Conversion factors

1 ft	=	0.305 m
1 mi	=	1.61 km
1 mph	=	0.447 m/s
1 y	=	3.16×10^7 s
1 u	=	1.66×10^{-27} kg
1 lb	=	4.45 N
1 dyn	=	10^{-5} N
1 erg	=	10^{-7} J
1 ft-lb	=	1.36 J
1 eV	=	1.60×10^{-19} J
1 cal	=	4.18 J
1 Btu	=	1.06×10^3 J
1 kWh	=	3.60×10^6 J
1 atm	=	1.01×10^5 Pa
1 cmHg	=	1.33×10^3 Pa
1 psi	=	6.90×10^3 Pa
1 u·c^2	=	932 MeV

Values are to three significant figures.

See Appendix D for further information.

PHYSICS
Part Two

Richard T. Weidner *Rutgers University*

in collaboration with
Michael E. Browne *University of Idaho*

Allyn and Bacon, Inc.
Boston • London • Sydney • Toronto

Developmental Editor: Jane Dahl

Production Editor: Mary Beth Finch

Copyright © 1985 by Allyn and Bacon, Inc., 7 Wells Avenue, Newton, Massachusetts 02159. All rights reserved. No part of the material protected by this copyright notice may be reproduced or utilized in any form or by any means, electronic or mechanical, including photocopying, recording, or by any information storage and retrieval system, without written permission from the copyright owner.

Library of Congress Cataloging in Publication Data

Weidner, Richard T.
    Physics.

    Includes index.
    1. Physics. I. Browne, Michael E. II. Title.
QC21.2.W425 1985   530   84-18576
ISBN 0-205-08082-0 (v. 2)

Printed in the United States of America.

10 9 8 7 6 5 4 3 2 1   89 88 87 86 85

Credits

The cover art is a special-effects photograph created by Michael Freeman, one of a series appearing in the May, 1983, issue of *Smithsonian,* as part of the article by James Trefil titled, "How the Universe Began."

    The picture shows schematically events occuring in the period from 10^{-10} s to 10^{-5} s after the "Big Bang," the event that marked the creation of the Universe about 15 billion years ago. The red clouds in the foreground symbolize background radiation; although still very hot, the Universe is cooling. The fragmented spheres, each with eighteen parts, represent quarks. Two quarks in the right foreground are interacting through the electromagnetic force, symbolized by magenta sparks, whereas two other quarks at left center interact through the weak force, symbolized by yellow sparks. Near the middle is a stringlike concentration of mass, shown as a loop; a future galaxy might condense around it. At the top we see quarks beginning to aggregate in groups of three to form protons and neutrons, and in groups of two for more exotic particles.

Figure 23-2: Courtesy of MIT Museum, Massachusetts Institute of Technology. **Figure 26-4:** Allyn and Bacon file photo. **Figure 28-2:** Burndy Library. **Figure 29-2:** Fermi National Accelerator Laboratory. **Figure 30-1:** Educational Development Center. **Figure 30-18:** Fermi National Acceler-

(Continued on Page A-27.)

Contents

Preface *vii*

23 Point Electric Charges *491*

- **23-1** Some Qualitative Features of the Electric Force *491*
- **23-2** Coulomb's Law *495*
- **23-3** Further Characteristics of Electric Charge *498*
- **23-4** Electric Field Defined *501*
- **23-5** Electric-Field Lines *505*

Summary *507*

Problems and Questions *508*

24 Continuous Distributions of Electric Charge *512*

- **24-1** Electric Field for Three Simple Geometries *512*
- **24-2** Uniform Electric Field *516*
- **24-3** Electric Flux *519*
- **24-4** Gauss's Law *521*
- **24-5** Electric Field and Charged Conductors *525*
- **24-6** An Electric Dipole in an Electric Field (Optional) *530*
- **24-7** General Proof of Gauss's Law (Optional) *531*

Summary *533*

Problems and Questions *534*

25 Electric Potential *539*

- **25-1** The Coulomb Force as a Conservative Force *539*
- **25-2** Electric Potential Defined *541*
- **25-3** Electric Potential for Point Charges *543*
- **25-4** Electric Potential Energy *545*
- **25-5** Equipotential Surfaces *547*
- **25-6** Relations between V and **E** *549*
- **25-7** Electric Potential and Conductors *552*

Summary *554*

Problems and Questions *555*

26 Capacitance and Dielectrics *560*

- **26-1** Capacitance Defined *560*
- **26-2** Capacitor Circuits *564*
- **26-3** Dielectric Constant *567*
- **26-4** Energy of a Charged Capacitor *569*
- **26-5** Energy Density of the Electric Field *571*

26-6 Electric Polarization and Microscopic Properties (Optional) *572*
Summary *574*
Problems and Questions *575*

27 Electric Current and Resistance *582*

27-1 Electric Current *582*
27-2 Current and Energy Conservation *585*
27-3 Resistance and Ohm's Law *585*
27-4 Resistivity *587*
27-5 RC Circuits *589*
27-6 Electric Resistance from a Microscopic Point of View *591*
Summary *592*
Problems and Questions *593*

28 DC Circuits *598*

28-1 EMF *598*
28-2 Single-Loop Circuits *601*
28-3 Resistors in Series and Parallel *603*
28-4 DC Circuit Instruments *606*
28-5 Multiloop Circuits *609*
Summary *611*
Problems and Questions *612*

29 The Magnetic Force *619*

29-1 Magnetic Field Defined *620*
29-2 Magnetic-Field Lines and Magnetic Flux *622*
29-3 Charged Particle in a Uniform Magnetic Field *625*
29-4 Charged Particles in Uniform **B** and **E** Fields *628*
29-5 Magnetic Force on a Current-Carrying Conductor *630*
29-6 The Hall Effect *632*
29-7 Magnetic Torque on a Current Loop *633*
29-8 Magnetic Dipole Moment *635*
Summary *631*
Problems and Questions *638*

30 Sources of the Magnetic Field *643*

30-1 The Oersted Effect *643*
30-2 The Magnetic Force between Current-Carrying Conductors *646*
30-3 The Magnetic Field from a Current Element; the Biot-Savart Relation *647*
30-4 Gauss's Law for Magnetism *651*
30-5 Ampère's Law *652*
30-6 The Solenoid *656*
30-7 Magnetic Materials *658*
Summary *661*
Problems and Questions *661*

31 Electromagnetic Induction *666*

31-1 Induced Currents and EMF's *667*
31-2 EMF in a Moving Conductor *671*
31-3 Lenz's Law *675*
31-4 Faraday's Law and the Induced Electric Field *677*
31-5 Eddy Currents *680*
31-6 Diamagnetism *681*
Summary *683*
Problems and Questions *683*

32 Inductance and Electric Oscillations *689*

32-1 Self-Inductance Defined *689*
32-2 The *LR* Circuit *694*
32-3 Energy of an Inductor *696*
32-4 Energy of the Magnetic Field *697*
32-5 Electrical Free Oscillations *697*
32-6 Electrical-Mechanical Analogs *702*
32-7 Mutual Inductance (Optional) *704*
Summary *705*
Problems and Questions *706*

33 AC Circuits *710*

33-1 Some Preliminaries *710*
33-2 Series *RLC* Circuit *713*
33-3 Phasors *719*

33-4 Series *RLC* Circuit with Phasors 721
33-5 RMS Values for AC Current and Voltage 724
33-6 Power in AC Circuits 725
33-7 The Transformer 729
Summary 730
Problems and Questions 731

34 Maxwell's Equations 737

34-1 The General Form of Ampère's Law 737
34-2 Maxwell's Equations 741
34-3 Electromagnetic Waves from Maxwell's Equations (Optional) 745
Summary 749
Problems and Questions 749

35 Electromagnetic Waves 751

35-1 Basic Properties of Electromagnetic Waves 751
35-2 Sinusoidal Electromagnetic Waves and the Electromagnetic Spectrum 752
35-3 Energy Density, Intensity, and Poynting Vectors 754
35-4 Electric-Dipole Oscillator 757
35-5 The Speed of Light 759
35-6 Radiation Force and Pressure, and the Linear Momentum of an Electromagnetic Wave 762
35-7 Polarization 765
Summary 768
Problems and Questions 769

36 Ray Optics 772

36-1 Ray Optics and Wave Optics 772
36-2 The Reciprocity Principle 774
36-3 Rules of Reflection and Refraction 775
36-4 Reflection 777
36-5 Index of Refraction 779
36-6 Refraction 781
36-7 Total Internal Reflection 784
Summary 785
Problems and Questions 786

37 Thin Lenses 792

37-1 Focal Length of a Converging Lens 792
37-2 Ray Tracing to Locate a Real Image 795
37-3 Ray Tracing to Locate a Virtual Image 798
37-4 Diverging Lens 799
37-5 Lens Combinations 801
37-6 The Lens Maker's Formula 804
37-7 Lens Aberrations 807
37-8 Spherical Mirrors (Optional) 808
Summary 810
Problems and Questions 811

38 Interference 816

38-1 Superposition and Interference of Waves 816
38-2 Interference from Two Point Sources 818
38-3 More on Interference from Two Point Sources 822
38-4 Young's Interference Experiment 824
38-5 The Diffraction Grating 826
38-6 Coherent and Incoherent Sources 829
38-7 Reflection and Change in Phase 830
38-8 Interference with Thin Films 832
38-9 The Michelson Interferometer 834
Summary 835
Problems and Questions 836

39 Diffraction 841

39-1 Radiation from a Row of Point Sources 841
39-2 Single-Slit Diffraction 844
39-3 The Double Slit Revisited 846
39-4 Diffraction and Resolution 847
39-5 X-Ray Diffraction 850
39-6 $I(\theta)$ for Single Slit (Optional) 853
39-7 The Diffraction Grating Revisited (Optional) 855
Summary 856
Problems and Questions 857

40 Special Relativity 860

40-1 The Constancy of the Speed of Light 860
40-2 Relativistic Velocity Transformations 861

- 40-3 Space and Time in Special Relativity 865
- 40-4 Relativistic Momentum 873
- 40-5 Relativistic Energy 875
- 40-6 Mass-Energy Equivalence and Bound Systems 878
- 40-7 The Lorentz Transformations (Optional) 880

Summary 881
Problems and Questions 882

41 Quantum Theory 886

- 41-1 Quantization 886
- 41-2 Photoelectric Effect 887
- 41-3 X-Ray Production and Bremsstrahlung 891
- 41-4 Compton Effect 892
- 41-5 Pair Production and Annihilation 895
- 41-6 Matter Waves 897
- 41-7 Probability Interpretation of the Wave Function 899
- 41-8 Complementarity Principle 902
- 41-9 Uncertainty Principle 903
- 41-10 The Quantum Description of a Confined Particle 908

Summary 911
Problems and Questions 913

42 Atomic Structure 917

- 42-1 Nuclear Scattering 917
- 42-2 The Hydrogen Spectrum 920
- 42-3 Bohr Theory of Hydrogen 921
- 42-4 The Four Quantum Numbers for Atomic Structure 927
- 42-5 Pauli Exclusion Principle and the Periodic Table 933
- 42-6 The Laser 937

Summary 941
Problems and Questions 941

Appendixes

- A International System of Units A-1
- B SI Prefixes for Factors of Ten A-2
- C Physical Constants A-2
- D Conversion Factors A-4
- E References A-5

Answers to Selected Problems A-7

Index A-15

Preface

- *To concentrate on basic principles* and skip much of the rest *The aim*
- *To be persuasive,* with crystalline clarity on the fundamentals, no cutting of corners, no superficial explanations, lots of worked examples that illustrate and apply basic ideas, a large variety and number of problems and questions for students to work
  - *To be informal,* with a conversational tone and interesting asides
- And then, *to stop.*

That's what I've tried to do in this calculus-based introductory physics textbook for students of engineering and science.

If you're going to do more than just name topics and give a superficial once-over to each, if the book is to be small enough to be carried around easily, and if students taking introductory calculus-based physics are to have at least some time for other courses and interests, you simply cannot cover all of the topics that, in some ideal world, would be "nice" to include. Something's got to give. *Coverage*

I've concentrated on the fundamental topics, the topics that every budding engineer or scientist simply has to know, and treated these basic ideas in much more detail than is typical in introductory textbooks. Why, for example, is that factor $\cos\theta$ in the definition of work? What do heat and work mean at the microscopic level? These are the sorts of questions that I've not brushed over. And dealing thoroughly with the most basic concerns pays off, in my opinion, in at least two ways—students find the physics to be easier, and they even come to like it.

But which are the truly basic topics? Before this work was begun, I had the benefit of a survey conducted by the publisher and the opinions of several hundred college instructors. Of course, they did not agree in detail. But on some central questions they reached substantial convergence: treat fewer topics thoroughly, and don't attempt to be encyclopedic. Or, don't cover a lot, uncover a little. Although the final choices reflect my own predilections, the topics included here are those that most physics instructors said were most important. This meant dropping entirely some really nice items; probably every instructor will find one or more favorites missing. As a scan of the table of contents will show, this is pretty much straight physics, with a strong emphasis on classical physics. There is little on the history of physics, little on experimental details, little on trendy items. The panels are intended to add a touch of spice to what is basically a meat-and-potatoes diet.

The sequence of topics is pretty much canonical. Here are specific instances in which the text departs from tradition: *Departures from convention*

- Chapter 1 is short (barely 4 pages). It does just three things—tells the student to be on the look-out for the "Message of Physics" (that things hang together, and the basic laws of physics are simple); cautions the student that this is "textbook physics" with all of the limitations that implies; and, in the fashion of a preface for the student, tells "What's Where".

• Chapter 2, mostly on the properties of vectors, also includes the statics of a particle to illustrate vector properties applied to forces. This item may, of course, be treated at a later stage.

• Rotation progresses from the simple to the more sophisticated as follows: first equilibrium of a rigid body (Chapter 12), then rotational dynamics (Chapter 13), and finally angular momentum (Chapter 14).

• The first chapter on electrostatics deals with point charges (Chapter 23) and the next one with continuous distributions of electric charges, including Gauss's law (Chapter 24).

• Topics in modern physics are condensed into three (longish) chapters — special relativity (Chapter 40), Quantum Physics (Chapter 41), and Atomic Structure (Chapter 42). This is what most instructors surveyed said that they wanted.

Antecedents This book is in some respects a lineal descendant of earlier texts of the same publisher that I co-authored with Robert L. Sells. But the changes relative to these earlier works is substantial enough as to make this effectively a new book.

Optional sections Instructors may wish to skip sections marked *optional*. Some are on detailed proofs (the parallel-axis theorem, for example) that are good to have for the record but may not get much class attention. The results of an optional section are not required in later sections, and they are not included in chapter summaries nor in the problem and question sets.

Use of mathematics Very little calculus is needed in the first several chapters, and the calculus that does appear there is pretty simple. Elementary calculus is a co-requisite. The dot and cross products are introduced where they are first needed for the physics (work and torque, respectively). Unit vectors, introduced in separate sections, are not used much in problems and can easily be skipped.

Panels The *panels* are short items that appear on a colored background and are sprinkled throughout the text. They are stories about physics and physicists, asides, extras, independent of the text development. Their style is informal, even breezy, but their subject matter is serious. The chapter location, title, and (in parentheses) the topics for panels in Part Two of the text are these:

Chapter 23 "Famous Physicist, Founding Father" (biographical sketch, Benjamin Franklin)
Chapter 26 "Knowing the Connections" (how physics connects disparate items)
Chapter 30 "And the Beat Goes On" (electromagnetism viewed from different levels of sophistication)
Chapter 31 "How Does Physics Advance?" (strategies for progress in physics)
Chapter 34 "The Odd Couple" (biographical sketches of Faraday and Maxwell, and the Maxwell demon)
Chapter 35 "Everything Was Big but the Particles" (1984 Nobel Prize in Physics)
Chapter 36 "Using a Point Source" (geodesy with radiointerferometry)
Chapter 38 "Hey, Phenomenal!" (biographical sketch, Thomas Young)
Chapter 40 "The Italian Navigator Has Just Landed" (biographical sketch, Enrico Fermi)

Chapter 41 "Particles, Fields" (the electromagnetic interaction from the point of view of quantum field theory)

Chapter 42 "Aha, That Did It!" (biographical sketch, Wolfgang Pauli)

Surely every physics instructor has had the following experience. You're fed up with working yet one more inclined-plane problem, you're afraid that the engineering students are getting a distorted picture of what's really important in physics, so you launch into a discussion of, say, the complementary roles of theory and experiment in physics. After just a few minutes have gone by, and you silently congratulate yourself on the apt phrase you've used, the telling point you've made, someone near the back of the room asks in an exasperated tone, "Is this going to be on the test?" It is hard to present these crucial ideas to intensely goal-oriented freshman engineering students by a frontal attack. What the panels attempt is to do this by guile.

Of worked examples there are many, on the average about one per section. *Examples* They stress insight, not computation. Skipped steps are only in simple algebra, never in the explanation. I should think that a typical student can read and understand the examples strictly on his own.

It's SI all the way. Well, almost. English and cgs units and their relation to SI *Units* units are given, usually in footnotes. In a very few instances English units are used in problems where to do otherwise would seem highly contrived (whoever, for example, heard of the power of an automobile engine given in kW, rather than hp?) In thermal physics I've eschewed calories.

The summaries at the chapter ends all have the same format with entries *Chapter summaries* organized under four main headings: *Definitions, Units, Fundamental Principles,* and *Important Results.* Deliberately telegraphic, the summaries are intended simply to remind students, of the most essential items, not tell them everything they must know.

There are many Problems (P) and Questions (Q) at the chapter ends. As *Problems and questions* defined here, a problem requires some computation, algebraic or numerical; a question does not. Multiple-choice items are most often marked Q. Each P or Q is identified by level of difficulty:

$$\text{Easy} \equiv \cdot \qquad \text{Medium} \equiv : \qquad \text{Hard} \equiv \vdots$$

Of course, people may disagree on the level of difficulty, but I am pretty confident that no one will think that an item marked easy should have been called hard, or conversely. A student who has the fundamental ideas straight should be able readily to work easy Ps and Qs.

Problems and Questions are also identified by the section number and title. Answers to odd-numbered Ps and Qs are given in the back of the book. So-called *supplementary problems* are found at the very ends of some chapters; not identified by section or difficulty, these problems involve some interesting quirk or applied aspect.

Professor Michael E. Browne of the University of Idaho is identified as collaborator in this work because he produced a large fraction of the problems and questions.

Acknowledgments

I am especially grateful for the comments and criticisms on some portions of the manuscript by my long-time partner and friend, Professor Robert L. Sells of State University College (SUNY), Geneseo. By mutual agreement, Duke is not co-author in this work. His special style and insights have influenced me over many years.

The following professors of physics have reviewed at least portions of the manuscript and made especially valuable comments and suggestions:

Paul A. Bender, Washington State University; George H. Bowen, Iowa State University; Jack Brennan, University of Central Florida; Keith H. Brown, California State Polytechnic University, Pomona; Roger W. Clapp, University of South Florida; Roger Creel, University of Akron; John E. Crew, Illinois State University; Harriet H. Forster, University of Southern California; Simon George, California State University, Long Beach; George Goedecke, New Mexico State University; George W. Greenlees, University of Minnesota; Alvin W. Jenkins, Jr., North Carolina State University; Mohan Kalelkar, Rutgers University; Clement J. Kevane, Arizona State University; Brij M. Khorana, Rose-Hulman Institute of Technology; Sung K. Kim, Macalester College; David Markowitz, University of Connecticut; W. F. Parks, University of Missouri, Rolla; Philip C. Peters, University of Washington; R. L. Place, Ball State University; Marllin Simon, Auburn University; F. B. Stumpf, Ohio University; George A. Williams, University of Utah; John S. Zetts, University of Pittsburgh at Johnstown; E. J. Zimmerman, University of Nebraska, Lincoln; Earl Zwicker, Illinois Institute of Technology.

These physicists have made valuable contributions in checking answers and solutions to problems and questions and in other aspects of manuscript preparation:

Professor Roger W. Clapp, Jr., University of South Florida, Tampa
Professor A. Douglas Davis, Eastern Illinois University
Professor William T. Franz, Randolph-Macon College
David Klesch, Michigan State University
Dr. Arthur E. Walters, freelance consultant
Jake Zwart, University of Waterloo, Ontario

I have had a happy association over many years with Allyn and Bacon, Inc. and I am particularly appreciative of the special efforts of the following people in helping to see this work reach completion:

John Gilman, Vice President
Gary Folven, Editor-in-Chief
James M. Smith, Science Editor
Jane Dahl, Developmental Editor
Mary Beth Finch, Production Editor

Above all I am grateful to the students who were not satisfied with facile, stock answers and who kept pressing me for simpler, more persuasive explanations.

The manuscript was effectively typed by Allegra L. Cushing and Cynthia Sells.

I accept responsibility for all errors, and I should like to be informed of those residual infelicities that escaped earlier detection.

R. T. W.

Point Electric Charges

23

23-1 Some Qualitative Features of the Electric Force
23-2 Coulomb's Law
23-3 Further Characteristics of Electric Charge
23-4 Electric Field Defined
23-5 Electric-Field Lines
     Summary

All the known forces in physics arise from four fundamental interaction forces: the strong nuclear force, the electromagnetic and the closely related weak interaction force, and the gravitational force. The nuclear and the weak interaction forces are important only within the atomic nucleus, in certain collisions between nuclei, and in the decay of unstable elementary particles. The familiar gravitational force is important only when one of two interacting objects has a mass comparable to that of a planet. This leaves the *electromagnetic force.* Except for the force due to gravity, all the forces of ordinary experience—the restoring force of a stretched string, the tension in a cord, the force between colliding automobiles—indeed, all the forces acting between the atomic nucleus and its surrounding electrons, or between atoms in molecules, are ultimately electromagnetic in origin. Electromagnetic forces dominate the interactions of all systems from the size of atoms to the size of planets.

The term *electromagnetism* emphasizes that electric and magnetic phenomena are not separate or unrelated. Both electrical and magnetic effects are a consequence of the property of matter called *electric charge.*

23-1 Some Qualitative Features of the Electric Force

The distinctively new concept introduced in electromagnetism is *electric charge.* First, how do ordinary objects become charged? Whenever any two

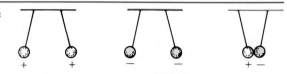

Figure 23-1. Charged objects suspended from insulated strings. Like charges repel, unlike charges attract.

dissimilar nonmetallic objects are brought into intimate contact—for example, glass rubbed with silk—and then separated, they show a mutual attraction, which far exceeds their gravitational attraction. Such objects are said to be electrically charged, and the presence of electric charge on each object is responsible for their interaction through an electric force. We assume that you are familiar with this effect and the qualitative experiments that establish the following fundamental facts about these electric interactions:

- *Like charges repel* each other.
- *Unlike charges attract* each other.

In this chapter we deal with electrostatics, the study of electric charges at rest. Electric charges in motion, besides interacting by the electric force, also interact through the so-called magnetic force. In the next several chapters we deal only with those situations in which the speeds of any charges are so much smaller than the speed of light (3.0×10^8 m/s) that the magnetic force is negligible compared with the electric force.

All electrical phenomena exist because fundamental particles in physics may have the property of electric charge. Thus, the electron has a negative charge, the proton a positive charge, and the neutron a zero charge. The use of the algebraic signs $+$ and $-$ to denote the two kinds of charge is appropriate, since combining equal amounts (to be defined precisely below) of positive and negative charges produces a zero electric force on an object. As we know, the nucleus of an atom consists of protons and neutrons bound together by the nuclear force within a volume whose length dimensions are never much greater than 10^{-14} m; the nucleus is surrounded by electrons that are bound to it by the electric force. An atom is electrically neutral as a whole when the number of electrons surrounding the nucleus equals the number of protons in the nucleus.

In solids, atoms are closely packed, their nuclei being separated from one another by distances of the order of 10^{-10} m. In *conductors,* of which many metals are common examples, most of the electrons are bound to their parent nuclei and remain with them. But in a conductor, typically one electron per atom is a *free electron.* A free (or conduction) electron, although bound to the conductor as a whole, may wander throughout the interior of the material and can easily be displaced by external electric forces. In *insulating,* or *dielectric,* materials on the other hand, all atomic electrons are *bound,* to a greater or lesser degree, to their parent nuclei. Electrons are removed from an insulating material or added to it only with the expenditure of significant energy. Examples of common conductors are metals, liquids having dissociated ions (electrolytes), the earth, moist air, and the human body. Good insulators are very often transparent materials: plastics, glass, and a vacuum, which is a perfect insulator. The best electrical conductors are better than the worst conductors (or best insulators) by enormous factors, up to 10^{20} (this conductivity ratio is given precise quantitative meaning in Chapter 27). Lying between these extremes are

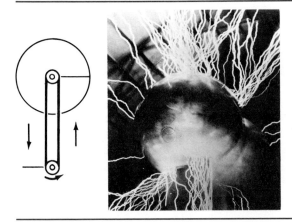

Figure 23-2. (a) Schematic diagram for a Van de Graaff generator, a device for separating charge. The moving belt continuously carries charge from sharp points near it at the lower roller to points on the interior of a spherical conductor. (b) Photograph of a Van de Graaff accelerator.

the so-called semiconductors; their conductivity is intermediate between conductors and insulators. Silicon is an example of a semiconductor. In a semiconductor, only a very small fraction of the electrons are free.

Qualitative electrostatic effects can be understood on the basis of the atomic model and of the properties of conductors and dielectrics. Suppose two unlike dielectrics are rubbed together. Electrons at the interface will leave the material to which they are less tightly bound for the other material, to which they become more tightly bound. On separation, one object carries excess electrons and is negatively charged, and the other has a deficiency of electrons and is positively charged. More generally, whenever a large-scale object has its electrical neutrality disturbed by losing electrons, the object is positively charged. Similarly, a negatively charged object has excess electrons. "Charging" an object consists simply in adding or subtracting electrons. Of course, when one type of charge is produced on an ordinary object, the other type must appear in equal amounts on a second object. The charging of any large-scale body results from the separation of charged particles (see Figure 23-2).

A charged body has acquired or lost electrons, and it is natural sometimes to speak of the net charge *on* a body. Of course the body acquires (or loses) not only the charge of electrons added to (or removed from) it but also the mass of the added (or removed) electrons. The additional mass is usually so trivial as to be negligible.

A simple device for detecting electric charge, often used in classroom demonstrations of electrostatic effects, is the *electroscope*. See Figure 23-3. The angular deflection from alignment with the vertical of a light pivoted conducting rod and the vertical conducting rod to which it is connected measures the charge on the electroscope. When charge is added to the conducting plate at the top of the vertical rod by touching it, the charge spreads, because of mutual repulsion, over the entire conductor. The pivoted rod is repelled by like charge on the fixed conducting rod. An electroscope is not a precise charge-measuring device; and we shall later learn of other devices through which charge magnitude is measured indirectly but far more precisely.

When a charged object touches an initially neutral conductor, the conductor acquires some of the object's net charge. It is possible, however, to charge a conductor without ever touching it directly with a charged object. The procedure is known as *charging by induction*. The steps are shown in Figure 23-4.

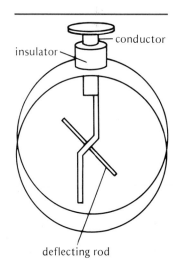

Figure 23-3. An electroscope that shows its state of net electric charge by the deflection of the pivoted conducting rod.

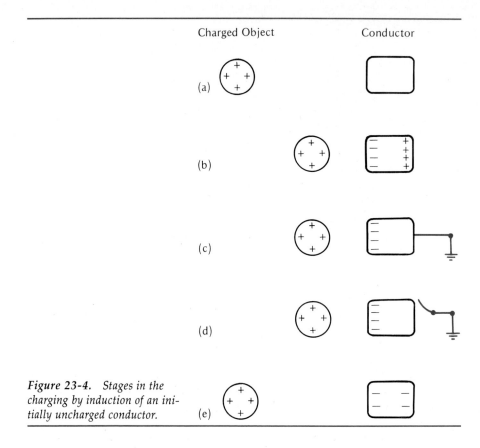

Figure 23-4. Stages in the charging by induction of an initially uncharged conductor.

(a) The charged object (here chosen as positive) is far away, and the conductor is electrically neutral, both as a whole and on all its surfaces.

(b) The charged object comes close to the conductor's left side. Electrons within the conductor are attracted to and move readily to the left side. This leaves an equal amount of positive charge on the conductor's right side. Bringing the charged object close to the conductor has merely separated positive from negative charge on the conductor; strictly, electrons have simply shifted left. If the charged object were removed at this stage, the situation would return to that shown in part (a).

(c) The conductor is *grounded*. It is connected by a conducting wire to the largest and most readily accessible good conductor nearby—the entire earth. (The electrical symbol for ground is ⏚.) Now the conductor has effectively become a part of a much larger conductor, the earth, and the positive charge that had been located on the conductor's right side is now distributed over the entire earth. The earth, because it is such a large conductor and acquires a relatively modest amount of positive charge from the conductor, is still effectively neutral. The negative charge does not leave the left side, however. It is held there by the attractive force from the positive charge on the nearby object.

(d) The conducting wire is disconnected and things remain as in (c).

(e) The charged object is taken away. Because of the mutual repulsion of electrons, the negative charge that had been concentrated on the left side is now distributed over the entire conductor. Note that the conductor has a charge that is *opposite* in sign to that on the charging object.

23-2 Coulomb's Law

The basic relation for the electric force between a pair of electrically charged particles is *Coulomb's law*. Consider Figure 23-5, where two point charges q_1 and q_2 (assumed to be of the same sign) are separated by distance r. Each charge repels the other by the electric interaction. The principal features, based on experiment, are these:

- The *magnitudes* of the Coulomb force on each charge are the *same*; their directions are *opposite*:

$$\mathbf{F}_{2\,on\,1} = -\mathbf{F}_{1\,on\,2}$$

- The electric force is a *central* force. It lies on the line connecting the two point charges.
- The magnitude of the electric force is given by

$$F_e = \frac{kq_1q_2}{r^2} \qquad (23\text{-}1)$$

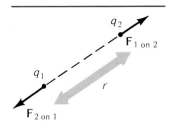

Figure 23-5. *Two point electric charges interacting through the Coulomb force.*

where k is constant.*

- Equation (23-1) applies only to *point charges*. This means that the physical size of charge q_1 or q_2 must be small compared with their separation distance r. In this chapter we consider point charges only. We consider continuous distributions of charge in the next chapter; the procedure for dealing with a charge distribution is simply to imagine it subdivided into a collection of effective point charges, and to apply (23-1) to each pair of charges, summing the contributions by the techniques of integral calculus.
- That the electric force is inverse-square with distance was first established by C. A. de Coulomb (1736–1806). He used a torsion balance like that Cavendish used later in studying the gravitational force (Section 15-3). It is now known through indirect experiments (Section 24-4) that the exponent of r is 2 to within a few parts in 10^{16}. Indeed, Coulomb's law is perhaps the most

---

* Equation (23-1) is a *scalar* equation; it gives the *magnitude* of the electric force on either charged particle. If the signs of q_1 and q_2 are included in (23-1), the sign we compute for F_e tells whether the force is repulsive or attractive: positive F_e for like charges and repulsion, negative F_e for unlike charge and attraction.

The *direction* of the electric force can be incorporated in a *vector* equation. Let radius vector $\mathbf{r}_{1\,to\,2}$ have its tail at q_1 and its head at q_2. Then the vector force on q_2 can be written as

$$\mathbf{F}_{1\,on\,2} = \frac{kq_1q_2\mathbf{r}_{1\,to\,2}}{r_{1\,to\,2}^3}$$

When q_1 and q_2 have the same sign, $\mathbf{F}_{1\,on\,2}$ is in the same direction as $\mathbf{r}_{1\,to\,2}$, corresponding to a repulsive force; the direction of $\mathbf{F}_{1\,on\,2}$ is opposite to $\mathbf{r}_{1\,to\,2}$ when q_1 and q_2 have opposite signs. Note that an additional factor $r_{1\,to\,2}$ appears in the denominator to compensate for the magnitude of $\mathbf{r}_{1\,to\,2}$ introduced into the numerator.

We can write the vector relation a little differently by using a unit vector. Let $\mathbf{u}_{1\,to\,2}$ be a unit vector parallel to $\mathbf{r}_{1\,to\,2}$. Then,

$$\mathbf{F}_{1\,on\,2} = \frac{kq_1q_2\mathbf{u}_{1\,to\,2}}{r_{1\,to\,2}^2}$$

Force $\mathbf{F}_{2\,on\,1}$ is, of course, just the negative to $\mathbf{F}_{1\,on\,2}$. As a practical matter, it is usually best to compute just the magnitude of the electric force from Coulomb's law and indicate the direction of the force by a vector in a diagram.

thoroughly tested basic relation in physics. It works down to separation distances as small as 3×10^{-18} m, about one-thousandth the size of a proton.

- In the SI system of units, the unit for the scalar quantity called electric charge is the *coulomb* (abbreviated C). The definition of the coulomb is a bit complicated. First, by definition, a net charge of 1 C passes through the cross section of an electric conductor when an electric current of one ampere (1 A) exists in the conductor for one second: $1 \text{ C} \equiv (1 \text{ A})(1 \text{ s})$. The ampere is defined in turn by the magnetic force between two current-carrying conductors (Section 30-4).

With the coulomb unit for charge defined as given above, the constant in the Coulomb-law relation has the value

$$k = 8.987\ 55 \times 10^9 \text{ N} \cdot \text{m}^2/\text{C}^2$$

For most computations, it is satisfactory to use the rounded value, $k \approx 9.0 \times 10^9 \text{ N} \cdot \text{m}^2/\text{C}^2$.

The units assigned to k assure that with q_1 and q_2 both given in coulombs and r in meters, the force is in newtons. Thus, with two point charges each of 1 C separated by 1 m, the electric force on each is

$$F_e = (9.0 \times 10^9 \text{ N} \cdot \text{m}^2/\text{C}^2)(1 \text{ C})(1 \text{ C})/(1 \text{ m})^2 = 9.0 \times 10^9 \text{ N}$$

As static charges go, one coulomb is enormous. A laboratory device of ordinary size would usually not have a charge larger than about 10^{-6} C = 1 μC ($\mu \equiv micro \equiv 10^{-6}$); this might be the charge on the sphere of a classroom Van de Graaff generator. A small conductor — say, a dime or a paper clip — touched to the terminal of a 9-V battery picks up a charge no larger than about 10^{-11} C = 10 pC (p $\equiv pica \equiv 10^{-12}$).

- The constant k of Coulomb's law* is also written as

$$k \equiv \frac{1}{4\pi\epsilon_0} \tag{23-2}$$

where ϵ_0 is called the *electric permittivity of the vacuum*. The zero subscript indicates that as written in (23-1) and (23-2), Coulomb's law applies to electric charges in a *vacuum*. As we shall see in later chapters, it is sometimes preferable to write equations related to Coulomb's law using the constant ϵ_0 instead of k.

- Two fundamental forces in physics — the electric and the gravitational — are similar in some ways. Both are *central, conservative, inverse-square forces*. Because of this, many of the concepts and relations developed for gravity carry over unchanged into electricity. But there are also important differences. For one thing, there are *two* kinds of electric charge, but only one kind of gravitational charge (more often called gravitational mass). Electric charges may attract or repel; gravitational charges attract only. Further, there is no such thing as a gravitational conductor or shield. Another important difference is the relative magnitudes of these forces. The electric force is immensely larger than the gravitational force; for example, between an electron and

---

* Constant k relates to the *electric* interaction between charged particles. We shall encounter later (Chapter 32) a second constant k for the *magnetic* interaction between charged particles. To keep straight which is which, we shall later designate the constant associated with Coulomb's law as k_e and the constant associated with the magnetic interaction as k_m.

proton the electric attraction is 10^{39} times greater than the gravitational attraction (Example 23-3).

At the atomic level the gravitational force is so very small, compared with the electric force, that it can be ignored entirely. How is it then—if the electric force is so much greater than the force of gravity—that the only force a person is aware of in ordinary experience is gravity? It is that there are equal amounts of opposite charge in ordinary materials. The strong attraction between the opposite charges makes ordinary materials electrically neutral. With the electric force balanced, only the far weaker gravitational force remains.

- The *superposition principle* for forces holds for the coulomb interaction. Suppose that a charge q_3 is in the presence of two other charges q_1 and q_2, as shown in Figure 23-6. Experiment shows that the force on q_3 is just the *vector sum* of the separate forces on it from q_1 and q_2. Said differently, the force between any two charges is independent of the presence of other charges; to find the resultant force, we merely add the individual forces as vectors. The superposition principle, although simple, is not self-evident; it is a result of observation for electric interactions. As we shall see, its consequences are many and important.

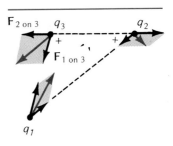

Figure 23-6. Three interacting point charges. The principle of superposition applies for Coulomb forces.

Example 23-1. What is the electric force exerted on each of two identical small conductors each with a charge of 0.10 μC and separated by 1.0 cm?

$$F_e = \frac{kq_1q_2}{r^2} = \frac{(9.0 \times 10^9 \text{N} \cdot \text{m}^2/\text{C}^2)(1.0 \times 10^{-7} \text{ C})^2}{(1.0 \times 10^{-2} \text{ m})^2} = 0.90 \text{ N}$$

We expect then, that electric forces between charged objects will be obvious when charges are of the order of 1μC and the charged objects are separated by the order of 1 cm.

Example 23-2. Two small identical conducting spheres first attract one another. Then the conductors are touched together and brought back to the same separation distance as initially. The conductors now repel one another with the same force magnitude as initially. What was the ratio of the original charge magnitudes?

Let the initial charge magnitudes be q_A and q_B. Since the spheres attract initially, the charges must be of opposite sign. The net charge magnitude initially of the two spheres together is $q_A - q_B$. After the spheres have touched, the net charge is shared equally between the identical conducting spheres. The magnitude of the charge of each sphere is now $\frac{1}{2}(q_A - q_B)$.

The magnitude of the initial attractive force is

$$F = \frac{kq_Aq_B}{r^2}$$

The magnitude of the final repulsive force at the same separation distance is

$$F = \frac{k[(q_A - q_B)/2]^2}{r^2}$$

Equating the two forces yields

$$q_Aq_B = \left(\frac{q_A - q_B}{2}\right)^2$$

which can be written as a quadratic equation in q_A/q_B:

$$\left(\frac{q_A}{q_B}\right)^2 - 6\left(\frac{q_A}{q_B}\right) + 1 = 0$$

whose two roots are

$$\frac{q_A}{q_B} = 3 + \sqrt{8}, \ 3 - \sqrt{8}$$

There should be just one answer to the question, but there are two different roots to the quadratic equation. Actually, the two values are equivalent, since one is the reciprocal of the other: $3 + \sqrt{8} = 1/(3 - \sqrt{8})$.

23-3 Further Characteristics of Electric Charge

How is Charge Magnitude Measured? Suppose that you observe two charges, q_A and q_B, not of the same magnitude, attracting one another. Which is the larger charge? With only two charges, you cannot tell, since the force magnitudes on the two charges are the same. In fact, you can't tell which of the two is positive and which is negative.

It takes a *third* charge. Suppose that this third charge is positive, and we first bring q_A close to it and measure the force F_A on it, as shown in Figure 23-7. With a repulsive force on q_A, we know that q_A is also *positive*. Now suppose that q_A is taken away, and q_B is brought to the same location to interact with our positive third charge. The force F_B on q_B is attractive, as shown in Figure 23-7(b). We conclude that q_B is negative. Moreover, the ratio of the force magnitudes gives us by *definition* the ratio of the charge magnitudes:*

$$\frac{q_A}{q_B} \equiv \frac{F_A}{F_B} \qquad (23\text{-}3)$$

For the situation shown in Figure 23-7, we have $|F_A| = 2\,|F_B|$; therefore, $|q_A| = 2|q_B|$.†

Electron Charge The electron's charge is designated by e; this is also the magnitude of the charge of any other ordinary charged "elementary" particle. The best current value for e is

$$e = 1.602\ 189\ 2 \times 10^{-19} \text{ C}$$

The direct measurement of the electronic charge e, as first made in the Millikan experiment of 1909, is discussed in Problem 25-7.

An electron's charge is extremely small. One microcoulomb, a fairly typical charge by laboratory standards, corresponds to an excess or deficiency of 6×10^{12} electrons. Thus, we can ordinarily assume electric charge to be infinitely divisible and continuous and ignore its essential "graininess." We can, for example, imagine a negatively charged surface to have charge spread

---

* The procedure used here to define electric-charge magnitude is exactly like the one used earlier to define the ratio of the magnitudes of gravitational mass (Section 15-7).

† Does the magnitude of the charge of a particle depend on its speed? The evidence that a particle's charge always has the same magnitude, independent of speed, comes from a simple observation: All atoms that are expected to be electrically neutral because they have equal numbers of electrons and protons actually have no net electric charge, despite the fact that the electrons differ drastically in their speeds in the various elements.

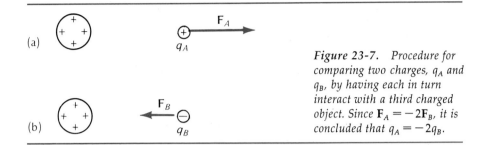

Figure 23-7. Procedure for comparing two charges, q_A and q_B, by having each in turn interact with a third charged object. Since $\mathbf{F}_A = -2\mathbf{F}_B$, it is concluded that $q_A = -2q_B$.

continuously over it, rather than concern ourselves with the actual finite number of discrete electrons, acting as point charges, that reside on it.

When electrons are transferred to or from a laboratory object being charged, the difference in mass is trivial. For example, when an object acquires a charge of 1 μC (or 6×10^{12} electrons), its mass changes by only $(9.1 \times 10^{-31}$ kg/electron$)$ $(6 \times 10^{12}$ electrons$) = 5 \times 10^{-18}$ kg.

Charge Conservation Electric charge is conserved. According to the fundamental law of conservation of charge, the net charge, or *algebraic sum of the charges, in any isolated system is constant*. This simple but important experimental result is illustrated very simply by the processes in which *two* neutral objects become charged. Electrons are transferred from one object to the other, and the result is one object with positive charge and the second with an equal amount of negative charge. No violation of charge conservation has ever been observed.

Electric-charge conservation does not imply, however, that charged particles can be neither created nor destroyed; it implies only that the creation of one of positive charge must be accompanied by the creation of another with an equal amount of negative charge. An important example is the phenomenon of *pair production*. When a sufficiently energetic photon, an electrically neutral particle of electromagnetic radiation, enters a closed container, as in Figure 23-8, it may be annihilated when close to an atom, and in its stead two particles appear, an electron with charge $-e$ and a positron with charge $+e$. The *net* charge within the container has not changed.

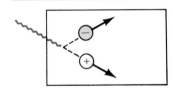

Figure 23-8. A photon enters a closed chamber and produces an electron-positron pair.

The positron, identical with an electron in all respects except for the sign of its charge (and the consequences of the difference in sign), is called the *antiparticle* of the electron. The electron and positron are but one example of a particle-antiparticle pair. Other examples are the proton and antiproton (charges $+e$ and $-e$, respectively) and more exotic elementary particles such as the π^+ meson and the π^- meson. Just as particle-antiparticle pairs can be created, a particle and its antiparticle can *annihilate* each other, producing two or more photons, or pairs of other particles. No matter what processes take place within a system—whether charge transfer between bodies in contact, nuclear transformations, creation of matter, or annihilation of particles—the total charge is always conserved. With the conservation of electric charge, the *classical* list of conservation laws is complete. They refer to the conservation of linear momentum (Section 7-2), angular momentum (Section 14-6), mass (Section 5-2), energy (Section 10-5), and now, electric charge.

Famous Physicist, Founding Father

When Benjamin Franklin (1706–1790) arrived in Europe in 1757, he was already world-renowned, not as a representative of the American colonies, but as a physicist. For his fundamental contributions to physics, Franklin had received honorary degrees from Harvard, Yale, and the College of William and Mary, and his book on electricity was then in its third edition.

He was an extraordinarily accomplished experimentalist, who constructed special apparatus to perform a variety of experiments that clarified many aspects of electrostatics. His "single fluid" theory eliminated the necessity of speaking of two distinct types of electric charge. Thus, an object with an excess of the basic charge, or fluid, was "positive," and another object with a deficiency was "negative."

The idea of electric-charge conservation originated with Franklin. He clarified the behavior of capacitors and the role of the dielectric. He showed that unbalanced charge resides on the outside of conductors and is especially concentrated at points. He explained what happens in the process we now call charging by induction. Anticipating electrons as the mobile carriers of electric charge, he believed that electric charge consisted of "extremely subtle particles." And Franklin showed in the famous kite and related experiments that electricity is not merely a laboratory curiosity but also the basis for such a large-scale natural phenomenon as lightning; he showed further that the lightning stroke typically goes from earth to the cloud, instead of the reverse. Franklin gave the sensible advice, "Never take shelter under a tree in a thunder gust [a thunderstorm]," and he invented the lightning rod. Other inventions: bifocals, the rocking chair, the Franklin stove, Daylight Saving Time.

Franklin's scientific interests ranged widely. He devised a thermometer to measure temperatures to depths in water of 100 ft. He studied cloud formation. He advanced arguments against the idea that light consisted of particles. He performed experiments to measure thermal conductivity. He advocated the caloric theory of heat. He wrote on such subjects as lead poisoning, gout, the physiology of sleep, deafness.

And not the least, Franklin had quite a hand in founding a nation.

Charge Quantization The net electric charge of any object is just the algebraic sum of the charges of the particles that make up the object. All elementary particles, although they may differ greatly in mass and other properties, are found to have just one of *three possible charge values:*

$$+e, \quad 0, \quad \text{or} \quad -e$$

where e is the magnitude of the charge of an electron.* For example, the charges of the electron and positron are $-e$ and $+e$, respectively; and the charges of the three kinds of mesons, the π^+, π^0, and π^-, are $+e$, 0, and $-e$, respectively. The neutron and its antiparticle, the antineutron, have charges of exactly zero.

Since every charged object is nothing more than a collection of particles, the only possible values of the total charge Q of any object are integral multiples of e:†

---

* One exception: The very short-lived exotic particle Δ^{++} has the charge $+2e$.

† There is indirect but nevertheless compelling evidence that some of the so-called elementary particles actually consist of still smaller parts called *quarks*. A quark has a fractional electric charge: $+\frac{2}{3}e$, $-\frac{2}{3}e$, $+\frac{1}{3}e$, or $-\frac{1}{3}e$. A proton, for example, is believed to consist of two "up" quarks, each with a charge of $+\frac{2}{3}e$, together with a "down" quark with charge $-\frac{1}{3}e$. A neutron consists of one up quark with charge $+\frac{2}{3}e$ and two down quarks, each with a charge $-\frac{1}{3}e$. A positive pion π^+ consists of an up quark (charge $+\frac{2}{3}e$) plus an "antidown" quark (charge $+\frac{1}{3}e$).

Quarks come in six types: up, down, charm, strange, top, and bottom. Furthermore, each quark comes in one of three "colors": red, blue, green. Finally, for each quark there exists an anti-quark. Adding up all possibilities gives a total of 36 distinct varieties.

$$Q = \pm Ne \quad \text{where } N = 0, 1, 2, \ldots \quad (23\text{-}4)$$

Electric charge is said to be *quantized*. It appears only as *integral* positive and negative *multiples* of the charge of the electron, and no others; see Figure 23-9.

The discreteness, or granularity, of electric charge is evident only through subtle experiments, simply because most charged objects have a charge very much larger than e; that is, the integer N in (23-4) is typically very much larger than 1. Charge quantization shows up in chemistry in the chemical idea of atomic number; this integer is merely the total number of electrons (or protons), and hence, the total negative (or positive) charge in a neutral atom. Charge quantization is also implied in valence, which can have only an *integral* value.

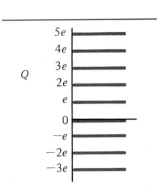

Figure 23-9. Charge quantization. The only possible values of any charge Q are integral multiples of the electron charge e.

Example 23-3. What is the ratio of the electric to the gravitational force between a proton and an electron?

The gravitational force (15-1) is $F_g = Gm_1m_2/r^2$, and the electric force is $F_e = ke^2/r^2$, where e is the charge magnitude for both electron and proton. The force ratio for any separation distance r is

$$\frac{F_e}{F_g} = \frac{ke^2}{Gm_1m_2}$$

$$= \frac{(9.0 \times 10^9 \text{ N} \cdot \text{m}^2/\text{C}^2)(1.6 \times 10^{-19} \text{ C})^2}{(6.7 \times 10^{-11} \text{ N} \cdot \text{m}^2/\text{kg}^2)(1.7 \times 10^{-27} \text{ kg})(9.1 \times 10^{-31} \text{ kg})} \approx 10^{39}$$

(The electron and proton masses come from Appendix C.)

23-4 Electric Field Defined

The concept of field—here the electric field, later the magnetic field—is absolutely central to electromagnetism. Indeed, electromagnetism can be said to be the study of electric and magnetic fields and their relation to electric charge. At the first level, the field concept makes it easier to visualize, to graph, and to compute electromagnetic interactions; at a more fundamental level, the fields themselves are endowed with such physical properties as energy, momentum, and angular momentum.

To see why it is useful to define an electric field, consider first Figure 23-10(a). Here we have some point charges—we shall call them the *source charges*—fixed in position. Still one more charge, a *test charge* q_t, is brought to point P. The resultant electric force ΣF_e acting on q_t is found by applying Coulomb's law to its interaction with each of the source charges and adding the individual forces as vectors.

To make the circumstances more specific, suppose that the test charge happens to be $q_t = +1 \ \mu\text{C}$ and the resultant force on it at P is found to be $F = 1 \times 10^{-2}$ N in the direction east, as shown in Figure 23-10(b). Now suppose we take away the $+1$-μC charge and replace it by a charge of $-1 \ \mu$C, again at point P, as shown in Figure 23-10(c). What is the force on the -1-μC charge? There is no need to repeat detailed computations. We can immediately say that with the charge magnitude unchanged but the sign reversed, the resultant force is still 1×10^{-2} N, but in the direction west. Similarly, if the charge magnitude is doubled so that we have $q_t = +2 \times 10^{-6}$ C, as in Figure 23-10(d), the force is also doubled to 2×10^{-2} N in the direction east. In short, if we know the resultant electric force on some *one* charge at P, we can immedi-

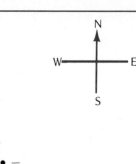

Figure 23-10. *Source charges on the left produce the resultant electric force ΣF_e on test charge q_t.*

ately find the force on *any other* charge at P. Clearly, the electric force per unit positive charge at point P is

$$\frac{\Sigma F_e}{q_t} = \frac{1 \times 10^{-2} \text{ N east}}{+1 \times 10^{-6} \text{ C}} = 1 \times 10^4 \text{ N/C east}$$

This is the electric field **E** at P.

More formally, the *electric field* **E** at point *P* arising from source charge(s) *fixed* in position is given by

$$\mathbf{E} \equiv \frac{\Sigma \mathbf{F}_e}{q_t} \quad (23\text{-}5)$$

where $\Sigma \mathbf{F}_e$ is the resultant electric force from the source charges on test charge q_t at point *P*. More succinctly, the electric field is the electric force per unit positive charge.

Said a little differently, the resultant electric force acting on charge q_t at a location where the electric field is **E** is given by

$$\Sigma \mathbf{F}_e = q_t \mathbf{E}$$

The electric field is a vector whose magnitude and direction may change from one point in space to another. As we shall see, we should think of the electric field as existing at point *P* even though there may be no actual charge q_t at this location to feel an electric force.

Why must the source charges be fixed in position? If the charges creating the electric field were not nailed down, then bringing an additional test charge q_t to *P* would produce a force on the source charges that might shift their positions and therefore *change* the electric field they create. This consideration is especially important for a conductor. Suppose that we have an uncharged conductor. Clearly, it produces zero electric field at all external locations. But if a test charge is brought near the conductor, this additional charge will redistribute the free electrons in the conductor, as shown in Figure 23-4, and the conductor may then produce electric fields different from zero at external locations. Rather than insist that the source charges remain fixed, we could instead imagine test charge q_t to be infinitesimally small. Then the forces that q_t would produce on the source charges would be so small that the locations of the source charges would not be changed.

Sometimes the quantity **E** is referred to as the intensity, or strength, of the electric field. Here we shall call it simply the electric field. From its definition, we see that the units for **E** are newtons per coulomb (N/C); we shall see later (Section 25-6) that equivalent (and more common) units are volts per meter.

What is the electric field a distance *r* from a single point charge q_s? With a test charge q_t a distance *r* from q_s, we have, from Coulomb's law,

$$F = \frac{kq_s q_t}{r^2}$$

But $\mathbf{E} = \mathbf{F}/q_t$. Therefore

$$E = \frac{kq_s}{r^2} \quad (23\text{-}6)$$

The magnitude of the electric field from a single point charge falls inversely with the square of the distance *r* from it. The direction of the electric field is along the line connecting q_s and q_t. For *positive* q_s, then, **E** is radially *outward* from q_s; for a negative q_s, **E** is radially *inward*.* See Figure 23-11. The electric field for a single point charge is especially important, because any more com-

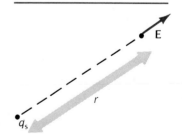

Figure 23-11. Source charge q_s produces electric field **E** at a distance *r*.

---

* A vector equation for **E** is $\mathbf{E} = kq_s\mathbf{r}/r^3$.

plicated charge distribution can always be regarded as a collection of point charges with (23-6) applying to each one.

The electric-field concept gives a new way of looking at the interaction between two charges q_a and q_b. We can say that charge q_a acts as a source of, or creates, an electric field $\mathbf{E}_a$ that surrounds it. Charge q_b, immersed in this field, is then subject to an electric force $\mathbf{F}_{a\,on\,b}$. That is,

$$q_a \text{ creates } \mathbf{E}_a; \qquad \mathbf{F}_{a\,on\,b} = q_b \mathbf{E}_a$$

(Alternatively, we can interpret the electric force of q_b on q_a by saying that q_b generates electric field $\mathbf{E}_b$ and that q_a immersed in the field $\mathbf{E}_b$ then experiences the force $\mathbf{F}_{b\,on\,a} = q_a \mathbf{E}_b$.)

This two-stage process—the production of the field by one charge and the response to the field by the second charge—may seem at first sight to be pedantry. But there is physical justification for the field concept, that is, for visualizing electric interactions as taking place via the electric field. For one thing, when q_a is moved, so that its separation distance from q_b is changed, q_b is not subject to a different force *instantaneously*. Rather, q_b continues to be influenced by the original force (therefore, electric field) for the time required for light to travel from a to b. That is, disturbances in the electric field arising from accelerated charges are propagated at the *finite* speed of light. This is no mere accident. As we shall see, light consists of electric (and magnetic) fields traveling through space.

Example 23-4. Point charges $+q$ and $-q$ are separated by a distance d. What is the electric field of the two charges at point P in Figure 23-12? (Point P is on a line perpendicular to the line separating the two charges and at a distance d above charge $+q$.)

The electric field from charge $+q$ has the magnitude [from (23-6)] of

$$E_+ = \frac{kq}{d^2}$$

with $\mathbf{E}_+$ along the $+y$ direction.

Charge $-q$ is a distance $\sqrt{2}d$ from point P. The magnitude of its electric field at P is

$$E_- = \frac{kq}{(\sqrt{2}d)^2} = \frac{kq}{2d^2}$$

Field $\mathbf{E}_-$ is at angle of $45°$ below the x axis in the third quadrant, as shown in Figure 23-12(b).

To find the vector sum of $\mathbf{E}_+$ and $\mathbf{E}_-$, we use their rectangular components. The resultant electric field $\mathbf{E}$ has the components

$$E_x = -E_- \cos 45° = -\frac{kq}{2d^2}\left(\frac{1}{\sqrt{2}}\right)$$

$$E_y = E_+ - E_- \sin 45° = \frac{kq}{d^2} - \frac{kq}{2d^2}\left(\frac{1}{\sqrt{2}}\right)$$

so that

$$E = \sqrt{E_x^2 + E_y^2} = \frac{kq}{d^2}\left(\frac{5}{4} - \frac{1}{\sqrt{2}}\right)^{1/2} = 0.74\,\frac{kq}{d^2}$$

The direction of $\mathbf{E}$ relative to the negative x axis [see Figure 23-12(c)] is given by

$$\tan \phi = \frac{E_y}{E_x} = \frac{1 - 1/(2\sqrt{2})}{1/(2\sqrt{2})} = \tan 61°$$

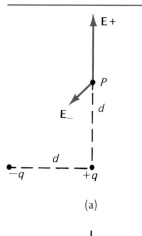

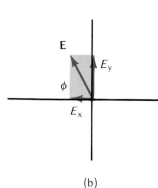

Figure 23-12. (a) Point charges $+q$ and $-q$ produce respective electric fields $\mathbf{E}_+$ and $\mathbf{E}_-$ at point P. (b) The resultant field $\mathbf{E}$ as the vector sum of its x and y components.

Example 23-5. Charge $+q$ is at the origin. Charge $+3q$ is on the x axis at $x = d$. See Figure 23-13. Where is the electric field of the two charges zero?

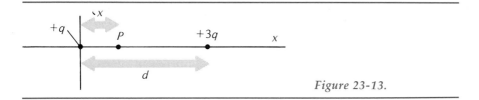

Figure 23-13.

Apart from being an infinite distance from the two charges, the point P at which $E = 0$ must be on the x axis. Only along this line can the forces of $+q$ and $+3q$ on a third charge cancel. The coordinate of P is x.

Equating the magnitudes of the electric fields produced at P by the two charges separately then gives

$$kq/x^2 = 3kq/(d-x)^2$$

where we have used the fact that charge $+3q$ is at distance $d - x$ from point P.

The equation above can be recast in the form

$$2(x/d)^2 + 2(x/d) - 1 = 0$$

which is a quadratic equation whose roots are

$$\frac{x}{d} = -\tfrac{1}{2} + \tfrac{1}{2}\sqrt{3},\ -\tfrac{1}{2} - \tfrac{1}{2}\sqrt{3}$$

or

$$\frac{x}{d} = 0.366,\ -1.37$$

The answer we want is $x/d = 0.366$. This corresponds to a location *between* the two charges, where the two *positive* charges contribute electric fields in *opposite* directions with equal magnitudes. The second root, $x/d = -1.37$ corresponds to a location to the *left* of both charges; there the two charges contribute electric fields of the same magnitude but in the same direction (to the left).

23-5 Electric-Field Lines

How can you represent the electric field of a point charge? Since **E** is a vector, one way is to use several vectors, as shown in Figure 23-14(a), with the tail of each vector at the location where **E** is specified. For a positive charge, the vectors are all radially outward. The magnitude of an **E** vector is inversely proportional to the square of the distance r from the charge.

An equivalent way of mapping the electric field is shown in Figure 23-14(b). Here we have uniformly spaced, outwardly directed, straight *electric-field lines* radiating from the positive point charge. (Strictly, the lines go outward in three dimensions.) Clearly, at any point the direction of the field line is also the direction of **E**.

What about the magnitude of **E**, which falls off with distance as $1/r^2$ does? Imagine the point charge to be surrounded by a sphere of radius r. All the same field lines will penetrate the sphere's surface, whatever its radius. The sphere's surface area ($4\pi r^2$) *is proportional to* r^2, so that the number of electric-field lines per unit area falls off with $1/r^2$ in exactly the same way as the magnitude of **E**.

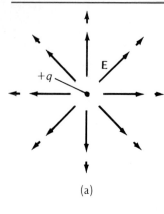

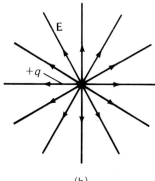

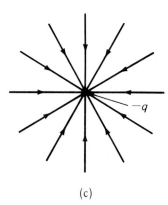

Figure 23-14. (a) Electric-field vectors for a positive point charge q. (b) Electric-field lines for a positive point charge and (c) electric-field lines for a negative point charge.

We can use the number of electric-field lines per unit area to indicate the magnitude of **E** at any point. It turns out (the general proof is given in Section 24-7) that the density of electric-field lines can always be used to represent the magnitude of the electric field. What clearly works for a single point charge, works for any charge distribution.

The electric-field lines surrounding a negative point charge are shown in Figure 23-14(c). Since the direction of **E** always gives the direction of the electric force on a *positive* charge placed in the field, the electric-field lines here are radially *inward* and converge on a negative charge.

Sketches of electric-field lines are especially useful because we can tell at a glance what will happen to still one more charge introduced into the field. A positive charge will be accelerated in the direction of **E**; a negative charge will be accelerated in the opposite direction. Important properties of electric-field lines representing the electric field **E** are these:

- Field lines **E** originate from positive charges and terminate on negative charges. Furthermore, as will be proved in Chapter 24, **E** lines are always continuous.
- At any point, the direction of an **E** line gives the direction of the electric field at that point.
- The density of **E** lines at *any* point—the number of lines per unit transverse area—is proportional to the magnitude of the electric field at that point. Therefore, the electric field becomes weaker where the **E** lines diverge and stronger where they converge.
- Electric-field lines give the direction of the *force* that will act on a unit positive charge introduced into the field. The field lines do not portray the paths or velocities of charged particles, although particle velocities, displacements, and paths can be computed from a knowledge of **E**. Only a particle's *acceleration* is along (or opposite to) the direction of **E**.
- The lines representing **E** are a useful fiction; the electric field itself is real, or as real as any other measurable physical property.

Figure 23-15 shows electric-field configurations produced by two point charges. In Figure 23-15(a) we have two separated point charges of the same magnitude but opposite sign—an *electric dipole* (see also Section 24-6). All field lines originating from the positive charge terminate on the negative charge. Point P corresponds to the location for which **E** was computed in Example 23-4. Indeed, the electric field at any location can be computed in exactly the fashion of that example. Note that close to either point charge the field lines are essentially those of a single point charge, since the second charge is then relatively far away and uninfluential. It is important to appreciate that Figure 23-15 shows the electric force acting on a *third* positive charge introduced into the presence of the two charges creating the electric field.

In Figure 23-15(b), with two separated positive charges of the same magnitude, the location midway between the two charges is one at which $\mathbf{E} = 0$, and there are no field lines there. All field lines diverge outward. Viewed from afar, the field lines approximate those of a *single* positive charge of $+2q$. If the two charges were both negative and of the same magnitude, then the configuration of the field lines would not change, but the arrows would be reversed, with the field lines then converging on the negative charges.

In Figure 23-15(c), we have charges $-q$ and $+3q$. Three times as many field lines originate from the $+3q$ charge as go to the $-q$ charge. In Figure 23-15(d)

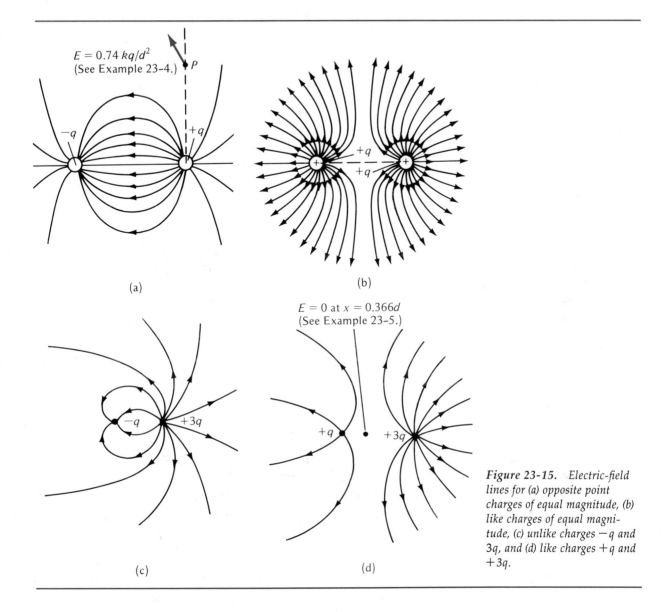

Figure 23-15. Electric-field lines for (a) opposite point charges of equal magnitude, (b) like charges of equal magnitude, (c) unlike charges $-q$ and $3q$, and (d) like charges $+q$ and $+3q$.

the spot where $\mathbf{E} = 0$ (computed in detail in Example 23-5) is closer to the $+q$ charge than the $+3q$.

Summary

Definitions

Electric field $\mathbf{E}$: the resultant electric force $\Sigma \mathbf{F}_e$ per test charge q_t,

$$\mathbf{E} \equiv \frac{\Sigma \mathbf{F}_e}{q_t} \qquad (23\text{-}5)$$

Electric-field lines may be used to present the vector field $\mathbf{E}$.

Units

Electric charge: Coulomb (C)
Electric field: Newtons per coulomb (N/C)

Fundamental Principles

Coulomb's law: for the electric force F_e between point

electric charges q_1 and q_2 separated by distance r

$$F_e = \frac{kq_1q_2}{r^2} \tag{23-1}$$

where

$$k = \frac{1}{4\pi\epsilon_0} \simeq 9.0 \times 10^9 \text{ N} \cdot \text{m}^2/\text{C}^2$$

Electric charge conservation: the net electric charge of an isolated system is constant.

Important Results

Electric charge is quantized; the observed charge magnitude is always an integral multiple of the electronic charge, $e = 1.60 \times 10^{-19}$ C.

Problems and Questions

Section 23-1 Some Qualitative Features of the Electric Force

· **23-1 Q** When a positively charged rod is brought near a negatively charged electroscope, the pivoted conductor will
(A) deflect more
(B) deflect less
(C) become positively charged.
(D) stay the same.

· **23-2 Q** A balloon that has been rubbed on your sweater and then pressed to a wall will often stick there. Which of the following is the best explanation?
(A) Rubbing removes a surface layer of grease, allowing the balloon to come close enough to the wall that air pressure holds it there.
(B) Rubbing the balloon charges it electrostatically, and this charge on the balloon induces an opposite charge on the wall. The attraction between the induced charge and the charge on the balloon holds the balloon to the wall.
(C) A wall typically has a net electric charge on it, and rubbing the balloon charges it electrostatically. If the wall happens to have a charge opposite to that on the balloon, the balloon will stick.
(D) Rubbing the balloon causes moisture to condense on it, and surface tension causes the balloon to stick to the wall.
(E) Rubber molecules form weak polymers with molecules in the wall.
(F) Rubbing the balloon surface causes it to become slightly conducting. When the balloon is touched to the wall, electrons flow from the balloon to the wall. This sets up an electric field that bonds the balloon weakly to the wall.

· **23-3 Q** A charged particle is fired at right angles toward an electrically neutral flat conducting surface. Does the particle speed up or slow down as it approaches the surface?

: **23-4 Q** Figure 23-4 shows how a single, initially uncharged conductor is given a charge by the process of "charging by induction." The positively charged object could actually be used over and over again to give an indefinitely large number of initially uncharged conductors neg-

ative charges. Does this violate energy conservation? Where does the energy come from?

: **23-5 Q** Two objects can exert an electric force on each other
(A) only if they are both conductors.
(B) only if they are both insulators.
(C) only if each carries a net nonzero charge.
(D) only if each contains some electrons.
(E) even if only one of the objects has a net charge.

: **23-6 Q** An uncharged metal sphere hangs from an insulating thread. If a positively charged glass rod is brought near the sphere (but not touching), what will happen?
(A) The sphere will be attracted to the rod.
(B) The sphere will be repelled by the rod.
(C) The sphere will experience no net force since it is electrically neutral.
(D) The sphere will acquire a net charge.

: **23-7 Q** Two identical negative charges are placed along the x axis. Midway between them is placed a small positive test charge. What can be said about the stability of the test charge?
(A) It is stable for motion along any axis.
(B) It is stable for motion only along the x axis.
(C) It is stable only for motion along the y axis.
(D) It is stable for motion along any direction perpendicular to the z axis.
(E) It is not stable for motion in any direction.

Section 23-2 Coulomb's Law

· **23-8 Q** Two charged spheres of the same size and mass are suspended by threads. Initially they attract each other. Then they touch and are electrically repelled. One can then conclude that before the spheres touched
(A) both were positively charged.
(B) both were negatively charged.
(C) the spheres had charges of equal magnitude and opposite sign.
(D) the spheres had charges of unequal magnitude and opposite sign.
(E) None of the above conclusions can be drawn.

· **23-9 Q** Two identical spheres hold charges of different

magnitudes. After these spheres are allowed to touch they will always
(A) repel each other.
(B) attract each other.
(C) sometimes attract and sometimes repel depending on the circumstances.
(D) be electrically neutral.
(E) attract at small distances and repel at large distances.

· **23-10 P** Two point charges experience an attractive force of 4 N when they are separated by 1 m. What force do they experience when their separation is (a) 0.5 m? (b) 2 m? (c) 10 m?

: **23-11 P** Charges q_1, q_2, q_3, and q_4 are placed at the corners of a square of side 20 cm (Figure 23-16). Determine the magnitude and direction of the force on q_1, given $q_1 = q_2 = q_3 = 5 \, \mu C$ and $q_4 = -5 \, \mu C$.

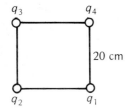

Figure 23-16. Problem 23-11.

: **23-12 P** Three charges are placed as shown in Figure 23-17. What is the force on each charge?

Figure 23-17. Problem 23-12.

: **23-13 P** A charge of $+2 \, \mu C$ is placed at the origin and a charge of $-4 \, \mu C$ is placed at $x = 6$ cm. Where can a third charge of $+1 \, \mu C$ be placed (not at infinity) so that the electric force on it will be zero?

: **23-14 P** Charges $+2Q$, $+2Q$, and $-Q$ are placed at the corners of an equilateral triangle of side a. Determine the magnitude and direction of the force on each charge.

: **23-15 P** Three charges are placed as follows: $+Q$ at $x = a$, $+Q$ at $x = -a$, and $-Q$ at $y = +a$. Determine the force on the charge $-Q$.

: **23-16 P** Five charges $+Q$ are placed in a line with equal spacing a between adjacent charges. Determine the force on each.

: **23-17 P** Two small spherical conductors are separated by a distance d, which is large compared with their radii. How should a fixed amount of charge Q be distributed between them to maximize the force of repulsion between them?

: **23-18 P** Two small spherical conductors, each of mass 10 gm, are suspended from two threads attached to a common point. The threads are each 60 cm long. A charge Q is placed on one conductor. This conductor is then touched to the second conductor, which was initially uncharged. The two then repel each other. When they come to rest, each string makes an angle of 37° with the vertical. What is the initial charge Q?

: **23-19 P** Charges $+Q$ and $-Q$ are placed at the corners of a cube of side a so that nearest-neighbor charges all have opposite signs. Determine the magnitude and direction of the force on a charge $+Q$ and on a charge $-Q$.

: **23-20 P** Charge Q_1 is placed at $z = 1 \times 10^{-10}$ m and charge Q_2 at $z = -1 \times 10^{-10}$ m. Determine the force on a test charge q placed at $x = 1 \times 10^{-8}$ m and at $x = 2 \times 10^{-8}$ m. (a) First take $Q_1 = Q_2 = 1.6 \times 10^{-19}$ C. (b) Then take $Q_1 = -Q_2 = 1.6 \times 10^{-19}$ C. This combination of charges is called an *electric dipole*. Note that the force due to it decreases with increasing distance much more rapidly than the force due to two like charges (which are effectively a "monopole"). (c) Calculate the ratio of the force at 10^{-8} m and at 2×10^{-8} m in each case.

: **23-21 P** Two equal positive charges $+Q$ are placed on the x axis at $x = +a$ and $x = -a$. Where should a charge q be placed on the z axis to experience the maximum force?

Section 23-3 Further Characteristics of Electric Charge

· **23-22 P** A rubber rod that has been rubbed with cat's fur acquires a charge of -4×10^{-10} C. How many electrons were transferred from the fur to the rod?

· **23-23 P** Two electrons are initially at rest with a separation of 2 cm. What acceleration will they experience when released?

· **23-24 P** The electron and the proton in a hydrogen atom are separated by about 5.3×10^{-11} m. (a) What is the magnitude of the electric force between them? (b) What is the ratio of the electric force to the gravitational force of attraction?

· **23-25 P** In an electroplating process, the ions Ag⁺ deposit a total charge of 10 C. To what mass of silver does this correspond? (Atomic mass of silver, 108.)

· **23-26 P** A cell membrane has a typical thickness of 10^{-8} m. What is the force of attraction between ions Na⁺ and Cl⁻ separated by this distance?

· **23-27 Q** A negative electric charge
(A) interacts only with positive charges.
(B) interacts only with negative charges.
(C) interacts with both positive and negative charges.
(D) may interact with either positive or negative charges, depending on the circumstances.
(E) can always be subdivided into two equal negative electric charges.

· **23-28 P** The 92 protons in the uranium nucleus have an

average separation of 2.5×10^{-15} m. What is the electric force of repulsion between two adjacent protons? (Another force, the so-called strong force, or nuclear force, pulls them together.)

· **23-29 P** The fission of $^{236}_{92}$U can produce the fragments $^{146}_{56}$Ba and $^{90}_{36}$Kr. These two nuclei have charges $+56e$ and $+36e$ respectively. Determine the Coulomb force acting on each just after their formation when their centers are separated by 1.6×10^{-14} m.

: **23-30 P** In the Bohr model of the hydrogen atom, an electron is a point charge orbiting the proton, which is the atom's nucleus. For an electron moving in a circular orbit of radius 5.3×10^{-11} m, what is the frequency of revolution if its Coulomb attraction to the proton provides the required centripetal force?

: **23-31 P** One theory of the expanding universe imagines that a hydrogen atom, instead of being electrically neutral, carries a slight positive charge (that is, the charge on the nucleus is slightly greater than the charge on the orbiting electron). This causes an electric force of repulsion between hydrogen atoms that counteracts the gravitational attraction. If these two forces are in balance, what value of $\Delta q/q$ is required, where $-q =$ charge on the electron and $q + \Delta q =$ charge on the hydrogen nucleus (the proton)? In other words, Δq is the net charge on the atom.

: **23-32 P** A conducting sphere with a diameter of 1.0 cm has a net charge of only -1.0 pC. Imagine for simplicity that the excess electrons are uniformly spread in a single layer over the sphere's outer surface. What is the approximate distance between adjacent electrons?

: **23-33 P** What fraction of the electrons in the earth would have to be removed from it and placed on the moon so that the electrostatic attraction between the earth and moon would have the same magnitude as their gravitational attraction? (To simplify the computation make the approximation that the nucleus of any element has equal numbers of protons and neutrons, each with a mass of 1.7×10^{-27} kg.)

Section 23-4 Electric Field Defined

· **23-34 P** A spherical drop of latex of mass 2×10^{-4} gm is suspended in a vertically downward electric field of 600 N/C. What is the net charge on the droplet?

· **23-35 P** A dust particle of mass 2×10^{-6} kg has a charge of 3 μC. What vertical electric field is required to suspend the particle against the downward force of gravity?

· **23-36 P** The surface of the earth carries a negative charge, and this charge gives rise to an electric field of about 30 N/C downward at an elevation of 1.5 km. (a) What net charge would a very small aircraft of only 500 kg mass have to carry to be supported against the force of gravity? (b) To how many extra electrons would this correspond?

: **23-37 Q** In a perfect conductor, charge carriers can move freely. Because of this property one is led to conclude that in electrostatic equilibrium
(A) the net charge on the conductor is distributed uniformly throughout its volume.
(B) there is a constant nonzero electric field throughout the volume of the conductor.
(C) the electric field is zero everywhere throughout the volume of the conductor.
(D) a conductor cannot carry a net charge.

: **23-38 P** Charge $+Q$ and $-2Q$ are separated by a distance d. Where is the electric field due to them zero?

: **23-39 P** A charge of $+4$ μC is 2 cm from a charge of $+6$ μC. Where is the electric field of the two charges zero?

: **23-40 Q** Charges $+Q$ and $-2Q$ are placed as shown in Figure 23-18. Near which of the points indicated is the electric field due to these charges most likely to be zero?

Figure 23-18. Problem 23-40.

: **23-41 Q** Charges Q, Q, and $-2Q$ are placed at the corners of an equilateral triangle, as shown in Figure 23-19.

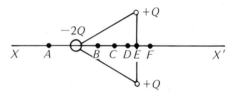

Figure 23-19. Problem 23-41.

The point along the line XX' where the electric field is most likely to be zero is
(A) A
(B) B
(C) C
(D) D
(E) E
(F) F
(G) None of the above, since the field vanishes only at infinity.

: **23-42 P** The electric field between the electrodes of the gas discharge tube used in a certain neon sign has a magnitude of 5×10^4 N/C. What acceleration will a neon ion experience? A neon ion's mass is 3.3×10^{-26} kg; its charge is $+e$?

: **23-43 Q** The electric field at a point in space near electric charges is the
(A) force per unit charge acting on a small test charge placed at the point.
(B) work done per unit test charge against the force vectors in carrying a charge from infinity to the point.

(C) electric force at the point.
(D) work done against electric forces in carrying a test charge from infinity to the point.

: **23-44 Q** You have a collection of small spheres. You can adjust the sign and magnitude of the charge on each, and the mass of each, to whatever values you wish. Can you then, using only electrostatic forces, place the spheres so that they would be in stable equilibrium?

: **23-45 Q** The electric field lines arising from two charges Q_1 and Q_2 are shown in Figure 23-20. From this drawing we can see that
(A) the electric field could be zero at P_1.
(B) the electric field could be zero at P_2.
(C) both Q_1 and Q_2 have the same sign.
(D) $|Q_1| > |Q_2|$.
(E) none of the above is true.

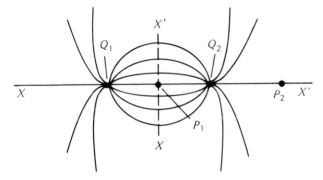

Figure 23-20. Problem 23-45.

: **23-46 P** A charge $+2Q$ is a distance d from a charge $-Q$. (a) Where is the electric field from the two charges zero? (b) Sketch the electric-field lines from the charges.

: **23-47 P** Charges $+Q$ are placed at two adjacent corners of a square of side a, and charges $-Q$ are placed at the other two corners. (a) Determine the magnitude and direction of the electric field at the center of the square. (b) Sketch the electric-field lines.

: **23-48 P** Which of the following is the most accurate statement?
(A) *Electric field* and *electric force* are two terms with the same meaning.
(B) Electric-field lines never cross.
(C) Electric-field lines indicate the direction but not the magnitude of the electric field.
(D) An electric field line at a given point indicates the direction an electron placed at that point would start to move if released from rest.
(E) An electric field is strictly a theoretical construct without physical reality.

: **23-49 Q** Charges $+q$, $+Q$, and $-Q$ are placed at the corners of an equilateral triangle of side a. (a) Determine approximately where the electric field is zero. (b) Sketch some of the electric field lines.

Supplementary Problem

23-50 An *electrostatic precipitator* is a device for removing dust particles from a smoke stack. It works by giving an electric charge to dust particles in a smoke stack, deflecting them in an electric field, and then collecting the charged particles on an electrode with the opposite charge. Suppose a dust particle rises through the precipitator with a constant velocity of 5 m/s, and it is given a specific charge of 10^{-5} C/kg. What horizontal electric field would deflect the dust particle 0.5 m horizontally as it ascends 20 m through the stack?

24 Continuous Distributions of Electric Charge

24-1 Electric Field for Three Simple Geometries
24-2 Uniform Electric Field
24-3 Electric Flux
24-4 Gauss's Law
24-5 Electric Field and Charged Conductors
24-6 An Electric Dipole in an Electric Field (Optional)
24-7 General Proof of Gauss's Law (Optional)
Summary

In continuing with electric fields, we see first how a continuous distribution of electric charge can be subdivided into a collection of effective point charges, and their individual contributions added, to yield a resultant electric field. Through the concepts of electric flux and Gauss's law, we consider an alternative procedure for finding the electric field from charge distributions. We examine special properties for the electric fields produced by charged conductors. The characteristics of electric dipoles in an electric field and the rigorous proof of Gauss's law are given in optional sections.

24-1 Electric Field for Three Simple Geometries

Here we consider the electric field produced by three particularly simple geometrical arrangements of electric charge.*

- A single point charge.
- A uniform, infinite straight line of continuous charge.
- A uniform, infinite flat sheet of continuous charge.

* The results of Section 24-1 are also derived in Section 24-4, using Gauss's law.

More complicated charge distributions can frequently be thought of as superpositions of simple distributions.

Point Charge We have already found the electric-field magnitude to be

$$\text{Point charge:} \quad E = \frac{kq}{r^2} \propto \frac{1}{r^2} \qquad (23\text{-}6)$$

The field **E** is radial (outward for positive q, inward for negative q) and its value falls off inversely with the square of the distance r from the charge.

Line of Charge Electric charge is distributed uniformly along the length of an infinitely long straight line. The constant charge per unit length, or *linear charge density*, is represented by λ (the Greek letter L for "linear"); SI units for λ are coulombs per meter.

The line of charge lies along the x axis. See Figure 24-1(a). We want to find the direction and magnitude of **E** at point P (shown on the y axis), which is a perpendicular distance r from the charged wire. It is easy to show, on the basis of *symmetry alone*, that **E** must have the following characteristics:

- **E** is perpendicular to the charged wire (outward for positive charge).
- **E** has the same magnitude at all points the same distance from the wire.
- **E** is uniformly distributed in angular position in a plane transverse to the wire [in Figure 24-1(a), the yz plane].

The kind of proof to be used here will appear several times in this chapter and subsequent ones. We assume the contrary proposition and then show that it would lead to a contradiction and therefore cannot be true.* Consider the first assertion, that **E** is perpendicular to the wire. Suppose, for the sake of argument, that **E** at some point were tipped toward the right. This would mean that the side to the right of P differed from the left side. But this cannot be true because the wire is *uniformly* charged throughout and extends to *infinity* in both directions. The two other properties of **E** follow from symmetry considerations likewise.

To deal with a *continuous* charge distribution, we simply imagine it subdivided into effectively infinitesimal *point* charges dq; then we add, in vector fashion, the contribution $d\mathbf{E}$ to the resultant electric field **E** from all the point charges in the continuous distribution.†

We concentrate on the charge element dq lying within the infinitesimal length element dx [Figure 24-1(b)], where $dq = \lambda dx$ at the coordinate x. This effective point charge is a distance R from point P, and along line R it produces an electric field $d\mathbf{E}$, whose magnitude is

$$dE = \frac{k \, dq}{R^2} = \frac{k\lambda \, dx}{R^2}$$

---

* This type of argument, originated by the ancient Greeks, is officially known as *reductio ad absurdum*, "reduction to an absurdity."
† By "point" here we mean, strictly, a very small volume. The volume must be very small compared with overall dimensions of the charged wire, yet large enough to contain many atoms. If the volume were truly submicroscopic, the electric field would fluctuate violently as we approached individual electrons or nuclear particles and receded from them.

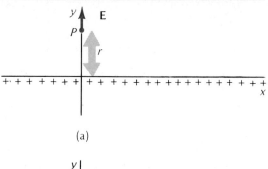

Figure 24-1. (a) The electric field **E** at point P a distance r from a uniformly charged infinite straight line. (b) The electric field d**E** at point P contributed by the element dx of an infinite, uniformly charged wire.

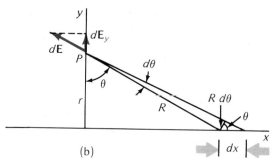

We know that only the transverse component dE_y will contribute to the resultant field (the longitudinal component dE_x is cancelled by an equal but opposite component from a charge element of the same distance from the origin on the *negative x* axis). Therefore,

$$dE_y = dE \cos \theta = \frac{k\lambda \, dx \cos \theta}{R^2}$$

We have three variables—x, R, and θ. The integration we must carry out is simplest with θ as the variable. To eliminate R and x, consider the geometry of Figure 24-1(b), where

$$R = \frac{r}{\cos \theta} \quad \text{and} \quad dx = \frac{R \, d\theta}{\cos \theta}$$

Using these relations to eliminate x and R in the equation above, we get

$$dE_y = \frac{k\lambda \cos \theta \, d\theta}{r}$$

To find the resultant field E from an infinitely long wire, we integrate θ from $-\pi/2$ to $+\pi/2$:

$$E = \int dE_y = \frac{k\lambda}{r} \int_{-\pi/2}^{\pi/2} \cos \theta \, d\theta$$

Line of charge: $$E = \frac{2k\lambda}{r} \propto \frac{1}{r} \qquad (24\text{-}1)$$

The magnitude of **E** falls off inversely as the *first* power of the distance r from the uniformly charged wire.

SECTION 24-1 Electric Field for Three Simple Geometries   515

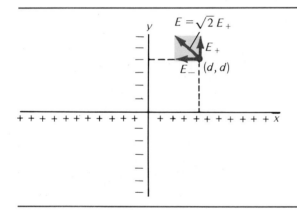

Figure 24-2.

Example 24-1. There is a uniform *positive* linear charge density λ along the x axis and a uniform *negative* charge density of the same magnitude along the y axis. What is **E** at the coordinates (d, d)?

See Figure 24-2, where the electric fields, $\mathbf{E}_+$ and $\mathbf{E}_-$, contributed separately by the positive and negative lines of charge, are shown. We exploit the superposition principle for electric forces, and now for electric fields; the resultant field is simply the vector sum of the contributions from all the parts. In the present problem, we simply find the electric fields separately from the two infinite lines of charge and then add them as vectors to find the resultant **E**. From (24-1), with $r = d$, we have

$$E_+ = \frac{2k\lambda}{d}$$

Similarly,

$$E_- = \frac{2k\lambda}{d}$$

The resultant field **E** has the magnitude

$$E = \sqrt{2}\, E_+ = \frac{2\sqrt{2}\, k\lambda}{d}$$

and makes an angle of 45° with the negative x axis, as shown in Figure 24-2.

The lines of charge in this example could certainly *not* be on two long and initially uniformly charged *conducting* wires, even if somehow they could be kept from touching at the origin. Charges on one wire would influence the charge distribution on the other, and neither could remain uniform.

Infinite Surface of Charge   The infinite plane surface is uniformly charged, with constant *surface charge density* σ (Greek s for surface); in SI units, σ would be given in coulombs per square meter.

Again, the direction of **E** is determined by the geometry of a uniform, charged infinite plane; **E** must be perpendicular to the surface (outward for positive charge). See Figure 24-3(a). Suppose it were not, so that **E** at some point was tipped, say, to the right. Then, the region to the right would have to be somehow different from the left. But this would violate the assumption of *uniform* charge over an *infinite* plane. Furthermore, the magnitude of **E** can depend, at most, on the perpendicular distance to the plane, not on location along the plane.

To sum all contributions to **E**, we could imagine the plane subdivided into small patches $dx\,dy$, each effectively a point charge. Still more simply, however,

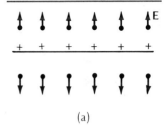

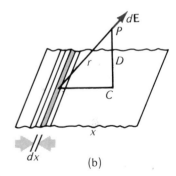

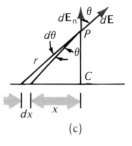

Figure 24-3. Electric field from a uniformly charged infinite flat sheet. (a) On the basis of symmetry, **E** must be perpendicular to the sheet. (b) and (c) An infinitely long charged strip of width dx produces field $d\mathbf{E}$ at point P.

we can think of the plane as a collection of infinitely long, uniformly charged, and parallel wires. In Figure 24-3(b), the field $d\mathbf{E}$ at point P a distance D from the surface is contributed by one wire of width dx and a distance x from point C on the plane directly below P. Over a length L along one wire, there is a charge λL; we can also think of this charge as spread over an area $L\,dx$ and having the magnitude $\sigma L\,dx$. Therefore, $\lambda L = \sigma L\,dx$, or $\lambda = \sigma\,dx$. One strip contributes an electric field $d\mathbf{E}$, whose magnitude is, from (24-1),

$$dE = \frac{2k\lambda}{r} = \frac{2k\sigma\,dx}{r}$$

Only the normal component dE_n of $d\mathbf{E}$ will not be cancelled, and we have

$$dE_n = dE\cos\theta = \frac{2k\sigma\,dx\,\cos\theta}{r}$$

Again, the integration is most easily carried out with θ as the variable. From the geometry of Figure 24-3(c), we have

$$dx = \frac{r\,d\theta}{\cos\theta}$$

Using this relation in the one above it, we then get

$$dE_n = 2k\sigma\,d\theta$$

We integrate θ from $-\pi/2$ to $+\pi/2$ to include all strips covering the infinite plane:

$$E = \int dE_n = 2k\sigma \int_{-\pi/2}^{+\pi/2} d\theta = 2\pi k\sigma$$

Plane of charge: $E = 2\pi k\sigma$ (24-2)

The electric field is *constant* in magnitude; for a truly infinite plane of uniform charge, **E** does not depend on the distance from the plane.*

24-2 Uniform Electric Field

The simplest electric-field configuration is this: **E** is *uniform, constant,* or *homogeneous*. The direction and magnitude of **E** is the same at all points in a uniform field, and electric-field lines are straight, uniformly spaced, and parallel.

As we have seen in Section 24-1, an infinite, uniformly charged sheet produces a uniform field. But even a uniformly charged sheet of finite size produces an effectively constant **E** at points that are close to the sheet and relatively far from the edge of the sheet. It is simply that if we are very close to a flat sheet of finite size, the sheet appears to be effectively infinite.

Now let us find the electric field produced by two uniformly charged sheets with the same surface charge density but of opposite sign. We merely superpose the fields produced by each sheet separately. First, with a single positively

---

* We shall later find it advantageous to replace the constant k by $1/4\pi\epsilon_0$. Then (24-2) can be written as $E = \sigma/2\epsilon_0$.

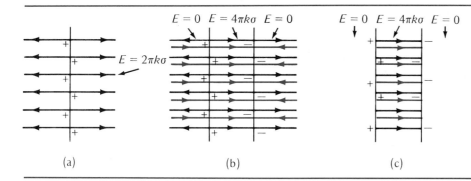

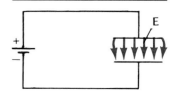

Figure 24-4. (a) Electric field of a single uniformly charged sheet. (b) Electric fields of two parallel uniformly charged sheets of opposite sign. (c) Resultant electric field of part (b).

Figure 24-5. A uniform electric field **E** between parallel charged conducting sheets connected to the two terminals of a constant-voltage source such as a battery.

charged sheet, a uniform outward electric field of magnitude $E = 2\pi k\sigma$ is produced on *both sides* of the sheet, as shown in Figure 24-4(a). Of course, for a negatively charged sheet, field **E** is toward the surface on both sides of the charged sheet. Now, consider the net field from two parallel charged sheets with *opposite* surface charges, both of the same magnitude σ, as in Figure 24-4(b). Between the sheets, the resultant field has equal contributions from the two surfaces. The resultant field between the sheets is $E = 4\pi k\sigma$. Outside this region, the fields from the two oppositely charged surfaces cancel and $E = 0$ [see Figure 24-4(c)].

The most common way to produce a uniform **E** in the laboratory is with two parallel conducting sheets or plates connected to the oppositely charged terminals of a battery (or some other source maintaining a constant "voltage"). See Figure 24-5. If the separation distance between plates is small compared with the width of either plate, the electric field is very nearly uniform at interior points, far from the edges of the plates. A fringing of the electric field, or a bending of **E** lines, takes place at the plate edges.

What is the motion of a particle with charge q and mass m in a uniform field **E**? The electric force $\mathbf{F}_e$ on the particle is

$$\mathbf{F}_e = q\mathbf{E}$$

The force is constant in both magnitude and direction. The particle's acceleration **a**, also constant, is given by

$$\mathbf{a} = \frac{\mathbf{F}_e}{m} = \frac{q}{m}\mathbf{E}$$

We already know that a particle or projectile in a uniform gravitational field has a constant acceleration **g** downward and traces out, in general, a parabolic path (Chapter 4). So too, a charged particle in a uniform **E** field traces out a parabolic path. All the relations derived for a projectile apply also to the electrically charged particle. We merely replace **g** by the acceleration $(q\mathbf{E}/m)$. Of course, a *positively* charged particle has a constant acceleration *in* the direction of **E**; a *negatively* charged particle is accelerated in the direction *opposite* to **E**.

Example 24-2. An electron with charge e, mass m, and initial horizontal velocity $\mathbf{v}_0$ is fired horizontally between two parallel, oppositely charged conductors, producing

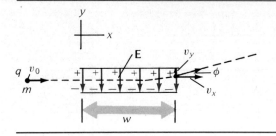

Figure 24-6. A charged particle is fired into the uniform electric field between two oppositely charged parallel plates.

a constant, vertically downward electric field **E**. See Figure 24-6. The electron travels a horizontal distance w in the uniform electric field; the gravitational force on the electron is negligible compared with the electric force on it. What are the electron's velocity components (v_x and v_y) as it emerges from the plates?

While the electron is in the uniform field, it has the following acceleration components:

$$a_x = 0$$

$$a_y = +\frac{qE}{m}$$

Note that the acceleration is *upward*; we have a *negative* charge in a *downward* **E**.

The velocity components of the electron as it leaves the uniform field are

$$v_x = v_0 = \frac{w}{t}$$

$$v_y = a_y t = \left(\frac{qE}{m}\right)t$$

where t is the time the electron spends between the plates. Along the horizontal, the electron has a constant velocity component v_0 and travels a horizontal distance w. Eliminating t between the two equations above gives

$$v_y = \left(\frac{qE}{m}\right)\left(\frac{w}{v_0}\right)$$

The angle ϕ at which the electron emerges is then specified by

$$\tan\phi = \frac{v_y}{v_x} = \frac{qEw/mv_0}{v_0} = \frac{qEw}{mv_0^2}$$

After emerging from the plates, the electron coasts in a straight line.

Precisely the arrangement shown in Figure 24-6 can be used to deflect an electron beam in a CRT (cathode ray tube, or oscilloscope), using electrostatic deflection. The vertical position of electrons striking the fluorescent screen is controlled by horizontal charged plates; the horizontal position, by vertical charged plates. Note that for small ϕ (with $\tan\phi \approx \phi$), the deflection angle $\phi \approx qEw/mv_0^2$ is directly proportional to the electric field E between the plates and inversely proportional to the electron's kinetic energy. The displacement of electrons on the screen is a direct measure of the electric field applied to the deflecting plates.

Example 24-3. An electron with an initial speed of 3.0×10^6 m/s ($\frac{1}{100}$ the speed of light, a relatively low speed for an electron) is fired in the same direction as a uniform electric field with a magnitude of 100 N/C. How far does the electron travel before being brought to rest momentarily and turned back?

The constant force on the electron is opposite to its initial velocity $\mathbf{v}_0$. The acceleration magnitude is $a = F_e/m = -eE/m$. If the electron travels a distance x before coming to rest, we have

$$x - x_0 = \frac{v^2 - v_0^2}{2a} \tag{3-10}$$

$$x = -\frac{v_0^2}{2a} = \frac{v_0^2 m}{2eE}$$

$$= \frac{(3.00 \times 10^6 \text{ m/s})^2 (9.11 \times 10^{-31} \text{ kg})}{2(1.60 \times 10^{-19} \text{ C})(100 \text{ N/C})} = 0.26 \text{ m}$$

The last equation can also be written as

$$\tfrac{1}{2}mv_0^2 = (eE)x = F_e\, x$$

and we can describe the situation above in terms of work and kinetic energy as follows. An electric force $F_e = eE$ does negative work on the electron in slowing it to rest as the electron's kinetic energy drops from $\tfrac{1}{2}mv_0^2$ to zero.

24-3 Electric Flux

Gauss's law, our principal concern in Section 24-4, is a statement about electric flux, so we first define this quantity.

Imagine a closed surface of arbitrary shape — a so-called Gaussian surface; we can think of it as an imaginary balloon that can be pushed into any shape we wish. We focus on some very small patch of a Gaussian surface, as shown in Figure 24-7(a). An infinitesimally small patch of surface is effectively flat, and we can represent both its orientation and the size of its area by a vector $d\mathbf{S}$. The element of surface vector $d\mathbf{S}$ is *perpendicular* to the patch and points *outward* from the closed surface. The magnitude of $d\mathbf{S}$ represents the area of the patch. Clearly, the orientation of $d\mathbf{S}$ will differ from one part of the Gaussian surface to another, as shown in Figure 24-7(b).

Now consider the electric field $\mathbf{E}$ penetrating some surface element $d\mathbf{S}$, as shown in Figure 24-8. (What is responsible for $\mathbf{E}$ does not concern us at the moment.) Electric field $\mathbf{E}$ makes an angle θ with $d\mathbf{S}$; equivalently, $\mathbf{E}$ makes angle θ with the outward normal of the surface patch. Since $d\mathbf{S}$ is vanishingly small, $\mathbf{E}$ can be taken to be constant over $d\mathbf{S}$. By definition, the *electric flux* through the surface element is

$$d\phi_E \equiv E \cos\theta\, dS$$

In words, the electric flux is the component $E \cos\theta$ of the electric field parallel to $d\mathbf{S}$ (or perpendicular to the surface) multiplied by area dS. We can express the same thing more compactly by using the dot product (Section 9-4) of vectors $\mathbf{E}$ and $d\mathbf{S}$:

$$d\phi_E \equiv \mathbf{E} \cdot d\mathbf{S} \tag{24-3}$$

Figure 24-9 shows three particularly simple situations for finding $d\phi_E$:

(a) $\phi = 0$; $\mathbf{E}$ is parallel to $d\mathbf{S}$; and $d\phi_E = +E\, dS$.
(b) $\phi = 90°$; $\mathbf{E}$ is perpendicular to $d\mathbf{S}$ (and lies in the surface); and $d\phi_E = 0$.
(c) $\phi = 180°$; $\mathbf{E}$ is opposite to $d\mathbf{S}$ (and points into the Gaussian surface); and $d\phi_E = -E\, dS$.

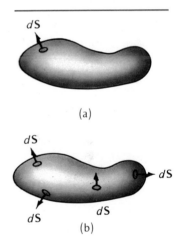

Figure 24-7. *(a) A closed Gaussian surface. Vector $d\mathbf{S}$ represents a small patch of area on the surface. (b) Every differential surface element $d\mathbf{S}$ is outwardly directed from the Gaussian surface.*

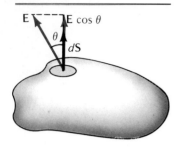

Figure 24-8. Electric field **E** at the location of surface element d**S**. The component of **E** perpendicular to the surface (and parallel to d**S**) is $E \cos \theta$.

Electric-field lines coming *out* of the surface correspond to *positive* electric flux; lines going *in* correspond to *negative* flux. When the field lines lie in the surface, there is no flux through that surface.

The magnitude of **E** can be imagined as represented by the number of electric-field lines per transverse cross section (Section 23-5); therefore, electric flux $d\phi_E$ is a measure of the number of electric-field lines coming through d**S**.*

A Gaussian surface can be chosen, we shall see, to have any shape. To make the calculation of electric flux easy, we shall always try to choose the shape so that:

- **E** is parallel to d**S**, and the flux is $d\phi_E = +E\, dS$.
- **E** is perpendicular to d**S**, and the flux is $d\phi_E = 0$.

To find the total electric flux ϕ_E over an entire closed Gaussian surface, we merely sum up the contributions from all surface elements, taking into account, of course, that both **E** and d**S** may vary in magnitude and direction from one point on the surface to another. The total electric flux is

$$\phi_E = \oint d\phi_E = \oint \mathbf{E} \cdot d\mathbf{S} \qquad (24\text{-}4)$$

The total flux $\oint \mathbf{E} \cdot d\mathbf{S}$ is sometimes referred to as the surface integral of vector field **E** over the closed Gaussian surface. The little circle on the integral sign is there simply to remind us that the integration is taken over the entire Gaussian surface. Although evaluating such an integral might, at first sight, seem formidable, we shall see that it is remarkably easy in situations with a high degree of geometrical symmetry.

Example 24-4. What shape of Gaussian surface will yield *zero* total electric flux in a *uniform* electric field?

Two orientations of a Gaussian surface are easily dealt with: **E** perpendicular to d**S**, and **E** parallel to d**S**. The Gaussian surface shown in Figure 24-10 — a right cylinder

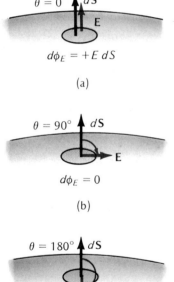

Figure 24-9. Electric flux $d\phi_E$ through surface element d**S** for three simple orientations of the electric field **E**: (a) **E** parallel to d**S** and $d\phi_E$ positive; (b) **E** perpendicular to d**S** and $d\phi_E$ zero; and (c) **E** anti-parallel to d**S** and $d\phi_E$ negative.

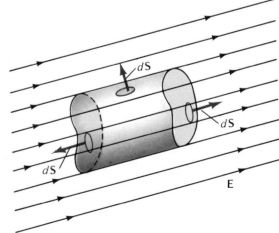

Figure 24-10. Electric flux through a cylinder aligned with uniform field **E**.

---

* Why call $\mathbf{E} \cdot d\mathbf{S}$ "flux," which implies something flowing? For the electric flux of the vector electric field, nothing flows. But consider the corresponding quantity for the streamlines of a fluid's velocity field (Sections 16-6 and 16-7). The flux $d\phi_v = \mathbf{v} \cdot d\mathbf{S}$ is actually the volume of fluid per unit time, or volume flux, flowing "out" through d**S**.

with ends of area S perpendicular to $\mathbf{E}$—satisfies these conditions. The flux through the end from which electric-field lines emerge is $+ES$. The flux at the other end is $-ES$. There is no flux through the side wall of the cylinder. The total electric flux is zero.

It turns out, as we shall see, that the electric flux is zero for a Gaussian surface of any shape in a uniform field. Indeed, the result is still more general. The total electric flux through a closed Gaussian surface of any shape is zero for *any* configuration of electric field—uniform or nonuniform—so long as there is no net electric charge inside the surface.

24-4 Gauss's Law

Elegant, powerful, neat. That's how Gauss's law is often described. With it, one can for simple geometries perform easily—sometimes with just one line of mathematics—calculations that would otherwise be long and complicated (like the one for the electric field from an infinitely long charged wire given in Section 24-1). The derivation of Gauss's law here is simple and special; a general proof is given in Section 24-7.

Let us find the total electric flux from a single point charge q (assumed to be positive for definiteness) under the simplest possible circumstances. We imagine q to be surrounded by a spherical Gaussian surface of radius r with q at its center. See Figure 24-11. The electric field $\mathbf{E}$ produced by the single point charge is radially outward and perpendicular to the spherical surface at each point. Finding the total electric flux is then particularly simple.

The magnitude of $\mathbf{E}$ at a distance r from q is

$$E = \frac{kq}{r^2} \qquad (23\text{-}6)$$

For reasons soon to be obvious, it is a good idea to replace the constant k in Coulomb's law by its equivalent.

$$k \equiv \frac{1}{4\pi\epsilon_0} \qquad (23\text{-}2)$$

Written in terms of ϵ_0, the permittivity of free space, the equation above for E becomes

$$E = \left(\frac{1}{4\pi\epsilon_0}\right)\frac{q}{r^2}$$

The total flux through the Gaussian sphere, with a surface area of $4\pi r^2$, is

$$\phi_E = \oint \mathbf{E} \cdot d\mathbf{S} = \frac{1}{4\pi\epsilon_0}\frac{q}{r^2}(4\pi r^2)$$

$$\phi_E = \frac{q}{\epsilon_0} \qquad (24\text{-}5)$$

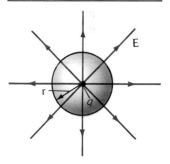

Figure 24-11. Point charge q at the center of a Gaussian surface of radius r.

That's the end of the derivation. The crucial point: distance cancels out, because the way in which the Coulomb force falls off with distance ($\propto 1/r^2$) is matched exactly by the way a spherical surface area grows with increased radius ($\propto r^2$).

What (24-5) says in words is this: The total electric flux ϕ_E through the

522   CHAPTER 24   Continuous Distributions of Electric Charge

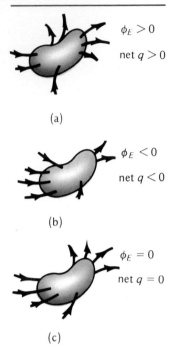

Figure 24-12. To find the net charge within a Gaussian surface, one can simply count the net number of **E** lines out of the surface: (a) positive electric flux, positive charge within; (b) negative electric flux, negative charge within; and (c) zero net flux, zero net charge within.

closed surface with charge q inside is equal to the value of that charge (including sign) divided by the permittivity constant ϵ_0. *Gauss's law* is just the statement that this simple result holds *in general*—the total electric flux through a closed surface of *any* shape is simply the *net* charge q enclosed *anywhere* within that surface divided by ϵ_0 (and does not depend on any charge that may be outside the Gaussian surface). Proving Gauss's law in general (Section 24-8) depends simply on invoking geometry and two fundamental characteristics of the Coulomb force: (a) it is exactly inverse-square; and (b) the superposition principle for forces applies.

Gauss's law says, in effect, that if we know what is happening to electric-field lines over a closed surface, we can deduce the net charge within. Recall that the flux out of a surface is a direct measure of the number of electric-field lines emerging through the surface. Then Gauss's law in graphical terms is illustrated in Figure 24-12 as follows:

- Net positive flux; more E lines out of the surface than in; net charge within positive.
- Net negative flux; more E lines in than out; net charge within negative.
- Zero net flux; same number E lines in and out; zero net charge within.

Since ϕ_E is proportional to the net charge q, and ϕ_E is also proportional to the number of **E** lines, it is proper to draw the number of electric-field lines from a charge that is proportional to q.

Gauss's law may be thought of as a fundamental statement about the properties of electric fields and the lines through which they are portrayed. It incorporates the inverse-square character of Coulomb's law, but it also has this to say about **E** lines:*

- **E** lines originate from positive charges.
- **E** lines terminate on negative charges.
- **E** lines are continuous.

Applied to charge distributions in general, Gauss's law involves a surface integral whose evaluation may be difficult. But for uniform charge distributions of high geometrical symmetry—a uniform, infinite line of charge, for example—it is a simple, almost trivial matter to compute the integral. The key is this; since the Gaussian surface can have any shape, we choose the shape to match the symmetry of the charge distribution.

Example 24-5 Electric Field of a Uniformly Charged Infinite Wire. We have already worked this problem (Section 24-1) by summing the contributions to the resultant electric field from infinitesimal charge elements. The result:

$$E = \frac{2k\lambda}{r} = \frac{\lambda}{2\pi\epsilon_0 r} \tag{24-1}$$

where λ is the constant linear charge density.

To apply Gauss's law, especially to choose the shape of the Gaussian surface to make applying Gauss's law easy, we exploit what we already know about the electric

---

* As treated here, Gauss's law applies to electric fields produced by electric charges at rest. But as we shall see (Chapter 31), electric fields can also be created by a changing magnetic flux. Such an electric field consists solely of loops (each **E** line ends on its own tail). Loops of **E** contribute nothing to the flux, and Gauss's law is valid for all situations.

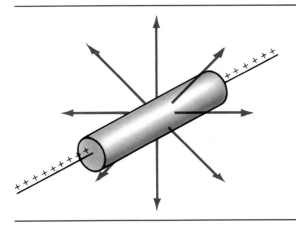

Figure 24-13. A uniform infinite line of charge with a coaxial cylindrical Gaussian surface.

field on the basis of symmetry alone. Arguments given in Section 24-1 (not repeated here) show that all **E** lines are perpendicular to the charged wire, uniformly spaced and radially outward from the wire in a plane perpendicular to the wire, and dependent only on the distance r to the wire. The shape of Gaussian surface that will make it easy to find flux ϕ_E is a right circular cylinder of radius r and length L whose axis coincides with the wire, as shown in Figure 24-13. The electric field at the ends of the cylinder lies in the planes of the ends; consequently the flux through the two ends is zero. This leaves only the cylindrical surface with a total area of $2\pi rL$; the **E** lines are perpendicular everywhere to this curved surface, and the flux through it is $E(2\pi rL)$. Therefore, the total flux through the Gaussian surface is

$$\phi_E = \oint \mathbf{E} \cdot d\mathbf{S} = E(2\pi rL)$$

The charge inside the cylinder of length L is λL. Therefore, Gauss's law gives

$$\phi_E = \frac{q}{\epsilon_0}$$

$$E(2\pi rL) = \frac{\lambda L}{\epsilon_0}$$

or

$$E = \frac{\lambda}{2\pi\epsilon_0 r}$$

precisely the result found before, but with almost no computation.

Example 24-6 Electric Field from a Uniformly Charged Infinite Sheet. We also worked this problem (Section 24-1) by summing contributions to E from elements of the surface. The result:

$$E = 2\pi k\sigma = \frac{\sigma}{2\epsilon_0} \qquad (24\text{-}2)$$

where σ is the constant surface charge density.

From the symmetry arguments given in Section 24-1, we know that **E** is everywhere perpendicular to the infinitesimally thin sheet of charge and could depend, at most, on the distance from the sheet. Again, the Gaussian surface is chosen for easy flux computation. It is a cylinder with its axis perpendicular to the sheet and its two ends, each of area A, at the same distance from the sheet, as shown in Figure 24-14. The **E** lines now lie in the curved cylindrical surface, and there is then no flux through this surface. We

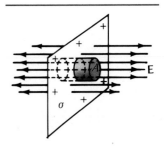

Figure 24-14. The cylindrical Gaussian surface used for computing the electric field from a uniformly charged sheet.

are left with contributions to ϕ_E from the two ends, where **E** is perpendicular to the surfaces and outward. We count a contribution to ϕ_E of EA from each of the two ends. The total charge within the cylindrical Gaussian surface is σA. From Gauss's law, we have

$$\phi_E = \frac{q}{\epsilon_0}$$

$$2EA = \frac{\sigma A}{\epsilon_0}$$

or

$$E = \frac{\sigma}{2\epsilon_0}$$

again in agreement with our earlier result.

Example 24-7 Spherical Shell of Charge. Here we have a total charge Q (assumed positive for definiteness) spread uniformly over a thin spherical shell of radius R. We shall find the electric field, using Gauss's law, at a distance r from the center of the shell with (a) $r > R$ and (b) $r < R$.

(a) By symmetry, we know that any electric field from the spherical shell of charge must be in the radial direction and depend, at most, on the distance r. Imagine the shell surrounded by a spherical Gaussian surface of radius r, as shown in Figure 24-15(a), so that **E** is at any point on the Gaussian surface perpendicular to the surface. Over the entire spherical surface, the flux ϕ_E is then $\phi_E = E(4\pi r^2)$, and Gauss's law yields

$$\phi_E = \frac{q}{\epsilon_0}$$

$$E(4\pi r^2) = \frac{Q}{\epsilon_0}$$

$$E = \left(\frac{1}{4\pi\epsilon_0}\right)\frac{Q}{r^2} = \frac{kQ}{r^2}$$

Anywhere *outside* the shell, the electric field is just like that from a single point charge Q at the center of the shell; **E** does not depend on the shell's radius R.

(b) To find **E** inside the spherical shell, we suppose that the Gaussian surface is a concentric spherical surface *inside* the shell of charge, as in Figure 24-15(b). Since *no charge* is enclosed by this surface, the electric flux through it must be *zero*. But if the shell of charge is truly spherically symmetric, any **E** must be radial and with the same magnitude at all points on the Gaussian surface. The net charge q inside is zero, so that Gauss's law gives

$$\phi_E = \frac{q}{\epsilon_0}$$

$$E(4\pi r^2) = 0$$

or

$$E = 0$$

for all points inside the shell. There is no electric field in the interior of a uniform sphere of charge, and therefore no electric force on any point electric charge placed inside a uniformly charged shell.

The results here are just like those found earlier (Section 15-9) for the gravitational effects of a uniform shell of mass: (a) equivalent on the outside to a point mass at the center of the shell; and (b) no gravitational force at all on the inside. The reason is that

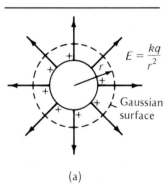
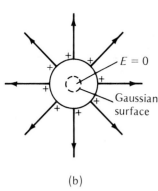

Figure 24-15. (a) Spherical shell of charge with a concentric Gaussian surface. (b) The concentric spherical Gaussian surface is now inside the shell of charge.

both Coulomb and gravitational forces are inverse-square. The special advantage of Gauss's law is seen if one compares the simple way (one line of mathematics) we arrived at the results for the spherical shell compared with the involved calculus derivation (Section 15-9).*

* Gauss's law can be formulated for the universal gravitational force as follows: $\int \mathbf{g} \cdot d\mathbf{S} = -4\pi Gm$. Here the gravitational field **g** is the gravitational force per unit mass, G is the universal gravitational constant, and m is the mass inside any Gaussian surface. A minus sign appears on the right side of the equation because any two masses attract one another whereas electric charges of the same sign repel one another.

24-5 Electric Field and Charged Conductors

Consider an electric conductor of any shape. It may carry a net charge, and there may be other charged objects nearby. If electrostatic equilibrium has been achieved, so that any net charges remain fixed in position, the following properties, summarized in Figure 24-16(a), apply to the charged conductor:

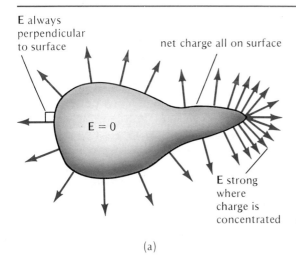

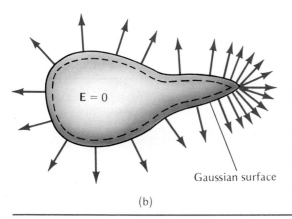

Figure 24-16. (a) Charged conductor of arbitrary shape in electrostatic equilibrium. (b) The Gaussian surface lies just inside the outer boundary of the charged conductor.

- **E** is *zero* at any point *inside* the conducting material.
- At the exterior surface, **E** is *perpendicular* to the surface at each point.
- Any *net charge* on the conductor is entirely at the *surface*.
- **E** is the strongest at those points on the surface at which the net charge is most heavily concentrated.

We prove each of these assertions in turn.

E = 0 on Interior of Conducting Material Suppose that **E** were not zero at some point within the conducting material. Then a free electron there would be subject to a force. The free electron would be accelerated, in contradiction to the assumption that the charged conductor is in electrostatic equilibrium. Therefore, **E** must be zero.

E Perpendicular to Surface Suppose that **E** were not perpendicular to an external surface. There would then have to be a component of **E** parallel to the surface. A charged particle at the surface would be accelerated, again in contradiction to electrostatic equilibrium. Therefore, **E** must be perpendicular to the surface.

Net Charge at Surface Here we exploit Gauss's law. We choose the Gaussian surface to lie just within the outer surface of a charged conductor of any shape, as shown in Figure 24-16(b). Since **E** = 0 at all points on the Gaussian surface, the net electric flux ϕ_E is zero. But from Gauss's law, if $\phi_E = 0$, then $q = 0$, and there is *no* charge (net) inside the charged conductor. If there is no charge inside the Gaussian surface, any net charge on the conductor must be *outside* the Gaussian surface, or on its surface. Within the conducting material, atoms are neutral; any excess or deficiency of electrons exists only in a thin layer of atoms at the conductor's surface. We saw earlier that the electric field within a *spherically symmetrical* shell of charge is zero. Our present result for charged conductors in equilibrium is more general; the interior field is zero for *any* shape, and for *any* charges on the exterior surface.

Our result—that quite apart from the electric fields or charges that may exist at the external surface of a conductor, there is *no net charge inside*—has profound consequences. It is, first of all, a result derived from Gauss's law, which was in turn derived from Coulomb's law. Testing by experiment to see whether there is actually no net charge inside a conductor amounts to testing indirectly whether the electric interaction between point charges is inverse-square. Indeed, a sensitive experimental test for net charge inside a conducting shell is by far the most sensitive test of the exponent of r in Coulomb's law. Many experimenters have done the experiment—Benjamin Franklin, Michael Faraday, Henry Cavendish. The most recent and precise tests, by E. R. Williams, J. E. Faller, and H. A. Hill in 1971, show that in the relation $F_e = kq_1q_2/r^n$, the exponent n differs from 2 by fewer than three parts in 10^{16}.

The traditional form of the experiment is the ice-pail experiment first performed by Faraday (who used an actual bucket for ice) in the mid–nineteenth century, and illustrated in Figure 24-17. The conductor is an ice pail, and the charge on its outer surface, indicated by the charge on the attached electroscope, is initially zero [Figure 24-17(a)]. A positively charged conducting object, suspended from an insulating thread, is introduced into the conductor. Negative charges appear on the inner surface and positive charges necessarily

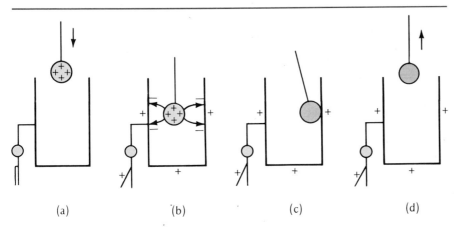

Figure 24-17. Stages in the transfer of charge from a charged object to the outside of a hollow conductor, whose state of charge is indicated by the attached electroscope. (a) The charged object is outside. (b) The charged object is within the conductor, and equal and opposite charges are found on the inside and outside of the conductor. (c) When the object is touched to the conductor's inside, it annuls the charge on the conductor's inner surface and the exterior charge is unchanged. (d) The object, now electrically neutral, is removed.

of the same magnitude on the outer surface of the pail [Figure 24-17(b)]. It is customary to say that the charged object "induces" charges on the inner and outer surfaces of the conductor. Then the introduced conducting object is touched to the inside of the bucket [Figure 24-17(c) and — this is the crucial point — the electroscope shows no change. The charged object introduced annuls *all* the charges on the inside of the conductor, so that at the end [Figure 24-17(d)], *all* the net charge is on the conductor's outer surface, *none* is on the inside.

Inside a conducting shell, there is *electric* shielding. No matter how strong the electric fields may be at or outside the surface of a conducting shell, inside the shell of whatever shape, a charge feels nothing. The interior is shielded completely from what goes on outside.

We can see, by considering Figure 24-18, why electric shielding works on a basis less formal than Gauss's law. Here we have a conducting shell first [Figure 24-18(a)] with no external electric field. On the interior of the conducting shell, $\mathbf{E} = 0$. Now an external $\mathbf{E}_{ext}$ field (from a point charge) is applied, as in Figure 24-18(b). Positive charge, and equal amounts of negative, appear on the surface of the conductor, and the electric field within is still zero. We can look at this in a different way, as shown in Figure 24-18(c); the external field $\mathbf{E}_{ext}$ still exists within the conducting shell but it is exactly cancelled by the electric field $\mathbf{E}_{sep}$ in the opposite direction arising from the separated charges on the surface of the conducting shell.

Dependence of E on Local Surface-Charge Density For a *spherical* conducting surface, the charge per unit area σ has the *same* value at all points on the surface. The electric field also has the same magnitude at all points on a charged spherical conductor.

Not so for any other shape of charged conductor. The surface-charge density σ and electric field $\mathbf{E}$ both may change in magnitude (and in sign) from

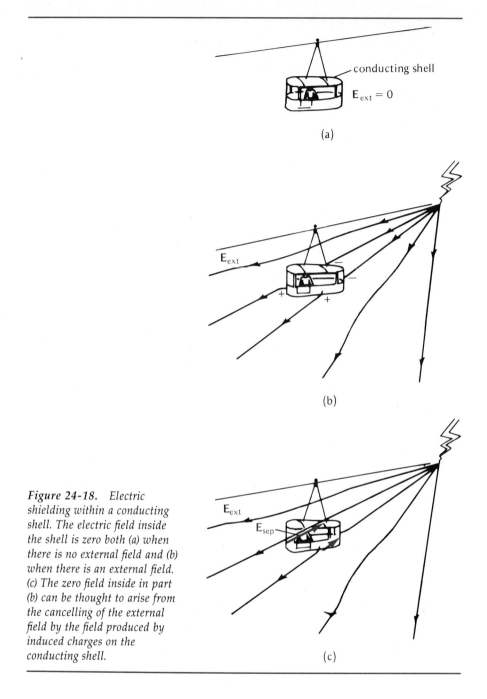

Figure 24-18. Electric shielding within a conducting shell. The electric field inside the shell is zero both (a) when there is no external field and (b) when there is an external field. (c) The zero field inside in part (b) can be thought to arise from the cancelling of the external field by the field produced by induced charges on the conducting shell.

point to point. (We shall show in detail in Section 25-7 that σ and **E** have maximum values at points, or at regions of small radius of curvature, and minimum values at flat portions of a conducting surface.) Here we derive the general relation between σ and **E**.

Consider the conducting surface of arbitrary shape shown in Figure 24-19. We concentrate on a portion of the surface so small that it can be regarded as flat. We know already that **E** on the outer surface is perpendicular to the surface and that **E** = 0 within the conductor. We choose as a Gaussian surface a small flat cylinder whose axis is parallel to the normal to the surface and whose ends lie, respectively, within the conductor and just outside its surface. We can regard both σ and **E** to be constant over the cylinder's surface area A.

The total charge enclosed by the Gaussian surface is $q = \sigma A$. The only contribution to the electric flux is through the outer surface of the cylinder, for which $\phi_E = EA$. The flux is zero over the inner surface (**E** is zero there) and over the curved cylinder surface (**E** is parallel to the surface there). Gauss's law then gives

$$\phi_E = \frac{q}{\epsilon_0}$$

$$EA = \frac{\sigma A}{\epsilon_0}$$

$$E = \frac{\sigma}{\epsilon_0} \qquad (24\text{-}6)$$

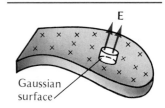

Figure 24-19. *Surface of a charged conductor with a Gaussian surface used to evaluate the electric field* **E**.

The magnitude of the electric field at any point of the surface of a charged conductor is directly proportional to the surface charge density at that point.

Example 24-8. A hollow conductor initially has a net charge of $+3Q$; all of this charge is, of course, on its outer surface. See Figure 24-20(a). Then a charged object of $+4Q$ is introduced into the interior of the conductor without touching the inner surface. What is the net charge on the inner surface and on the outer surface of the conductor?

When the $+4Q$ charge is introduced into the interior of the hollow conductor, it induces an equal but opposite charge of $-4Q$ on the *inner* surface of the conductor. (To satisfy yourself on this point, starting with fundamental principles, first choose a Gaussian surface lying entirely within the conductor material, as shown in Figure 24-20(b). The electric field **E** is zero all over this surface; the electric flux is zero from Gauss's law; and therefore the net charge within the Gaussian surface is zero.) Since the hollow conductor still has a net charge of $+3Q$ but now has $-4Q$ on its inner surface, the outer surface must have a net charge of $+7Q$.

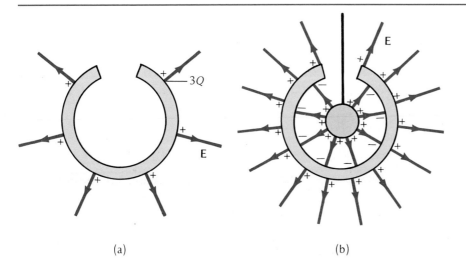

Figure 24-20.

Note again the effect of electric-field shielding. With the $+4Q$ charge on the inside and the $-4Q$ charge on the conductor's inner surface there *is* an electric field on the interior of the hollow conductor. But this field is produced solely by charges *within* the conductor, not at all by electric effects on the exterior of the conductor. Even if still another charged object were to be brought close to the exterior of the charged conductor, its outer surface would continue to have a net charge of $+7Q$ (even though this charge might be redistributed over the outer surface) but nothing inside the conductor would change.

24-6 An Electric Dipole in an Electric Field (Optional)

An electric dipole consists of two point charges of equal magnitude but opposite sign separated by a distance d. As a whole, the dipole is electrically neutral. It is useful to define the *electric dipole moment* $\mathbf{p}$ as

$$\mathbf{p} \equiv q\mathbf{d} \tag{24-7}$$

where q is the magnitude of either of the two charges and $\mathbf{d}$, a vector, is the displacement of the positive point charge relative to the negative point charge (see Figure 24-21). The magnitude of the electric dipole moment is simply qd.

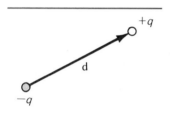

Figure 24-21. *An electric dipole.*

What happens when an electric dipole is placed in a uniform electric field $\mathbf{E}$? In Figure 24-22 we have an electric dipole $\mathbf{p}$ making an angle θ with the electric-field lines. Each of the two point charges is acted on by an electric force of magnitude qE. The two forces are in opposite directions. Consequently, the resultant force on the dipole is zero, and the dipole is not accelerated translationally in a *uniform* electric field.

There is, however, a resultant torque on the dipole. We use the general definition for the torque τ,

$$\boldsymbol{\tau} = \mathbf{r} \times \mathbf{F} \tag{12-5}$$

or in magnitude,

$$\tau = r_\perp F \tag{12-4}$$

It is convenient to choose the tail of the vector $\mathbf{d}$ as the axis for computing torques (actually, we get the same torque for any choice of axis). Then, from Figure 24-22, we see that

$$\tau = r_\perp F = (d \sin \theta)(qE) = pE \sin \theta \tag{24-8}$$

This relation can be written more compactly in vector form as

$$\boldsymbol{\tau} = \mathbf{p} \times \mathbf{E} \tag{24-9}$$

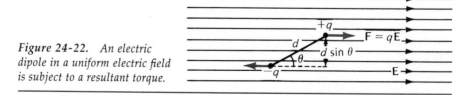

Figure 24-22. *An electric dipole in a uniform electric field is subject to a resultant torque.*

The torque on the dipole is zero when it is aligned with the uniform field and the maximum when it is at right angles ($\theta = 90°$) to the field.

What is the potential energy U of an electric dipole **p** in electric field **E**? The potential energy is equal to the work done, W, in reorienting the dipole. In general, the work done by the torque is

$$U = W = \int \tau \, d\theta \qquad (13\text{-}11)$$

Using (24-8), we have

$$U = \int_{\pi/2}^{\theta} pE \sin\theta \, d\theta = -pE \cos\theta \qquad (24\text{-}10)$$

The zero for measuring potential energy is chosen arbitrarily, and the lower limit of the integral was here chosen as $\frac{1}{2}\pi$ to give a simple final result. The zero for the potential energy then corresponds to the angle $\theta = 90°$. We can write (24-10), using the dot product, as

$$U = -\mathbf{p} \cdot \mathbf{E} \qquad (24\text{-}11)$$

The potential energy has its minimum ($U = -pE$) when the dipole is aligned with the field; and the maximum ($U = +pE$) corresponds to $\theta = 180°$.

The cross product of **p** with **E** gives the torque of an electric dipole in an electric field, and the dot product gives the potential energy. We can regard (24-9) and (24-11) as *defining* an electric dipole moment. We can attribute an electric dipole moment to *any* object that is subject to a torque and has potential energy of orientation when in a uniform electric field. For example, a neutral molecule may have an electric dipole moment that is reoriented in an electric field. Here there are *not* two separated *point* charges of equal magnitude and *opposite* sign; a neutral molecule has an electric dipole moment because its "centers" of positive and negative charge do not coincide. The water molecule, H_2O, has a permanent electric dipole moment, and is said to be a *polar molecule*, because the "end" of the molecule with the hydrogen atoms has a net positive charge while the "end" with the oxygen has a net negative charge. A *nonpolar molecule* has no permanent electric-dipole moment but acquires a dipole moment when immersed in an electric field. It is not necessary to specify the charge separation distance d for a polar molecule or any other electric dipole; indeed, it is not possible to give q and d separately, since it is their product qd that enters into the torque and potential-energy relations.

When an electric dipole is in a *nonuniform*, or inhomogeneous, electric field, the forces acting on the two opposite charges are *not* of equal magnitude; the dipole is then subject to a resultant force.

24-7 General Proof of Gauss's Law (Optional)

In Section 24-4, we found that the relation $\oint \mathbf{E} \cdot d\mathbf{S} = q/\epsilon_0$ holds for a *single* point charge q at the *center* of a *spherical* Gaussian surface. To prove Gauss's law in general, we shall show that this relation holds, first, for a single charge inside or outside a Gaussian surface of *any shape*, and then, for *any number* of point charges.

Consider Figure 24-23, where point charge q is at the center of a spherical Gaussian surface and also inside a closed surface of arbitrary shape. We con-

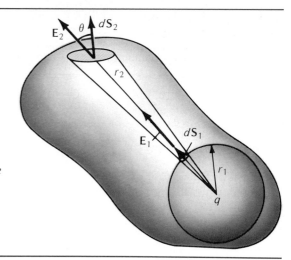

Figure 24-23. A point charge q surrounded by a volume of arbitrary shape and by a concentric sphere. The electric flux through the two shaded areas is exactly the same.

centrate on the small solid angle, or cone, that intercepts the surface element $d\mathbf{S}_1$, at a distance r_1 from q and also surface element $d\mathbf{S}_2$ at a distance r_2. We want to show that the electric flux is the same through these two matched surface elements. The electric fields at $\mathbf{r}_1$ and $\mathbf{r}_2$ are $\mathbf{E}_1$ and $\mathbf{E}_2$.

Note that $\mathbf{E}_2$ and $d\mathbf{S}_2$ are not parallel in general; the angle between them is θ. The flux $d\phi_2$ through $d\mathbf{S}_2$ is

$$d\phi_2 = \mathbf{E}_2 \cdot d\mathbf{S}_2 = E_2 \cos\theta\, dS_2$$

Area dS_2 is larger than dS_1, for two reasons: dS_2 is farther from q, and dS_2 is inclined relative to the direction of $\mathbf{E}_2$. From the geometry of Figure 24-22, we have that

$$dS_2 \cos\theta = \left(\frac{r_2}{r_1}\right)^2 dS_1$$

where the factor $(r_2/r_1)^2$ comes from the difference in distances and the factor $\cos\theta$ from the inclination of $d\mathbf{S}_2$.

The electric field from a point charge falls off inversely with the square of distance, so that

$$E_2 = \left(\frac{r_1}{r_2}\right)^2 E_1$$

Substituting the two equations immediately above in the relation for $d\phi_2$ then gives

$$d\phi_2 = E_2(\cos\theta\, dS_2) = E_1 dS_1 = d\phi_1$$

The flux is the *same* for this pair of matched surfaces. Indeed, the flux is the same for all pairs of matched surface elements. The total flux $\oint \mathbf{E}\cdot d\mathbf{S}$ has the same value (q/ϵ_0) for *any* shape of surface enclosing q.

Suppose now that a point charge is *outside* some arbitrary Gaussian surface, as in Figure 24-24. Nothing changes. For every straight electric-field line into the surface there is a corresponding line out. A charge outside the surface does not contribute to the net flux.

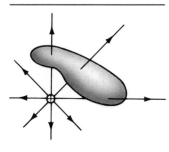

Figure 24-24. The total electric flux through a closed surface not enclosing a net charge is zero. $\oint \mathbf{E}\cdot d\mathbf{S} = 0$ for q outside.

Thus far we have proved Gauss's law for a *single* point charge. Now suppose that we have a collection of point charges—$q_1, q_2, q_3, \ldots$—at various locations. We apply Gauss's law to each charge in turn with, however, the *same* Gaussian surface for all. We then can write

$$\oint \mathbf{E}_1 \cdot d\mathbf{S}_1 = \frac{q_1}{\epsilon_0}, \quad \oint \mathbf{E}_2 \cdot d\mathbf{S}_2 = \frac{q_2}{\epsilon_0}, \ldots$$

where $\mathbf{E}_1$ is the field produced by q_1 alone, $\mathbf{E}_2$ by q_2 alone, and so forth.

Now we add all the equations above, to get

$$\oint (\mathbf{E}_1 + \mathbf{E}_2 + \mathbf{E}_3 + \cdots) \cdot d\mathbf{S} = \frac{q_1 + q_2 + q_3 + \cdots}{\epsilon_0}$$

We have added the several $\mathbf{E}$'s—*under* the integral sign—and it's all right because the same surface element applies to all of them. Furthermore, because of the superposition principle for electric forces, we *can* add as vectors the fields produced by separate charges to find the resultant electric field, $\mathbf{E} \equiv \mathbf{E}_1 + \mathbf{E}_2 + \mathbf{E}_3 + \cdots$, produced by the net algebraic charge $q = q_1 + q_2 + q_3 + \cdots$, *inside* the Gaussian surface.

The relation above is Gauss's law in its general form, which we usually write more compactly as

$$\phi_E = \frac{q}{\epsilon_0} \tag{24-5}$$

Summary

Definitions

Uniform electric field: $\mathbf{E}$ = constant, magnitude and direction.

Electric flux $d\phi_E$ through an outwardly directed surface element $d\mathbf{S}$, where the electric field is $\mathbf{E}$:

$$d\phi_E \equiv \mathbf{E} \cdot d\mathbf{S} \tag{24-3}$$

Gaussian surface: a closed surface of arbitrary shape.

Fundamental Principles

Gauss's law, a fundamental principle relating electric fields to electric charges and equivalent to Coulomb's law, states: For any closed Gaussian surface, the net electric flux $\phi_E = \oint \mathbf{E} \cdot d\mathbf{S}$ is equal to q/ϵ_0, where q is the algebraic net charge enclosed by the surface.

$$\phi_E = \frac{q}{\epsilon_0} \tag{24-5}$$

Important Results

E for various charge distributions

- Point charge:

$$E = \frac{kq}{r^2} \propto \frac{1}{r^2} \tag{23-6}$$

- Infinite uniform line of charge (linear charge density λ):

$$E = \frac{2k\lambda}{r} = \frac{\lambda}{2\pi\epsilon_0 r} \propto \frac{1}{r} \tag{24-1}$$

- Infinite uniformly charged sheet of charge (surface-charge density σ):

$$E = 2\pi k\sigma = \frac{\sigma}{2\epsilon_0} = \text{uniform } \mathbf{E} \tag{24-2}$$

- Spherical shell of charge Q: Outside of shell—$\mathbf{E}$ same as that from point charge Q at center of shell. Inside of shell—$\mathbf{E} = 0$.

E and Charged Conductors:

- Within conducting material, $\mathbf{E}$ is zero.
- At exterior surface, $\mathbf{E}$ is perpendicular to surface.
- Net charge is on conductor's surface.
- Magnitude of $\mathbf{E}$ at surface is proportional to the local charge density:

$$E = \frac{\sigma}{\epsilon_0} \tag{24-6}$$

Problems and Questions

Section 24-1 Electric Field for Three Simple Geometries

· **24-1 P** What is the electric field at a distance of 4 cm from a long straight wire carrying a linear charge density of $2 \ \mu C/m$?

· **24-2 P** What is the electric field at a distance of 1 mm from a $1 \text{ m} \times 1 \text{ m}$ sheet of aluminum that carries a charge of 2×10^{-6} C?

· **24-3 P** Consider a sphere having a radius of 4 cm and carrying a charge of 8 μC. What would the surface charge density be if the charge were distributed (a) uniformly over the surface? (b) Uniformly throughout the volume?

: **24-4 P** Charge Q is distributed uniformly along the z axis from $z = -a$ to $z = +a$. Determine the electric field at a point on the x axis.

: **24-5 P** An infinite insulating plane sheet of charge has a charge density of $+8 \times 10^{-9}$ C/m². At a distance of 10 cm from the plane is a point charge of $+0.01$ C. What is the magnitude and direction of the electric field at a point midway between the charge and the plane?

: **24-6 P** The earth carries a negative charge, which is created by lightning strikes. On a fair day, a typical value of the electric field at the earth's surface is 130 N/C. Assuming that the earth is a conductor, determine the surface charge density needed to create this field.

: **24-7 P** A small foam sphere of mass 1×10^{-4} kg carries a charge 2×10^{-9} C. It hangs from a thread 60 cm long, attached to a vertical conducting wall carrying a uniform surface charge of 2×10^{-6} C/m². What angle does the thread make with the vertical?

: **24-8 Q** Charge Q is distributed uniformly along the x axis from $-L$ to L (Figure 24-25). What is the electric field at the point $(0, L)$?

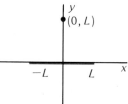

Figure 24-25. Question 24-8.

(A) $\dfrac{Q}{8\pi\epsilon_0} \displaystyle\int_{-L}^{L} \dfrac{dx}{(x^2 + L^2)^{3/2}}$

(B) $\dfrac{Q}{4\pi\epsilon_0} \displaystyle\int \dfrac{dx}{x^2}$

(C) $\dfrac{Q}{8\pi\epsilon_0} \displaystyle\int \dfrac{dx}{x^2}$

(D) $\dfrac{Q}{8\pi\epsilon_0 L} \displaystyle\int \dfrac{dx}{x^2 + L^2}$

(E) $\dfrac{Q}{4\pi\epsilon_0 L^2}$

(F) $\dfrac{Q}{\epsilon_0 \oint dS}$

: **24-9 P** Charge Q is distributed uniformly along the z axis from $z = -a$ to $z = +a$. Determine the electric field at all points on the z axis for $z > +a$ and $z < -a$.

: **24-10 P** Charge Q is distributed uniformly along a circular loop of wire of radius R. (a) What is the electric field on the axis of the loop a distance z from its center? (b) For what value of z is the field a maximum?

: **24-11 P** Charge Q is distributed uniformly throughout a thin disk of radius R. Determine the electric field at a point on the axis of the disk a distance z from the center. (*Hint:* Consider concentric rings of charge and find the field due to each. Integrate to find the total field.)

: **24-12 P** A rod of length L carries charge Q uniformly distributed along its length. Determine the electric field at point P indicated in Figure 24-26.

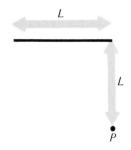

Figure 24-26. Problem 24-12.

: **24-13 P** Charge is distributed with uniform surface-charge density over the strip of the yz plane extending from $y = -a$ to $y = +a$. Determine the electric field at a point on the x axis.

: **24-14 P** A cylindrical shell of radius a and length L has a uniform surface-charge density σ (Figure 24-27). Show that the electric field at point P on the axis is

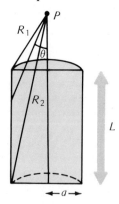

Figure 24-27. Problem 24-14.

$$E = \frac{\sigma}{4\pi\epsilon_0} \frac{1}{L} \left(\frac{1}{R_1} - \frac{1}{R_2} \right)$$

(*Hint:* To carry out the necessary integration, change variables from z to θ.)

Section 24-2 Uniform Electric Field

· **24-15 P** What charge must a water droplet of radius 1 mm carry if it is not to fall under the force of gravity when placed in a vertical electric field of 5000 N/C?

· **24-16 P** Three parallel insulating infinite sheets of charge have surface-charge densities of $\sigma_1 = +1$ nC/m², $\sigma_2 = +2$ nC/m², and $\sigma_3 = -3$ nC/m². Determine the magnitude and direction of the electric field in regions a, b, c, and d of Figure 24-28.

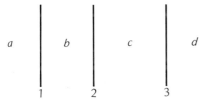

Figure 24-28. Problem 24-16.

: **24-17 P** In a CRT (cathode ray tube), a narrow beam of electrons is projected horizontally between two oppositely charged horizontal parallel plates. The electric field between the plates is 600 N/C. The beam consists of a line of electrons each separated by 1×10^{-5} m from its immediate neighbors. Calculate the force on an electron from (*a*) the electric field of the plates; (*b*) the electric field of a neighboring electron; (*c*) its weight.

: **24-18 P** Two large parallel metal plates, each of area A and separated by a small distance d, carry charges $+Q$ and $-2Q$ (distributed over both sides of each plate). Determine the electric field between the plates and outside the plates on each side.

: **24-19 P** An electron with an initial velocity of 2×10^6 m/s is slowed by a uniform electric field of 200 N/C. (*a*) How long does it take for the electron to come momentarily to rest? (*b*) How far does the electron travel before it comes to rest?

: **24-20 P** A uniform electric field exists between two parallel conducting plates. The magnitude of the surface-charge density on each plate is σ. A third conducting plate is now placed between the two charged plates, oriented parallel to them and one-fourth of the way from one plate to the other. (*a*) Determine the surface-charge density induced on each side of the third plate. (*b*) Does the result in (*a*) depend on how far the third plate is from either of the other two?

: **24-21 Q** A metal sphere is placed in the region between

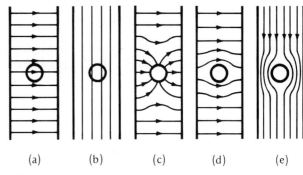

Figure 24-29. Question 24-21.

two oppositely charged parallel plates. Which of the sketches in Figure 24-29 best represents the electric field in this case?

: **24-22 P** Suppose you want to bend a beam of particles, each of mass m and charge q, through 90°. The particles are to exit at the same speed they enter the field. In practice this is often done using magnetic fields, but it can also be done with static electric fields. What is the electric field required in each of these two schemes: (*a*) The beam passes between two curved electrodes, which set up an electric field of constant magnitude directed in toward the center of curvature. See Figure 24-30. (*b*) An electric field constant in both magnitude and direction is used.

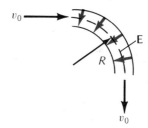

Figure 24-30. Problem 24-22.

Section 24-3 Electric Flux

· **24-23 P** What is the electric flux through a plane surface of area 2 m² inclined at an angle of 30° to a uniform electric field of 300 N/C?

: **24-24 P** Consider a cube of edge length L. What can you say about the outward electric flux through each face when a positive point charge Q is positioned (*a*) at the center of the cube? (*b*) inside the cube very near the center of one face? (*c*) outside the cube very near the center of one face?

: **24-25 P** A hemispherical surface of radius 4 cm is placed in a uniform electric field of 200 N/C. What is the maximum electric flux that can pass through the surface?

· **24-26 Q** Which of the following has the meaning closest to "electric flux through an open surface"?
(A) The flow of electric potential through the surface.

(B) The flow of electric charge through the surface.
(C) The total charge on a conducting body.
(D) The number of electric field lines passing through the surface.
(E) The net electric force acting on a charged object.

Section 24-4   Gauss's Law

· **24-27 P** A point charge Q is placed at the center of a cube of side a. What is the electric flux through one face of the cube?

· **24-28 Q** In using Gauss's law to find the electric field due to a charge distribution, which of the following is the most accurate statement?
(A) We should like a Gaussian surface over which **E** is constant and perpendicular to the surface.
(B) We should like to find a surface over all of which **E** is parallel to the surface.
(C) We first find the electric potential, then differentiate to find the electric field.
(D) We find it is easier to solve problems involving an array of point charges, rather than a continuous distribution of charge, such as a sphere.
(E) We try to find a Gaussian surface that will enclose zero net charge.

· **24-29 Q** In using the equation $\oint_s \mathbf{E} \cdot d\mathbf{S} = q/\epsilon_0$ to find the electric field, it is important to keep in mind which of the following?
(A) All charges have to be included in q.
(B) The surface S may be open or closed, depending on the particular problem.
(C) The surface S is the surface of the charge distribution of interest.
(D) The surface S should be chosen so that **E** and $d\mathbf{S}$ are either parallel or perpendicular over the surface.
(E) This law does not apply to conductors, where $\mathbf{E} = 0$.

· **24-30 Q** A Gaussian surface encloses zero net charge. Therefore,
(A) Gauss's law requires that the electric field be zero everywhere on the surface.
(B) the vector sum of all electric fields passing through the surface is zero.
(C) external charges cannot change the flux through the surface.
(D) the electric field is zero everywhere inside the surface.
(E) none of the above.

· **24-31 Q** Which of the following is an accurate statement about Gauss's law?
(A) It requires that if a closed surface contains no net charge, the electric field must be zero everywhere over the Gaussian surface.
(B) It requires that if the electric field is everywhere zero over a closed surface, the net charge enclosed by the surface must always be zero.
(C) It applies only to symmetric charge distributions.
(D) It applies only to continuous distributions of charges and not to point charges.

: **24-32 Q** (*a*) Suppose that charge is uniformly distributed over the surface of a spherical balloon. What can you say about the electric field inside and outside the balloon? (*b*) Suppose the balloon is then distorted into some sausagelike shape, still with constant surface-charge density. How does this change the field from case (a)?

: **24-33 Q** A point charge is placed at the center of a spherical Gaussian surface and the electric flux ϕ_E through this sphere is calculated. Which, if any, of the following would cause a change in the flux ϕ_E?
(A) The sphere is replaced by a cube of the same volume.
(B) The sphere is replaced by a cube of the same surface area.
(C) The point charge is moved off center of the original sphere but remains inside the sphere.
(D) The charge is moved just outside the original sphere.
(E) A second charge is placed near, and outside, the original sphere.

: **23-34 Q** In Gauss's law, $\epsilon_0 \oint_s \mathbf{E} \cdot d\mathbf{S} = q$
(A) The integral is always carried out over the surface of a conductor.
(B) The **E** in the integral is due only to the charge q.
(C) q is what we call a test charge.
(D) $d\mathbf{S}$ is an element of displacement (measured in meters).
(E) q is the charge lying on surface S.

: **24-35 P** Charge is distributed uniformly with density ρ throughout the volume of an insulating sphere of radius R. Determine the electric field at all points, inside and outside the sphere's surface.

: **24-36 P** A spherical shell of inner radius R_1 and outer radius R_2 has a uniform volume-charge density ρ. Determine the electric field everywhere.

: **24-37 P** Use Gauss's law for the gravitational field to find the gravitational field inside and outside the surface of a sphere of radius R and constant mass density. Sketch the gravitational field as a function of r.

: **24-38 P** Charge is distributed uniformly with density ρ throughout the volume of an infinite slab of insulator defined by the planes $z = +a$ and $z = -a$. Use Gauss's law to find the electric field at all points.

: **24-39 P** A long insulating cylinder of radius R has a uniform volume-charge density ρ. Determine the electric field inside and outside the cylinder.

: **24-40 P** A long cylinder of radius R_1 has a uniform volume charge density ρ (Figure 24-31). A long cylindrical void of radius R_2 runs parallel to the axis of the cylindrical charge; the center of the void is a distance a from the center of the cylinder of charge; and $R_1 > R_2$. Show that the elec-

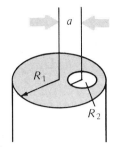

Figure 24-31. Problem 24-40.

tric field in the void is constant and determine its value. (*Hint:* Superimpose a positive and a negative charge distribution to arrive at the given geometry, and calculate separately the field due to each.)

: **24-41 P** An insulating slab of thickness t has a charge density given by

$$\rho(x) = \rho_0 \sin \pi x/t$$

where x is the distance from one face of the slab. Find E at points inside and outside the slab.

: **24-42 P** A spherically symmetric charge distribution of radius R gives rise to an electric field

$$E = \frac{q_0}{4\pi\epsilon_0}\left[\frac{(1-e^{-ar})}{r^2}\right] \quad \text{for } r \le R$$

(a) What is the charge density of the distribution? (b) What is the electric field for $r > R$? (c) Sketch $E(r)$ and $\rho(r)$ vs r.

: **24-43 P** Charge Q is distributed uniformly over an insulating spherical shell of radius R (Figure 24-32). A very small piece of the sphere is cut out and removed (along with the charge that was there). Show that the electric field on the outside of the sphere is $Q/4\pi\epsilon_0 R^2$ everywhere over the outside of the sphere except at the hole, where the field has the value $Q/2\pi\epsilon_0 R^2$. (*Hint:* Recall the idea of superposition of charge distributions.)

Figure 24-32. Problem 24-43.

: **24-44 P** In a simple model of the hydrogen atom, a point nucleus of charge $+e$ is assumed to be surrounded by a spherically symmetric cloud of negative charge $-e$ whose density is $\rho(r) = \rho_0 e^{-2r/a}$. Determine the electric field at a distance r from the nucleus for this model.

Section 24-5 Electric Field and Charged Conductors

· **24-45 P** Air will undergo electrical breakdown if subjected to an electric field in excess of about 3×10^6 N/C (the exact value depends on the pressure and humidity). What surface-charge density on a conductor will give rise to this field?

· **24-46 P** The electric field at a particular point on the surface of an irregularly shaped conductor is 5000 N/C. What is the surface-charge density at the point? If this cannot be determined, explain.

· **24-47 Q** Charge is placed on the lump of aluminum sketched in Figure 24-33. How will the charge be distributed on the object?

Figure 24-33. Problem 24-47.

(A) Uniformly throughout the volume.
(B) Uniformly over the surface.
(C) With greatest density near point C on the surface.
(D) With greatest density near point D in the interior.
(E) With greatest density near point E on the flat surface.

· **24-48 P** A sufficiently large electric field will cause an electric breakdown of air. For this reason, a given amount of charge on a spherical conductor of small radius is more likely to result in a "spark" than a larger sphere is. To see this, calculate the field at the surface of a conducting sphere when a charge of 12 μC is placed on a sphere of radius (a) 2 cm, (b) 20 cm, (c) 2 m.

· **24-49 Q** A spherical conducting shell carries a charge Q. The electric field in the space inside the shell is
(A) directed radially inward.
(B) zero
(C) of the same magnitude as at a point just outside the shell.
(D) dependent on the position inside the shell.

: **24-50 Q** Identify the incorrect statement.
(A) The electric field inside a charged conductor can never be zero.
(B) The electric field is a measure of the force per unit charge.
(C) Gauss's law is an alternative formulation of Coulomb's law.
(D) The electric field due to two charges can only be zero somewhere on the line joining them.
(E) All charges are made up of multiples of either the electron or the proton charge.

: **24-51 P** The surface of the earth carries a negative

charge that, it is believed, is maintained by negatively charged lightning striking the ground and by positive corona discharge. In clear weather, the electric field near the earth's surface is approximately 100 V/m. To what surface charge does this correspond?

: **24-52 P** Three concentric conducting spheres carry charges $+Q$ (on the inner sphere), $-2Q$ (on the middle sphere), and $+3Q$ (on the outer sphere). Determine the charge on each surface (inner and outer) of each sphere.

: **24-53 Q** Three concentric conducting spherical shells carry charges as follows: $+4Q$ on the inner shell, $-2Q$ on the middle shell, and $-5Q$ on the outer shell. The charge on the inner surface of the outer shell is
(A) zero.
(B) $4Q$.
(C) $-Q$.
(D) $-2Q$.
(E) $-3Q$.
(F) impossible to determine unless the radii of the shells are known.

: **24-54 P** A spherical conducting shell of inner radius 2 cm and outer radius 4 cm carries a charge of $+4~\mu C$. Determine the electric field at the following distances from the center: (a) 1 cm, (b) 3 cm, (c) 4 cm, (d) 8 cm.

Supplementary Problems

24-55 P An electrostatic precipitator is a device for removing dust particles from a smoke stack. The rising dust particles are given an electric charge, then they are deflected in an electric field, and finally they are collected on an electrode with the opposite charge before they can leave the stack. A typical precipitator produces an electric field between a cylindrical conductor and a thin coaxial conductor inside it. Choose reasonable order-of-magnitude values for the various parameters that would enter into the design of a precipitator and show that it can work effectively—that is, that charged dust particles can be precipitated out of the emerging smoke in a stack of reasonable dimensions.

24-56 P When a charge $+q$ is placed a distance d from a plane metal surface, a negative surface charge will be induced on the metal. We have seen that the resulting electric field lines must be perpendicular to the metal surface, and that the magnitude of the electric field at the metal surface is proportional to the surface charge induced there. Electric-field lines would be so distributed, entering the metal normal to the surface, if the field were due just to the charge $+q$ and a second, fictitious "image" charge, placed

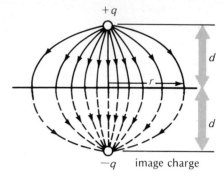

Figure 24-34. Problem 24-56.

as shown in Figure 24-34. (The charge $+q$ appears thus to be "reflected" in the mirror of the metal surface.) Using this method of images, determine (a) the electric field at the metal surface as a function of r; (b) the surface-charge density as a function of r; and (c) the total induced surface charge on the metal. Determine the total charge by integrating the surface-charge density over the entire infinite surface. You should find the total induced charge to be $-q$.

24-57 Q What accounts for the purple glow (the corona discharge) one sometimes sees around high-voltage apparatus, such as transformers?
(A) When an electric field is sufficiently intense, the field becomes visible. The purple color exists because it is nearest the UV, or high-energy, end of the spectrum.
(B) Intense electric fields accelerate stray electrons or ions to high speeds. These particles collide with air molecules, ionizing them. These fragments are in turn accelerated, forming an avalanche. Some of the ions and electrons recombine and in so doing emit a characteristic light.
(C) Electrons, which are normally invisible, can be seen when present in high concentrations. Generation of high voltage requires high charge concentration, and electrons are the carriers of electricity in metals.
(D) The intense electric fields cause nuclei to undergo fission, much as in a nuclear reactor, and as a result radiation is given off. This radiation is also known as Cerenkov radiation.
(E) Intense electric fields are known to focus light beams. The purple light seen is the result of focusing stray light that is usually unnoticed. The effect is dependent on light wavelength and is most pronounced for short wavelengths.

Electric Potential

25

25-1 The Coulomb Force as a Conservative Force
25-2 Electric Potential Defined
25-3 Electric Potential for Point Charges
25-4 Electric Potential Energy
25-5 Equipotential Surfaces
25-6 Relations between V and **E**
25-7 Electric Potential and Conductors
Summary

We have already found in our study of mechanics how much easier it is to analyze physical situations and solve problems with energy-conservation ideas than with forces. Energy is a scalar quantity; force, and such force-related concepts as the electric field, are vectors. We now concentrate on electric potential, a central concept relating to the energy of interacting electric charges.

25-1 The Coulomb Force as a Conservative Force

Is energy conserved for the Coulomb force? Is the Coulomb force a conservative force? We know that the electric force between point charges must be conservative because it is of the same form as the universal gravitational force between point masses; both are central and inverse-square forces. Since the gravitational force is conservative (Section 15-5), the electric force must also be conservative.

It is easy to check separately, however, that the net work done by the electric force between a pair of point charges is zero when one charge is moved around any closed path while the other charge remains at rest. Then the

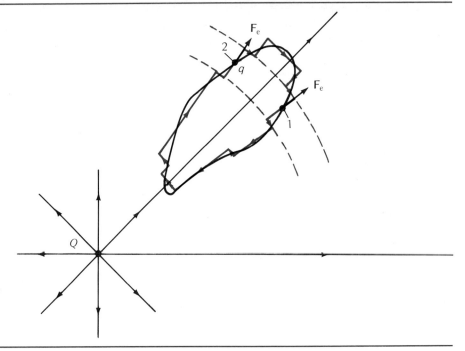

Figure 25-1. Charge q is moved in a closed path in the vicinity of fixed charge Q. Charge q is subject to an outward electric force $\mathbf{F}_e$ over the inward displacement at 1 and to the same electric force over the outward displacement at segment 2.

requirement for a conservative force,

$$\oint \mathbf{F} \cdot d\mathbf{r} = 0 \tag{9-14}$$

is met, and we can be sure that energy conservation applies to the Coulomb interaction.

Consider Figure 25-1. Charge Q is fixed in position and produces the electric field shown; charge q is moved around some arbitrary closed path. Actually, we imagine the path to consist of two types of small line segments:

- Circular arcs centered on Q.
- Radial lines pointing towards or away from charge Q.

We are to find the net work done on q by the electric force $\mathbf{F}_e$ on it from Q. (For definiteness we imagine both q and Q to be positive.) The results are simple:

- Along any circular arc, the force $\mathbf{F}_e$ is perpendicular to the displacement $d\mathbf{r}$, and the work done is zero.
- Along a radial line, every inward displacement, such as that at 1 in Figure 25-1, can be matched with a corresponding outward displacement along a radial line, as at 2 in Figure 25-1. The force magnitude is the same at the two locations; both 1 and 2 are the same distance from Q. The displacement magnitudes are also the same. But $\mathbf{F}_e$ and $d\mathbf{r}$ are antiparallel for the inward displacement and parallel for the outward. Positive work is done by $\mathbf{F}_e$ on q for an outward displacement; an equal amount of negative work is done for the matched inward displacement. The net work is zero, for this and every other matched pair of radial displacements.

The net work $\oint \mathbf{F}_e \cdot d\mathbf{r}$ done around any closed path is zero. Another way of saying the same thing is this: the net amount of work done on a charge by the Coulomb force as it moves from some initial location A to a final location B does not depend on the path followed between A and B, but *only on the end points.*

25-2 Electric Potential Defined

To see how electric potential is defined, consider Figure 25-2. Here charges Q_1, Q_2, Q_3, ... are fixed while test charge q_t is moved at constant speed from A to B by some external agent. The agent applies a force $\mathbf{F}_a$ that balances in magnitude and direction at each location the electric force $\mathbf{F}_e$ of the fixed charges on q_t. As q_t is moved from A to B, the work done on it by the applied force is designated $W_{A \to B}$. We know that it does not matter how q_t gets from A to B; the work $W_{A \to B}$ is the same for all paths between the same end points.

Then, by definition, the *electric potential difference* between A and B produced by charges Q_1, Q_2 ... is

$$V_B - V_A \equiv \frac{W_{A \to B}}{q_t} \tag{25-1}$$

In words, the electric potential difference at the final location (B) relative to the initial location (A) is the work done per unit positive test charge by an external agent that takes the charge at constant speed from the initial location to the final.

Now, if we know the work required to transport a *unit* positive charge from one location to another, we can immediately specify the work required to move *any* charge q at constant speed from A to B. From (25-1), it is

$$W_{A \to B} = q(V_B - V_A) \tag{25-2}$$

If q is a positive charge and the work done is also positive, then the electric potential V_B at point B is higher (more positive) than V_A at A.

In Figure 25-2, we had an external agent doing work on charge q_t without, however, changing its kinetic energy. Therefore, the potential energy of the entire system of charged particles must have increased. In fact, the change in electric potential energy between the two configurations, $U_B - U_A$, is just equal to the work done, $W_{A \to B}$. Therefore, we can relate the change in electric *potential*, $V_B - V_A$, between locations A and B to the change in the system's electric *potential energy*, $U_B - U_A$. From the relation above, it follows that

$$U_B - U_A = q(V_B - V_A) \tag{25-3}$$

In applying the energy-conservation principle, we are always concerned with energy *differences*; the choice of the zero of energy is arbitrary. Similarly, we are

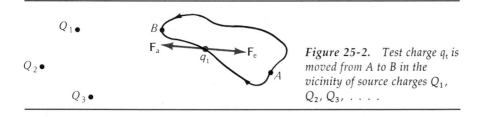

Figure 25-2. Test charge q_t is moved from A to B in the vicinity of source charges Q_1, Q_2, Q_3, ...

free to choose arbitrarily the zero for electric potential. The most common choices are these:

- The electric potential *zero* at an *infinite* distance from one or more charged particles.
- The electric potential of the *earth* (ground) *zero*, particularly for electric circuits.

With starting point A in Figure 25-2 chosen to have zero potential ($V_A = 0$), the electric potential at any point V_p can then be written from (25-1) as

$$V_p = \frac{W_p}{q_t} \tag{25-4}$$

where W_p is now the work required to move q_t at constant speed from the location where the electric potential is zero to p.

The SI unit for electric potential is the *volt* (abbreviated V), named for Alessandro Volta (1745–1827), the inventor of the voltaic pile (a primitive form of energy cell, or battery). By definition,

$$1 \text{ volt} \equiv 1 \text{ joule/coulomb}$$

$$1 \text{ V} \equiv 1 \text{ J/C}$$

Common derived units are the microvolt ($\mu V = 10^{-6}$ V), the millivolt (mV = 10^{-3} V), and the megavolt (MV = 10^6 V).

Electric potential V and electric potential energy U have nearly the same names. Although closely related, the two terms have different meanings and different units, and they must be carefully distinguished:

- *Electric potential V*, measured in volts, is *a property at a location in space of the source charges*. This algebraic number tells us how much work must be done per unit positive charge to move the charge at constant speed to the location where V is measured from the place where the electric potential is chosen to be zero. Equivalently, electric potential change is electric potential energy change per unit positive charge.
- *Electric potential* energy U, measured in joules, is *a property of a system of two or more charged particles*. This algebraic number tells us how much net work must be done to assemble the system.

Example 25-1. Points A, B, and C have the following electric potentials:

$$V_A = 0 \text{ V} \quad V_B = -9 \text{ V} \quad V_C = +3 \text{ V}$$

The way in which a 9-V and a 12-V battery can be connected, using conducting wire, to

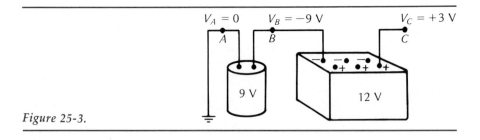

Figure 25-3.

SECTION 25-3 Electric Potential for Point Charges

Table 25-1

CHARGE MOVED, IN μC	FROM	TO	ROUTE	WORK DONE	COMMENT
+0.1	A	A	$A \to B \to C \to A$	0	$\oint \mathbf{F}_e \cdot d\mathbf{r} = 0$
+0.1	A	B	$A \to C \to B$	$-0.9\ \mu J$	$W_{A \to B} = q(V_B - V_A)$ $= (0.1\ \mu C)(-9\ V - 0)$
+0.1	B	A	$B \to A$	$+0.9\ \mu J$	$W_{B \to A} = -W_{A \to B}$
+0.2	B	C	$B \to A \to C$	$+2.4\ \mu J$	$W_{B \to C} = q(V_C - V_B)$ $= (0.2\ \mu C)[3\ V - (-9V)]$
-0.2	B	C	$B \to A \to C$	$-2.4\ \mu J$	Sign reversed for W, compared with entry above, because sign of q is reversed.
-0.2	A	B	$A \to B$	$+1.8\ \mu C$	$W_{A \to B} = (-0.2\ \mu C)(-9V - 0)$
-0.2	A	C	$A \to B \to C$	$-0.6\ \mu C$	$W_{A \to C} = W_{A \to B} + W_{B \to C}$ $= (+1.8 - 2.4)\ \mu J$

produce these electric potentials is shown in Figure 25-3. (A battery maintains a constant potential difference between its terminals. All points on a conducting wire are at the same potential.) How much work must be done to move the charge specified in the first column of Table 25-1 from the initial location to the final one, following the route specified? The answers, together with pertinent comments, are given in the last column of Table 25-1.

In all entries, we recognize that the net work done, which is independent of the route between the end points, p to r, is given by

$$W_{p \to r} = q(V_r - V_p)$$

25-3 Electric Potential for Point Charges

Here we derive the relation for the electric potential V at a distance r from a single point charge q. We choose $V = 0$ for $r \to \infty$.

Consider Figure 25-4, where q is fixed in position while test charge q_t is brought in along a radial line from infinity to its final location, a distance r from q. We compute the work done by the force $\mathbf{F}_a$ applied by an external agent who moves charge q_t at constant speed. The inward force $\mathbf{F}_a$ is matched in magnitude at each location by the oppositely directed electric force $\mathbf{F}_e$. The origin of the radius vector $\mathbf{r}$ is at the location of q. Therefore, $+dr$ represents an *outward* incremental radial displacement; an inward infinitesimal displacement is given by $-dr$.

The incremental work dW done by $\mathbf{F}_a$ as q_t moves inward over an infinitesimal displacement $d\mathbf{r}$ is

$$dW = \mathbf{F}_a \cdot d\mathbf{r} = (F_e)(-dr) = \frac{kqq_t}{r^2}(-dr)$$

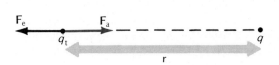

Figure 25-4. Test charge q_t is moved at constant speed relative to fixed charge q under the action of electric force $\mathbf{F}_e$ and applied force $\mathbf{F}_a$.

The total work done is

$$\int_\infty^r dW = -\int_\infty^r \frac{kqq_t}{r^2}\,dr = \frac{kqq_t}{r}$$

$$W = \frac{kqq_t}{r} \tag{25-5}$$

The electric potential V is, by definition, the work per unit charge, so that

$$V = \frac{W}{q_t} = \frac{kqq_t}{q_t r}$$

$$V = \frac{kq}{r} \tag{25-6}$$

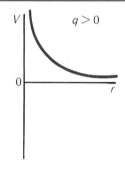

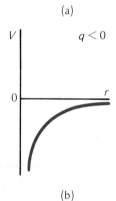

Figure 25-5. Electric potential V as a function of distance r from a point charge q that is (a) positive and (b) negative.

Electric potential is an algebraic quantity and can have positive and negative values. The potential from a single positive charge is positive; its magnitude increases as one approaches the charge. Negative charges produce negative potentials. See Figure 25-5, where the electric potential of a single (a) positive and (b) negative point charge is plotted as a function of distance r. For example, at a distance of 1.0 cm from a point charge of 0.1 μC, the electric potential is

$$V = \frac{kq}{r}$$

$$= \frac{(9.0 \times 10^9 \text{ N}\cdot\text{m}^2/\text{C}^2)(1.0 \times 10^{-7}\text{C})}{1 \times 10^{-2}\text{ m}}$$

$$= 9.0 \times 10^4 \text{V} = 90 \text{ kV}$$

where we have used the fact that newton meters per coulomb equal joules per coulomb equal volts.

What is the electric potential produced by two or more point charges? See Figure 25-6, where charge q_A is a distance r_A from point P, and similarly for charges q_B, q_C, ... The potential V at P is simply the algebraic sum of the potentials produced separately by individual point charges:

$$V = \frac{kq_A}{r_A} + \frac{kq_B}{r_B} + \frac{kq_C}{r_C} + \cdots$$

Writing this result more compactly, we get

$$V = \Sigma \frac{kq_i}{r_i} \tag{25-7}$$

Each charge produces its own potential at P. This result follows because the work done by q_A, q_B, q_C, ... on still another charge brought to P equals the sum of the work done by the electric force of each charge separately, a result that in turn follows from the superposition principle for the electric force. The electric potential for any collection of charge can be found simply, by applying (25-7). To find the *electric field* from a collection of charges may, on the other hand, be far more complicated—you would have to add *vectors*. Computing electric potential is easier because you add algebraic, *scalar* quantities.

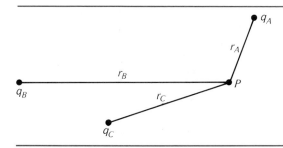

Figure 25-6. Point P has respective distances r_A, r_B, r_C, ... from point charges q_A, q_B, q_C, ...

Example 25-2. Two opposite point charges, each of magnitude q, are separated by a distance $2d$. For the point midway between the charges, what is (a) the electric potential? (b) the electric field? See Figure 25-7.

(a) The electric potential at the midpoint is

$$V = \frac{k(+q)}{d} + \frac{k(-q)}{d} = 0$$

(b) The *magnitude* of the electric field produced at the midpoint by either point charge alone is

$$E_{\text{one charge}} = \frac{kq}{r^2} = \frac{kq}{d^2}$$

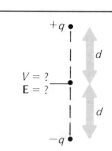

Figure 25-7.

The direction of **E** produced by both $+q$ and $-q$ is from $+q$ to $-q$ (downward in Figure 25-7). Therefore, the net electric field at the midpoint is twice the value from either charge separately, or

$$E = \frac{2kq}{d^2}$$

in the direction from $+q$ to $-q$.

The potential V is zero at the midpoint, but **E** is not. A zero value for V does not imply that **E** must also be zero at that location. Zero V at the midpoint means that no work need be done to bring a third charge from infinity to that location. Nonzero **E** at the midpoint means that an electric force would act on a third charge placed at that location.

25-4 Electric Potential Energy

Point charges q_1 and q_2 are separated by a distance r_{12}. What is their electric potential energy U_e?

We found in Section 25-3 that the work required to bring charge q_t at constant speed from infinity to a distance r from q is kqq_t/r [(25-5)]. This is then also the potential energy of the system. With $q = q_1$, $q_t = q_2$, and $r = r_{12}$ we have

$$U_e = \frac{kq_1q_2}{r_{12}} \qquad (25\text{-}8)$$

The significance of the potential-energy function is that it appears in the general energy-conservation principle. A system of two or more isolated but mutually interacting charged particles has a *constant* total energy E, where

$$E = K + U_e = \text{constant}$$

and K is the total kinetic energy of the particles.

The sign of the potential energy depends on the sign of q_1 and q_2. If the two charges have *like* signs and repel one another, U_e is *positive*. On the other hand, if the two charges have *unlike* signs and attract one another, U_e is negative.* There is nothing wrong with a negative potential energy. Suppose, for example, that a positive and negative charge are far separated and initially at rest. Then $K = 0$, $U_e = 0$, and $E = K + U_e = 0$. The total energy is initially zero and must remain so. But these charges attract one another and they will later have gained kinetic energy as they approach one another. Consequently, with $E = 0$ and $K > 0$, we must have $U_e < 0$.

The matter of signs arises often in this chapter. You can always deal properly with this matter by scrupulously assigning the right signs to q, V, and U. Still better, physical arguments will always confirm that the signs have been correctly assigned.

What is the total electric potential energy for a system with three or more charges? Let r_{12} be the distance separating q_1 and q_2, and r_{23} the distance between q_2 and q_3, and so on. Then the potential energy of the system is

$$\text{Total } U = U_{12} + U_{23} + U_{13} + \cdots$$
$$= \frac{kq_1q_2}{r_{12}} + \frac{kq_2q_3}{r_{23}} + \frac{kq_1q_3}{r_{13}} + \cdots \quad (25\text{-}9)$$

The system's total potential energy is just the sum of terms for all possible pairs of particles. This simple result, like that for gravity, is a direct consequence of the superposition principle for the Coulomb force.

Because the electric force is of the same form as the gravitational force, detailed results derived for gravity (Chapter 15) carry over to the corresponding situations involving charged particles. For example, a charged particle attracted electrically to an oppositely charged particle of far greater mass will orbit it in an elliptical path, and the Kepler rules for planetary motion will have exact counterparts.

Electric potential energy in the SI system is expressed in joules. For atomic particles or atomic systems, a more convenient energy unit is the *electron volt* (abbreviated eV). Consider an electron or any other particle with charge magnitude e that is moved across a potential difference ΔV of 1 V. From (25-3), the potential energy changes by ΔU, where

$$\Delta U = q\,\Delta V$$

Here, with $q = e$ and $\Delta V = 1$ V, we can write

$$\Delta U = (e)(1\text{ V}) = 1\text{ eV} = 1\text{ electron volt}$$

Alternatively, we can express the energy change in joules as

$$\Delta U = (1.602\,189\,2 \times 10^{-19}\text{J})(1\text{ V}) = 1.602\,189\,2 \times 10^{-19}\text{ J}$$

---

* The form of (25-8) for the electric potential energy of a pair of point electric charges is very much like the relation $U_g = -Gm_1m_2/r_{12}$, for the gravitational potential energy of a pair of point masses [(15-4)]. We simply have $-Gm_1m_2$ replaced by kq_1q_2. The difference in sign arises as follows. Two interacting masses (with both m_1 and m_2 positive) *attract* one another and the potential energy is *negative*; two interacting unlike charges (with one of q_1 and q_2 positive and the other negative) also attract one another, and the potential energy is again negative.

Therefore, the conversion relation for the electron volt can be written as

$$1 \text{ eV} \equiv 1.602\ 189\ 2 \times 10^{-19} \text{ J}$$

The electron volt can therefore be defined as the work done on a particle with charge e as it goes across a potential difference of 1 V. Equivalently, a particle of charge e will have its kinetic energy change by 1 eV when it is accelerated across a potential difference of 1 V.

Example 25-3. An electron (charge, $-e$) is released from rest at a distance of 0.2 nm = 2.0×10^{-10} from a proton (charge, $+e$) also initially at rest. What is the electron's kinetic energy when it is 1 nm from the proton?

We can regard the proton as remaining essentially at rest because it has a far greater mass than the electron. Only the electron gains kinetic energy. With subscript 1 denoting the initial state and subscript 2 the final state, we have from energy conservation

$$E = K_1 + U_1 = K_2 + U_2$$

$$= 0 + \frac{k(+e)(-e)}{0.2 \text{ nm}} = K_2 + \frac{k(+e)(-e)}{0.1 \text{ nm}}$$

$$K_2 = \frac{ke^2}{0.2 \text{ nm}}$$

$$= \frac{(9.0 \times 10^9 \text{ N} \cdot \text{m}^2/\text{C}^2)(1.60 \times 10^{-19} \text{ C})^2 (1 \text{ eV})}{(2.0 \times 10^{-10} \text{ m})(1.60 \times 10^{-19} \text{ J})} = 7.2 \text{ eV}$$

The result here is typical of subatomic particles separated by atomic dimensions; kinetic and potential energies are of the order of a few electron volts.

Example 25-4. Three charges $+Q$, $+Q$, and $-Q$ are brought to the corners of an equilateral triangle of side L. See Figure 25-8(a). What is the total energy required to assemble these charges?

The work W done in assembling the system equals the system's total potential energy. Therefore, from (25-9),

$$W = U = +\frac{kQ^2}{L} - \frac{kQ^2}{L} - \frac{kQ^2}{L} = -\frac{kQ^2}{L}$$

The total energy is *negative*. This means that the three charges form a *bound* system. The total U turns out to be equal to that of just a single pair of unlike charges. We can see that this must be the result from another angle. Imagine that first the charges $+Q$ and $-Q$ are separated by L and therefore have a potential energy $-kQ^2/L$. Then the third charge is brought to its final position along a line perpendicular to the line joining the first two charges; see Figure 25-8(b). This third charge is acted on by the electric field of the first two charges [see Figure 23-15(a)]. But that field is always perpendicular to the displacement of the third particle, and no work is required to move the third charge to its final location along this route (or any other)! From another point of view, the potential of the first two charges is zero at all points on the dashed line in Figure 25-8(b).

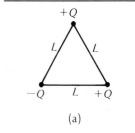

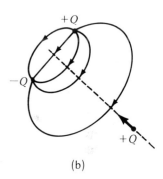

Figure 25-8. (a) A system consisting of three point charges at the corners of an equilateral triangle. (b) No work is required to bring the third charge midway between the other two charges (constituting an electric dipole) along the dashed path, or along any other path.

25-5 Equipotential Surfaces

Consider a single point charge q, taken to be positive for definiteness. The electric potential V at any distance r from the point charge is $V = kq/r$. All points at a fixed distance r from q are on a sphere of radius r, and all these points are at the *same* potential V. The concentric spherical surface about a point charge constitutes an *equipotential surface*. Indeed, we can imagine the point

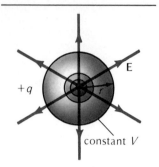

Figure 25-9. Electric-field lines and equipotential surfaces for a point charge.

charge to be surrounded by a number of concentric spherical surfaces; every point on a given surface has the same electric potential.

Now consider the electric field produced by a single positive point charge and its relation to equipotential surfaces. The electric field is radially outward and therefore perpendicular at every point to an equipotential surface. Furthermore, as one moves along the direction of **E**, radially outward for a positive charge, one moves to surfaces at progressively lower potentials. In a two-dimensional figure, the equipotential surfaces are represented by equipotential lines perpendicular to electric-field lines. See Figure 25-9.

More generally, an equipotential surface for any charge distribution is the locus of points all at the same electric potential. By definition, a charge can be moved along an equipotential surface with no work done on it by the electric field. Therefore, **E** must at every location be perpendicular to the constant V surface; otherwise, a component of **E** would lie on the V surface and do work on and change the energy of a charge lying on that surface. Moreover, the direction of **E** is always toward surfaces of *lower* potential.

The equipotential surfaces for a uniform electric field consist of flat planes perpendicular to **E**. See Figure 25-10.

Field lines and the associated equipotential lines for two equal but opposite electric charges (an electric dipole) are shown in Figure 25-11. The constant V loops are reminiscent of contour lines drawn on an ordinary map to indicate points at the same elevation. Just as one can walk in a loop around a mountain peak at constant elevation without doing work against the gravitational force, a

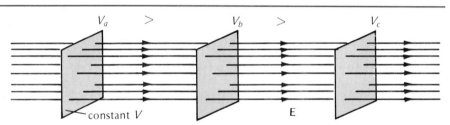

Figure 25-10. Electric-field lines and equipotential surfaces for a uniform electric field.

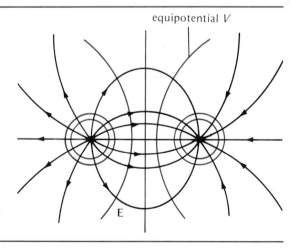

Figure 25-11. Electric-field lines **E** and the associated equipotential lines in the plane containing two point charges of equal and opposite charges.

charge can traverse an equipotential line with no work done on it. In the potential-hill analogy, where elevation represents electric potential, a positive point charge produces an infinitely high "spike" ($V = kq/r$), and a negative point charge an infinitely deep "hole in the ground." A uniform electric field is represented by an inclined plane. As will be proved in the next section, the electric field in the potential-hill model corresponds to the steepest slope between points at different elevation (or potential).

25-6 Relations between V and E

The electric properties of a region of space can be mapped in two different ways:

- By the *vector electric field* **E**, the *force* per unit positive charge, with **E** mapped by *electric field lines*.
- By the *scalar electric potential* V, the *work* per unit positive charge, with V mapped by *equipotential surfaces*.

What are the general relations between **E** and V? If we know **E** at some location, how does V vary as we go to nearby points? Or the converse question, if we know V over a small region of space, how do we find **E** there?

Consider first an important special situation: a *uniform electric field*. Electric-field lines and mutually perpendicular associated equipotential lines are shown in Figure 25-12. Suppose that a positive charge q is moved a distance d along the direction of **E** from an initial electric potential V_i to a final potential V_f. Charge q moves in the direction of **E**; therefore, V_i is higher than V_f. The change in potential from i to f is $\Delta V = V_f - V_i$; with $V_i > V_f$, the change ΔV is a negative.

The electric force on q has the magnitude $F = qE$ and the work done by this constant force over distance d is

$$W = Fd = qEd$$

The change in electric potential ΔV is also related to the work done by the electric force through (25-2):

$$W = -q\,\Delta V$$

(The minus sign here simply reflects the circumstance that, since ΔV is the potential *increase*, $-\Delta V$ is the potential *drop*.)

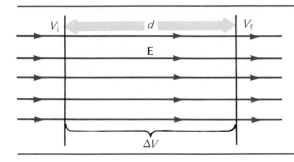

Figure 25-12. Drop in electric potential ΔV over a distance d along a uniform electric field **E**.

Equating the two relations above then yields

$$\Delta V = -Ed$$

or
$$E = -\Delta V/d \qquad (25\text{-}10)$$

The minus sign tells us that if we travel in the direction of **E** over a distance d, the electric potential *decreases*; the change ΔV is negative because the potential drops.

Equation (25-10) shows that appropriate units for the electric field are *volts per meter*, V/m. For example, a uniform electric field with a magnitude of, say, 100 V/m is one in which the electric potential drops by 100 V for each meter traversed in the direction of the electric-field lines (or a 1-V drop for each centimeter). It is easy to see that the two units for electric field, originally newtons per coulomb and now also volts per meter, are equivalent: V/m = (J/C)/m = (N·m/C)/m = N/C.

To get the general relations between V and **E**, we simply consider an infinitesimal displacement $d\mathbf{s}$ in any direction in any electric field **E**. The work done on q can be expressed either as

$$dW = \mathbf{F}_e \cdot d\mathbf{s} = q\mathbf{E} \cdot d\mathbf{s}$$

or
$$dW = -q\,dV$$

Again equating the two relations gives

$$dV = -\mathbf{E} \cdot d\mathbf{s} \qquad (25\text{-}11)$$

If we now integrate this relation from starting point 1 to final point 2, we have for the difference in electric potential between the two points:

$$\int_1^2 dV = -\int_1^2 \mathbf{E} \cdot d\mathbf{s}$$

$$V_2 - V_1 = -\int_1^2 \mathbf{E} \cdot d\mathbf{s} \qquad (25\text{-}12)$$

The quantity $\int \mathbf{E} \cdot d\mathbf{s}$ is described in mathematical language as the *line integral* of the vector field **E**. We see from (25-12) that if the electric field **E** is known as a function of position, the potential V can be computed by integration.

Displacement $d\mathbf{s}$ was considered above to be in an arbitrary direction. In what direction must $d\mathbf{s}$ point to produce the maximum change dV in electric potential? Obviously, $d\mathbf{s}$ must then be parallel to **E**. Under these circumstances we can write (25-11) as

$$E = -\frac{dV}{ds} \qquad (25\text{-}13)$$

where the derivative dV/ds is computed for an infinitesimal displacement *along the direction* in space for which V changes *most rapidly* with distance; **E** is in the direction for which V drops most rapidly. The quantity dV/ds is described in mathematical language as the *gradient* of the scalar function V; thus, (25-13) says that the static electric field at any point in space is the negative gradient of the electric potential in the neighborhood of that point. The electric potential can be expressed as a function of coordinates x, y, and z. Then the rectangular components of the electric field are

$$E_x = -\frac{\partial V}{\partial x}, \quad E_y = -\frac{\partial V}{\partial y}, \quad E_z = -\frac{\partial V}{\partial z} \qquad (25\text{-}14)$$

where the partial derivative $\partial V/\partial x$ of V with respect to x implies that variables y and z are held constant, and likewise for $\partial V/\partial y$ and $\partial V/\partial z$.

Example 25-5. As shown in Section 24-1, an infinite plane sheet of charge with a constant surface-charge density σ produces a uniform electric field. On both sides of the sheet, **E** is perpendicular to the sheet and has a magnitude

$$E = 2\pi k\sigma \qquad (24\text{-}2)$$

How does the electric potential vary with distance from the sheet?

For definiteness, assume the sheet to be positively charged and to lie in the yz plane. Then **E** points along the x axis, as shown in Figure 25-13.

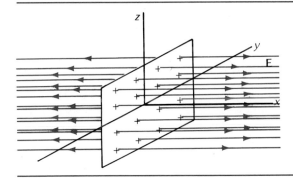

Figure 25-13. Electric field **E** from a uniformly charged infinite plane sheet.

The electric potential V is given in general in terms of **E** by (25-12):

$$V_2 - V_1 = -\int_1^2 \mathbf{E} \cdot d\mathbf{s}$$

We take displacement $d\mathbf{s}$ to be along the x axis, the only direction in which **E** has a nonzero component. The two relations above yield

$$V_x - V_0 = -\int_0^x 2\pi k\sigma \, dx$$

where we integrate variable x from 0 to x, and designate the arbitrarily chosen zero of potential (at $x = 0$) by V_0. Then,

$$V_x - V_0 = -2\pi k\sigma x$$

We see that the electric potential drops uniformly with distance x from the charged sheet.

Example 25-6. Figure 25-14 shows a uniformly charged ring of radius R and total charge Q. What is the electric field at P, a distance x from the ring's center measured outward along the symmetry axis?

One way to find **E** is to add *as vectors* the contributions to **E** from small charge elements dq distributed around the ring. An easier way is first to find the electric potential V at P by adding *as scalar quantities* the contributions from all charge elements and then applying (25-14).

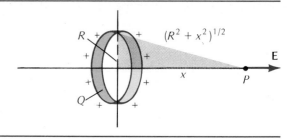

Figure 25-14. Electric field **E** from a uniformly charged ring of radius R.

Every charge element on the ring is the same distance $(R^2 + x^2)^{1/2}$ from P. The total electric potential at P is then

$$V = \Sigma \frac{kq_i}{r_i} = \frac{kQ}{(R^2 + x^2)^{1/2}}$$

Since V depends only on the variable x, the electric field has an x component only; **E** is along the symmetry axis. From (25-14), we have

$$E_x = -\frac{dV}{dx} = \frac{kQx}{(R^2 + x^2)^{3/2}}$$

For positive Q, field **E** points away from the ring.

We can check that the relation above is correct for two extreme values of x:

- $x = 0$, at the center of the ring; then $E_x = 0$. A point charge brought to the center of the uniformly charged ring would be surrounded equally in all directions by identical charges; at this location there is no force on a charge, and necessarily, **E** = 0.
- $x \to \infty$, far from the ring; then $E_x \simeq kQ/x^2$ with $R^2 \ll x^2$. Viewed from afar, the ring is effectively a point charge, and the field a distance x away is that of point charge Q.

25-7 Electric Potential and Conductors

Consider an insulated conductor with a net charge. We have already seen (Section 24-5) that after electrostatic equilibrium has been achieved:

- The net charge appears only on the conductor's outer surface.
- Inside the conducting material, **E** = 0.
- Just outside the conductor's surface, **E** at every location is perpendicular to the surface.

We can now add one more general property:

- *All points* on the charged conductor's surface and interior are at the *same electric potential*; the entire volume of the conductor is an equipotential volume. See Figure 25-15.

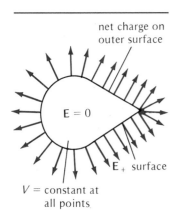

Figure 25-15. A charged conductor in electrostatic equilibrium.

The proof is straightforward. First, just outside the conductor's surface, **E** is perpendicular at all points to the surface, so that this surface must be an equipotential surface. Said a little differently, a charge can be carried along the surface with no work done on the charge because there is no component of **E** along the displacement of the charge. Indeed, the charge can be carried into the conducting material and around anywhere *within* it, and again no work is done

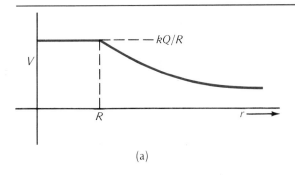

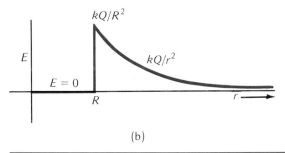

Figure 25-16. For a spherical conducting shell of radius R with charge Q, as a function of distance r from the center, (a) electric potential V and (b) electric field **E**.

on the charge, because inside $\mathbf{E} = 0$. The entire volume of an insulated charged conductor is equipotential.

Consider a spherical conducting shell of radius R with a net charge Q. How do the electric potential V and the electric field **E** each vary in magnitude with distance r from the center of sphere? We know that from the outside, a spherical shell of charge is equivalent to a single point charge at its center. Therefore,

$$\text{For } r \geq R: \quad V = \frac{kQ}{r}$$

$$E = \frac{kQ}{r^2}$$

At any point within the charged shell, the field **E** is zero, and therefore, the potential is the same as on the shell itself.

$$\text{For } r < R: \quad V = \frac{kQ}{R}$$

$$E = 0$$

Figure 25-16 shows V and E as functions of r.

Suppose that some well-designed Van de Graaff lecture-demonstration electrostatic generator has a spherical conducting shell of 15-cm radius at an electric potential of 250 kV. Then the charge on the sphere is, from the relation above,

$$Q = \frac{VR}{k} = \frac{(250 \times 10^3 \text{ V})(15 \times 10^{-2} \text{ m})}{(9.0 \times 10^9 \text{ N} \cdot \text{m}^2/\text{C}^2)} = 4.2 \ \mu\text{C}$$

and the electric field at the shell's surface is

$$E = \frac{kQ}{R^2} = \frac{V}{R} = \frac{250 \times 10^3 \text{ V}}{15 \times 10^{-2} \text{ m}} = 1.7 \times 10^6 \text{ V/m}$$

This value for E on the surface is close to the critical value, called the *dielectric strength* (equal to 3×10^6 V/m for dry air at atmospheric pressure), at this critical value, the very strong electric field starts to rip electrons from air molecules, ionizing the air. Any initially charged conductor will lose its charge when E at any point on the surface exceeds the dielectric strength of the medium in which the conductor is immersed.

The relation above, $E = V/R$, gives the magnitude of E at the surface of any spherical conductor of radius R with electric potential V. Suppose that several charged spherical conductors all have the same potential. With $E \propto 1/R$, then E is greatest for small R. A large charged sphere is less likely to discharge than a small one at the same potential. This inverse relation, $E \propto 1/R$, applies to any single charged conductor of arbitrary shape, all points of which are necessarily at the same potential. We can now interpret R to mean the local radius of curvature over any convex small region of the conductor's surface. The electric field is a minimum at a flat region (large R) and a maximum at a sharp point (small R). We saw earlier (Section 24-5) that the electric-field magnitude on the surface of a charged conductor is proportional to the local surface-charge density, $E \propto \sigma$. It follows then that surface charge is also concentrated at sharp points. See Figure 25-15. This result has important practical consequences: to avoid discharge, make a charged conductor's surface as flat and smooth as possible (a sphere has the flattest surface for a given volume); to encourage discharge, use sharp points (for example, the lightning rod).

Summary

Definitions

Electric potential (V): the electric potential difference between points A and B is the work done $W_{A \to B}$ by an applied force to move charge q_t at instant speed from A to B:

$$V_B - V_A = \frac{W_{A \to B}}{q_t} \quad (25\text{-}1)$$

Equipotential surface: locus of points at the same electric potential.

Units

Electric potential: Volt (V)

$$1 \text{ V} \equiv 1 \text{ J/C}$$

Energy: electron volt (eV)
1 eV = energy to move charge e across a potential difference of 1 V = 1.60×10^{-19} J

Electric field: Volt/meter (V/m)

$$1 \text{ V/m} = 1 \text{ N/C}$$

Important Results

The Coulomb force is a conservative force: $\oint \mathbf{F}_e \cdot d\mathbf{r} = 0$.

The electric potential V arising from point charge(s) q_i at distance(s) r_i:

$$V = \sum \frac{kq_i}{r_i} \quad (25\text{-}7)$$

The electric potential energy U_e of point charges q_1 and q_2 separated by distance r_{12}:

$$U_e = \frac{kq_1 q_2}{r_{12}} \quad (25\text{-}8)$$

Electric-field lines are perpendicular at every location to the associated equipotential surfaces and point to regions of lower electric potential.

Computing V from $\mathbf{E}$ gives

$$V_2 - V_1 = -\int_1^2 \mathbf{E} \cdot d\mathbf{s} \quad (25\text{-}12)$$

Computing $\mathbf{E}$ from V gives

$$E = -\frac{dV}{ds} \quad (25\text{-}13)$$

where the infinitesimal displacement $d\mathbf{s}$ is along the direction for which V decreases most rapidly with distance, also the direction of $\mathbf{E}$.

For a uniform $\mathbf{E}$:

$$E = -\frac{\Delta V}{d} \quad (25\text{-}10)$$

for a displacement d along the direction of $\mathbf{E}$.

All points on the surface and interior of a charged conductor in equilibrium are at the same potential.

At the surface of a charged conductor, $\mathbf{E}$ is inversely proportional to the radius of curvature of a convex region. Surface charge and electric field are large at sharp points of a charged conductor.

Problems and Questions

Section 25-2 Electric Potential Defined

· **25-1 Q** In comparing electric and gravitational forces, we find that electric potential is analogous to
(A) mass m.
(B) mgh.
(C) g.
(D) mg.
(E) gh.
(F) h.

· **25-2 P** An electron with initial velocity 4×10^6 m/s is accelerated through a potential difference of 100 V. What is its final velocity?

: **25-3 P** In a typical lightning flash, the potential difference between discharge points is about 10^9 V, and about 30 C of charge is transferred. How much ice could a lightning bolt melt if all the energy were used for that purpose?

: **25-4 Q** Which of the following is an accurate statement?
(A) A person raised to a potential of a few thousand volts will almost certainly suffer serious injury.
(B) It is possible for clouds to reach potentials of the order of half a million volts.
(C) By walking across a nylon rug, you cannot raise your body to a potential of more than a few millivolts.
(D) Electrostatic demonstration experiments are unaffected by humidity.
(E) It is now believed that our galaxy carries a net electric charge.

Section 25-3 Electric Potential for Point Charges

: **25-5 P** A charge of $+2\ \mu C$ is placed at the origin and a second charge of $-4\ \mu C$ at the point (2 m, 0). Determine the electric potential at the following points: (a) (1 m, 0) (b) (1 m, 1 m) (c) (−1 m, 0) (d) (0, 1 m).

: **25-6 P** Charge Q is distributed uniformly along the x axis from $x = 0$ to $x = L$. Determine the potential at a point $x > L$ on the x axis.

: **25-7 P** A beam of charged particles forms a cylindrical column of radius R. If the charge density is uniform throughout the beam, what is the potential (a) inside and (b) outside the beam?

: **25-8 P** Electric charge is distributed with constant density throughout an insulating sphere of radius R. Determine the electric potential inside and outside the sphere as a function of distance from the center.

Section 25-4 Electric Potential Energy

· **25-9 Q** An electron volt is a unit used to measure
(A) energy.
(B) potential difference.
(C) electric field.
(D) force.
(E) electric charge.

· **25-10 P** An electron passes a proton at a distance of 2×10^{-10} m. What is the minimum kinetic energy of the electron that will prevent its being bound to the proton?

: **25-11 P** Three charges q_1, q_2, and q_3 are placed at positions x_1, x_2, and x_3 on the x axis. What is the total potential energy of the charges if $q_1 = +1\ \mu C$, $x_1 = 0$; $q_2 = -2\ \mu C$, $x_2 = 2$ cm; $q_3 = +3\ \mu C$, $x_3 = 3$ cm?

: **25-12 P** Two 1-μC point charges are separated by 2 cm. How much work is needed to bring an electron from a point far away to the point midway between the charges?

: **25-13 P** An electron travels in a circular orbit of radius r with constant speed v around a stationary proton. The charge on the electron is $-e$ and on the proton is $+e$. The electron mass is m. The system's total energy is
(A) $\frac{1}{2} mv^2$
(B) $\dfrac{-k e^2}{r}$
(C) $\dfrac{mv^2}{2} + \dfrac{ke^2}{r}$
(D) $\dfrac{-ke^2}{2r}$
(E) $\dfrac{ke^2}{r}$

: **25-14 P** Four charges $+Q$ and four charges $-Q$ are placed at the corners of a cube of side a in such a way that nearest-neighbor charges have opposite signs. How much energy is needed to assemble this charge array?

· **25-15 Q** Point A is at an electric potential of -4.0 V and point B at an electric potential of $+6.0$ V. An electron passes

point A with a kinetic energy of 12.0 eV. When it reaches point B, its kinetic energy will be
(A) 2 eV.
(B) 6 eV.
(C) 18 eV.
(D) 22 eV.
(E) indeterminable unless we know the direction the electron was moving at point A.

: **25-16 P** Two fixed point charges, each $+Q$, are separated by a distance d. (a) Graph the potential energy of a third charge $+q$ as a function of position along the line joining the two fixed charges. (b) Show that for small displacements x from the midpoint, the potential energy E_p is parabolic, that is, $E_p \propto x^2$. (c) When the charge q is at the midpoint, is it in stable or unstable equilibrium with small displacements along the line joining the fixed charges? along a direction perpendicular to this line?

: **25-17 P** A proton initially very far away from a nucleus with charge $+Ze$ is projected toward the nucleus with velocity v_0. When the proton is a distance R from the nucleus, its velocity has decreased to $\frac{1}{2}v_0$. (a) When its velocity has dropped to $\frac{1}{4}v_0$, how far will it be from the nucleus? (b) What is the distance of closest approach to the nucleus?

: **25-18 P** An alpha particle (a helium nucleus with charge $+2e$) with an initial kinetic energy 10.0 MeV approaches a gold nucleus (charge $+79e$) along a path that would bring it within 2.0×10^{-14} m of the gold nucleus if it were not for the Coulomb repulsion. Assume that the heavy gold nucleus remains at rest. (a) What is the distance of closest approach of the two particles? (b) What is the kinetic energy of the alpha particle at the point of closest approach?

: **25-19 P** Assuming that the density of the earth is constant (and equal to M_e/V_e), determine how much energy would be needed to blow the earth apart, that is, to send each piece of it off to infinity. (Hint: Gravitational potential energy may be handled much like electric potential energy.)

: **25-20 P** Electrons are emitted from a cathode with negligible kinetic energy and accelerated through a potential difference of 500 V to an anode 20 cm from the cathode. Take the beam to have circular cross section of diameter 1 mm, and assume that a constant electric field acts on the electrons. What is the charge density in the beam as a function of the distance from the cathode?

Section 25-5 Equipotential Surfaces

· **25-21 Q** Equipotential lines, in a two-dimensional plot, are most nearly analogous to
(A) flow lines in a fluid.
(B) the tracks left by a charged particle moving in a cloud chamber.
(C) the path followed by a particle undergoing Brownian motion.
(D) the direction a stream would follow on a map.
(E) contour lines on a topographic map.
(F) a map of regions in a room in which there is no sound.

: **25-22 P** Charges of $+1$ nC and -1 nC are separated by 2 cm. Using approximate numerical methods, carefully sketch the equipotential lines that pass through points at distances of 8 mm, 9 mm, 10 mm, 11 mm, and 12 mm from the positive charge along the line joining the two charges.

Section 25-6 Relations between V and E

· **25-23 P** An electron is accelerated by an electric field of 2000 V/m over a distance of 2 cm. By how much, in electron volts and in joules, does its energy increase?

· **25-24 P** A 3-keV electron is shot through a hole in one of two parallel plate electrodes. A potential difference of 10,000 V is applied between the electrodes, as shown in Figure 25-17. How far past the first electrode will the electron go before stopping?

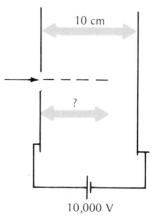

Figure 25-17. Problem 24-25.

· **25-25 P** In a given region, the electric field in the x direction has the constant value -100 V/m. If the potential at $x = 4$ cm is 6 V, what is the potential at $x = 10$ cm?

· **25-26 P** If the electric potential varies from 100 V to 150 V in a distance of 5 cm, what is the magnitude of the average electric field in this region?

: **25-27 P** An infinite conducting sheet carries a uniform charge density of 10^{-7} C/m². How far apart are equipotentials that differ by 5 V?

: **25-28 P** In a tandem Van de Graaff accelerator, a hydrogen ion is accelerated from rest through a potential difference of 10 MV. Then the two electrons are stripped from the ion and the resulting bare proton is accelerated back to ground potential through a potential difference of -10 MV. What is (a) the final kinetic energy and (b) the speed of the proton?

· **25-29 P** Sketched in Figure 25-18 is the potential arising from a hypothetical charge distribution. At which indicated point or points on the x axis does the electric field E_x (a) have its largest positive value? (b) have its largest negative value? (c) have the value zero?

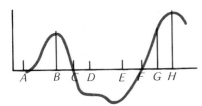

Figure 25-18. Problem 25-29.

: 25-30 Q Which of the following is the most accurate statement?
(A) The electric potential difference between two points equals the work that must be done in moving an electron from one point to the other.
(B) The presence of a positive point charge always tends to make the potential in its vicinity more positive.
(C) We associate electric potential with the slope of the electric field.
(E) The potential due to two point charges q and $-q$ will be zero at only one point in a finite region of space.
(F) More than one of the above is true.

· 25-31 P An electric field varies from 200 V/m at point A to 300 V/m at point B, and A and B are 0.5 m apart. Can you calculate the difference in potential between points A and B with only this information? If so, what is the value of the potential difference?

: 25-32 P The potential due to a point charge is $V = q/4\pi\epsilon_0 r$ where $r = \sqrt{x^2 + y^2 + z^2}$. Calculate electric-field components E_x, E_y, and E_z from this expression, and show that the expected result $\mathbf{E} = (q/4\pi\epsilon_0)(\mathbf{r}/r^3)$ is obtained.

: 25-33 Q Consider the electric-field lines and equipotentials arising from an array of positive and negative point charges. Which statement is true?
(A) The electric field does not vanish (except possibly at infinity).
(B) The electric field lines never cross.
(C) The equipotential surfaces never intersect.
(D) No work is required to move a test charge parallel to an electric field line.
(E) Both B and C above are true.

: 25-34 Q Which of the following statements are true?
(A) The electric potential arising from an array of point charges is the sum of the potentials arising from each individual charge.
(B) The electric field is zero where the potential is zero.
(C) The potential arising from a single point charge may vary from positive to negative in different regions.
(D) The force on a charged particle located on an equipotential surface is zero.
(E) All of the above are true.

· 25-35 P Charges $+Q$, $+Q$, and $-2Q$ are at the corners of an isosceles triangle. Make a qualitative sketch of some of the equipotentials and electric-field lines for this configuration.

: 25-36 Q Which of the following is a true statement?
(A) If the electric field is zero at a point, the electric potential will be zero there also.
(B) If the electric potential is zero at a point, the electric field must be zero there also.
(C) If the electric field is zero throughout some region of space, the electric potential must also be zero throughout that region.
(D) If the electric potential is zero throughout some region, the electric field must also be zero throughout the region.
(E) The electric field is large where the electric potential is large and small where the electric potential is small.

: 25-37 Q In Figure 25-19 are drawn some equipotential planes or electric-field lines. At which of the points indicated is the electric-field strength greatest?
(A) A.
(B) B.
(C) C.
(D) D.
(E) It is equally large at B and D.
(F) It is equally large at A and C.
(G) The question cannot be answered unless we know whether the lines are equipotentials or electric-field lines.

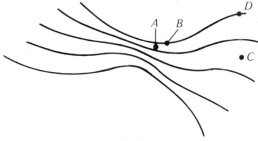

Figure 25-19. Question 25-37.

: 25-38 P Charge is uniformly distributed over a circular disk of radius a. (a) Show that the potential at a point on the symmetry axis a distance z from the center is

$$V = \frac{\sigma}{2\epsilon_0}(\sqrt{a^2 + z^2} - z)$$

(b) Show that this result gives the electric field on the axis as

$$E_z = \frac{\sigma}{2\epsilon_0}\left(1 - \frac{z}{\sqrt{a^2 + z^2}}\right)$$

(c) Consider what the limiting values for V and E are when $z \gg a$ and when $z = 0$. Do these results agree with what you would expect?

Section 25-7   Electric Potential and Conductors

· **25-39 Q** In learning about electrostatics, it is helpful to use the analogy between the description of electrical phenomena and a contour map. The surface of a piece of copper, then, would be analogous to
(A) the direction in which a stream flows on the map.
(B) a contour line.
(C) a point of high elevation on the map, such as a mountain peak.
(D) a point of low elevation on the map, such as the bottom of a well.
(E) a uniform slope.
(F) a road drawn on the map.

· **25-40 Q** Can there be a potential difference between two adjacent insulated conductors, each of which carries the same charge?
(A) No, since if two conductors carry the same charge, they must be at the same potential.
(B) No. All conductors are equipotential surfaces; hence they are all "grounded," that is, they are at zero potential.
(C) Yes. An example would be two spheres of different radii.
(D) A conductor is by definition electrically neutral; hence this question makes no sense, since one cannot have a "conductor carrying a charge."

· **25-41 Q** Two conductors are connected by a long copper wire. Thus
(A) each conductor must have the same charge.
(B) each conductor must be at the same potential.
(C) the electric field at the surface of each conductor is the same.
(D) one conductor always feels a force due to the other conductor.
(E) each conductor has zero net charge.

· **25-42 P** Trailing from the wings of a jetliner are pointed needles. When static charge accumulates on the airplane, a corona discharge occurs at these needles, thereby preventing dangerous sparks near the fuel tanks. If air breaks down in an electric field of 3×10^6 V/m, about what is the maximum potential to which the plane could be charged electrostatically? (The needles have a tip radius of 0.01 mm.)

: **25-43 Q** A long metal bar of square cross section is given a charge. Using qualitative reasoning, sketch a series of equipotential lines and electric-field lines. No calculation is necessary, but your drawing should be accurate for the areas close to the conductor and very far away from it.

· **25-44 Q** To understand new concepts, it sometimes helps to draw analogies with familiar ideas. In this sense one can think of a two-dimensional plot of equipotential surfaces as a contour map. In such a representation, a charged hollow copper sphere would be most like
(A) a volcano.
(B) a mountain lake.
(C) a very steep cliff.
(D) a sloping hillside of constant slope.
(E) a very narrow well, dug straight down far into the earth.

: **25-45 P** Four spherical raindrops, each of radius r, carry such a charge that the potential of each is V_0. They join to form a single large droplet. (a) What is the potential of the large droplet? (b) What is the change in electrostatic energy of the system when the droplets combine? How do you account for the change?

: **25-46 P** Calculate the work necessary to charge a conducting sphere of radius R with a total charge Q.

: **25-47 P** What charge must be placed on a conducting sphere to raise it to a potential of 2 MV (2×10^6 V) if the radius of the sphere is (a) 1 m? (b) 10 cm?
(c) Since less charge is needed to raise the smaller sphere to the required voltage, why are large spheres used in a Van de Graaff generator? Substantiate your answer by calculating the electric field at the surface of the sphere in each of the above cases. Note that a large electric field is more likely to cause electrical breakdown and also cause a more rapid leakage of charge because of migrating ions in the surrounding atmosphere.

: **25-48 Q** Why would one want to use a large spherical conductor at the high-voltage end of a Van de Graaff generator?
(A) Because the electric field is smaller at the surface of a large sphere than at the surface of a small sphere at the same potential.
(B) Because the larger sphere can store more charge.
(C) Because a given amount of charge placed on a large sphere will raise it to a higher potential than what would happen if the same charge were placed on a small sphere.
(D) Because with a large sphere it is easier to generate the corona discharge needed for acceleration of particles.
(E) The main reason for a large sphere is that this in turn allows one to use a wide belt to carry charge up to the sphere; hence charging can be done more rapidly and efficiently.

: **25-49 Q** Three hollow, conducting aluminum spheres of radius 1 m, 2 m, and 3 m are placed at the corners of an equilateral triangle of side 50 m. A charge of 400 μC is placed on the large sphere and then the spheres are connected by wires. What is the final charge on each sphere?

: **25-50 P** In a lecture demonstration, three conducting spheres, of radius 2 cm, 4 cm, and 6 cm, are used. The largest sphere is charged to a potential of 500 V. It is first touched to the smallest sphere and next it is touched to the intermediate sphere. What will be the potential of each sphere and the charge on each after this is done, each sphere then being separate from the others?

: **25-51 P** A Geiger counter uses two concentric metal cylinders of radius $R_1 = 0.1$ mm and $R_2 = 1.2$ cm; between the cylinders a voltage of 1000 V is applied. Incident radiation particles ionize the air between these electrodes, giving rise to a pulse of electric current (flow of charge) between them. (a) What is the electric field at the surface of the inner conductor? (b) What is the electric field at the surface of the outer conductor? (c) What is the linear charge density along each conductor?

Supplementary Problems

25-52 P The electrostatic charge that can build on the surface of a liquid drop (such as a raindrop or a droplet in an aerosol spray) can cause it to break up. This has many important consequences. Consider a spherical drop of radius R that can break into two identical smaller drops. The energy of the conducting drop consists of electrostatic energy and surface-tension energy, $E_s = 4\pi R^2 \sigma$, where σ = surface tension = 0.07 J/m² for water.

(a) Write an expression for the energy of the drop before it breaks up and after. When it has broken up, it consists of two small touching drops. The initial charge Q is shared equally between these two smaller drops. (b) The initial drop will break up when the energy of the two small touching drops is less than the total energy of the large drop. Show that the condition for instability is $Q^2/4\pi\epsilon_0 > 2.5\sigma R^3$. (c) Determine the maximum electric field that can exist on the surface of the drop before it breaks up.

(If you have never noticed that water droplets acquire a charge, try deflecting a fine spray from a faucet by using a comb you have charged by rubbing it on a sweater. This phenomenon is important in meteorology, spray painting, aerosol sprays, and ink jet printers.)

25-53 P In an ink jet printer, a thin stream of conducting ink is projected from a nozzle down the center of a cylindrical electrode. A high voltage is applied between the jet and the surrounding cylinder. The droplets acquire a net charge opposite to that of the cylinder. Electrostatic and surface-tension forces cause the jet to break into charged droplets. These droplets are then deflected by charged parallel metal plates, just as an electron beam is deflected in an oscilloscope.

(a) If the jet stream has radius R_j and the charging cylinder has radius R_c, what is the charge per unit length on the jet in terms of the voltage V between the jet and the cylinder? (b) If the jet breaks into droplets of radius $R_d = 2R_j$ how many droplets per second will be generated when $R_j = 2 \times 10^{-5}$ m and the jet velocity is 25 m/s? (c) What will be the charge per droplet if $R_c = 5$ mm and the charging voltage is $V = 500$ V? (d) Through what angle θ will the droplets be deflected when they pass through the deflecting plates shown in Figure 25-20? A deflecting voltage of 1000 V is used.

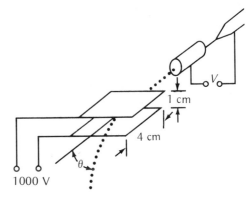

Figure 25-20. Problem 25-53.

25-54 P In an electrostatic dust precipitator, a long straight wire passes down the center of a cylindrical metal duct. A large potential difference V is maintained between the wire and the duct, with the duct positive. This causes a corona discharge near the wire, and the resulting free electrons tend to attach to dust particles that are then swept out to the outer duct wall. Periodically the voltage is shut off and the dust particles fall to the bottom of the duct, where they are collected. During some industrial processes (smelting, for instance) useful materials such as sulfur are recovered in this way while also reducing air pollution. If it is desired to produce a field of 3×10^6 V/m at the surface of the wire, what voltage must be applied if the radius of the wire is 1 mm and of the duct is 20 cm?

25-55 P The high-voltage electrode of a Van de Graaff generator is a metal sphere of radius 0.8 m. Electrical breakdown will occur in the pressurized gas in the machine at a field of 1.2×10^8 V/m. Charge leaks off the high-voltage electrode at a rate of 2×10^{-4} C/s from a variety of effects. This charge is replenished by a moving rubber belt 60 cm wide. (a) What is the maximum voltage to which the machine can be charged? (b) At what minimum speed must the belt be driven to replace the charge lost by leakage? (c) What then is the surface-charge density on the belt? (Hint: If the charge density on the belt is too great, breakdown will occur.) (d) What power is needed to drive the belt if frictional losses are neglected?

26 Capacitance and Dielectrics

26-1 Capacitance Defined
26-2 Capacitor Circuits
26-3 Dielectric Constant
26-4 Energy of a Charged Capacitor
26-5 Energy Density of the Electric Field
26-6 Electric Polarization and Microscopic Properties (Optional)
Summary

The capacitor is important for several reasons:

- It stores separated electric charge.
- It stores electric potential energy.
- It plays a crucial role in electric and electronic circuits, particularly for varying electric currents. (Anyone who has turned the dial to tune a radio has changed the capacitance of a variable capacitor.)

26-1 Capacitance Defined

A general type of capacitor is shown in Figure 26-1. It consists of two separated conductors—we call them *plates* whatever their shape—carrying opposite charges of equal magnitude Q with a potential difference V between them.* Since all points on the surface and interior of any one conductor in electrostatic equilibrium are at a single potential, V represents the potential difference between *any* point on one plate and *any* point on the oppositely charged plate.

---

* A potential difference is often represented by such a symbol as ΔV. For a capacitor, however, it is customary to write the potential difference simply as V.

The net charge of the entire capacitor is zero, with charges $+Q$ and $-Q$ on the two plates. To charge a capacitor clearly requires work. If it were possible to take just one electron from one uncharged plate and place it on the other plate, this electron would repel all other electrons brought to the negatively charged plate.

The charge magnitude Q on either plate is directly proportional to the potential difference V between the plates, so that we can write

$$Q = CV$$

where C is the proportionality constant for any particular configuration of capacitor plates.* Indeed, C is defined to be the *capacitance* of the capacitor:

$$C \equiv \frac{Q}{V} \qquad (26\text{-}1)$$

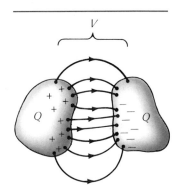

Figure 26-1. Most general form of capacitor; two conductors carrying equal but opposite charges of magnitude Q and having an electric potential difference V between them.

Capacitance is a quantitative measure of a capacitor's capacity for storing separated charges and therefore for storing electric potential energy. A relatively large C means that even a small V can produce a large charge magnitude Q. A capacitor is symbolized in electric circuit diagrams by ─┤├─ or by ─┤∈ ; the symbol ─⟋├─ denotes a variable capacitance.

From its definition in (26-1), we see that capacitance has the units coulombs per volt. A special name, the farad (abbreviated F) is given to this ratio:

$$1 \text{F} \equiv 1 \text{C/V}$$

(The capacitance unit is named after the famed nineteenth-century experimentalist in electromagnetism, Michael Faraday, who first investigated the effects of dielectric materials in capacitors.) As we shall see, a capacitance of one farad is enormous (simply because a charge of one coulomb is enormous). More common capacitor units are the microfarad (1 μF = 10^{-6} F) and the picofarad (1 pF = 10^{-12} F).

Any two oppositely charged conductors make up a capacitor. A practical capacitor, however, is small, has its oppositely charged plates easily insulated from one another, and is so constructed that external fields will not disturb the distribution of charge on the plates. These requirements are met by capacitors with parallel plates or with coaxial cylindrical conductors, and the symmetry of these geometries allows the capacitance to be computed easily.

Example 26-1 Parallel-Plate Capacitor. A capacitor with two parallel plates, each of area A and separated by a distance d, is shown in Figure 26-2. The plate separation d is small compared with the size of the plates, so that the electric field **E** is uniform and confined almost entirely to the region between the plates. Although fringing of the electric force lines at the boundaries is always present, we take it to be negligible. The two plates must be held apart by some insulating material, not only to prevent shorting of the capacitor but also to hold apart the oppositely charged plates

---

* That C is indeed a constant for any capacitor follows directly from the relation ($V = kq/r$) for the electric potential V of a single point charge q. At any particular location (any r), $V \propto q$. But any continuous distribution of charges over the capacitor plates consists basically of a collection of point charges with the proportionality relation $V \propto q$ holding for each point charge. The potential at any point is then proportional to *all* charges. Further, when the net charge on any conductor is changed, the relative distribution of charge on the conductor is not changed but depends only on the geometry of the conductor. Therefore, $Q = CV$, where C is a constant.

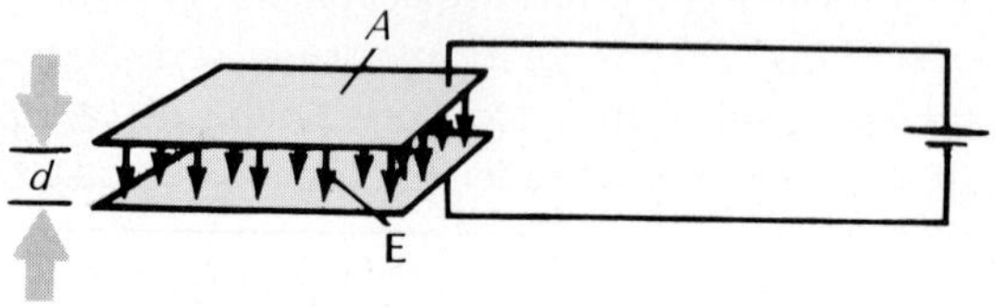

Figure 26-2. A parallel-plate capacitor with plate area A and plate separation distance d.

under the action of the attractive force between them. The insulator may be a dielectric material sandwiched between the plates, in which case the capacitance is enhanced over its value when a vacuum exists between the plates. Here we assume the plates to be in a vacuum; the effect of dielectric materials will be treated in Section 26-3.

The most direct way to charge a capacitor is simply to connect its plates with conducting wire to the terminals of a battery. The constant potential difference between the battery terminals then also becomes the same potential difference V across the capacitor.

To find the capacitance C of a parallel-plate capacitor we must know the charge Q on either plate in terms of the potential difference V. The link is the electric field E. Since the field is uniform over the plate separation distance d, we have

$$E = \frac{V}{d} \tag{25-10}$$

It was shown earlier (Section 24-5) that the electric field near the surface of any conductor is related to the local surface charge density σ by

$$E = \frac{\sigma}{\epsilon_0} = \frac{Q/A}{\epsilon_0} \tag{24-6}$$

where $\sigma = Q/A$ for a parallel-plate capacitor, with σ constant over the area A of each plate carrying charge Q. Using the relations above in the definition of C, we get

$$C = \frac{Q}{V} = \frac{Q}{Ed} = \frac{Q}{(Q/\epsilon_0 A)d} = \frac{\epsilon_0 A}{d} \tag{26-2}$$

where

$$\epsilon_0 = \frac{1}{4\pi k} = 8.854\,187\,8 \times 10^{-12} \; \text{C}^2/\text{N} \cdot \text{m}^2$$

As (26-2) shows, the capacitance is (1) proportional to the plate area and (2) inversely proportional to the plate separation. Both make sense on the basis of simple physical arguments: (1) the larger the plate area (for a given V and d), the larger the amount of charge that can be stored on the plates; (2) with Q and A fixed (and E thereby constant), bringing the plates closer together reduces the distance over which this field exists and hence reduces V.

Notice further from (26-2) that C equals the constant ϵ_0 multiplied by the ratio A/d, a quantity with the dimensions of the length. This result applies for all types of capacitors, and each capacitor's capacitance depends only on the geometry of the conductor. We can see that one farad of capacitance is enormous directly from (26-2). Suppose that each of the two plates has an area of 1.0 m² and that they are separated by 1.0 mm = 10^{-3} m. Then,

$$C = \frac{\epsilon_0 A}{d} = \frac{(8.85 \times 10^{-12} \; \text{C}^2/\text{N} \cdot \text{m}^2)(1.0 \; \text{m}^2)}{10^{-3} \; \text{m}} = 8.85 \times 10^{-9} \; \text{F}$$

or somewhat less than $\frac{1}{100}$ μF. With this separation distance, one would need square plates 10 km along a side to have a capacitance of 1 F.

Example 26-2 Coaxial Capacitor. A coaxial capacitor consists of two concentric cylindrical conductors of radius r_1 and r_2, as shown in Figure 26-3. A cylindrical capacitor of extended length, often constructed with a dielectric material filling the space between the conductors, is known as coaxial cable (colloquially, "coax" or sometimes merely "cable," as in "cable TV").

Each of the two cylindrical surfaces is an equipotential surface; between these exists a potential difference $V_1 - V_2$. For definiteness, the inner conductor is taken to be positive, so that **E** is radially outward and potential V_1 is higher than V_2. To compute the capacitance, we shall use the relation giving the electric field E in terms of the charge per unit length λ on either conductor and then find the potential difference from (25-12):

$$V_2 - V_1 = -\int_1^2 \mathbf{E} \cdot d\mathbf{r}$$

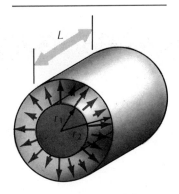

Figure 26-3. A coaxial cylindrical capacitor.

It is easy to see that the electric field between the two cylindrical shells is exactly the same as that from a single infinite, straight, uniformly charged wire at the axis of the cylinders with the same linear charge density λ as either cylindrical shell. The infinite charged wire produces a radial field **E** whose magnitude is given by $E = 2k\lambda/r$ (Section 24-1); equipotential surfaces, necessarily at right angles to the radial field **E**, are also right circular cylinders. (That the field between the conductors of the coaxial capacitors is $E = 2k\lambda/r$ can also be deduced directly from Gauss's law; one chooses a concentric cylindrical Gaussian surface lying between the two conductors.)

When $E = 2k\lambda/r$ is substituted into the relation above, we get

$$V_2 - V_1 = -2k\lambda \int_{r_1}^{r_2} \frac{dr}{r} = -2k\lambda \ln \frac{r_2}{r_1}$$

With $\lambda = Q/L$, where Q is the charge on one conductor of length L, we find the capacitance per unit length C/L to be

$$\frac{C}{L} = \frac{Q}{(V_1 - V_2)L} = \frac{1}{2k \ln(r_2/r_1)} = \frac{2\pi\epsilon_0}{\ln(r_2/r_1)} \qquad (26\text{-}3)$$

For example, any air-filled coaxial cable with $r_2 = 2r_1$ has a capacitance per unit length of $1/(2)(9.0 \times 10^9 \text{ N} \cdot \text{m}^2/\text{C}^2) \ln 2 = 80$ pF/m.

Figure 26-4 shows several types of commercial capacitors.

Figure 26-4. Various types of capacitors. The capacitor at back left is a variable capacitor of the type used to tune a radio receiver; in the middle back is an old-fashioned Leyden jar, the original capacitor, named for the town in the Netherlands where it first originated.

26-2 Capacitor Circuits

Electric circuits composed entirely of capacitors are not very interesting. They are worthy of attention, however, because they illustrate fundamental concepts common to all circuits.

First, these fundamental principles:

- *Electric charge conservation.* For our immediate purposes this implies that the *net* charge of an isolated conductor does not change, although the charge may be redistributed.
- *Energy conservation.* For an equilibrium distribution of electric charges, the net work done by electric forces on a point charge carried around a closed loop (a circuit) is zero, $\oint \mathbf{F}_e \cdot d\mathbf{r} = 0$. Another way of saying the same thing is that the *net change* in *electric potential* around any circuit is *zero,* or in still other words, the sum of all of electric potential drops ΣV around any circuit is zero:

$$\sum_{a \to a} V = 0 \qquad (26\text{-}4)$$

A drop in potential over one portion of the circuit must be compensated for by an equal potential rise somewhere else in the circuit.

Consider the circuit of Figure 26-5(a) with a battery (symbolized by ⊣⊢) connected across a single capacitor. If we travel along any closed loop, not necessarily coinciding with the conducting wires connecting the battery to the capacitor, the net change in potential is zero. This of course holds also for the special route shown in Figure 26-5(b), one that coincides with the conducting wire:

- From *a* to *b*, a potential *rise* in going from the negative to positive terminal of the battery.
- From *b* to *c*, along a conductor in electrostatic equilibrium, no change in potential.
- From *c* to *d*, from the positive to the negative plate of the charged capacitor, a potential drop.
- From *d* to *a*, again along a conductor in equilibrium, no change in potential.

Here we have

$$\Sigma V = 0$$
$$V_{ab} + V_{bc} + V_{cd} + V_{da} = 0$$
$$V_{ab} + 0 + V_{cd} + 0 = 0$$
$$V_{cd} = -V_{ab}$$

The potential *drop* across the capacitor, V_{cd}, is equal to the negative potential *rise,* $-V_{ab}$, across the battery. The potential drops across the two are the same.

We now consider capacitors in two simple circuit arrangements:

- *In series* [Figure 26-6(a)]: in going from *A* to *B*, we encounter in turn each of the circuit elements *a, b, c.*
- *In parallel* [Figure 26-6(b)]: there are alternative ("parallel") routes from *A* to *B*, each one containing one circuit element.

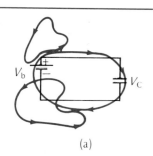

(a)

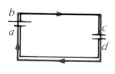

(b)

Figure 26-5. Circuit consisting of a battery and a capacitor. (a) For electrostatic equilibrium, the net change in electric potential around any closed loop (shown with color lines) is zero. (b) A loop abcda around the circuit passing along the conducting wires.

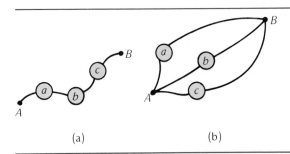

Figure 26-6. Circuit elements a, b, and c connected (a) in series and (b) in parallel.

Capacitors in Series Capacitors C_1, C_2, C_3 are connected in series and attached across battery terminals as shown in Figure 26-7(a). We wish to find the value of C of the single capacitor that is equivalent to this group of capacitors in the sense that it has the same charge as all series capacitors taken together when connected to the same battery.

From Figure 26-7(a), we see that the *net* charge on the single conductor shown within the color loop must remain zero; this section is isolated electrically from everything else. The battery separates the charges on this conductor; equal amounts of positive and negative charge appear on the plates of adjoining capacitors. Thus, all capacitors in series have the *same charge*:

$$Q_1 = Q_2 = Q_3$$

The potential across the battery V_b is the same as the total drop across the capacitors in series,

$$V_b = V_1 + V_2 + V_3 = \frac{Q_1}{C_1} + \frac{Q_2}{C_2} + \frac{Q_3}{C_3}$$

where we have used $V_1 = Q_1/C_1$, and similarly for C_2 and C_3.

The relation may be written

$$V_b = Q\left(\frac{1}{C_1} + \frac{1}{C_2} + \frac{1}{C_3}\right)$$

where we have used Q to represent any one of $Q_1 = Q_2 = Q_3$. With a single equivalent capacitor C across the battery terminals, the potential difference

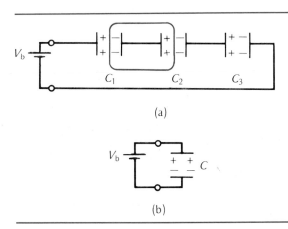

Figure 26-7. (a) Capacitors in series; the net charge within the color loop is zero. (b) The equivalent single capacitor.

Knowing the Connections

What is this experimenter doing in the physics lab?

As the figure shows, two oppositely charged plates are attracting one another; the electric force on the upper charged plate is balanced by a weight placed on the pan on the other side of the beam balance. In this fairly simple arrangement, we have, basically, two electrically charged objects attracting one another. The formula for the electric force on a plate ($F = Q^2/2\,\epsilon_0 A$, where A is plate area) is a little different from Coulomb's law for a couple of point charges, but here also the force depends on essentially the same quantities: the magnitude of charge, the dimensions, and the constant ϵ_0. So what is being measured?

There a few possible interpretations:

- The experimenter might be merely checking that the relation for the force is right, a pretty dull experiment.

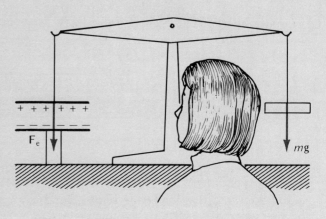

- She might be comparing an electric force with a gravitational force. Now that could be interesting. We can see from the setup above that the electric interaction is enormously larger than the gravitational interaction. After all, it takes the entire earth pulling

across its plates, $V = Q/C$, would equal V_b. Therefore,

$$\frac{Q}{C} = Q\left(\frac{1}{C_1} + \frac{1}{C_2} + \frac{1}{C_3}\right)$$

$$\frac{1}{C} = \frac{1}{C_1} + \frac{1}{C_2} + \frac{1}{C_3} = \Sigma \frac{1}{C_i} \qquad (26\text{-}5)$$

For capacitors in series, the reciprocal of the single equivalent capacitance is the sum of the reciprocals of the individual capacitances. As a consequence, the equivalent capacitance is always *less* than the smallest of the series capacitors. For example, with four capacitors, each of 20 μF and connected in series, the equivalent capacitance is 5 μF.

Capacitors in Parallel For capacitors connected in parallel, see Figure 26-8(a). What single capacitor, as in Figure 26-8(b), is equivalent for this group of capacitors? The charges and potential differences for C_1, C_2, and C_3 are again designated, respectively, Q_1, Q_2, Q_3 and V_1, V_2, V_3. By taking potential differences along the closed loops shown by color lines in the figure, we find that the

Figure 26-8. (a) Capacitors in parallel; the net potential difference around any color loop is zero. (b) The equivalent single capacitor.

on the object on the right-hand pan to balance the electric force between two objects with very modest amounts of charge.

- The experimenter might be—as surprising as this might seem—measuring the speed of light! Not directly, of course. But, as we shall later see (Section 34–3), light is an electromagnetic phenomenon, and its speed through empty space is related directly to ϵ_0. Know one and you can compute the other.

That's the kind of remarkable thing you can do when you know the connections that physics reveals. When you know basic physics—that things hang together and how they hang together—you have gained not only extraordinary insight, but also power.

We've already seen examples of how quantities that might at first seem absolutely inaccessible can be computed indirectly from basic relations, and we shall see many more.

- What's the mass of the earth? Certainly you can't measure it directly. But if you know the law of universal gravitation and the value of the gravitational constant G, then computing the earth's mass is almost ridiculously easy (Section 15–3 and Example 15–1).

- At the other extreme, what's the mass of an electron? Again, any direct measurement is ruled out. But if we know how electric and magnetic fields influence a moving charged particle, we can set up what amounts to an obstacle course for electrons (Section 29–4). If an electron makes it all the way through the course, we can compute its mass directly.

The list could go on. The conclusion is simple but important: when you know how things fit together in the physical universe—when you know basic physics—then you can do astonishing things with it.

And that, of course, is why, if you're a student in engineering, physics is almost certainly a required course. To do the challenging and difficult job of applying basic scientific ideas to practical use—to be an effective engineer—you must first have straight the fundamental insights that physics provides.

---

potential difference is the same across each capacitor as across the battery terminals. Parallel capacitors have the *same potential difference*; that is,

$$V_b = V_1 = V_2 = V_3$$

The circuit consists, effectively, of only two conductors attached to the battery terminals (all the upper plates with their connecting wires, and all the lower plates with theirs). Therefore, the total charge Q held on a single capacitor equivalent to those in parallel is given by

$$Q = Q_1 + Q_2 + Q_3$$

or

$$Q = C_1 V_1 + C_2 V_2 + C_3 V_3 = V_b(C_1 + C_2 + C_3)$$

But $C = Q/V_b$. Therefore,

$$C = C_1 + C_2 + C_3 = \Sigma\, C_i \tag{26-6}$$

For parallel capacitors, the single equivalent capacitance equals the sum of the individual capacitances. The equivalent capacitance always exceeds the largest capacitance in parallel.

26-3 Dielectric Constant

First we describe how the introduction of a dielectric material changes the capacitance of a capacitor. Then (Section 26-6) we discuss in qualitative terms why a dielectric changes a capacitor's capacitance.

The dielectric constant κ of an insulating, or dielectric, material can be defined in terms of the capacitance C of a capacitor whose plates are immersed in a vacuum and the capacitance C_d of the same capacitor but with dielectric material filling the region between the plates:

$$\kappa \equiv \frac{C_d}{C} \qquad (26\text{-}7)$$

The capacitance always *increases* with the introduction of dielectric. $C_d > C$, and κ is always *greater than one*.

Table 26-1 lists dielectric constants and dielectric strengths (explained below) at room temperature for several materials. Note that κ for air is very close to 1.00, which is, by definition, the dielectric constant for a vacuum.

Since a capacitor's capacitance changes with the introduction of a dielectric, other associated properties of the capacitor may also change. Here we distinguish two simple circumstances:

• Capacitor maintained at a *constant potential difference* V (by being attached to a battery). See Figures 26-9(a) and (b). Then from the definitions,

$$\text{Same } V: \quad \kappa = \frac{C_d}{C} = \frac{Q_d/V}{Q/V} = \frac{Q_d}{Q}$$

The *charge* Q_d on a plate *increases* with the introduction of a dielectric; the *electric field* $E = V/d$ between the plates is *unchanged*. The additional charge is supplied by the battery.

• Capacitor maintained at *constant charge* Q (by being detached from other circuit elements). See Figures 26-9(c) and (d). Again from the definitions,

$$\text{Same } Q: \quad \kappa = \frac{C_d}{C} = \frac{Q/V_d}{Q/V} = \frac{V}{V_d} = \frac{E}{E_d}$$

Both the potential difference V_d and electric field E_d are reduced from their respective values for a vacuum.

The *electric permittivity* ϵ of a dielectric material is defined by

$$\epsilon = \kappa \epsilon_0 \qquad (26\text{-}8)$$

With $\kappa = 1$, we have $\epsilon = \epsilon_0$, the permittivity of the vacuum.* To generalize any relation we have derived so as to include filling by a dielectric material, we simply replace ϵ_0 by ϵ. Thus, from (26-2) the capacitance of a parallel-plate capacitor filled with dielectric is $C_d = \kappa C = \kappa \epsilon_0 A/d = \epsilon A/d$.

Dielectric materials serve essential functions in actual capacitors. They insulate the plates from one another, and they enhance the capacitance by a factor κ. An ordinary "paper capacitor" is formed by sandwiching thin paper between two metallic foils and rolling an essentially parallel-plate capacitor into cylindrical shape.

A capacitor can maintain its insulating properties only up to some maximum electric field E_{max}, called the *dielectric strength* of the insulating material. Above this electric field, electric discharge, or arcing, occurs through the di-

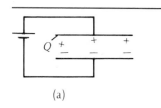

(a)

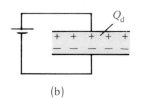

(b)

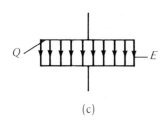

(c)

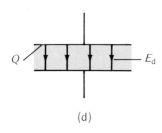

(d)

Figure 26-9. Charged capacitor maintained at constant potential difference (a) before and (b) after dielectric is introduced. Insulated charged capacitor (c) before and (d) after dielectric is introduced.

---

* The term *permittivity* is used for ϵ and ϵ_0 because it gives a measure of the degree to which a dielectric "permits" an external electric field to pass through it. More on this in Section 26-6.

Table 26-1

MATERIAL	DIELECTRIC CONSTANT κ	DIELECTRIC STRENGTH E_{max} AT ROOM TEMP. (V/m)
Air (1 atm)	1.00059	30×10^6 3×10^6
Air (100 atm)	1.0548	—
Germanium	16	—
Mylar	3.1	—
Plexiglas	3.4	40×10^6
Pyrex	5.6	15×10^6
Quartz	3.8	8×10^6
Paraffined paper	2	40×10^6
Mica	5	200×10^6
Barium titanate	1200	300×10^6

electric. Therefore, a dielectric material is specified by the maximum electric field E_{max} (for example, 200 kV/mm for mica and 40 kV/mm for paraffined paper) as well as by its dielectric constant (see Table 26-1). The dielectric strength of the insulating material determines in turn the maximum potential difference that can be applied to the capacitor terminals without breaking down the dielectric material. Commercial capacitors are characterized by two numbers: capacitance in farads, and maximum allowable potential difference to avoid breakdown.

Example 26-3. A parallel-plate capacitor is half-filled with a dielectric, as shown in Figure 26-10(a). What is its capacitance?

The half-filled capacitor is equivalent to two capacitors in parallel, since they have a common potential difference. One capacitor is completely filled with dielectric and has a capacitance C_d; the other is empty and has a capacitance C_e, shown in Figure 26-10(b). The two capacitances are, from (26-2) and (26-8),

$$C_d = \frac{\kappa \epsilon_0 A}{2d} \quad \text{and} \quad C_e = \frac{\epsilon_0 A}{2d}$$

where A represents the plate area of the whole capacitor. Then, from (26-6),

$$C = C_d + C_e = \frac{\epsilon_0 A}{2d}(1 + \kappa)$$

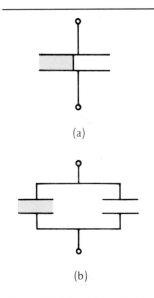

Figure 26-10. (a) Parallel-plate capacitor half-filled with dielectric; (b) the equivalent circuit of part (a).

26-4 Energy of a Charged Capacitor

To charge a capacitor, work is required. After the first small charge is placed on one plate, this charge repels other charges of like sign subsequently added. We want to find the total work U_e needed to charge a capacitor to a final potential difference V_f with a final charge Q on each plate.

The potential difference between the plates, always proportional to charge Q, grows from zero to V_f as the capacitor is being charged. On the average, the work per unit charge, or potential difference between the plates, is $\frac{1}{2}V_f$, so that the work done in putting a total charge Q on each plate is just

$$U_e = Q(\tfrac{1}{2}V_f) = \tfrac{1}{2}QV_f$$

The same result follows from summing the contributions to the work $dW = V\, dq$ done on charge dq as it goes across a potential difference V. We have

$$U_e = \int_0^Q V\, dq = \int_0^Q \frac{q}{C}\, dq = \frac{Q^2}{2C}$$

where U_e is the electric potential energy of the separated charges on the capacitor plates. From the definition, $C = Q/V_f$, the total energy can be written in three equivalent forms as

$$U_e = \frac{Q^2}{2C} = \frac{1}{2} CV_f^2 = \frac{1}{2} QV_f \tag{26-9}$$

Example 26-4. The spherical conducting shell on a Van de Graaff electrostatic generator has a radius of 15 cm, and it is charged to a potential (relative to ground as zero) of 250 kV. (a) What is the capacitance of the spherical shell? (b) What is the energy of the charged conducting shell?

(a) We can regard an isolated spherical conductor as one plate of a capacitor in which the other plate is the earth, effectively an infinite distance away and at zero potential.

The potential of a spherical conductor of radius R with charge Q is

$$V = \frac{kQ}{R}$$

so that its capacitance is

$$C = \frac{Q}{V} = \frac{R}{k}$$

The capacitance is proportional to the conductor's radius. For a 15-cm conducting shell, we have

$$C = \frac{R}{k} = \frac{15 \times 10^{-2}\text{ m}}{9.0 \times 10^9 \text{ N} \cdot \text{m}^2/\text{C}^2} = 17 \times 10^{-12}\text{ F} = 17 \text{ pF}$$

(b) The energy of the charged sphere is

$$U_e = \tfrac{1}{2} CV^2 = \tfrac{1}{2}(17 \times 10^{-12}\text{ F})(250 \times 10^3\text{ V})^2 = 0.53 \text{ J}$$

roughly equal to the kinetic energy of a $0.1 =$ kg object after it falls from rest over a distance of 0.5 m.

Example 26-5. A capacitor C_1 is charged initially with a potential difference V_i, as shown in Figure 26-11(a). Then this capacitor is connected to an initially uncharged capacitor C_2, as shown in Figure 26-11(b). What is (a) the final potential difference across each capacitor? (b) the ratio of the final to the initial potential energy of the capacitors?

(a) The charge Q_i initially on C_1 is

$$Q_i = C_1 V_i$$

After the switch has been closed, this charge is shared between the two capacitors

$$Q_i = Q_1 + Q_2$$

where Q_1 and Q_2 are respectively the final charges on C_1 and C_2. But $Q_1 = C_1 V_f$ and $Q_2 = C_2 V_f$, so that the relation can be written

$$C_1 V_i = C_1 V_f + C_2 V_f$$

and the final potential difference is

$$V_f = V_i\left(\frac{C_1}{C_1 + C_2}\right)$$

We see that V_f is *always less* than V_i.

(b) The energy U_i initially of C_1 is, from (26-9),

$$U_i = \tfrac{1}{2}C_1 V_i^2$$

and the energy U_f finally on the two capacitors is

$$U_f = \tfrac{1}{2}C_1 V_f^2 + \tfrac{1}{2}C_2 V_f^2$$

The ratio of final to initial energy is then

$$\frac{U_f}{U_i} = \frac{\tfrac{1}{2}(C_1 + C_2)V_f^2}{\tfrac{1}{2}C_1 V_i^2} = \frac{(C_1 + C_2)V_f^2}{C_1 V_i^2}$$

Using the result for part (a) to eliminate V_f/V_i, we can write the energy ratio as

$$\frac{U_f}{U_i} = \frac{C_1}{C_1 + C_2}$$

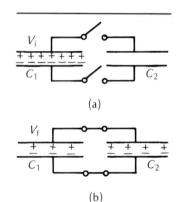

Figure 26-11.

For any capacitors, the final potential energy of the charged capacitors is less than the initial energy. Where does the energy go? It is dissipated mostly as thermal energy, and to a lesser degree as radiation, as the charges redistribute themselves over the conductors when the switch is closed, as will be discussed in later chapters.

26-5 Energy Density of the Electric Field

A charged capacitor has electric potential energy because work is required to assemble the charges at the plates. We can say the same thing differently: A charged capacitor has energy because an electric field has been created in the region between its plates. In this view we ascribe energy to the electric field itself. (To avoid double counting, we include either the electric potential energy of separated charges, or now its equivalent, energy in the electric field, but not both.)

Using (26-9) and (26-2) for a parallel-plate capacitor, we have for the energy

$$U_e = \frac{1}{2}CV_f^2 = \frac{1}{2}\left(\frac{\epsilon_0 A}{d}\right)(Ed)^2 = \frac{1}{2}\epsilon_0 E^2 (Ad)$$

We neglect fringing and suppose that the uniform field **E** is confined entirely to the volume Ad between the plates. Then the electric energy per unit volume, the *electric energy density* u_e, is

$$u_e = \frac{U_e}{Ad} = \frac{1}{2}\epsilon_0 E^2 \qquad (26\text{-}10)$$

for an electric field of magnitude E in a vacuum. Note that the energy density is proportional to the *square* of the electric field. Although derived here for the specific case of a parallel-plate capacitor (with negligible fringing), the result above gives the electric energy density for any electric field.

We now have two ways of describing the energy associated with electric charges:

- The potential energy associated with the relative positions of the charges.
- The electric field created by the charged particles.

Actually, the more fundamental view is to identify energy with the field itself. The reason is that we *can* have a *field without an electric charge* even though we cannot have a charge without an associated field. As we shall see, an electric field (and also a magnetic field) may become detached from the charge or charges that created it and propagate through empty space as an electromagnetic wave. Such unattached fields traveling at the speed of light have such physical attributes as energy, linear momentum, and angular momentum.

26-6 Electric Polarization and Microscopic Properties (Optional)

In this section we discuss qualitatively how the microscopic properties of dielectric materials are responsible for the increase in a capacitor's capacitance when a dielectric material is inserted between the plates. More broadly, we are concerned with the behavior of dielectric materials in an external electric field.

Recall first what happens when insulating material with dielectric constant κ is inserted into an isolated and initially charged capacitor (Section 26-4). (For simplicity, we suppose the capacitor to have parallel plates.) The capacitance is increased by the factor κ. Since charge Q is unchanged, the *potential difference* between the plates, $V_d = Q/C$, must *decrease* (by the factor κ). The uniform *electric field*, $E_d = V_d/d$, between plates separated by distance d must also *decrease* with the introduction of dielectric. Since the electric field between the charged plates is reduced, even though the free charge, $+Q$ and $-Q$, on each of the two plates is unchanged, we must conclude that the dielectric material between the charged plates itself produces an electric field that cancels at least partially the electric field between the plates.

How this occurs is shown in Figure 26-12. In Figure 26-12(a), there is no

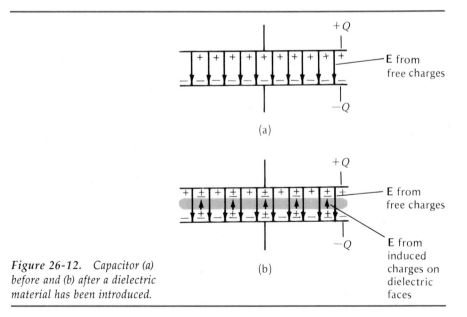

Figure 26-12. Capacitor (a) before and (b) after a dielectric material has been introduced.

dielectric within the capacitor and the electric field, shown downward, arises solely from the free charges $+Q$ and $-Q$ on the capacitor plates. In Figure 26-12(b), a dielectric slab has been inserted between the plates of the capacitor, still with free charges $+Q$ and $-Q$. (For clarity, a gap is shown between each of the capacitor plates and the nearby face of the dielectric.)

Under the influence of the electric field between the capacitor plates, the dielectric has become *polarized*. That is to say, negative charges appear on its upper face and positive charges on its lower face. The charges *on the capacitor plates* are called *free charge*; they are able to move freely within a conductor. The charges *on the dielectric faces* are called induced charges, or *polarization charges*; they are induced by the external electric field and appear on the dielectric *surfaces* only. The magnitude of the induced charge on either dielectric face is less than Q. We see from Figure 26-12(a) that the electric field from the free charges is downward whereas the electric field (within the dielectric) from the induced charges is upward. The net effect: reduction of the electric-field magnitude between the capacitor plates.

The question now is how the microscopic properties of dielectric materials, under the influence of an external electric field, produce polarization charges on the dielectric faces.

First, we must distinguish between two general types of molecules, polar molecules and nonpolar molecules. A *polar* molecule has a *permanent electric dipole* moment (Section 24-6); although electrically neutral as a whole, its "centers" of positive and negative charge do not coincide. For example, in the polar molecule NaCl, the end with the sodium ion is positive while the end with the chlorine ion is negative. On the other hand, a *nonpolar* molecule has *no electric dipole* moment *in the absence of an external field*.

What happens when a nonpolar dielectric material is placed in an external electric field? The field polarizes the molecules, that is, it displaces the electrons (on the average) in the direction opposite to that of **E**, as shown in Figure 26-13. The external electric field induces electric dipole moments in the molecules, shifting the center of negative charge of electrons relative to the positive charge at the nucleus. This polarization is manifest, however, only at the surface of the

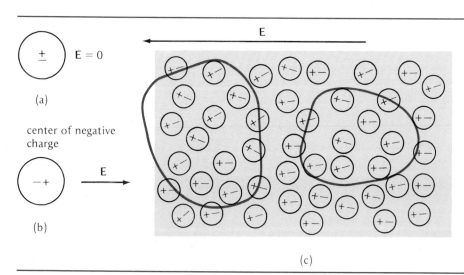

Figure 26-13. (a) In the absence of an external electric field, the centers of positive and negative charge in a molecule coincide. (b) An external electric field separates the centers of positive and negative charge. (c) A dielectric material polarized by an external electric field. A net charge appears at the surface of the dielectric material, but not within the material.

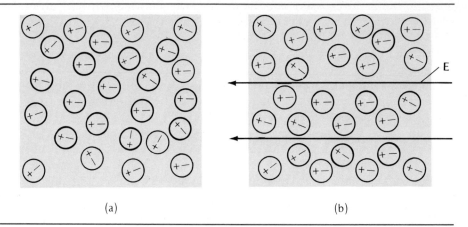

Figure 26-14. (a) Polar molecules in the absence of an external field. (b) Polar molecules in the presence of an external field.

dielectric. As the figure shows, for any small volume lying entirely *within* the dielectric material the positive and negative charges, although not coincident, appear in equal magnitudes. This does not happen however, when we choose a small volume that encloses one external surface of a dielectric. A net charge appears at the surface. The degree of polarization, measured by the amount of induced charge at the dielectric surface, is typically proportional to the magnitude of the field.

In the absence of an external electric field, a polar dielectric, one with molecules having a permanent electric dipole moment **p**, shows no net polarization. Because of the thermal agitation within the material, the molecular dipoles are randomly oriented, as shown in Figure 26-14(a). When an external field **E** is applied, each dipole is subject to a torque $\tau = \mathbf{p} \times \mathbf{E}$ [(24-9)] that tends to align the dipole moments with the external field (Figure 26-14(b)). The polarizing influence of the external field competes with the depolarizing influence of thermal motion. As the temperature is lowered, the polarization of the dielectric increases for a given external field. In addition to aligning the permanent dipoles, the external field induces electric dipole moments in the molecules, just as for nonpolar dielectrics. Again, the polarization of the dielectric produces net charges only at the surface of the material.

Summary

Definitions
    Capacitance:
$$C \equiv Q/V \qquad (26\text{-}1)$$
where Q is the charge magnitude on each of the two plates between which a potential difference V exists.
    Dielectric constant:
$$\kappa \equiv C_d/C \qquad (26\text{-}7)$$
where C_d is the capacitance of a capacitor filled with dielectric material with dielectric constant κ, and C is the capacitance of the same capacitor in a vacuum.
    Dielectric strength: the maximum electric field that can be applied to a dielectric material without electrically breaking down the material.

Units
    Farad, the unit for capacitance:
$$1\text{ F} \equiv 1\text{ C}/\text{V}$$

Fundamental Principles
    For any circuit in electrostatic equilibrium,

• From energy conservation: the net change in electric potential around any circuit is zero.

$$\Sigma V = 0 \qquad (26\text{-}4)$$

- From charge conservation: the net charge on any isolated circuit element is constant.

Important Results
Parallel-plate capacitor—plate area A, plate separation d:

$$C = \frac{\epsilon_0 A}{d} \qquad (26\text{-}2)$$

Coaxial cylindrical capacitor—inner radius r_1, outer radius r_2, length L:

$$\frac{C}{L} = \frac{2\pi\epsilon_0}{\ln(r_2/r_1)} \qquad (26\text{-}3)$$

Equivalent capacitance C for capacitors (C_i) in *series* (same Q):

$$\frac{1}{C} = \Sigma \frac{1}{C_i} \qquad (26\text{-}5)$$

Equivalent capacitance C for capacitors (C_i) in *parallel* (same V):

$$C = \Sigma C_i \qquad (26\text{-}6)$$

Energy U_e of a capacitor charged with final charge Q and final potential difference V_f:

$$U_e = \frac{Q^2}{2C} = \frac{1}{2}CV_f^2 = \frac{1}{2}QV_f \qquad (26\text{-}9)$$

Energy density (energy per unit volume) u_e of electric field E:

$$u = \tfrac{1}{2}\epsilon_0 E^2 \qquad (26\text{-}10)$$

Problems and Questions

Section 26-1   Capacitance Defined

· **26-1 Q** One farad is the same as one
(A) ohm/meter.
(B) coulomb/volt.
(C) volt/coulomb.
(D) joule/coulomb.
(E) ampere·second.

· **26-2 P** (a) What is the capacitance of an air-filled parallel-plate capacitor of plate area 30 cm × 30 cm (about 1 ft²) and plate separation 1 mm? (b) Air breaks down at about 3×10^6 V/m. What is the maximum voltage that could be applied to such a capacitor before breakdown will occur?

· **26-3 Q** Show that ϵ_0 has units of farads per meter.

: **26-4 Q** The capacitance between two conductors depends on
(A) the charge on them.
(B) the potential difference between them.
(C) the electric field between them.
(D) the energy stored between them.
(E) none of the above.

: **26-5 Q** Two large parallel copper plates carry charges $+Q$ and $-Q$ respectively. They are separated by a small distance of 10 mm. They differ in potential by 12 V. If the plates are moved together slowly to a spacing of 6 mm, the potential difference between them will
(A) increase.
(B) decrease.
(C) remain unchanged.
(D) increase or decrease according to what is chosen as the "zero" of potential.

: **26-6 P** What is the capacitance of an isolated conducting sphere of radius R? (*Hint:* If some charge were on the sphere, electric-field lines would be directed out radially. They would end on the other "plate" of this capacitor. Where is the other plate?)

: **26-7 P** The first precision measurement of the charge of an electron was made by Robert A. Millikan in 1909 in his famous *oil-drop experiment*. In this experiment an oil drop having an excess or deficiency of no more than a few electrons is in the uniform vertical electric field of a parallel-plate capacitor. An oil drop of charge q, radius r, and velocity v is subject to three forces in the vertical direction: (1) its weight, (2) the electric force qV/d, where V is the potential difference between plates separated by d, and (3) a resistive force, $F_r = -6\pi\eta r v$, where η is the viscosity of the medium (for air, $\eta = 1.82 \times 10^{-5}$ N·s/m²).

In its simplest form the experiment can be done as follows: (a) the charged drop is brought to rest by balancing its weight against an upward electric force; (b) the drop is allowed to fall at a constant measured velocity with the electric field turned off, so that the weight is now balanced against an upward resistance force. Part (b) yields the drop's radius and therefore its weight; part (a) then allows q to be computed.

Suppose that some particular oil drop falls at constant speed over a vertical distance of 1.00 mm (as viewed through a microscope) in a time interval of 50 s. (a) What is the radius of the drop? (b) Suppose that this drop had a net charge of $-2e$. What electric potential difference would have to be applied across the capacitor plates, separated by 1.75 mm, in order to bring the drop to rest? (Oil density, 850 kg/m³)

: **26-8 P** Show that the expression for the capacitance of a coaxial capacitor approaches that for a parallel-plate capacitor as $r_1 \to r_2$. Use the approximation $\ln(1+x) \simeq x$ if $x \ll 1$, and write $r_2 = r_1 + \delta$, where $\delta \ll r_1$.

: 26-9 Q Is it possible for a single conductor to have capacitance? Which of the following statements best answers and explains this question?
(A) No. A single conductor can have no capacitance, because although charge can be placed on it, there is no second conductor against which to measure a potential difference.
(B) No. A single conductor can have no capacitance because the unit of capacitance, the farad, is specifically defined for the case of two capacitors.
(C) No. A single conductor can have no capacitance, because a conductor is electrically neutral and can carry no net charge.
(D) Yes. A single conductor can have capacitance, because one part of the conductor can be considered one plate and another part the second plate.
(E) Yes. A single conductor can have capacitance, since capacitance is only a measure of the ability to store charge and is unrelated to other factors such as potential difference or the proximity of other conductors.
(F) Yes. A single conductor can have capacitance. The "second plate" of the capacitor may be thought of as the equipotential surface at infinity.

: 26-10 P Consider two isolated conductors of arbitrary shape, one carrying charge $+Q$ and the other $-Q$. In general, the electric field between them might be complicated, as suggested in Figure 26-15. Suppose that along one of the field lines, say the one labeled XX', the electric field intensity varies according to

$$E = E_0\left(1 - \frac{x}{2L}\right)Q$$

This is the electric field at a point P a distance x measured along the line XX', and L is the length of this line. The quantity E_0 is a constant, which depends on the shape and separation of the two conductors. What is the capacitance of these two conductors?

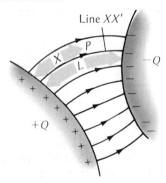

Figure 26-15. *Problem 26-10.*

Section 26-2 Capacitor Circuits

· 26-11 P Three capacitors (1-pF, 2-pF, and 4-pF) are to be connected to a 120-V power supply. Determine the potential difference across each and the charge on each when (*a*) all three are connected in series; (*b*) all three are connected in parallel.

· 26-12 P Determine the equivalent capacitance of the combinations of capacitors shown in Figure 26-16.

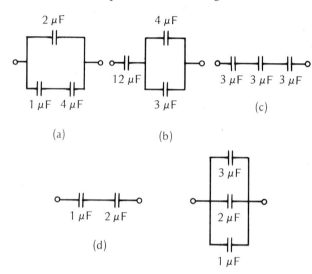

Figure 26-16. *Problem 26-12.*

: 26-13 P A capacitor of known capacitance C_1 is charged to a potential difference V_1. It is then connected across an uncharged capacitor of unknown capacitance C_2. The final potential difference across the two capacitors is V_2. Show that the capacitance of the unknown capacitor is

$$C_2 = \left(\frac{V_1 - V_2}{V_2}\right)C_1$$

: 26-14 Q A sheet of aluminum foil is placed between the plates of a parallel-plate capacitor as shown in Figure 26-17. How will this affect the capacitance?

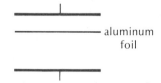

Figure 26-17. *Question 26-14.*

(A) C will increase.
(B) C will decrease.
(C) C will not change.
Suppose that the aluminum foil is then connected to one of the plates with a wire. How will this affect the capacitance?

(D) C will increase.
(E) C will decrease.
(F) C will not change.

: **26-15 P** An air-filled variable capacitor shown in Figure 26-18, of the type used to tune a radio, is made by varying the overlapping area between two sets of plates through rotation of one set of plates (attached to the knob you turn). Such a capacitor consists of n adjacent plates, half of which are attached to one pole and half to the other, with a gap d between adjacent plates. Show that the capacitance is $C = (n-1)\epsilon_0 A/d$, where A is the overlapping area of a plate.

Figure 26-18. Problem 26-15.

: **26-16 P** A metal slab of thickness t and area A is inserted into a parallel-plate capacitor of area A and separation d, as shown in Figure 26-19. (a) What is the new capacitance between the plates? (b) Does the result in (a) depend on where the slab is placed? Explain.

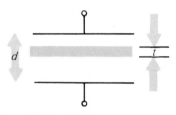

Figure 26-19. Problem 26-16.

: **26-17 Q** Suppose that a capacitor is connected to a battery. Is it true that each plate will receive a charge of the same magnitude? If so, why? If not, why not?
(A) Each plate will receive equal amounts of charge only if they are of the same area.
(B) A capacitor is designed so that all the electric-field lines emanating from one plate terminate on the other plate. From Gauss's law, we then know that there is an equal amount of charge on each plate.
(C) No, each plate will not receive an equal amount of charge, whether or not they have equal areas. The charge on a plate depends on the shape of the plate as well as on its area.
(D) Yes, each plate receives the same charge because each is at the same potential.

: **26-18 P** (a) Determine the capacitance of two concentric spherical conductors of radius R_1 and R_2. (b) Show that if $R_2 = R_1 + \delta$, the result from (a) reduces to the formula for the capacitance of a parallel-plate capacitor, where $\delta \ll R_1$.

: **26-19 P** Consider the circuit shown in Figure 26-20 (a) What is the charge on each capacitor and the potential difference across each? (b) Suppose that the 4-μF capacitor breaks down electrically and becomes a conductor. What then would be the potential differences across each capacitor and the charge on each?

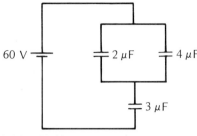

Figure 26-20. Problem 26-19.

: **26-20 P** Two capacitors (2-μF and 4-μF) are each charged to a potential difference of 100 V and then joined in parallel, with plates of opposite polarity connected. What is the final potential difference across each capacitor, and how much charge is on each?

: **26-21 P** The capacitors shown in Figure 26-21 are initially uncharged. The switch is first thrown to position 1, then to position 2. What is the final charge on each of the three capacitors?

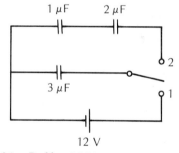

Figure 26-21. Problem 26-21.

: **26-22 P** What is the equivalent capacitance of the five identical capacitors shown in Figure 26-22? (*Hint:* From symmetry, what is the charge on the middle capacitor?)

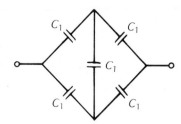

Figure 26-22. Problem 26-22.

: **26-23 P** How many different values of capacitance can

you get by connecting three identical capacitors in various ways? What are the values in terms of C_0, the value of an individual capacitance?

26-24 P Consider two spheres of radius R_1 and R_2 whose centers are separated by a large distance d, where $d \gg R_1, R_2$. What is the capacitance of this system? (*Hint:* Imagine charge $+Q$ on one and $-Q$ on the other. Write down the potential of each due to both charges. The difference in potential will be found to be proportional to Q; and from the constant of proportionality, one can deduce C.)

26-25 P Three capacitors C_1, C_2, and C_3 are given charges Q_1, Q_2, and Q_3. They are then connected in series, with plates of opposite polarity being joined. What is the final charge on each capacitor?

Section 26-3   Dielectric Constant

· 26-26 P What is the capacitance per unit length of a coaxial cable filled with polyethylene (dielectric constant 2.3) if the inner-conductor radius is 2 mm and the outer-conductor radius is 7 mm?

26-27 P One-fourth of the area of a parallel-plate capacitor is filled with material with dielectric constant κ that extends from one plate to the other. What is the capacitance of this capacitor if its capacitance with no dielectric is C_0?

26-28 Q Consider the three arrangements of capacitors shown in Figure 26-23. When a dielectric slab is inserted into one capacitor, the capacitance between the terminals X and Y will increase
(A) only in case A.
(B) only in case B.
(C) only in case C.
(D) in all three cases, independent of whether or not X and Y are connected to a battery.
(E) in all three cases, provided X and Y are connected to a battery.
(F) in all three cases, provided X and Y are not connected to a battery.

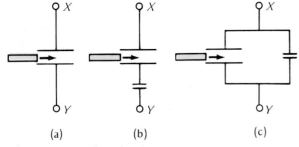

Figure 26-23.   *Question 26-28.*

26-29 Q An air-filled parallel-plate capacitor is charged by a 12-V battery. The capacitor is then disconnected from the battery. What will happen if a dielectric slab is then slid between the plates of the capacitor?
(A) The amount of free charge on the metal plates of the capacitor will change.
(B) The potential difference between the plates will decrease.
(C) The electric field in the region between the plates will increase.
(D) The capacitance of the capacitor will be unchanged.
(E) The potential difference between the plates and the electric field in the region between the plates will both be unchanged.

26-30 Q Suppose that while a capacitor is connected to a battery a dielectric slab is placed between the plates.
(A) The capacitance will not be affected, since the battery holds the potential difference fixed.
(B) More charge will flow on to the plates from the battery.
(C) Charge will flow from the capacitor plates back into the battery.
(D) The effective capacitance will be reduced.
(E) The potential difference between the plates will increase because of induced charges on the surface of the dielectric.

26-31 Q An air-filled parallel-plate capacitor is charged by a battery and then disconnected from the battery. A dielectric slab is then placed between the plates. (*a*) Will the charge on the plates increase, decrease, or remain unchanged? (*b*) Will the potential difference between the plates increase, decrease, or remain unchanged? (*c*) Will the capacitance increase, decrease, or remain unchanged? (*d*), (*e*), (*f*) Answer the above questions for the case in which the capacitor remains connected to the battery.

26-32 P (*a*) The earliest capacitors were called Leyden jars. They were glass jars coated inside and outside with metal. To estimate the capacitance of the jars used, calculate the capacitance of a cylinder that is 30 cm tall and has a diameter of 20 cm; its ends are flat, and the whole is made of glass that is 3 mm thick. The dielectric constant of glass is about 5. (*b*) If the dielectric strength of glass is 15×10^6 V/m, what is the maximum potential difference that can be maintained across the glass? (*c*) What is the maximum charge that can be stored in this capacitor?

26-33 Q You are designing a capacitor to be used in an experiment for which a particular value C_0 of capacitance is required. The capacitor must be capable of operating up to a given maximum voltage V_m. You decide to use a pair of parallel metal plates immersed in an oil bath. Unfortunately, you find that on your first test run, the plates are oversized and will not fit in the container with the oil. What should you do to redesign the capacitor?
(A) Drain out the oil and use the capacitor as air-filled.
(B) Increase the area of the plates and move them closer together.

(C) Increase the area of the plates and move them farther apart.
(D) Decrease the area of the plates and move them farther apart.
(E) Decrease the area of the plates and move them closer together.

: 26-34 P Suppose that you want to design a 1-μF oil-filled capacitor to be used at 5000 V. Oil with a dielectric constant of 5.0 and dielectric strength of 15×10^6 V/m is to be used. As a safety factor, the electric field is to be limited to 80 percent of the breakdown value. What is the minimum amount of oil required, expressed in liters?

: 26-35 P A parallel-plate capacitor is charged by a battery and then disconnected. A slab of dielectric is then inserted between the plates, part way in, as shown in Figure 26-24. The plates are horizontal, and friction is negligible. What will happen when the slab is released?
(A) It will shoot straight through the plates and fly out on the other side.
(B) It will be pushed out of the capacitor.
(C) It will be drawn into the capacitor and oscillate back and forth horizontally.
(D) It will be drawn into the capacitor and come to rest with its midpoint at the midpoint of the plates.

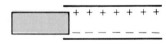

Figure 26-24. Problem 26-35.

: 26-36 P A parallel-plate capacitor is filled with equal volumes of materials with dielectric constants κ_1 and κ_2. What is the capacitance if the plate area is A and the plate separation is d, for the two arrangements shown in Figure 26-25?

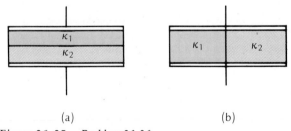

Figure 26-25. Problem 26-36.

: 26-37 P An air-filled coaxial capacitor has an inner radius of 1 cm and an outer radius of 2 cm. What is the minimum volume (per unit length of the capacitor) of material of dielectric constant 3 that can be added to the space between the plates to double the capacitance? Sketch where the dielectric material should be placed. (*Hint:* You can do this if you think about how a dielectric works. It is most effective if placed in the strongest electric field.)

: 26-38 Q Which of the following is the best explanation of how (or why) a dielectric increases the capacitance of a capacitor?
(A) The electric field acting on the dielectric sets up elastic stresses. This means that more energy is stored in the capacitor as elastic potential energy, and hence the capacitance is increased (since $U = \tfrac{1}{2}CV^2$).
(B) Free charge can flow from the dielectric on to the plates of the capacitor. This means that the battery must provide less charge; hence it does less work, and the capacitance is increased.
(C) Charge displacement in the dielectric tends to reinforce the electric field due to the charge on the plates, thereby increasing the potential difference across the capacitor and increasing the capacitance.
(D) Induced charge on the surface of the dielectric sets up electric fields that tend to cancel the field due to charge on the plates. This means that more charge must be placed on the plates to cause a given potential difference, and more coulombs per volt means more capacitance.
(E) None of the above is correct. A dielectric *decreases* capacitance.

Section 26-4 Energy of a Charged Capacitor

· 26-39 P Suppose that one square centimeter of a muscle-cell membrane has a capacitance of 8×10^{-7} F and that a potential difference of 70 mV exists across it (these are typical values). How much energy is stored in one square centimeter of membrane?

· 26-40 P What is the ratio of the stored energy in capacitors C_1 and C_2 when they are connected (a) in parallel? (b) in series? See Figure 26-26.

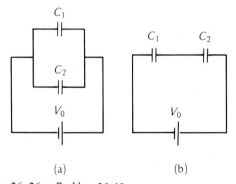

Figure 26-26. Problem 26-40.

: 26-41 P While the capacitors shown in Figure 26-27 are connected to a 12-V battery, a slab of dielectric constant 3 is inserted into the 4-μF capacitor. Determine for each capacitor the (a) potential difference, (b) charge, and (c) stored energy before and after the slab is inserted. The slab completely fills the capacitor.

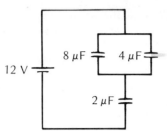

Figure 26-27. Problem 26-41.

: **26-42 P** (a) How should three capacitors C_1, C_2, and C_3 be connected to provide the maximum energy storage when connected to a single power supply? The capacitor C_1 is 1-μF rated at 500 V, C_2 is 4-μF rated at 100 V, and C_3 is 2-μF rated at 300 V. (b) What is the maximum stored energy?

: **26-43 P** The potential drop across the plates of a capacitor has decreased to one-half of its initial value. By what factor has the energy stored in the capacitor been reduced?

: **26-44 P** Show that a plate of a parallel-plate capacitor experiences a force $F = q^2/(2\epsilon_0 A)$ due to the electric field between the plates. (Hint: Consider the work done and the gain in energy when the separation of the plates is increased by a small amount.)

: **26-45 P** Integrate the energy density $\frac{1}{2}\epsilon_0 E^2$ over the volume of a cylindrical capacitor, to show that the energy stored in the capacitor is $Q^2/2C$.

: **26-46 Q** Two identical capacitors are connected in series to a battery. While they are connected to the battery, a dielectric slab is inserted into one of the capacitors. What happens—increases, decreases, or remains the same—to each of the following when this is done?
  (a) Charge on upper capacitor. (b) Charge on lower capacitor. (c) Potential difference across upper capacitor. (d) Potential difference across lower capacitor. (e) Total energy stored in the two capacitors. (f) Total capacitance.

: **26-47 P** An air-filled parallel plate capacitor in Figure 26-28 has capacitance C_0. It is charged to a potential difference V_0. A slab of dielectric constant κ is inserted a distance x into the capacitor. As a function of x, what is (a) the voltage across the plates? (b) the energy stored in the capacitor?

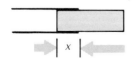

Figure 26-28. Problem 26-47.

: **26-48 P** An air-filled parallel-plate capacitor has capacitance C_0. It is charged to a potential difference V and disconnected from the battery. A dielectric slab (which just fills the capacitor) is released from rest at the edge of the capacitor and allowed to be drawn into the capacitor. What is the maximum velocity the slab can achieve if its mass is m?

: **26-49 P** Einstein showed that energy and mass are equivalent and related by the expression $E = mc^2$. It was once speculated that the rest mass of an electron might be just the energy associated with the electric field created by the electron. What is the radius of a conducting sphere with charge e whose total electric field energy is just mc^2, where m is the mass of the electron? (As far as we know the electron acts essentially like a point object, and its radius, if it has one, is less than 10^{-18} m.)

Section 26-5 Energy Density of the Electric Field

· **26-50 P** The energy content of 1 liter of gasoline is about 4.4×10^7 J. What would be the volume of a mica-filled capacitor that would store this much energy?

: **26-51 P** Calculate the energy stored in the electric field of charge Q when the charge is distributed (a) uniformly throughout the volume of a sphere of radius R; (b) uniformly over the surface of a sphere of radius R. (c) Compute the ratio of the result of (a) to the result of (b) and explain qualitatively the reason for such a result.

: **26-52 P** A spherical conductor of radius R carries a charge Q. (a) What is the radius of an imaginary concentric sphere that would enclose half the stored energy in the electric field surrounding the charged sphere? (b) At what radius does the potential due to the sphere drop to half its value at the surface of the sphere. Compare this result with that found in (a) and comment on the relation.

: **26-53 P** A parallel-plate capacitor of area A and separation d contains a dielectric slab of dielectric constant κ, area A, and thickness αd, where $\alpha < 1$. A potential difference V_0 is established between the plates of the capacitor. Determine (a) the total energy stored in the capacitor and the fraction of the energy stored in the air gap and in the dielectric; (b) the potential difference across the air gap and across the dielectric. (c) Evaluate each of the above for the case $\alpha = \frac{1}{2}$, $\kappa = 2$, $V_0 = 100$ V, $A\epsilon_0/d = 2 \times 10^{-6}$ F.

: **26-54 P** Consider a spherical conductor of surface charge density σ. Since the charges try to repel each other, they give rise to an outward "electric pressure." (a) Show that the electric pressure is $P = \frac{1}{2}\epsilon_0 E^2$ = electric field energy density. Consider first the work that would have to be done to compress the radius of the sphere by a small amount δr. Then equate this to the increase in the energy stored in the electric field (because then there would be slightly more space around the smaller sphere than initially). (b) Would it be possible to keep a conducting balloon spherical with no air inside just by charging it? Remember that the surrounding air will break down if E exceeds 3×10^6 V/m. What pressure in atmospheres could one create in this way?

Supplementary Problems

26-55 P A useful model of a nerve fiber (an axon) treats it as a cylindrical coaxial capacitor filled with myelin, which has a dielectric constant of 7. For a frog axon, typical values of the inner and outer radii are 5×10^{-6} m and 7×10^{-6} m. A potential difference of approximately 70 mV exists across the dielectric layer in the resting state. (*a*) What is the capacitance per meter of such an axon? (*b*) What is the surface-charge density on the outer surface? (*c*) Assuming that a molecule occupies about 25×10^{-20} m^2, determine how many electrons per molecule the surface-charge density found in (*b*) corresponds to.

26-56 P Energy storage is very important. The "energy density" measured in joules per kilogram is an important parameter, for example, in the design of an electric car. How to store the electrical energy needed by such a car is the main problem in the development. For example, whereas gasoline provides about 4.4×10^7 J/kg, a lead-acid battery stores only about 1.8×10^5 J/kg. Could one do better with a capacitor than with a battery? Calculate the energy density (in joules per kilogram) for a mica-filled capacitor charged almost to breakdown. The density of mica is about 2900 mg/m^3. Assume the plates contribute negligible mass and volume to the capacitor.

26-57 P Capacitors play an essential role in microelectronics technology. An individual capacitor can act as a storage cell for one bit of information (charged = yes = 1; uncharged = no = 0). In a silicon semiconductor memory chip, for example, a pair of conducting plates (combined with a transistor) is a capacitor storage cell. One large-scale integrated memory chip with RAM (random access memory) has 16,384 bits spread in a single layer over an area less than 0.5 cm on an edge. (*a*) Make reasonable assumptions to show that the size of one capacitor plate is of the order of 20 microns (1 micron = 1 μm = 10^{-6} m). (Recently developed storage cells are as small as about 1 micron). (*b*) Using 12 as the dielectric constant for the silicon dioxide insulator between the plates, show that the capacitance of the storage cell is of the order of 50 fF (where f = femto = 10^{-15}). (*c*) If the storage cell is fully charged with a potential difference of 10 V, what is the net charge on one plate in multiples of the charge of an electron? (*d*) Assuming that all electrons on the negatively charged plate are spread uniformly over the plate, what is the average distance between adjacent electrons? (*e*) What is the energy required to charge all 16,384 storage cells in the chip? [For more on "Microelectronics Memories" see the article with that title by David A. Hodges in *Scientific American* **237**, 130 (September, 1977).]

27 Electric Current and Resistance

27-1 Electric Current
27-2 Current and Energy Conservation
27-3 Resistance and Ohm's Law
27-4 Resistivity
27-5 RC Circuits
27-6 Electric Resistance from a Microscopic Point of View
    Summary

Thus far we have studied electrostatics — electric charges at rest. Now we turn to electric currents — electric charges in motion. Electric resistance and resistivity are defined. A circuit consisting of a capacitor and resistor is discussed. Finally, we give a brief and qualitative account of the microscopic properties influencing a material's resistivity.

27-1 Electric Current

Electric charges in motion constitute an *electric current*. By definition, the current i through a chosen surface is the total *net* charge dQ passing through that surface divided by the elapsed time dt:

$$i = \frac{dQ}{dt} \tag{27-1}$$

Electric current is always given in terms of charges passing through a surface bounded by a loop; for an ordinary conducting wire the loop would ordinarily be the wire's circumference.

The direction of the so-called *conventional current* is taken to be that in which positive charges move. Thus, protons (charge, $+e$) traveling to the right constitute a current to the right, while electrons (charge, $-e$) traveling to the right constitute a current to the left. To obtain the total current i through a loop, we must always count the *net* charge passing through per unit time. For example, if positive ions move to the right through a surface and at the same time negative ions move to the left through the same surface, as might happen for currents in liquids, *both* types of ions contribute to an electric current to the right; see Figure 27-1.

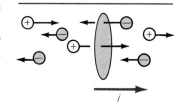

Figure 27-1. Both the positive charges to the right and the negative charges to the left through the loop contribute to a conventional current i to the right.

The SI unit of current is the *ampere* (abbreviated A). The unit is named after André M. Ampère (1775–1836), who made significant contributions to understanding the relation between an electric current and its associated magnetic field. One ampere corresponds to a net charge of 1 C passing through a chosen surface in the time of 1 s:*

$$1 \text{A} \equiv 1 \text{ C/s}$$

Whereas the coulomb is an uncommonly large amount of charge by laboratory standards, laboratory currents of a few amperes are typical. Related units are the milliampere (1 mA = 10^{-3} A) and the microampere (1 μA = 10^{-6}A).

A direct current (dc) is one in which net positive charges move in one direction only (but *not* necessarily at a constant rate). For an alternating current (ac), on the other hand, the current changes direction periodically.

How is the current i related to the properties of the charge carriers? Suppose that a current consists of n identical charge carriers per unit volume drifting at a speed v_d along a single direction. See Figure 27-2. Each carrier has a positive charge q. In the time dt a charge carrier advances a distance $v_d \, dt$, and all the charge carriers within a cylindrical volume of length $v_d \, dt$ and cross-sectional area A will have passed through one end of the cylinder. By the same token, an equal number of charge carriers will have entered the cylindrical column through the other end. The total charge dQ passing through the area A in time dt is the charge per unit volume nq multiplied by the cylinder's volume $Av_d \, dt$:

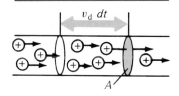

Figure 27-2. Charges drifting at speed v_d through a cylinder of cross-sectional area A in the time dt.

$$dQ = nqAv_d \, dt$$

Consequently, the current i contributed by these charge carriers is

$$i = \frac{dQ}{dt} = nqAv_d \tag{27-2}$$

The charge carriers need not move at a *constant* velocity. They may have a very complicated motion. The free electrons in an ordinary conductor are in random thermal motion at speeds of the order of 10^6 m/s. The drift speed v_d, on the other hand, is the average displacement per unit time along the direction of the current. As we shall see the drift speed is typically *very* small compared with the thermal speed.

The most common device used for measuring current is an *ammeter*; its operation depends on the torque produced by a magnet on a current-carrying coil of wire (Section 29-6).

---

* Strictly, the coulomb is defined as one ampere-second and the ampere itself, one of the base units in the SI system, is defined in terms of the magnetic interaction between a pair of current-carrying conductors. See Appendix A and Section 30-2.

Example 27-1. A current of 1.0 A exists in a conducting copper wire whose cross-sectional area is 1.0 mm². What is the average drift speed of the conduction electrons under these conditions?

In copper, as in other typical metallic conductors, there is approximately one free (conduction) electron per atom. We can find the density n of charge carriers by computing the number of atoms per unit volume from Avogadro's number N_A (the number of atoms per mole), the atomic mass m (the number of grams per mole), and the mass density ρ_m. Clearly,

$$n = \frac{\text{charge carriers}}{\text{volume}} = \left(\frac{1 \text{ charge carrier}}{\text{atom}}\right)\left(\frac{\text{atoms}}{\text{volume}}\right)$$

$$= \left(\frac{1 \text{ charge carrier}}{\text{atom}}\right)\left(\frac{\text{atoms}}{\text{mole}}\right)\left(\frac{\text{moles}}{\text{kilogram}}\right)\left(\frac{\text{kilograms}}{\text{volume}}\right)$$

or

$$n = (1)(N_A)\left(\frac{1}{m}\right)(\rho_m)$$

$$= \left(\frac{1 \text{ ch car}}{\text{atom}}\right)(6.0 \times 10^{23} \text{ atoms/mol})\left(\frac{1}{64 \times 10^{-3} \text{ kg/mol}}\right)(9.0 \times 10^3 \text{ kg/m}^3)$$

$$= 8.4 \times 10^{28} \text{ charge carriers/m}^3$$

Then from (27-2), we have

$$v_d = \frac{i}{nqA}$$

$$= \frac{1.0 \text{ A}}{(8.4 \times 10^{28} \text{ m}^{-3})(1.6 \times 10^{-19} \text{ C})(1.0 \times 10^{-6} \text{ m}^2)}$$

$$= 7.4 \times 10^{-5} \text{ m/s} = 0.074 \text{ mm/s}$$

The free electrons drift through the conductor with an average speed that is actually less than 0.1 mm/s, a remarkably low speed. It should *not* be inferred, however, that when such a current is established at one end of a conducting copper wire it takes almost 10 s for the signal to travel a mere 1 mm. The speed at which the electric field driving the free electrons is established is close to the speed of light. One must distinguish here between the speed with which the charged particles drift and the speed at which the signal is propagated, just as one must distinguish between the speed (possibly very low) at which a liquid drifts through a pipe and the much higher speed at which a change in pressure is propagated along the pipe. The *drift* speed of conduction electrons is much less than the *random* thermal speeds of electrons at any finite temperature. (We shall explore this point further in Section 27-5).

Equation (27-2) can be written in an alternative form by introducing the *current-density vector* **j**, whose magnitude is

$$j = \frac{i}{A} = nqv_d \tag{27-3}$$

and whose direction is determined by the direction of the velocity vector $\mathbf{v}_d$ of the charge carriers:

$$\mathbf{j} = nq\mathbf{v}_d \tag{27-4}$$

with **j** parallel to $\mathbf{v}_d$.

To see the utility of the current-density vector, consider Figure 27-3. Here vector d**S** is a small element of a surface through which charges flow. We find

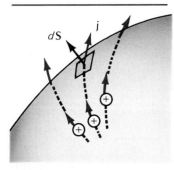

Figure 27-3. Current-density vector **j** and an element of surface d**S**.

the electric-charge flux, or electric current di through dS, by multiplying the component of $\mathbf{j}$ along the normal to the surface by the area dS; that is,

$$di = \mathbf{j} \cdot d\mathbf{S}$$

The total current i through the surface S is

$$i = \int_S \mathbf{j} \cdot d\mathbf{S} \tag{27-5}$$

When $\mathbf{j}$ and $d\mathbf{S}$ are parallel and the surface is flat, (27-5) reduces to (27-3).

27-2 Current and Energy Conservation

Here we set down the fundamental relation for the rate at which electric energy is transferred to any device through which charges flow. Our arrangement is altogether general; some energy source—for example, a battery, a solar cell, or a gasoline generator—maintains a potential difference $V_{ab} = V_a - V_b$ across the terminals of what we call simply *the load* and through which current i passes. See Figure 27-4. The load might be an electric motor, a lamp, or a heater. All that matters is that current i goes through the load while potential difference V_{ab} is maintained across its terminals. The input potential V_a is higher than the output potential V_b, so that conventional current is *in* at a and *out* at b. (Electrons would go "uphill" in potential, from b to a.)

When positive charge dQ goes through the load, the work dW done on it by electric forces associated with potential difference V_{ab} is

$$dW = (dQ)V_{ab}$$

Since $i = dQ/dt$,

$$dW = iV_{ab}\, dt$$

and the rate of doing work on the load dW/dt, or delivering power P to it, is

$$P = \frac{dW}{dt} = iV_{ab} \tag{27-6}$$

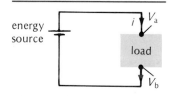

Figure 27-4. An energy source delivering energy to a load.

The instantaneous power is just the product of i and V_{ab}. Note that power P has units of watts: $(A)(V) = (C/s)(J/C) = J/s = W$.

If i and V_{ab} are constant, so that P is also constant with time, the total work W done on the charges transported through the load over a time interval t is

$$W = iV_{ab}t$$

If i and V vary with time, the work is

$$W = \int P\, dt = \int iV\, dt$$

27-3 Resistance and Ohm's Law

Figure 27-5 shows voltage-current plots for three dissipative current elements: (a) a gas in a closed container with two electrodes, (b) a semiconducting device, and (c) a homogeneous solid conductor. Only for the conductor is the V-i

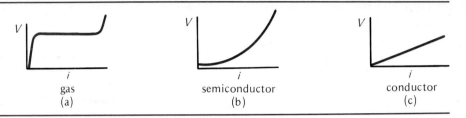

Figure 27-5. Voltage-current characteristics for (a) a gas, (b) a semiconducting device, and (c) a conductor obeying Ohm's law.

relation simple: a straight line. The ratio of V to i for any purely dissipative circuit element is *defined* to be the device's *electrical resistance* R

$$R \equiv V/i \qquad (27\text{-}7)$$

The experimental finding that R, so defined, is a *constant* for a given pair of connections to a conductor (maintained at constant temperature), so that R is independent of the magnitude of either i or V, is known as *Ohm's law*:

Ohm's law: $\quad R = $ constant for metallic conductors $\qquad (27\text{-}8)$

Georg S. Ohm (1787–1854) found that R was a constant for a large variety of materials and for an enormous range of currents and potential differences (a factor of 10^{10} for some materials).

The SI unit for resistance is the ohm (abbreviated by Greek capital omega, Ω), and it is defined by

$$1\,\Omega \equiv 1\text{ V/A}$$

Related units are the megohm (1 MΩ = 10^6 Ω) and the microhm (1 $\mu\Omega$ = 10^{-6} Ω). The reciprocal of resistance is called electrical *conductance*. The official SI unit for conductance is the *siemens* (abbreviated S):

$$1\text{ S} \equiv 1\text{ A/V} = 1/\Omega = 1\text{ mho}$$

The siemens unit is seldom used in practice; the common but unofficial unit for conductance is the reciprocal ohm, or mho.

In circuit diagrams, a resistor is depicted by the symbol —⋀⋀—; and a variable resistor, or rheostat, is represented by —⋀⋀— or —⋀⋀—. A dissipative circuit element whose resistance is not independent of V (and therefore also not independent of i) is known as a nonohmic or nonlinear resistor (its V-i plot is not a straight line).

The power delivered to any circuit element is iV. When that circuit element is a resistor the power dissipated can also be written

$$P = iV = i^2 R = \frac{V^2}{R} \qquad (27\text{-}9)$$

That the thermal energy dissipated per unit-time in a resistor (the so-called $i^2 R$ loss) is, in fact, exactly equal to the electric energy delivered to it was first established in the nineteenth century by the historic experiments of James Joule. The effect is sometimes referred to as Joule heating.

Example 27-2. A transistor radio operates at 10 mW with a 9-V dry cell. (a) What is the current into and out of the radio? (b) What is the radio's electric resistance?

(a) $\quad i = \dfrac{P}{V} = \dfrac{10 \times 10^{-3}\text{ W}}{9.0\text{ V}} = 1.1\text{ mA}$

(b) $R = \dfrac{V}{i} = \dfrac{9.0 \text{ V}}{1.1 \times 10^{-3} \text{ A}} = 8.2 \text{ k}\Omega$

27-4 Resistivity

Ohm's law holds for a conductor of any shape and with the leads attached to any two points. Consider, however, the simple arrangement in which a potential difference is established between the ends of a cylindrical conductor of length L and cross-sectional area A, as in Figure 27-6. For a given material, experiment shows that the resistance for this simple geometrical configuration is directly proportional to the length and inversely proportional to the cross-sectional area. We may write the resistance as

$$R = \rho \frac{L}{A} \qquad (27\text{-}10)$$

where the constant ρ, called the resistivity, is a property of the material of which the conductor is made but does not depend on the conductor's physical shape. Resistivity carries the units ohm meters ($\Omega \cdot$m).

Table 27-1 lists resistivities for several materials, both good and poor conductors. Note the great range in resistivities; the best conductors, metals, are better than the worst conductor by a factor of about 10^{24}. (The best *thermal* conductors are also metals; the low thermal and electrical resistivity arises from the free electrons in metallic conductors.) From (27-10), we see that the resistance of a cube 1 m along an edge, with a current between opposite faces maintained at different potentials, is equal numerically to the material's resistivity in ohm meters.

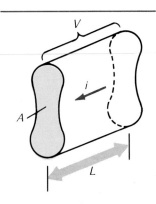

Figure 27-6. A potential difference V is maintained across the ends of a cylindrical conductor of length L and cross-sectional area A through which a current i passes.

Table 27-1.

MATERIAL	RESISTIVITY (MICROHM · CM)	TEMPERATURE COEFFICIENT OF RESISTIVITY AT 20°C (C°$^{-1}$)
Metal		
Aluminum	2.824	0.0039
Brass	7	0.002
Constantan	49	0.00001
Copper (annealed)	1.724	0.0039
Gas Carbon	5,000	-0.0005
Manganin	44	0.00001
Nichrome (trade mark)	100	0.0004
Silver	1.59	0.0038
Steel (manganese)	70	0.001
Tungsten (drawn)	5.6	0.0045
Insulator, semiconductor		
Germanium	4.3×10^7	
Silicon	2.6×10^{11}	
Wood (maple)	4×10^{19}	
Mica	9×10^{21}	
Quartz	5×10^{24}	

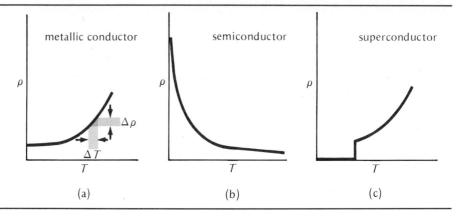

Figure 27-7. Variation in electrical resistivity ρ with temperature T for (a) a metallic conductor, (b) a semiconductor, and (c) a superconductor.

The reciprocal of the resistivity ρ is called the *conductivity* σ:

$$\sigma \equiv \frac{1}{\rho} \tag{27-11}$$

Figure 27-7 shows how the resistivity of three types of materials varies with temperature: (a) an ordinary metallic conductor, (b) a semiconductor, and (c) a superconductor.

The resistance, and therefore also the resistivity, of metallic conductors increases with temperature. Over a temperature range ΔT small enough for the curve in Figure 27-7(a) to be approximated by a straight line, the change in resistivity $\Delta\rho$ is proportional to ΔT. The fractional change in resistivity $\Delta\rho/\rho$ can be written as

$$\frac{\Delta\rho}{\rho} = \alpha \, \Delta T \tag{27-12}$$

where α is a constant (over the small temperature range) called the *temperature coefficient of resistance*. Values of α for some selected conductors are shown in Table 27-1. Copper at room temperature has $\alpha = 0.003\,93\,(C°)^{-1}$; this means that the resistivity of copper increases by 0.39 percent for an increase in temperature of one Celsius degree. The change in resistance with temperature can be used as the basis of a resistance thermometer. Note that the alloy constantan (60 percent copper, 40 percent nickel), with an extremely small value of α, has a resistivity that is nearly constant over a range of temperatures.

The resistivity of semiconductors [Figure 27-7(b)] decreases with a temperature rise, primarily because the number of charge carriers increases as the temperature is raised. Temperature coefficients of resistivity are negative for these materials, of which common examples are carbon, silicon, and germanium.

The behavior of superconductors [Figure 27-7(c)] is extraordinary. Below a certain critical temperature (below 7.175 K for lead), the resistivity of a superconductor drops to *zero* — not merely a small resistivity, but exactly zero. A superconductor offers no resistance to the flow of electric charge and dissipates no energy. Once a superconducting current is established in a superconducting loop, it persists indefinitely. Currents of several hundred amperes induced in a superconducting lead ring have been observed to persist with no measurable diminution for several years! Although bizarre, the phenomenon of superconductivity is now well understood on the basis of the quantum theory.*

Example 27-3. A current of 1.0 A is maintained in a copper wire 10 m long and having a cross-sectional area of 0.25 mm². (a) What is the potential difference between its ends? (b) What is the magnitude of the electric field within the wire driving electrons? (c) At what distance from a proton is the electric field from it of the same magnitude as the driving electric field?

(a) From Ohm's law and the definition of resistivity, we have

$$V = iR = i\rho \frac{L}{A}$$

$$= (1.0 \text{ A})(1.72 \times 10^{-8} \Omega \cdot \text{m}) \left(\frac{10 \text{ m}}{0.25 \times 10^{-6} \text{ m}^2} \right)$$

$$= 0.69 \text{ V}$$

Ordinarily the potential difference along short lengths of connecting wire in a circuit is small enough to be taken as zero.

(b) The electric field produced within the wire by the potential difference across its ends is constant in magnitude along the length of the wire, so that

$$E = \frac{V}{L} = \frac{0.69 \text{ V}}{10 \text{ m}} = 0.069 \text{ V/m}$$

(c) The electric field at a distance r from a point charge is $E = kq/r^2$ [(23-6)]. Therefore, the distance from a proton, with $q = 1.6 \times 10^{-19}$ C, at which $E = 0.069$ V/m is

$$r = \sqrt{\frac{kq}{E}} = \sqrt{\frac{(9.0 \times 10^9 \text{ N} \cdot \text{m}^2/\text{C}^2)(1.6 \times 10^{-19} \text{C})}{0.069 \text{ V/m}}} = 0.14 \text{ mm}$$

Compared with the electric forces acting on a conduction electron from other charged particles nearby (atoms in a solid might be separated by ~10^{-7} mm), the electric force producing the steady current is relatively feeble.

27-5 RC Circuits

Consider the circuit shown in Figure 27-8. It consists of an initially charged capacitor C connected across a resistor R. The capacitor has an initial charge magnitude Q_0 on each plate, and the potential difference between the plates is initially $V_0 = Q_0/C$.

It is easy to describe qualitatively what happens immediately after the switch is closed. Free electrons move counterclockwise around the circuit and through the resistor until finally the capacitor is electrically neutral. While charges are moving, a current i exists in the circuit; the current is zero when charge Q and potential difference V of the capacitor reach zero. At the same time, the electric-potential energy associated with the opposite charges on the capacitor plates is dissipated into thermal energy as charges move through the resistor.

---

* Actually, we are not surprised—but perhaps should be—to learn that electrons within an atom can remain in motion indefinitely without dissipating energy. The reason for this microscopic behavior, as well as for the macroscopic phenomenon of superconductivity, lies with quantum theory.

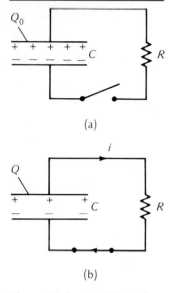

(a)

(b)

Figure 27-8. An RC circuit. (a) Capacitor initially charged; (b) capacitor discharging through the resistor.

We wish to find how charge Q and current i vary with time. The net change in potential around the circuit is zero at each instant of time. Going clockwise around the loop, we have a potential drop iR across the resistor and a potential drop $-Q/C$ across the capacitor. (We go from negative to positive plate through the capacitor—a potential *rise*). Therefore,

$$iR - \frac{Q}{C} = 0$$

The current at each instant is

$$i = -\frac{dQ}{dt}$$

The minus sign indicates that positive current corresponds to a *decrease* in charge Q on the capacitor plates. The first equation can then be written

$$-R\frac{dQ}{dt} - \frac{Q}{C} = 0$$

This differential equation holds at any instant. Rearranging terms and recognizing that R and C are independent of time, we have

$$\int_{Q_0}^{Q} \frac{dQ}{Q} = -\frac{1}{RC}\int_0^t dt$$

where we are integrating charge from its initial value Q_0 to any later value Q at time t and integrating time from zero to time t:

$$\ln\frac{Q}{Q_0} = -\frac{t}{RC}$$

$$Q = Q_0 e^{-t/RC} \qquad (27\text{-}13)$$

The charge Q on each capacitor plate decreases exponentially with time, as shown in Figure 27-9(a). The rate of decay is controlled by the quantity RC, the *time constant* τ of the RC circuit:

$$\tau \equiv RC \qquad (27\text{-}14)$$

As (27-13) shows, when $t = \tau = RC$, then $Q = Q_0/e$. The constant RC is the time elapsing until the capacitor's charge (and also potential difference) is $(1/e)$, or 37 percent, of its initial value. For example, in a circuit with $R = 1.0$ MΩ and $C = 1.0$ μF, the time constant is $\tau = (1.0 \times 10^6 \text{ V/A})(1.0 \times 10^{-6} \text{ C/V}) = 1.0$ s.

How does current vary with time? Using (27-13), we have

$$i = -\frac{dQ}{dt} = \left(\frac{Q_0}{RC}\right) e^{-t/RC}$$

But the initial current is $i_0 = V_0/R = Q_0/RC$. The equation above can then be written

$$i = i_0 e^{-t/RC} \qquad (27\text{-}15)$$

The current, zero until the switch is closed, rises abruptly to i_0 and then decays exponentially with time, again with characteristic time constant $\tau = RC$, as shown in Figure 27-9(b). The rate of energy dissipation in the resistor is i^2R,

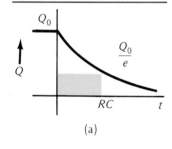

(a)

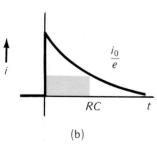

(b)

Figure 27-9. Variation with time of (a) the capacitor charge and (b) the current for an RC circuit.

Figure 27-10. *A sawtooth wave, with the sloped straight-line portion derived from an RC circuit.*

proportional to the *square* of i. Therefore, the rate at which energy is dissipated in the resistor (also the rate at which the charged capacitor loses energy) is more rapid than the decay in Q or i. The time constant for energy decay is $\frac{1}{2}RC$.

An *RC* circuit provides a means of producing a *sawtooth wave*, a variation in voltage with time shown in Figure 27-10. The change in potential across the capacitor or resistor with time is, as we have seen, exponential. But a small piece of an exponential line is effectively a *straight* line. The linear variation in voltage with time is obtained by allowing a capacitor to discharge (or charge) in an *RC* circuit for a time short compared with τ and then be abruptly recharged back to its initial state. The potential difference across V and R can thereby be made to vary as shown in Figure 27-10. Sawtooth waves are used frequently in electric circuits. For example, the steady and repeated progression of an electron beam across the face of an oscilloscope (or TV picture tube) can be produced by a sawtooth wave applied to horizontal deflecting plates.

27-6 Electric Resistance from a Microscopic Point of View

When there is no external electric field, the free, or conduction, electrons of a conductor are in random thermal motion. They move at relatively high speed ($\sim 10^6$ m/s), collide with one another and with the atoms of the conductor. In such a collision, a free electron is just as likely to gain as to lose kinetic energy. On the average, its kinetic energy is unchanged, and it undergoes no net displacement within the conductor.

Now consider what happens to free electrons in a conductor across which an electric potential difference is maintained. The simple cylindrical conductor in Figure 27-11 of length L and cross-sectional area A has a constant potential difference V between its ends. A constant electric field $\mathbf{E}$, pointing along the length of the conductor from high to low potential, then exists within the conductor, with $E = -V/L$. Free electrons still make frequent collisions, and are driven by this field. Electron drift is responsible for the current in the conductor. When a conductor is in *electrostatic* equilibrium, $\mathbf{E} = 0$ everywhere within a conductor, but when charges flow, and a current exists, in the conductor the electric field can no longer be zero.

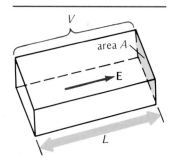

Figure 27-11. *Constant electric field $\mathbf{E}$ established on the interior of a conductor by electric potential difference ΔV.*

The approximate behavior of free electrons in a conductor under the influence of a driving electric field is illustrated by the mechanical model shown in Figure 27-12. Here balls (the free electrons) accelerate down a series of inclines driven by gravitational force (the driving electric field). Each ball gains kinetic energy as it goes down a ramp, but essentially all this kinetic energy is lost when the ball collides inelastically with a bumper (an atom in the conductor). After each collision, the ball starts from rest again and makes further inelastic collisions. At the bottom of the zigzag ramps (the resistor), each ball has the

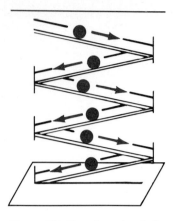

Figure 27-12. A mechanical model for a resistor: marbles rolling down a succession of inclined planes.

same kinetic energy it started with at the top—zero. All the energy gained in accelerating down the ramps becomes thermal energy in collisions. For any ball to go down the ramp again, it must be lifted back to the top (a battery, or other energy source) and its gravitational potential energy (electric potential) must be increased.

As the parenthetic entries above imply, the crucial aspects of conduction are these:

- *Inelastic* collisions with atoms of the conductor lattice.
- After each collision, each electron starts over again, essentially with zero *drift velocity.*
- Between collisions, electrons are driven at *constant acceleration* by an external electric field.

Superimposed on this drift is random thermal motion. Indeed, the average drift velocity acquired by free electrons is so very small compared to their average thermal speed that the time between collisions is effectively independent of the applied electric field.

Written as $V = iR$, Ohm's law is a macroscopic relation. We wish to express Ohm's law in microscopic form, that is, in terms of the driving electric field **E** and the current density **j** at any point within a conductor. For the uniform cylindrical conductor of Figure 27-11, we have $V = EL$ from (25-10) and $R = \rho L/A$ from (27-9). Therefore

$$V = iR$$

$$EL = i\rho \frac{L}{A}$$

Since the current density has the magnitude $j = i/A$, the relation above can be written

$$\mathbf{E} = \rho \mathbf{j} \qquad (27\text{-}16)$$

Equation (27-16) is written in vector form, since the charge carriers have drift velocities in the direction of the electric field.

Written in terms of conductivity $\sigma = 1/\rho$, Ohm's law is

$$\mathbf{j} = \sigma \mathbf{E} \qquad (27\text{-}17)$$

Although derived here for the special case of a uniform field, (27-16) and (27-17) are general relations that hold at any point within any conducting ohmic material.

Summary

Definitions

Electric current i:

$$i \equiv \frac{dQ}{dt} \qquad (27\text{-}1)$$

where net charge dQ passes through a chosen surface in time dt.

Current density j with current i passing transversely through area A:

$$j = \frac{i}{A}$$

Resistivity ρ of the material in a cylindrical conductor of resistance R, length L, and cross-sectional area A:

$$\rho \equiv \frac{RA}{L} \qquad (27\text{-}10)$$

Electric conductivity σ of a material with electric resistivity ρ:

$$\sigma = \frac{1}{\rho} \qquad (27\text{-}11)$$

Thermal coefficient of resistivity α for a material whose resistivity changes by the fractional amount $(\Delta\rho/\rho)$ for a temperature change ΔT:

$$\alpha \equiv \frac{\Delta\rho/\rho}{\Delta T} \qquad (27\text{-}12)$$

Units
Resistance: 1 ohm = $1\Omega \equiv 1$ V/A
Conductance: 1 siemens = 1 S = 1 A/V = 1 mho

Important Results
The power P delivered to a load across which the electric potential difference is V_{ab} and carrying current i:

$$P = iV_{ab} \qquad (27\text{-}6)$$

Ohm's law states that resistance R of a conductor is independent of current i and potential difference V:

$$R \equiv \frac{V}{i} = \text{constant} \qquad (27\text{-}7)$$

For microscopic application, Ohm's law is written

$$\mathbf{E} = \rho \mathbf{j} \qquad (27\text{-}16)$$

or

$$\mathbf{j} = \sigma \mathbf{E} \qquad (27\text{-}17)$$

where $\mathbf{E}$ is the driving electric field and $\mathbf{j}$ the current density at some point in a conductor with resistivity ρ and conductivity σ.

When a charged capacitor C is discharged through a resistor R (an RC circuit), the charge on either capacitor plate decays exponentially with time according to

$$Q = Q_0 e^{-t/\tau} \qquad (27\text{-}13)$$

where the time constant is

$$\tau = RC \qquad (27\text{-}14)$$

Problems and Questions

Section 27-1 Electric Current

· **27-1 P** How much charge passes in 2 min through a junction through which a steady current of 5 A exists?

· **27-2 P** In a typical lightning bolt, about 1 C of negative charge (a huge amount) is delivered to the earth in about 4×10^{-5} s. What current does this represent?

· **27-3 P** What is the current density in a copper wire of diameter 1.2 mm in which a current of 14 A flows?

· **27-4 P** A plasma is a neutral mixture of negatively charged electrons and positive ions, such as a very hot gas in a star. Attempts to create thermonuclear reactions like those in the sun are important in energy-source research and development. For the needed high temperatures, the plasma is often heated by passing a current through it. In one such device (a toroidal stellarator), the current is carried primarily by electrons whose density is $10^{19}/m^3$. What is the drift velocity required to produce a current density of 10^5 A/m²?

· **27-5 P** A typical current used in a household circuit is 10 A. Estimate the drift velocity of the carriers in this case, assuming that No. 14 copper wire (diameter 0.163 mm) is used.

· **27-6 P** In aluminum there are 1.8×10^{29} conduction electrons per cubic meter. What is the drift velocity in an aluminum wire of diameter 1.2 mm in which a current of 2 A flows?

: **27-7 P** A neutral beam has equal proton and electron densities, each $6.0 \times 10^6/cm^3$. Both electrons and protons have energies of 12 eV, with protons moving to the right and electrons moving to the left. If the cross-sectional area of the beam is 2 mm², what are the net (a) current and (b) current density in the beam?

: **27-8 P** Silver, with a density of 10.5 gm/cm³, has about one conduction electron per atom. In a silver wire of diameter 0.1 mm, what is (a) the density of conduction electrons (in electrons/m³)? (b) the drift velocity of the electrons when a current of 1 A exists in the wire?

: **27-9 P** What current is required in an electroplating process if 3 mg of copper (atomic mass 64) is to be deposited in 5 min? A copper sulfate solution containing the ions Cu^{2+} is used as the electrolyte.

: **27-10 P** What mass of gold (atomic mass 197) will be deposited in an electroplating process in which the ions Au^+ carry a current of 10 A for 5 minutes?

: **27-11 P** An experimenter wants to deposit a layer of silver of thickness 0.02 mm over a microwave component of area 4 cm². With a solution of Ag^+ and a current of 2 A, how long will it take to deposit the desired amount of silver? (Silver density, 10.5 gm/cm³; atomic mass, 108).

: **27-12 P** In a simple model of the hydrogen atom, a single electron circles a fixed proton in an orbit of radius 0.5×10^{-10} m under the action of the Coulomb force from the proton. (a) What is the frequency of rotation of the electron? (b) What current is associated with the electronic motion?

Section 27-2 Current and Energy Conservation

· **27-13 Q** In buying a battery you may find that one of the specifications provided is the "ampere-hour" rating. This is a measure of

(A) the maximum current the battery can provide.
(B) the maximum potential difference the battery can provide.
(C) the maximum electric charge the battery can deliver.
(D) the maximum power the battery can deliver.
(E) the maximum electrical energy the battery can deliver.

· 27-14 Q The kilowatt is a unit used to measure
(A) electric current.
(B) potential difference.
(C) resistivity.
(D) electrical energy.
(E) the rate at which work is done or energy is transferred.
(F) the rate of creation of electrical charge.

· 27-15 Q For you to be electrocuted,
(A) your body must acquire a large net electric charge.
(B) your body must be raised to a high electrical potential.
(C) a current must pass through your body.
(D) a potential difference of at least 1000 V must be applied across your body.
(E) you must be in contact with a dc source of voltage.
(F) at least part of your body must be wet.

· 27-16 Q A certain new, freshly charged car battery is rated at "12 volts, 83 ampere hours." This means
(A) each terminal has an initial charge of 83 C.
(B) the battery's internal resistance is 0.14 Ω.
(C) the battery will be dead after 83 h.
(D) the maximum current the battery can deliver is 83 A.
(E) the energy stored in the battery is 3.6×10^5 J.

: 27-17 P A 4.0-MeV proton beam has a density of $2.5 \times 10^{11}/m^3$ and an area of 2.0 mm³. (a) What is the current in the beam? (b) What is the power of the beam (that is, the energy delivered by the beam per second)? (c) How many years would it take such a beam to deposit 1 gm of protons?

: 27-18 P Electrons are emitted from a hot cathode at the rate of 5×10^{18} per second. They have negligible initial kinetic energy, but are accelerated by a potential difference V to an anode, where they are collected. Calculate (a) the current when $V = 100$ V; (b) the power dissipated at the anode when $V = 100$ V; (c) the current when $V = 200$ V; (d) the power when $V = 200$ V.

Section 27-3 Resistance and Ohm's Law

· 27-19 P A 100-Ω resistor is rated at 2.0 W. (a) What is the maximum potential difference that can be applied across it? (b) What maximum current can it carry?

· 27-20 P What current does a hair drier draw if it is rated at 800 W and is intended for use with a 120-V supply?

· 27-21 Q If the length and diameter of a wire of circular cross-section are both doubled, the resistance
(A) is unchanged.
(B) is doubled.
(C) is increased fourfold.
(D) is halved.
(E) None of the above.

· 27-22 Q Which of the following is most likely to cause you to suffer electrical injury?
(A) A large electric charge accumulates on your body.
(B) Your body is raised to a very high electric potential.
(C) A large electric current passes through your body.
(D) The resistance of your body becomes very large.
(E) The cells in your heart become polarized.

· 27-23 Q Suppose that a bird lands on a bare 14,000-V power line. What is likely to happen?
(A) It will be cooked.
(B) It will not be hurt because its feet are good insulators.
(C) It will not be hurt because it has a high body resistance.
(D) It will not be hurt because there is no potential difference across its body.
(E) It will not be hurt because the power line cannot deliver enough charge to cause injury.
(F) None of the above is true.

· 27-24 P A typical 20-A circuit in a home uses No. 12 copper wire with a resistance of 5.2×10^{-3} Ω/m. For such a 2-wire conductor of length 20 m, calculate (a) the potential drop when 20 A flows; (b) the power dissipated in the wire.

: 27-25 P An immersion electric heater used in the lab will change the temperature of 400 ml of water by 40C° in 5 min when operated with a 120-V supply. (a) What power does the heater deliver? (b) What current does it draw?

: 27-26 P A typical cost of electricity in the United States is $0.08 per kW·h. At this rate, assuming 100 percent efficiency, how much does it cost to (a) cook a piece of toast using a 1-kW toaster for 3 min? (b) heat a bathtub full of water (about 80 L) enough to increase the temperature 30°C?

: 27-27 P To minimize the danger of fire due to overheating, building codes limit the current in No. 10 (diameter 2.59 mm) rubber-coated copper wire to 25 A. Observing this condition, determine for a 100-m length of such wire (a) the wire resistance; (b) the current density; (c) the potential drop; (d) the electric field; (e) the rate at which heat is dissipated in the wire.

: 27-28 P The circuit breaker in a typical household circuit will open (and thus disconnect the circuit) if the current exceeds 15 A. Suppose that a person tries to plug the following appliances into a 120-V outlet, starting first with the lamp and then the following items in order. Which appliance will cause the circuit breaker to open?
(A) 150-W lamp (plugged in first).
(B) 800-W hair drier (plugged in second).
(C) 40-W radio.
(D) 700-W television set.

(E) 150-W typewriter.
(F) 50-W calculator.
(G) 100-W video game.

: **27-29 P** When a lamp is connected to a 12-V battery, it dissipates energy at the rate of 10 W. Assuming that the lamp resistance does not change, determine the rate at which the lamp will dissipate energy when connected to a 6-V battery? Assume that the batteries have negligible resistance.

: **27-30 P** A fully charged battery is rated at 12 V and 10 A·h. It is connected to a 5-Ω resistor. Assuming that the battery has negligible internal resistance, determine the total energy dissipated in the resistor by the time the battery has been completely discharged, or is "dead."

: **27-31 P** A radiant heater that is used to speed the drying of sheet rock mortar in cold weather is rated at 5000 W when the voltage supply is 120 V. (*a*) What is the resistance of the heating filaments? (*b*) What current does the heater draw? (*c*) How many kilocalories are generated in one hour by the heater? (*d*) If the change in resistance of the heating element with temperature is neglected, what would be the power output if the voltage were to drop by 10 percent to 108 V? In view of the actual variation of resistance with temperature for a metal filament, would the actual power be greater than the value calculated here, or less?

: **27-32 P** A fuse, such as that used in a car or a stereo set, is a conductor with a low melting point; it will thus get hot and melt if too much current exists in it. Such devices are used to limit the current in a circuit, and thereby prevent overheating from too much current.

When current exists in the fuse, its temperature will rise until the thermal losses equal the power P dissipated in the fuse. The temperature rise will generally be proportional to the power dissipation P, $\Delta T = aP$, where a is a constant that can be determined experimentally for a given geometry. As the fuse gets hot, its resistance will increase according to $R = R_{20}(1 + \alpha \Delta T)$. Show that if α is large enough, the fuse will "blow" (ΔT will become very large) when the current reaches the value $I = 1/\sqrt{\alpha a R_{20}}$.

Section 27-4 Resistivity

· **27-33 Q** Which of the following is the most accurate statement?
(A) *Resistance* and *resistivity* have the same meaning.
(B) The resistivity of a given conductor depends on its size and shape and on the material of which it is composed.
(C) A piece of copper has lower resistivity than a piece of iron.
(D) A thick copper wire has greater resistance than a thin copper wire of the same length.
(E) Resistance in a metal conductor is due to the "back" internal electric fields that the conduction electrons exert on each other.

· **27-34 P** What length of No. 18 copper wire (diameter = 0.040 in.) will have a resistance of 1 Ω?

· **27-35 P** What is the resistance of a copper bar of cross-sectional area 2 mm² and length 3 m at 20°C?

: **27-36 P** Consider an aluminum wire and a copper wire, each of the same length and resistance. What is the ratio of the weight of the copper wire to the weight of the aluminum wire? (Aluminum density, 2.7 gm/cm³; copper density, 8.9 gm/cm³.)

· **27-37 P** Nichrome is an alloy used as a heater wire in constructing electric heaters and furnaces. What resistance should such a heater have at 20°C if it is to dissipate 1000 W at 800°C when operated from a 120-V supply? (Assume that the temperature coefficient of resistivity of nichrome remains equal to 0.0004 C°⁻¹ over the entire temperature range.)

: **27-38 P** At what temperature will the resistance of a copper conductor be twice its value at 20° C, assuming its temperature coefficient of resistivity to remain constant? If you neglect the thermal expansion of the wire, will this affect seriously the accuracy of your result?

: **27-39 P** Carbon resistors find many applications in electronics. Consider such a resistor whose resistance at 20°C is 5.00 Ω. The mass of the resistor is 50 gm, its specific heat capacity is 0.10 kcal/kg°C, and its temperature coefficient of resistivity is $-5.0 \times 10^4/\text{C}°$ (note that resistance *decreases* when temperature is increased). A fixed potential difference of 12 V is applied to the resistor at $t = 0$. (*a*) If all the electric power dissipated goes into heating the resistor, how long will it take for the resistor to reach 40°C? (*b*) What will be the current through the resistor at 40°C?

: **27-40 P** A power station delivers 1 MW of power at voltage V to a transmission line made of aluminum cables of diameter 6 cm (Figure 27-13). The power is to be delivered to a town 80 km distant. The resistivity of aluminum is 2.8×10^{-6} Ω·cm.

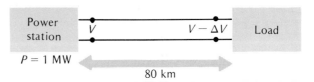

Figure 27-13. Problem 27-40.

(*a*) What is the resistance of the line? (Note that two wires are used.) (*b*) What is the voltage drop along the line in terms of the voltage V and the output power P at the power station? (*c*) If the voltage drop along the line is to be no greater than 4 percent of V, what minimum value of V is required? (*d*) What power is dissipated in the line? (*e*) Explain qualitatively why, for a given power P, line losses can be reduced by using high line voltage.

27-41 P An axon membrane has a typical resistance conductance of 5 mho/m² and a thickness of 7.5×10^{-9} m. (a) What is the resistivity of the membrane? (b) Suppose that the membrane conductivity is due to pores of diameter 8×10^{-10} m filled with conducting fluid of resistivity 0.16 $\Omega \cdot$m, with the bulk of the membrane material of very high resistivity. Approximately how many pores per meter squared would be required to give the observed resistance? (c) Estimate the spacing of the pores.

27-42 P What is the resistance of a conductor having resistivity ρ and the shape of a slightly tapered truncated cylindrical cone (see Figure 27-14)? Assume that the current density is uniform across any cross section.

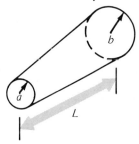

Figure 27-14. Problem 27-42.

Section 27-5 RC Circuits

· **27-43 P** A charged 2-μF capacitor is discharged through a 2-M Ω resistor. (a) What is the time constant for the decay? (b) After how many time constants will the energy stored in the capacitor decrease to half its initial value?

· **27-44 Q** The capacitor shown in Figure 27-15 is initially uncharged. The leads XX' are connected to the vertical axis of an oscilloscope, so a plot of the potential difference V_R across the resistance is obtained as a function of the time after the switch is closed. Which graph best describes what one would observe?

27-45 Q In working with electronic instrumentation, it is often necessary to use fast switching circuits. In such applications it is important to take care that the instrumentation and circuitry does not decrease the speed of response of the instruments. Which of the following is most likely to help attain this aim?

(A) Keep the capacitance in the circuits small.
(B) Keep the resistance in the circuits large.
(C) Keep the voltages used large.
(D) Keep the Joule heating small.
(E) Work in a room with very low humidity.

27-46 Q In a lecture demonstration intended to show the time needed for charging and discharging a capacitor, the circuit shown in Figure 27-16 was used. The switch was thrown to connect the capacitor to the dc power supply for charging, and it was then thrown to discharge the capacitor. The time to discharge was observed to be less than the time for charging. How do you best explain this result?

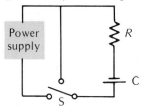

Figure 27-16. Question 27-46.

(A) Circuit analysis shows that it takes longer to charge a capacitor through a given resistor R than to discharge the capacitor through the same resistance.
(B) The potential difference across the capacitor aided the discharging, whereas it opposed the charging.
(C) The power supply had appreciable internal resistance.
(D) In discharge, the potential difference across R added to the potential difference across the capacitor, whereas this was not the case during charging.
(E) The capacitor was not fully charged when the switch was thrown; hence one would not expect the charging and discharging times to be equal.

27-47 P A parallel-plate capacitor is filled with mica and has the following characteristics:

Plate separation $d = 2 \times 10^{-2}$ cm
Plate area = 200 cm²
Dielectric constant = 6
Resistivity of mica = 1.7×10^{12} $\Omega \cdot$m

(a) What is the capacitance of the capacitor? (b) What is the leakage resistor of the capacitor? (c) How long will it

Figure 27-15. Question 27-44.

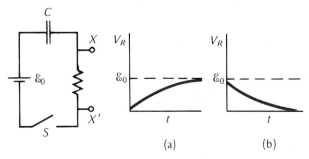

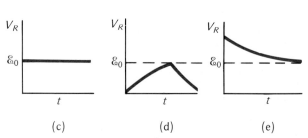

(a)   (b)   (c)   (d)   (e)

take for the voltage on the capacitor to drop to half its initial value?

: **27-48 P** A 4-μF capacitor is charged to 100 V. After 24 h, it is found that the voltage has decreased to 98 V. (a) What is the leakage resistance of the capacitor? (b) What will the voltage be 72 h after it has first been charged?

: **27-49 P** The region between the plates of a parallel-plate capacitor is filled completely with a material having resistivity ρ and dielectric constant κ. Show that the time constant for this capacitor is given by $(\rho\epsilon)^{1/2} = (\rho\kappa\epsilon_0)^{1/2}$, which is independent of the dimensions of the capacitor plates. The result is, in fact, a general one: all capacitors filled between their plates have a time constant that depends only on the filling material's resistivity and dielectric constant.

: **27-50 P** Using energy conservation, show that the differential equation describing the time variation in the energy U_C of a charged capacitor in an RC circuit is given by $dU_C/dt = i^2R$.

: **27-51 P** Show that when a charged capacitor is discharged through a resistor, the total energy initially stored in the capacitor, $Q^2/2C$, equals the energy dissipated in the resistor, $\int_0^\infty i^2R\,dt$.

: **27-52 P** A neon lamp has the following interesting property, which makes it useful in electronic circuits. The lamp consists of a small glass envelope filled with neon gas. There are two electrodes in the envelope. If the voltage applied to the electrodes exceeds 70 V, the neon becomes ionized and the resistance between the electrodes drops essentially to zero. The gas discharge is evident from the orange light given off. If the voltage drops below 70 V, the discharge stops in a few milliseconds and the tube is again an insulator. (Assume that the few milliseconds after the voltage drops below 70 V is enough time for the capacitor to be discharged completely, inasmuch as the circuit's resistance is then effectively zero.)

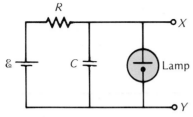

Figure 27-17. Problem 27-52.

(a) Determine the periodic frequency of the voltage between terminals X and Y for the *relaxation oscillator* shown in Figure 27-17. $V = 100$ V; $R = 100$ KΩ; $C = 2$ μF. (b) Sketch the voltage between X and Y as a function of time.

Supplementary Problems

27-53 P A common type of memory unit in a computer chip is a capacitor, whose state of charge corresponds to one bit of information. See supplementary problem 26-57. Even though the silicon dioxide between the plates of a memory capacitor is an excellent insulator, the charge on the plates does leak away, and it is therefore necessary to regenerate, or "refresh," the stored charge periodically. Indeed, the recharging must be done every 2 or 3 milliseconds in a typical microcomputer chip if the information stored in memory is not to be lost. Given that the dielectric constant of silicon dioxide is 12, compute the order of magnitude of the resistivity of silicon dioxide. (The result given in Problem 27-49 may be useful.)

27-54 P A diode is a device that allows current flow in only one direction. The circuit shown in Figure 27-18 is a *staircase generator*. A negative square pulse is applied to the input. This allows a current to flow through diode D_1, charging C_1. When the pulse ends, the positive charge that was on the right plate of C_1 can pass through D_2 and accumulate on C_2. The process is then repeated. Again C_1 is charged, and when the voltage drops again to zero, once more charge is passed on from C_1 to C_2.

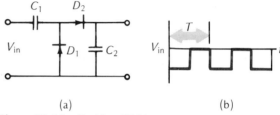

Figure 27-18. Problem 27-54.

(a) Plot output voltage vs time for $C_1 = 0.1$ μF, $C_2 = 1$ μF, $V_{in} = -2$ mV, $T = 1$ ms. From your result you will see why this is called a staircase generator. (b) If an ammeter is connected across the output terminals, a current I will flow in it, where $I = Q_2 f$. Here Q_2 is the charge that accumulates on C_2 every time a pulse of voltage V_{in} is applied to the input terminals, and f is the frequency of the input pulses. In this form, the circuit is a rudimentary frequency meter. Show that the frequency of the input pulses is $f = I/C_1 V_{in}$.

27-55 P An incandescent lamp uses a tungsten filament heated to a temperature of about 3000°C. A lamp operated with voltage V dissipates power P as radiated light and infrared radiation, where P is related to the Kelvin temperature T of the filament by $P = \sigma T^4$ (σ is a known constant). The resistance varies with temperature as $R = R_{29}(1 + \alpha\,\Delta T)$. Note that $\Delta T \approx T$ since the operating temperature of about 3000 K is so high. For tungsten $\alpha = 4.5 \times 10^{-7}$ per Celsius degree.

(a) What is the ratio of the filament resistance at 3000 K to the resistance at room temperature? (b) Show that the temperature T varies with the voltage V as $T \propto V^{2/5}$. (c) Show that the power varies with V as $P \propto V^{8/5}$ approximately. (d) How much does the power output increase if the voltage is increased by 10 percent?

28 DC Circuits

28-1 EMF
28-2 Single-Loop Circuits
28-3 Resistors in Series and Parallel
28-4 DC Circuit Instruments
28-5 Multiloop Circuits
Summary

28-1 EMF

Consider the simple dc circuit of Figure 28-1: resistor R connected across battery terminals. The positive battery terminal is a and the negative terminal b. Potential V_a is higher than V_b. The battery maintains the constant potential difference $V_a - V_b = V_{ab}$ across its terminals.

Positive charges enter the resistor at a, go downhill in potential, and leave at b. (Strictly, electrons enter the resistor at b, go uphill in potential, and leave at a.) The connecting wires have so small a resistance that the potential drop across them is negligible compared with that across R. Every time charge leaves any one point in the circuit, it is immediately replaced by other charge of the same magnitude. Consequently, there is no net change in charge at any point in the circuit, and each point in the circuit maintains a constant electrostatic potential.

Suppose that we follow a positive charge from a, through the resistor to b, and then through the battery back to a. From a to b the charge is acted on by an electric field within the conductor. It thereby gains energy, but because of frequent inelastic collisions with lattice atoms, it disposes of this electric energy as thermal energy in the conductor and emerges at point b with exactly the same energy it had at a. The charge now enters the negative battery terminal

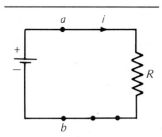

Figure 28-1. A simple dc circuit.

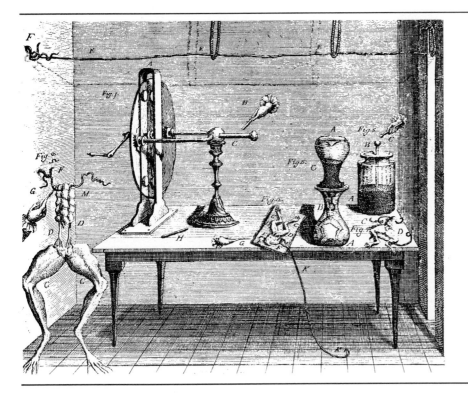

Figure 28-2. Luigi Galvani (1737–1798) first observed the existence of current through the nerve of a frog's leg. This engraving shows the frog legs hanging by the nerves (at left). When the electrostatic machine on the table revolved, or the Leyden jar at the right-hand end of the table was discharged, Galvani observed that the legs jerked when a scalpel touched the nerve. Current electricity, also once called "animal" electricity, was first thought to be a phenomenon entirely separate from static electricity.

and is transported through the battery back to the positive battery terminal; that is, the positive charge must move from a point at the potential V_b to the higher potential V_a. Clearly, this is not possible if the only force acting on the charge, as it passes through the interior of the battery, is the electric force derived from the potential difference (the force associated with the positive and negative charges residing at the battery terminals). There must exist, in addition to this electric force, a force not derived from the electrostatic charge distribution, which somehow drives the positive charge *uphill* in potential from V_b to V_a. In an electrochemical cell, or battery, the force that drives the charges against the opposing electrostatic force is sometimes called the *chemical force*.

When any two dissimilar metals are immersed in a conducting medium, a potential difference is found to exist between them; this is the most rudimentary form of an electrochemical cell.* The chemical reactions that are the origin of this potential difference and the details of what goes on within a battery lie in the area of electrochemistry and will not be dealt with here. Suffice it to say, however, that the potential difference has its origin in the differences in the binding energy of electrons to atoms of different types. In this sense, the nonelectrical, or chemical, forces are ultimately electrical in origin, while still being non-electrostatic. See Figure 28-2.

* The term *battery* means, strictly, a battery of electrochemical cells; for example, a 12-V automobile battery consists of six 2-V cells.

We may characterize any energy source that is capable of driving charges around a circuit against opposing potential differences as an *emf* (pronounced "ee-em-eff"). This is the abbreviation for *electromotive force*, a term so misleading (emf is *not* a force) that we shall hereafter refer to it simply as the emf, or symbolically, as $\mathcal{E}$.* By the $\mathcal{E}$ of a battery is meant the energy per unit positive charge gained (or the work per unit charge done by the chemical forces within a battery) in the transfer of a charge within the battery from the negative to the positive terminal:

$\mathcal{E}$ = work per unit positive charge done by the *chemical forces*

Since energy per unit charge, or joules per coulomb, is equivalent to volts, an emf, like a potential difference, is expressed in volts.

Some energy is always dissipated within any actual battery. One accounts for this energy loss by ascribing an *internal resistance r* to the battery, where the rate at which energy is dissipated within the battery is i^2r. A battery with emf $\mathcal{E}$ and internal resistance r, connected across a load resistor R, is represented by the circuit diagram shown in Figure 28-3. Since the emf is by definition the energy per unit charge gained from the battery, the time rate at which nonelectric chemical potential energy is transformed into electric energy, or the power associated with the emf, is $\mathcal{E}i$. Some of the electric energy is dissipated to thermal energy within the battery at the rate i^2r; the remaining energy is delivered to the load resistance at the rate $iV_{ab} = i^2R$. Therefore, energy conservation gives

$$\mathcal{E}i = i^2r + i^2R = i(ir + iR)$$

$$\mathcal{E} = i(r + R)$$

Since iR is the potential difference V_{ab} across the load resistor and also across the battery terminals, the equation above can be written

$$V_{ab} = \mathcal{E} - ir \tag{28-1}$$

Thus, the potential difference across the battery terminals is the battery's emf less the potential drop across its internal resistance. The potential difference V_{ab} across the battery terminals will equal the emf $\mathcal{E}$ only if the current i is zero, but as we see from (28-2), this will happen only when the load resistor R is infinite. Said differently, the potential difference appearing across the battery terminals on *open circuit* ($R = \infty$) is the battery's emf. A *short circuit* corresponds to $R = 0$.

The emf of a particular type of cell depends only on the chemical identity of its parts. As a battery ages or loses its "charge" (strictly, a battery loses only its internal chemical potential energy), the internal resistance increases but the emf remains unchanged. When the battery is "discharged," V_{ab} goes to zero as r increases, even for a relatively small load resistance R.

An electrochemical cell, or battery, is but one example of a device characterized by an emf. Other arrangements that can convert nonelectric energy to electric energy are these:

- A *thermocouple* consisting of two dissimilar conducting wires connected together in a circuit, with the two junctions maintained at different tempera-

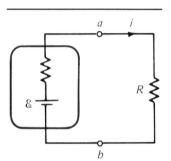

Figure 28-3. The color line shows what is within the battery: an emf and an internal resistance.

---

* The term *electromotance* is sometimes used for this quantity.

tures. An electric current then exists in such a circuit, driven by a *thermoelectric emf*.

- A *photovoltaic cell*, exemplified by a photographic exposure meter, where visible light strikes a sensitive material and generates an electric current.
- An electric generator, with mechanical energy transformed into electric energy.
- Electromagnetic induction effects, where an emf is produced by a changing magnetic flux (to be treated in Chapter 31).

Example 28-1. A transistor radio runs on a battery rated 9 V and 500 mA·h, and costing $2. (a) What is the battery's energy content? (b) What does this energy cost in cents per kilowatt-hour?

(a) The quantity 500 mA·h is electric *charge*, the product of current and time. The total electric energy delivered by the electrochemical cell is

$$\mathcal{E}Q = (9 \text{ V})(500 \text{ mA} \cdot \text{h}) = 4.5 \text{ W} \cdot \text{h} = 4.5 \times 10^{-3} \text{ kW} \cdot \text{h}$$

Not all this energy is delivered to the load, however. As the battery ages and its internal resistance increases, an increasingly larger fraction of the total energy is dissipated within the battery. (A fresh battery might have an internal resistance much less than 1 Ω. After a typical zinc-carbon battery has been connected to a 1000-Ω load for 4 h per day for two weeks, its internal resistance has grown to about 1000 Ω, and the potential drop across its terminals has dropped to half the initial value.)

(b) The total energy cost, only a portion of which goes to the load, is

$$\frac{\$2}{4.5 \times 10^{-3} \text{ kW} \cdot \text{h}} = 4.4 \times 10^4 \text{ ¢/kW} \cdot \text{h}$$

Compare this with a typical cost of 10¢/kW·h for household electricity. One pays dearly for the convenience of portable batteries.

28-2 Single-Loop Circuits

We wish to find the general relation between the emf's in a circuit and the potential differences. Consider, as a specific example, the circuit of Figure 28-4, which contains two batteries with emf's $\mathcal{E}_1$ and $\mathcal{E}_2$ and three resistors r_1, r_2, and R connected in series (r_1 is the internal resistance of $\mathcal{E}_1$ and r_2 of $\mathcal{E}_2$). The arrows associated with the two batteries, pointing from the negative to the positive terminal in both cases, indicate the directions of the electric field associated with the nonelectrostatic forces; they give, so to speak, the "directions" of the emf. The direction, or sense, of the current i is precisely the same at each point in the circuit loop because of electric-charge conservation, which implies that electric charge does not pile up or become depleted at any one point in the circuit.

Consider how energy conservation applies to the circuit. Proceeding in a clockwise direction around the loop, in the direction chosen for current i, we simply match the total energy delivered *by* the energy sources with the total energy delivered *to* the circuit elements (here, resistors). The power delivered *by* the batteries is

$$\mathcal{E}_1 i + \mathcal{E}_2 i = i\Sigma\mathcal{E}$$

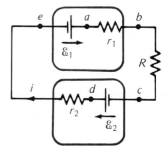

Figure 28-4.

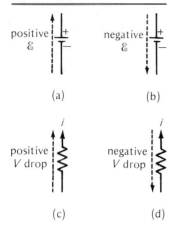

Figure 2-5. Sign conventions for emf's and potential drops. The dashed arrows indicate the direction in which the circuit element is traversed. (a) The emf is taken as positive if the battery is traversed from the negative to the positive terminal. (b) The emf is negative if traversed from the positive to the negative terminal. (c) The potential drop is positive if the resistor is traversed in the same direction as the current. (d) The potential drop is negative if the resistor is traversed in the direction opposite to that of the current.

We take both terms on the left side to be positive, since the directions of $\mathcal{E}$ and i are the same for both batteries. The total power delivered *to* the circuit elements is

$$iV_{ab} + iV_{bc} + iV_{de} = i\Sigma V$$

Note especially that ΣV includes only the potential differences across circuit elements, not the potential rises across the batteries.

Equating power delivered by the energy sources to the power delivered to circuit elements, we have

$$\Sigma \mathcal{E} = \Sigma V \qquad (28\text{-}2)$$

This is the fundamental relation for solving single-loop dc circuits. One must be careful about signs in applying this relation. Note that $\Sigma \mathcal{E}$ is the *algebraic* sum of the emf's in the circuit. An emf is taken as positive if the battery is transversed from the negative to the positive terminal; but the emf must be taken as negative if the battery is traversed in the other direction. See Figures 28-5(a) and (b). Similarly, ΣV is the *algebraic* sum of potential *drops* across circuit elements. Any V is taken as positive if we traverse the circuit element from one potential to a lower potential in the direction of the conventional current; V is negative if we traverse the circuit element in the direction opposite that of the current. See Figures 28-5(c) and (d).

Example 28-2. Two batteries are connected in opposition, as shown in Figure 28-6; the emf's and internal resistances are 18.0 V and 2.0 Ω, 6.0 V and 1.0 Ω, respectively. (a) What is the current in the circuit? (b) What is the potential difference across the battery terminals? (c) At what rate does the discharging battery charge the charging battery? (d) At what rate is energy dissipated as heat in the 6.0-V battery?

(a) We decide to traverse the circuit of Figure 28-6 in the *clockwise* sense. Moreover, we take this clockwise sense as the direction of the current i, knowing that the current will be clockwise, since its direction will be controlled by the emf of the larger, 18.0-V battery. But it is not necessary to do so. We may choose the current direction arbitrarily, and then if it is not correct, the current will appear *negative* in the solution; that is, the current will have been shown actually to exist in the sense opposite to that chosen initially.

With the circuit traversal and current directions taken as shown in Figure 28-6, we have

$$\Sigma \mathcal{E} = \Sigma V$$
$$18V - 6V = 2i + 1i$$

Note that the emf of the 6-V battery must be assigned a minus sign, since we pass through this battery from the positive to the negative terminal. The potential drops across the two resistors, $2i$ and $1i$, are both positive, since we traverse each resistor in the same direction as the current. Solving for the current i in the relation above gives

$$i = \frac{12.0 \text{ V}}{3.0 \text{ Ω}} = 4.0 \text{ A}$$

(b) Within the 18.0-V battery there is a potential drop across the 2-Ω internal resistance of

$$V = iR = (4.0 \text{ A})(2.0 \text{ Ω}) = 8.0 \text{ V}$$

Therefore, across the 18.0-V battery terminals the potential difference is 18.0 V − 8.0 V = 10.0 V.

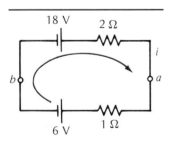

Figure 28-6.

If a 10.0-V potential difference exists across the 18-V battery terminals, this same potential difference must also appear across the 6.0-V battery terminals. Along the lower path from a to b in Figure 28-6, we have a potential drop (4.0 A)(1.0 Ω) = 4.0 V across the internal resistance and a *rise* of 6.0 V from the emf, giving a total potential drop of (4.0 + 6.0)V = 10.0 V.

(c) The rate at which any emf delivers energy is $\mathcal{E}i$. If $\mathcal{E}i$ is positive—that is, if the emf and current are both in the same direction—chemical energy from the battery is delivered *to* other circuit elements. On the other hand, if $\mathcal{E}$ and i are of opposite sign, as for the 6-V battery here, energy is delivered by other sources *to* it. Thus, the 18-V battery delivers energy (that is, loses chemical potential energy) at the rate

$$\mathcal{E}i = (18.0 \text{ V})(4.0 \text{ A}) = 72 \text{ W}$$

The 6-V battery has energy delivered to it (that is, it gains chemical potential energy) at the rate $\mathcal{E}i = (6.0 \text{ V})(4.0 \text{ A}) = 24 \text{ W}$. The battery with the larger emf is being discharged at the rate of 72 W, while the smaller battery is being charged at the rate of 24 W. What happens to the difference, 72 W − 24 W = 48 W? It must be dissipated in the two resistors.

(d) The rate at which energy is dissipated within the 6-V battery is

$$P = i^2R = (4.0 \text{ A})^2(1.0 \text{ Ω}) = 16 \text{ W}$$

and the rate at which energy is dissipated within the 12-V battery is $P = i^2R = (4.0 \text{ A})^2(2.0 \text{ Ω}) = 32 \text{ W}$. The total power dissipated in the resistors, 16 W + 32 W = 48 W, is just the difference between the power delivered *by* the discharging battery (72 W) and the power delivered *to* the charging battery (24 W).

28-3 Resistors in Series and in Parallel

Series and parallel arrangements of circuit elements were defined in Figure 26-6.

Series Resistors What is the equivalent resistance of resistors in series? By the equivalent resistance is meant the resistance of that single resistor which, when it replaces the separate resistors, does *not* change the current drawn from the energy source.

Clearly, in Figure 28-7, the same current i passes through each resistor, so that

$$V = V_1 + V_2 + V_3$$
$$= iR_1 + iR_2 + iR_3 = i(R_1 + R_2 + R_3)$$

where V_1, V_2, and V_3 are the potential drops across resistors R_1, R_2, and R_3, and V is the potential difference across the battery terminals. The same current i will exist in the circuit when the single equivalent R replaces the group in series:

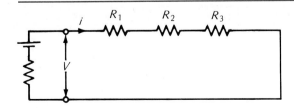

Figure 28-7. Resistors in series.

$$V = iR$$

Comparing the two equations above gives

$$\text{Resistors in series:} \quad R = R_1 + R_2 + R_3 = \Sigma R_i \quad (28\text{-}3)$$

The equivalent resistance is the sum of the separate resistances in series. The *same current* exists in each series resistor; the potential drop is the same across the entire group as across the single equivalent resistance.

Parallel Resistors Recall first that in a parallel arrangement of circuit elements [Figure 26-6(b)] a single conductor splits into two or more conductors, so that there are alternative ("parallel") routes between two points in a circuit. All the arrangements of resistors shown in Figure 28-8 are effectively the *same*.

We can see that the qualitative effect of adding resistors in parallel is to reduce the effective total resistance, since every additional parallel resistor provides an additional path for charges to flow from the input to the output location.

What is the equivalent resistance of resistors shown in Figure 28-8? We designate the potential drops across the resistors R_1, R_2, and R_3 as V_1, V_2, and V_3. The input current is i, and the currents through the resistors are i_1, i_2, and i_3. Current i is divided into three currents through the resistors:

$$i = i_1 + i_2 + i_3$$
$$= \frac{V_1}{R_1} + \frac{V_2}{R_2} + \frac{V_3}{R_3}$$

where $i_1 = V_1/R_1$, and so on.

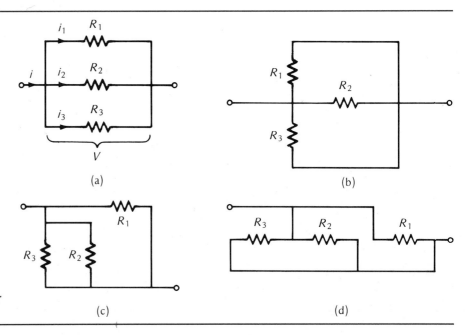

Figure 28-8. All four resistor circuits are the same circuit.

In traversing any loop in Figure 28-8(a), we see that the potential differences all have the *same* value V:

$$V = V_1 = V_2 = V_3$$

The equation above can then be written as

$$i = V\left[\frac{1}{R_1} + \frac{1}{R_2} + \frac{1}{R_3}\right]$$

A single equivalent resistor R replacing the parallel resistors must satisfy the relation

$$i = \frac{V}{R}$$

Comparing the last two equations gives, finally,

Resistors in parallel: $\quad \dfrac{1}{R} = \dfrac{1}{R_1} + \dfrac{1}{R_2} + \dfrac{1}{R_3} = \Sigma \dfrac{1}{R_i}\quad$ (28-4)

The reciprocal of the equivalent resistance equals the sum of the reciprocals of the separate parallel resistances. The equivalent resistance is always less than the smallest of the parallel resistances. The essential feature of parallel connections is that all elements have the *same potential difference*. It follows that for any two resistors in parallel, the ratio of the currents is inverse to the ratio of the respective resistances. For example, with 1.0- and 3.0-Ω resistors in parallel, the 1.0-Ω resistor always has three times the current of the 3.0-Ω resistor, which ensures that the potential difference across both is always the same.

Example 28-3. Consider the circuit of Figure 28-9. Find (a) the current in the battery, (b) the current in the 3.0-Ω resistor, (c) the potential difference across the 6.0-Ω

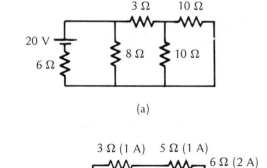

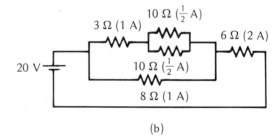

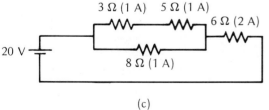

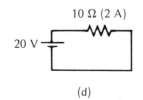

Figure 28-9. A circuit (a) and its evolution into progressively simpler equivalent forms (b) to (d). The currents in the various resistors (shown in parentheses) are found by starting with part (d) and working backward to (a).

resistor, and (d) the rate at which thermal energy is dissipated in the 8.0-Ω resistor. The currents shown in parentheses are not given, but calculated.

We first recognize that any of the questions that can be asked about a circuit like that in Figure 28-9(a) (potential differences, power dissipation, and the like) require that we first find the current through each resistor. Other quantities then can be easily computed. To find the current through each resistor, we must first reduce the complex of resistors, through the rules for combining resistors in series and in parallel, until we are left with a single equivalent resistor connected across the battery.

Figures 28-9(b) to (d) show the evolution of the circuit, Figure 28-9(a), into progressively simpler forms. (The numerical values of the resistances have been so chosen here that the computations can, without difficulty, be carried out in one's head.) We see that the current through the single equivalent resistor is 2 A. Now we work backward, through Figures 28-9(c), (b), and (a), in turn, to find the current in each resistor, Here we use the facts that the current through all series resistors is the same, and that the potential difference across all parallel resistors is the same. Of course, as soon as we find the current through a given resistor, we can immediately compute the potential drop iR across it and the power $i^2R = iV$ dissipated in it. In this way we have, finally:

(a) Current through battery = 2 A.
(b) Current through 3-Ω resistor = 1 A.
(c) Potential difference across 6-Ω resistor = 12 V.
(d) Power dissipated in the 8-Ω resistor = 8 W.

28-4 DC Circuit Instruments

Here we consider the essential features of four instruments common in dc circuits:

- Ammeter, for measuring electric current.
- Voltmeter, for measuring electric potential difference.
- Wheatstone bridge, for comparing resistances.
- Potentiometer, for comparing emf's.

Ammeter and Voltmeter Let us be clear on how the ammeter and voltmeter are connected in a circuit to measure current and potential difference. Figure 28-10(a) shows a simple circuit of a battery and resistor with a current i. When an ammeter (symbolized by —(A)—) is connected in series with the resistor to measure the current, as in Figure 28-10(b), it changes the circuit so that the measured current i' is less than i, the original current. If the resistance R_a of the ammeter is *much less* than R, then $i' \simeq i$. In Figure 28-10(c), a voltmeter (symbolized by —(V)—) is connected in parallel across the resistor. Again the current is changed. The current i'' through the battery is now greater than i. Only if the voltmeter resistance is *large* compared with R will $i'' \simeq i$.

Ideally, no current or potential difference is altered by connecting ammeters or voltmeters in a circuit; an ammeter resistance is so low that the potential difference across it is negligible, and a voltmeter resistance is so high that the current through it is negligible. Actually, any measuring instrument, electrical or otherwise, interferes somewhat with the quantity to be measured. In designing an ammeter or voltmeter, we want to minimize these perturbing effects, or at least to be aware of them.

Ordinary ammeters and voltmeters both use a pivoted needle whose angular position registers the current or potential difference. The needle is attached to a coil through which current passes, and this coil is immersed in the magnetic field of a magnet. Such a device is known as a *galvanometer* (an instrument we shall discuss in more detail in Section 29-7). A galvanometer (symbolized by —Ⓖ—) registers small currents. For example, a galvanometer showing full-scale deflection for a current of 1.0×10^{-6} A = 1.0 μA and having a resistance of 1,000 Ω might be typical (very much more sensitive galvanometers give full-scale deflection for currents as small as 10^{-12} A). Such a galvanometer, when placed into a circuit, will certainly register a full-scale deflection when 1.0 μA passes through it; at the same time, there will exist a potential difference across its terminals

$$V = iR = (1.0 \times 10^{-6} \text{ A})(1.0 \times 10^3 \, \Omega) = 1.0 \times 10^{-3} \text{ V} = 1.0 \text{ mV}$$

If this 1.0-mV potential difference is small compared with potential differences existing across other circuit elements, the galvanometer's influence is negligible.

By itself, such a galvanometer could be used as a microammeter to measure currents up to 1 μA, or as a millivoltmeter to measure potential differences up to 1 mV.

Suppose, however, that we are to construct an ammeter that registers 1.0 A full-scale, using this galvanometer. To do this, we connect a very small resistance R_p in parallel with the galvanometer, as shown in Figure 28-11; said differently, one "shunts" the galvanometer with a small resistance R_p. What is R_p? Nearly all the current, 1.0 A, through the ammeter will go through the shunt resistance R_p since only 1.0 μA is permitted through the galvanometer itself. Moreover, the potential differences across the galvanometer and its shunt must both be 1.0 mV at full-scale deflection. It follows that

$$R_p = \frac{V}{i} = \frac{1.0 \times 10^{-3} \text{ V}}{1.0 \text{ A}} = 1.0 \times 10^{-3} \, \Omega$$

The total resistance of the ammeter is also close to 10^{-3} Ω.

Suppose, now, that we use this same galvanometer to construct a voltmeter registering 10 V full-scale. To construct a voltmeter from a galvanometer, we place a high resistance R_s in series with it, as shown in Figure 28-12. Since the galvanometer alone can have a potential difference of only 1.0×10^{-3} V across its terminals, the potential difference V across the resistor R_s must be essentially 10 V, when the current i through it is 1.0×10^{-6} A. Therefore,

$$R_s = \frac{V}{i} = \frac{10 \text{ V}}{1.0 \times 10^{-6} \text{A}} = 1.0 \times 10^7 \, \Omega = 10 \text{ M}\Omega$$

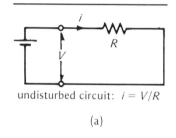

undisturbed circuit: $i = V/R$

(a)

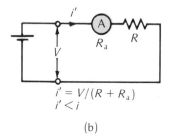

$i' = V/(R + R_a)$
$i' < i$

(b)

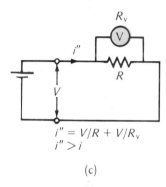

$i'' = V/R + V/R_v$
$i'' > i$

(c)

Figure 28-10. A circuit (a) undisturbed, (b) with an ammeter in series with the resistor, and (c) with a voltmeter in parallel with the resistor.

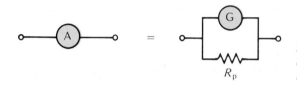

Figure 28-11. An ammeter is a galvanometer in parallel with a small resistance.

Figure 28-12. A voltmeter is a galvanometer in series with a large resistance.

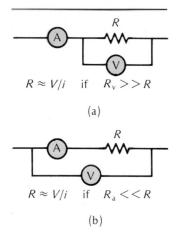

$R \approx V/i$ if $R_v \gg R$

(a)

$R \approx V/i$ if $R_a \ll R$

(b)

Figure 28-13. Two arrangements for measuring the resistance R with an ammeter and a voltmeter: (a) voltmeter across R; (b) ammeter in series with R.

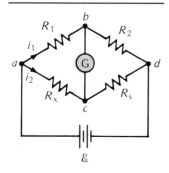

Figure 28-14. A Wheatstone bridge circuit.

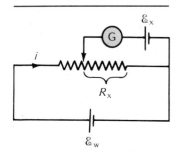

Figure 28-15. A potentiometer circuit for comparing emf's.

The voltmeter has so large an internal resistance that it produces little perturbation of any circuit with elements whose resistance is small compared with 10 MΩ.

A simple way of measuring resistance is shown in Figure 28-13(a); one measures the current i through the device with an ammeter and the potential difference V across it with a voltmeter, and then applies Ohm's law, $R = V/i$. Strictly, however, the current measurement is too high, since the ammeter of Figure 28-13(a) registers the current through both the resistor and the voltmeter. You could correct this by placing the ammeter inside the connections to the voltmeter, as shown in Figure 28-13(b). The ammeter then reads the current through R alone, but the voltmeter reads the potential difference, not across R alone, but across the resistance and ammeter.

Wheatstone Bridge The difficulties mentioned above are eliminated when resistance is measured, or more properly compared, with a known standard resistance R_s, using the bridge circuit devised by C. Wheatstone (1802–1875). See Figure 28-14. In its simplest form, the circuit consists of four resistors, a battery, and a sensitive galvanometer. The values of R_1, R_2, and R_s are all known; R_x is the unknown resistance. Like the ordinary beam balance, which indicates equal masses on its two pans when the needle shows no deflection from the vertical, the Wheatstone bridge is a *null instrument*. With a given unknown resistance R_x, the resistors R_1, R_2, and R_s are so adjusted that the galvanometer registers no current. Then points b and c are at the same potential. The current i_1 through resistor R_1 is the same as the current through R_2; likewise, R_x and R_s carry the current i_2. The potential differences V_{ab} and V_{ac} are equal, as are V_{bd} and V_{cd}. It follows that

$$V_{ab} = i_1 R_1 = V_{ac} = i_2 R_x$$
$$V_{bd} = i_1 R_2 = V_{cd} = i_2 R_s$$

Eliminating i_1 and i_2 from these relations yields

$$\frac{R_x}{R_s} = \frac{R_1}{R_2} \tag{28-5}$$

With the bridge balanced, the unknown resistance R_x is computed in terms of the standard resistance R_s and the ratio R_1/R_2, using (28-5).

Potentiometer Suppose that the emf of a battery is to be measured. A voltmeter placed across the battery terminals reads the *approximate* emf. The voltmeter reading V is always less than the true emf, since the potential difference appearing across a battery's terminals is the emf $\mathcal{E}$ less the potential drop ir across the battery's internal resistance r:

$$V = \mathcal{E} - ir \tag{28-1}$$

The emf $\mathcal{E}$ and potential difference V can be the same only if no current passes through the battery.

The potentiometer circuit permits the emf of a battery to be measured under the condition that the battery current is actually zero. Strictly, the potentiometer permits an unknown emf $\mathcal{E}_x$ to be compared with the precisely known emf $\mathcal{E}_s$ of a *standard cell* by a *null* method. The circuit is shown in Figure 28-15. A so-called working battery with emf $\mathcal{E}_w$, which need not be known but must be greater than $\mathcal{E}_x$ (or $\mathcal{E}_s$), maintains a constant current i through the resistor. The adjustable tap (see Figure 28-15 where the left end of the galvanometer wire connects to the resistor) is set so that the current through the sensitive galvanometer is zero. The potential drop across the galvanometer must be zero, and the potential drop across the internal resistance of the unknown battery is then also zero. Thus, the total potential drop across the branch in the circuit containing the unknown battery equals the battery's emf $\mathcal{E}_x$. But this is also the potential drop iR_x across the resistor from the tap to the right end. Therefore, $\mathcal{E}_x = iR_x$.

Now if the unknown battery is replaced by the standard cell and the resistor tap is again adjusted for balance (zero galvanometer current), we have $\mathcal{E}_s = iR_s$. Here R_s is the corresponding resistance from the tap to the end of the resistor and, because there is no current through the galvanometer, the current is still the same i. Eliminating i from these relations yields

$$\frac{\mathcal{E}_x}{\mathcal{E}_s} = \frac{R_x}{R_s} \qquad (28\text{-}6)$$

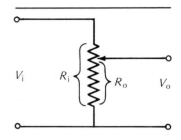

Figure 28-16. A potentiometer used as a voltage divider.

From this equation we see that comparing emf's with a potentiometer circuit consists in comparing resistances (or if the adjustable resistor is a wire of uniform cross section, in comparing lengths).

The term *potentiometer* is used in another sense in electric circuits—that of a *voltage divider*; see Figure 28-16. An input voltage V_i is applied across a variable resistor of total resistance R_i with a center tap. The output voltage is V_o, and R_o represents the resistance between the tap and the lower end of the resistor. Clearly, for negligible current in the output circuit, $V_o/V_i = R_o/R_i$.

28-5 Multiloop Circuits

Some circuits with more than one current loop, such as the one shown in Figure 28-9, can be solved simply by applying the rules for combining resistances in series and in parallel. Generally, however, this is not possible. Consider, for example, the relatively simple circuit shown in Figure 28-17. It is a circuit with *three* loops (a left inside loop, a right inside loop, and an outside loop going all the way around the circuit). There is no way to reduce this multiloop circuit into one involving a single battery and resistor.

One general procedure for solving multiloop circuit problems is expressed by *Kirchhoff's rules*. These two rules are simply statements, in the language of electric circuits, of the fundamental *conservation laws* of (1) *electric charge*, and (2) *energy*.

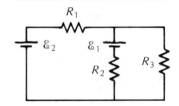

Figure 28-17. A simple multiloop circuit.

Junction Rule Consider a *junction*, a point in the circuit where three or more conducting wires are joined, as shown in Figure 28-18. Currents in the several wires are labeled $i_1, i_2, i_3,$ and i_4; current i_1 is into the junction and the other currents are out of the junction. Net charge cannot accumulate or be depleted at a junction. Therefore, from electric-charge conservation we know that the net charge per unit time into any junction must equal the net charge per unit time out of the junction. The net current into the junction equals the net current out of the junction. For the situation shown in Figure 28-18, this implies that

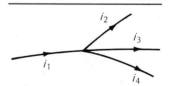

Figure 28-18. A circuit junction. The total current into the junction is zero.

$$i_1 = i_2 + i_3 + i_4 \quad \text{or} \quad i_1 - i_2 - i_3 - i_4 = 0$$

The negative terms appearing in the right-hand relation above may be interpreted as representing *negative* currents *into* the junction, which are equivalent to positive currents *out* of the junction. A general formulation of the junction theorem for currents is, then,

$$\Sigma i = 0 \tag{28-7}$$

where it is understood that currents into a junction are identified as positive and currents out of a junction as negative. The current rule is also known as *Kirchhoff's first rule*.

Loop Rule Recall the general relation

$$\Sigma \mathcal{E} = \Sigma V \tag{28-2}$$

which says that the algebraic sum of the emf's $\Sigma \mathcal{E}$ equals the algebraic sum of the potential drops ΣV around the loop. Equation (28-2) is merely an expression of the energy conservation law. The left-hand side represents the energy per unit charge supplied by energy sources in the circuit loop, and the right-hand side represents the energy per unit charge delivered to circuit elements around the loop. Equation (28-2) is known as the *loop equation*, or *Kirchhoff's second rule*.

Remember the conventions to be observed in applying (28-7) and (28-2):

- Having decided arbitrarily on the sense in which a particular loop will be traversed (clockwise or counterclockwise), we never reverse this sense through any circuit element around the loop.
- We choose the direction for any unknown current arbitrarily, but observe the sign convention denoting currents into or out of a junction in using (28-7).
- We take the emf of a battery to be *positive* when it is traversed from the *negative to the positive* terminal, and we take a potential drop across a resistor to be *positive* when the resistor is traversed in the *same* direction as that of *current* flow.

Example 28-4. Find the currents for the circuit of Figure 28-17, shown again in Figure 28-19. Currents in the *three branches* i_1, i_2, and i_3 are assigned directions arbitrarily. The three currents are the unknowns to be solved for. From (28-7), we have

$$i_1 - i_2 - i_3 = 0 \tag{A}$$

This is one equation in the three unknown currents. Applying (28-2) to each of the

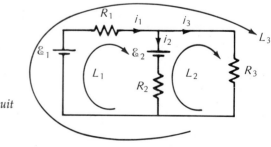

Figure 28-19. Three circuit loops (L_1, L_2, and L_3) for applying Kirchhoff's rules.

loops in Figure 28-19 gives three more equations; it is readily apparent, however, that one of these equations is not independent of the other two. A procedure for removing redundancies of this sort is to choose only loops in which one always goes in the same direction as the current. In Figure 28-19, for example, we can choose loops L_1 and L_3. Applied to loop L_1, Equation (28-2) yields

$$\mathcal{E}_1 - \mathcal{E}_2 = R_1 i_1 + R_2 i_2 \qquad \text{(B)}$$

Applied to loop L_3, it gives

$$\mathcal{E}_1 = R_1 i_1 + R_3 i_3 \qquad \text{(C)}$$

We now have three simultaneous linear equations (A, B, C) in the three unknowns (i_1, i_2, and i_3). Solving for the currents is straightforward. The results are

$$i_1 = \frac{R_3(\mathcal{E}_1 - \mathcal{E}_2) + R_2 \mathcal{E}_1}{R_1 R_2 + R_1 R_3 + R_2 R_3}$$

$$i_2 = \frac{R_3(\mathcal{E}_1 - \mathcal{E}_2) - R_1 \mathcal{E}_2}{R_1 R_2 + R_1 R_3 + R_2 R_3}$$

$$i_3 = \frac{R_2 \mathcal{E}_1 + R_1 \mathcal{E}_2}{R_1 R_2 + R_1 R_3 + R_2 R_3}$$

The equation above shows that current i_3 is always positive; it has the direction shown in Figure 28-19. On the other hand, i_1 and i_2 can be either negative or positive, depending on the relative sizes of the emf's and resistances. With negative current, the actual direction is opposite that chosen in Figure 28-19.

Solving any multiloop circuit, however complicated, means applying Kirchhoff's rules and solving some simultaneous linear equations. One can work this in reverse, using electric circuits to solve linear equations. The emf's and resistances in a circuit are made to correspond to the parameters of the linear equations; solving for the unknowns then consists merely in measuring the currents in the various branches of the circuit. This is one simple example of an *analog computer* in which one studies the physical behavior of a system obeying a well-known mathematical relationship, to solve for mathematical unknowns.

Summary

Definitions

Emf: the work done per unit positive charge by an energy source.

Important Results

Around any loop

$$\Sigma V = \Sigma \mathcal{E} \qquad (28\text{-}2)$$

The emf $\mathcal{E}$ is positive if the source is traversed from negative to positive terminal. The potential drop V across any circuit element is positive if the element is traversed in the same direction as the conventional current through it.

Resistors in *series*, all with the *same current*:

Equivalent single resistor $R = R_1 + R_2 + R_3 + \cdots$

$$(28\text{-}3)$$

Resistors in *parallel*, all with the *same potential difference*:

$$\frac{1}{R} = \frac{1}{R_1} + \frac{1}{R_2} + \frac{1}{R_3} + \cdots \qquad (28\text{-}4)$$

Ammeter: registers current, connected in series with circuit element, low resistance.

Voltmeter: registers potential difference, in parallel with circuit element, high resistance.

Wheatstone bridge: null instrument for comparing resistances.

Potentiometer: null instrument for comparing emf's.

Multiloop circuits (Kirchhoff's rules):

- Junction rule:

$$\Sigma i = 0 \qquad (28\text{-}7)$$

where i is positive into a junction.

- Loop rule:

$$\Sigma \mathcal{E} = \Sigma V \qquad (28\text{-}2)$$

with the conventions given above.

Problems and Questions

Section 28-1 EMF

· 28-1 P The manufacturer's specifications for a 12-volt storage battery rate the maximum current the battery can deliver as 80 A. What is the internal resistance of the battery?

· 28-2 P The internal resistance of a 1.5-V zinc-carbon battery is found to be 1.5 Ω. What is the maximum current it can deliver?

· 28-3 P A 6-V car battery provides a current of 4 A to a load for 8 min. By how much is the chemical energy of the battery decreased during this time?

· 28-4 P The open-circuit voltage of a particular solar cell is 2.2 V, and the short-circuit resistance (that is, a load resistor with $R_L = 0$) is 1.1 A. (a) What is the internal resistance of the battery? (b) What is the emf of the battery?

: 28-5 P Suppose that a load resistor R_L is connected across the terminals of a real battery. The power dissipated in the load resistor
(A) will be independent of the value of R_L.
(B) will increase when R_L is increased.
(C) will decrease when R_L is increased.
(D) may increase or decrease when R_L is increased, depending on the initial value of R_L.
(E) may decrease or stay the same, depending on the initial value of R_L.
(F) None of the above is true.

: 28-6 Q In some of our considerations, we have neglected any internal resistance of a source of emf, such as a battery. In fact, batteries have appreciable internal resistance. This internal resistance can be taken into account by including a resistor R_i in series with the battery, as shown in Figure 28-20. In this case, one could deduce that the potential difference between the battery terminals would
(A) be independent of the load resistance R_L.
(B) increase as the load resistance R_L is increased.
(C) decrease as the load resistance R_L is increased.
(D) either increase or decrease as the load resistance is varied, depending on whether the load resistance is larger or smaller than the internal resistance.
(E) drop to zero if the load resistance were to exceed the internal resistance.

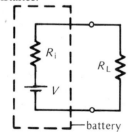

Figure 28-20. Question 28-6.

: 28-7 P When a resistor is connected to a battery, a current of 2.0 A exists. When an additional 15 Ω is added to the circuit in series, the current drops to 0.4 A. What is the emf of the battery?

: 28-8 Q A load resistance R_L is connected to a 6-V battery whose internal resistance is 0.1 Ω. The battery is rated at 120 A·h. Which of the following is the most accurate statement?
(A) The maximum current the battery can deliver is 30 A.
(B) The potential difference across the load will be a maximum when $R_L = 0.1$ Ω.
(C) The potential difference between the battery terminals will be a maximum when $R_L = 0$.
(D) The maximum power the battery can deliver to the load is 90 W.
(E) The maximum time for which the battery can deliver current without being recharged is 2 h.

Section 28-2 Single-Loop Circuits

· 28-9 P A battery with an emf of 12 V and an internal resistance of 1.0 Ω is connected to a load resistor of 2 Ω. How much power is delivered to the load?

· 28-10 P A 12-V car battery is rated at 200 A·h. If we assume that the potential difference stays constant and the internal resistance does not rise within the time for delivering the rated charge, for how many hours can the battery light a 50-W light bulb?

· 28-11 P Two batteries are connected as shown in Figure 28-21. (a) What current flows in each battery? Indicate the direction with an arrow. (b) At what rate is heat generated in the 6-Ω resistor?

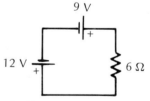

Figure 28-21. Problem 28-11.

· 28-12 P A current of 2 A flows in the 4-Ω resistor in

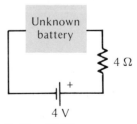

Figure 28-22. Problem 28-12.

Figure 28-22. What are the possible emf's of the unknown battery?

: **28-13 P** Two batteries are connected in series to a load resistor. A current of 4.0 A exists. The polarity of one battery is then reversed by reconnecting it, and the current drops to 1.0 A. One battery has an emf of 12 V. What is the emf of the other?

: **28-14 P** (a) Show that a battery of emf $\mathcal{E}$ and internal resistance r will deliver maximum power to a load resistor R_L when $R_L = r$. (b) Plot the power delivered to the load as a function of R_L/r.

: **28-15 P** Show that the resistance of the infinite network shown in Figure 28-23 is $(1 + \sqrt{3})R$.

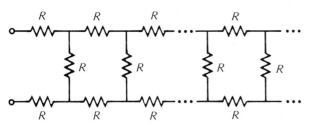

Figure 28-23. Problem 28-15.

Section 28-3 Resistors in Series and Parallel

· **28-16 P** Find the equivalent resistance of each combination of resistors shown in Figure 28-24.

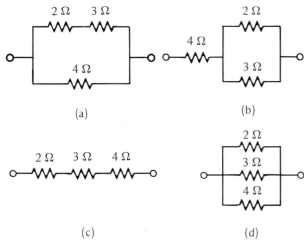

Figure 28-24. Problem 28-16.

· **28-17 Q** When several resistors are connected in parallel,
(A) the power dissipated in each is the same.
(B) the current through each is the same.
(C) the potential difference across each is the same.
(D) the net electric charge on each is the same.

· **28-18 P** What is the resistance between terminals X and Y for the network in Figure 28-25?

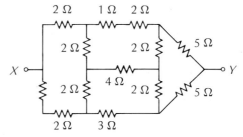

Figure 28-25. Problem 28-18.

· **28-19 P** What is the equivalent resistance of the network shown in Figure 28-26?

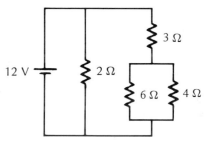

Figure 28-26. Problem 28-19.

· **28-20 P** What is the potential drop across the 4-Ω resistor in the circuit of Figure 28-27?

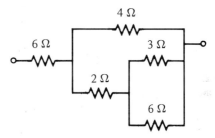

Figure 28-27. Problem 28-20.

· **28-21 P** When the switch in the circuit of Figure 28-28 is closed, what change will occur in the current in the 1-Ω resistor?

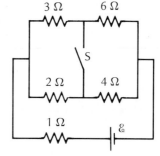

Figure 28-28. Problem 28-21.

(A) There will be no change.
(B) The current will increase.

(C) The current will decrease.
(D) Whether or not the current will increase depends on the value of the battery emf.

: **28-22 P** A car battery that is about to fail has an emf of 11.6 V and an internal resistance of 0.02 Ω. It is providing current to a 2-Ω load. A second battery, of emf 12.6 V and internal resistance 0.01 Ω, is connected in parallel to it to help it out. Determine (a) the circuit that describes the situation; (b) the current in each battery and in the load; (c) the total power delivered by the 12.6-V battery and the power delivered to the load; (d) the power delivered by the 11.6-V battery. (*Hint:* Pay attention to the sign of the power and the physical meaning of it.)

: **28-23 P** Four heating coils, each rated at 150 W at 120 V, are to be operated in combination from a 120-V supply. What different power outputs can be obtained by connecting them in various arrangements?

: **28-24 P** What is the electric potential at point P in the circuit of Figure 28-29?

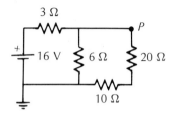

Figure 28-29. Problem 28-24.

: **28-25 Q** If the circuit were broken at point P in Figure 28-30,
(A) the current in R_1 would not change.
(B) the potential difference between point X and ground would increase.
(C) the current provided by the battery would increase.
(D) the emf provided by the battery (assumed to have no internal resistance) would change.
(E) the current in R_4 would increase.

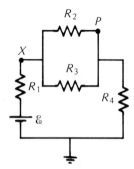

Figure 28-30. Problem 28-25.

: **28-26 Q** If the circuit shown in Figure 28-31 were broken at point P,
(A) the current in R_1 would not change.
(B) the current in R_2 would increase.
(C) the current in R_6 would decrease.
(D) the potential difference between G and P would decrease.
(E) the potential difference between G and Q would change.

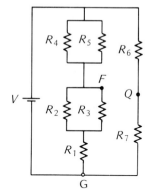

Figure 28-31. Question 28-26.

: **28-27 P** (a) Find the equivalent resistance of the network shown in Figure 28-32. (b) What potential difference applied to the terminals will result in a current of 1 A in the 6-Ω resistor?

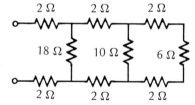

Figure 28-32. Problem 28-27.

: **28-28 P** A three-way light bulb has two filaments connected as shown in Figure 28-33. It is possible to apply 120 V across connections ab, ac, or bc. (a) What resistances are required for a 150 W-75 W-50 W lamp? (b) What power outputs would be obtained if $R_1 = 48$ Ω and $R_2 = 144$ Ω?

Figure 28-33. Problem 28-28.

: **28-29 P** Twelve identical resistors, each of resistance R, are connected to form a cube, with each resistor on an edge

(Figure 28-34). Determine the equivalent resistance between corners A and D.

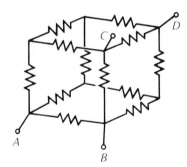

Figure 28-34. Problem 28-29.

: **28-30 Q** Sketched in Figure 28-35 is a circuit that has been designed so that the potential difference between points X and Y is 10 V. In trying to find the source of a difficulty that has arisen, you measure the potential between X and Y and find a difference of 15 V. You suspect that this is due either to a short circuit (from two bare wires touching) or to an open circuit (perhaps from a faulty solder joint). In the sketch, possible short circuits are indicated as dashed lines. Which of the following possibilities, if any, could account for your finding of $V_{XY} = 15$ V?
(A) Short circuit between T and Q.
(B) Short circuit between P and Z.
(C) Break in circuit at Q.
(D) Break in circuit at Z.
(E) Break in circuit at P.
(F) More than one of the above (indicate which).
(G) None of the above.

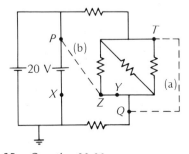

Figure 28-35. Question 28-30.

: **28-31 Q** Ordinary household light bulbs are meant to be connected in parallel to a 120-V supply. Suppose, however, that a 50-W lamp and a 100-W lamp are connected in series to a 120-V source. Which of the following statements best describes what will then happen?
(A) Both will be brighter than normal, with the 100-W lamp brighter than the 50-W lamp.
(B) Both will be dimmer than normal, with the 100-W lamp brighter than the 50-W lamp.
(C) The 100-W lamp will be brighter than normal, and the 50-W lamp will be dimmer than normal.
(D) The 100-W lamp will not be so bright as the 50-W lamp.
(E) Both lamps will still have their normal brightness.

: **28-32 P** What is the resistance of the network shown in Figure 28-36? (*Hint:* Use symmetry.)

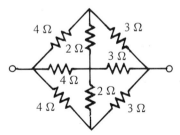

Figure 28-36. Problem 28-32.

: **28-33 P** Thermal energy is to be supplied to an experimental apparatus under changing conditions that may cause a change in the heater resistance R. It is desired to have the power dissipated in R independent (as much as possible) of the value of R. This can be achieved with the circuit shown in Figure 28-37. Determine the relation between R_1, R_2, and R if the power dissipated in R is independent, to first order in R, of R. The applied voltage V_0 is constant.

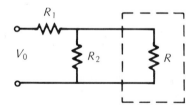

Figure 28-37. Problem 28-33.

Section 28-4 DC Circuit Instruments
· **28-34 P** The Wheatstone bridge shown in Figure 28-38 is balanced (there is no current in the galvanometer) when the variable resistance is 240 Ω. What is the value of the unknown resistance?

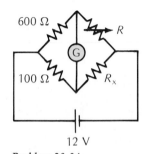

Figure 28-38. Problem 28-34.

· **28-35 Q** To say that a circuit is "grounded" at point P means that

(A) point P in the circuit is connected to the negative terminal of a battery.
(B) no current will flow through point P.
(C) point P is connected to a very large conductor, whose potential we arbitrarily choose to be zero volts.
(D) point P is electrically insulated from all other points, just as the earth is so insulated.
(E) point P is the point at which charge drains out of the circuit.

· **28-36 P** A galvanometer that has a resistance of 1000 Ω gives full-scale deflection for a current of 1.0 mA. (a) What shunt resistance is needed to construct an ammeter that deflects full-scale for a current of 100 A? (b) What length of No. 24 copper wire (diameter 0.0201 in., resistance 28.4 Ω per 1000 ft) would be needed to make the shunt resistor?

: **28-37 Q** In the circuit in Figure 28-39, the black box contains only resistors. If I make some changes inside the box and observe that the voltmeter reading decreases, then it is possible that I
(A) added some resistors.
(B) removed some resistors.
(C) changed resistors from parallel to series connection.
(D) decreased the current from the battery.
(E) All of the above are correct.

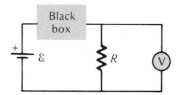

Figure 28-39. Question 28-37.

: **28-38 P** Shown in Figure 28-40 are two ways of measuring an unknown resistance R. If the resistance of the ammeter is 0.12 Ω and the resistance of the voltmeter 20 000 Ω, for what values of R will the resistance be given as V/i to within 1 percent under either arrangement?

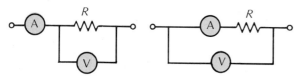

Figure 28-40. Problem 28-38.

: **28-39 P** A galvanometer has a resistance of 60 Ω and deflects full-scale for a current of 400 μA. Show how to use it to construct (a) a voltmeter that deflects full-scale for a voltage of 100 V; (b) an ammeter that deflects full-scale for a current of 5 A.

: **28-40 P** A galvanometer with a resistance of 800 Ω shows full-scale deflection when a current of 5 mA is passed through it. How would you use this galvanometer to construct (a) an ammeter registering 10-A full-scale deflection? (b) a voltmeter registering 500-V full-scale deflection?

: **28-41 P** The output of the voltage divider circuit shown in Figure 28-41 is to be measured with voltmeters of varying resistance. What will be the reading when the meter resistance is (a) 100 Ω? (b) 1 kΩ? (c) 50 kΩ? (d) 1 MΩ?

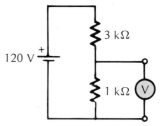

Figure 28-41. Problem 28-41.

: **28-42 P** A multirange ammeter is constructed using a galvanometer that gives a full-scale deflection when 1.0 mA flows in it (Figure 28-42). The galvanometer resistance is 1000 Ω. What are the values of the resistors (a) R_a? (b) R_b? (c) R_c?

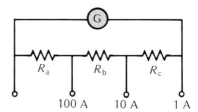

Figure 28-42. Problem 28-42.

: **28-43 P** A multirange voltmeter is constructed using a galvanometer that gives full-scale deflection for a current of 0.8 mA (Figure 28-43). The resistance of the galvanometer is 1200 Ω. What are the values of the resistors (a) R_a? (b) R_b? (c) R_c?

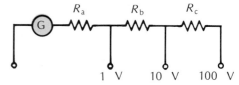

Figure 28-43. Problem 28-43.

: **28-44 P** Determine the ratio V/V_0 for the potential divider circuit shown in Figure 28-44.

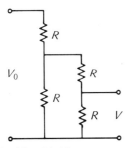

Figure 28-44. Problem 28-44.

28-45 Q If the circuit shown in Figure 28-45 is broken at point P,
(A) the power delivered by the battery will increase.
(B) the current in the ammeter, I, will increase.
(C) the current in the ammeter will not change.
(D) the potential difference read by the voltmeter, V, will increase.
(E) the potential difference read by the voltmeter will decrease.
(F) the potential difference read by the voltmeter will not change.
(G) The answer cannot be determined without knowing R_x.
(H) More than one of the above is true.

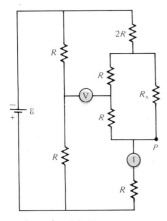

Figure 28-45. Question 28-45.

Section 28-5 Multiloop Circuit

28-46 Q For any network with current, it is always true that
(A) the sum of the currents around a closed loop is zero.
(B) the sum of the currents entering a junction equals the sum of the currents leaving the junction.
(C) the sum of the emf's around any closed loop is zero.
(D) the sum of the internal-resistance drops around any closed loop is zero.
(E) the total current, including all branches, is zero.

28-47 P What is the value of the emf $\mathcal{E}_1$ (Figure 28-46)?

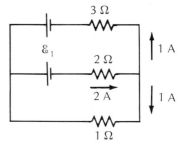

Figure 28-46. Problem 28-47.

28-48 P What is the potential difference between points X and Y in the circuit of Figure 28-47?

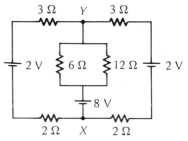

Figure 28-47. Problem 28-48.

28-49 P Suppose you have three 1.5-V flashlight batteries, each with internal resistance 1 Ω. What is the maximum power you can deliver to a load resistor, and how should the batteries be connected to do this when the load resistance has the value (a) 6 Ω? (b) 2 Ω? (c) 0.5 Ω?

Supplementary Problems

28-50 P A resistor network consists of an infinite square array of resistors, as shown in Figure 28-48. The resistance between any two adjacent junctions is R. What is the equivalent total resistance between two adjacent terminals, such as A and B in Figure 28-48? (*Hint:* the only easy way to solve this problem is to use two powerful ideas—symmetry and superposition. Superposition implies that the actual arrangement in which charge is injected into A and extracted

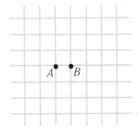

Figure 28-48. Problem 28-50.

from B may be considered as equivalent to taking the superposition of two situations: (a) injection of charge into A and collection at infinity and (b) extraction of charge from B with injection from infinity.)

28-51 P One of a pair of telephone wires is shorted out to ground in a storm. The pair of wires is 20 km long and runs through rough and inaccessible country. See Figure 28-49. The resistance of a single wire is 39 Ω/km. To determine the location of the short, the line crew must join the ends of the wires at one end of the line and establish a bridge circuit at the other. The bridge is balanced when $R_1 = 62\ \Omega$ and $R_2 = 86\ \Omega$. How far from point P has the short occurred?

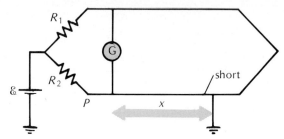

Figure 28-49. Problem 28-51.

The Magnetic Force

29

29-1 Magnetic Field Defined
29-2 Magnetic-Field Lines and Magnetic Flux
29-3 Charged Particle in a Uniform Magnetic Field
29-4 Charged Particles in Uniform **B** and **E** Fields
29-5 Magnetic Force on a Current-Carrying Conductor
29-6 The Hall Effect
29-7 Magnetic Torque on a Current Loop
29-8 Magnetic Dipole Moment
   Summary

The *fundamental* magnetic effect is this: An electric charge in motion produces a force on a second moving charge in addition to the electric (Coulomb) force. This velocity-dependent force between electrically charged particles is the *magnetic force*.

As we have seen, the electric force between charged particles may be thought to act through the intermediary of the electric field as follows:

- Charge q_a creates an electric field **E** at the site of charge q_b.
- Charge q_b, finding itself in field **E**, is acted on by an electric force.

In similar fashion, the magnetic interaction between charged particles is most readily described as taking place via a magnetic field:

- *Moving* charge q_a creates a *magnetic field* **B** at the site of moving charge q_b.
- *Moving* charge q_b, finding itself in field **B**, is acted on by a *magnetic force*.

The magnetic force is more complicated than the electric force since the magnetic force depends on the velocities of the two interacting charges as well as the charge magnitudes and signs and the charge separation distance. For this reason, it is useful to separate the magnetic interaction into two parts:

- One moving charge creates a magnetic field (the subject of Chapter 30).
- The magnetic field acts on a second moving charge (the subject of this chapter).

29-1 Magnetic Field Defined

The basic relation for the magnetic force $\mathbf{F}_m$ on a particle with charge q moving at velocity $\mathbf{v}$ in a magnetic field $\mathbf{B}$—the relation on which all else in this chapter is based—is this:

$$\mathbf{F}_m = q\mathbf{v} \times \mathbf{B} \qquad (29\text{-}1)$$

This equation involves the cross product (Section 12-2) of vectors $\mathbf{v}$ and $\mathbf{B}$; therefore, magnetic force $\mathbf{F}_m$ has the following properties:

- *Direction of the magnetic force.* The force $\mathbf{F}_m$ is perpendicular to the plane containing $\mathbf{v}$ and $\mathbf{B}$. More specifically, the direction of $\mathbf{F}_m$ is given by the right-hand rule for cross products: imagine $\mathbf{v}$ to be turned through the smaller angle θ by the right-hand curled fingers to align with $\mathbf{B}$; then the right thumb points in the direction of $\mathbf{F}_m$. See Figure 29-1, which shows the force direction for a *positive* charge; with a negatively charged particle, the direction of $\mathbf{F}_m$ is reversed.

- *Magnitude of the magnetic force.* From the definition of the cross product, we have

$$F_m = qv\,B\sin\theta = qv_\perp B \qquad (29\text{-}2)$$

where θ is the smaller angle between $\mathbf{v}$ and $\mathbf{B}$. The component of $\mathbf{v}$ perpendicular to $\mathbf{B}$ is $v_\perp = v \sin\theta$.

Equation (29-1) and Figure 29-1 say it all. The properties of the magnetic force and the definition of the magnetic field come from experiments in which charged particles are observed as they move through a magnetic field. (How the magnetic field is produced is not our concern here; it is treated in Chapter 30.) For example, one might study the deflection of a beam of electrons in an oscilloscope under the influence of a permanent magnet.

Characteristics of $\mathbf{F}_m$ and $\mathbf{B}$, all implicit in (29-1), are as follows:

- $F_m \propto q$. When a magnetic force acts, it is *proportional to charge magnitude.* Replacing a positive charge by a negative will, with other things unchanged, reverse the direction of $\mathbf{F}_m$.
- $F_m \propto v_\perp$. Only the *velocity component perpendicular* to $\mathbf{B}$ influences the magnitude of $\mathbf{F}_m$.
- $\mathbf{F}_m \perp \mathbf{v}$. When a magnetic force acts, it is always *perpendicular to the particle's velocity* and deflects the moving charged particle.
- $\mathbf{F}_m = 0$ for $\mathbf{v} \parallel \mathbf{B}$. A particle of either sign fired along the direction of $\mathbf{B}$ (or opposite to $\mathbf{B}$) is subject to no magnetic force. Indeed, the line of the magnetic field $\mathbf{B}$ is defined as that line in space along which a charged particle can be fired without being subject to a magnetic force.
- $\mathbf{F}_m$ has its maximum magnitude for $\mathbf{v} \perp \mathbf{B}$. With $\theta = 90°$ in (29-2), we have

$$F_{m,\text{max}} = qv_\perp B$$

or with $\mathbf{v} \perp \mathbf{B}$,

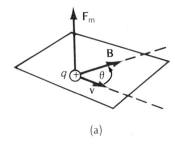

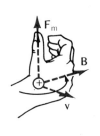

Figure 29-1. Magnetic force $\mathbf{F}_m$ on a particle with electric charge q and velocity $\mathbf{v}$ in magnetic field $\mathbf{B}$. (a) Relative orientation of the vectors; (b) the right-hand rule for $\mathbf{F}_m = q\mathbf{v} \times \mathbf{B}$.

$$B \equiv \frac{F_{m,max}}{qv} \qquad (29\text{-}3)$$

This relation *defines* the *magnitude* of **B** as the maximum magnetic force per unit charge divided by the particle speed.

- The direction of **B** comes from the right-hand rule in $\mathbf{F}_m = q\mathbf{v} \times \mathbf{B}$. With the *line* of **B** — the *undirected* line — defined as that for zero force, there are *two* possible opposite directions that might be chosen for **B**. The one universally used is shown in Figure 29-1 and summarized by the right-hand rule for cross products.*

Magnetic force $\mathbf{F}_m$ and magnetic field **B** *differ greatly* from their electric counterparts:

- An *electric force* $\mathbf{F}_e$ acts on charge q, whatever the state of its motion. The magnitude of $\mathbf{F}_e$ in field **E** depends only on the charge magnitude, $F_e = qE$. The direction of $\mathbf{F}_e$ is parallel (or antiparallel) to **E**.
- A *magnetic force* acts only on a moving charge q that has a velocity component perpendicular to **B**. The magnitude of $\mathbf{F}_m$ depends on the perpendicular velocity component $v_\perp$, as well as the charge q. The direction of $\mathbf{F}_m$ is *never* along **B**.

Equation (29-1) implies that magnetic field **B** is a vector. We have already discussed how to assign magnitude and direction to **B**. But a directed quantity is a true vector only if it obeys the rules of vector addition found to hold for displacement vectors (Section 2-1). Experiment shows that **B** is indeed a vector in the following sense. Two or more magnetic fields from separate sources acting simultaneously on a charged particle produce a magnetic force given by $\mathbf{F}_m = q\mathbf{v} \times \mathbf{B}$ if by **B** is meant the *vector sum* of the separate fields. Said a little differently, magnetic fields follow the superposition principle.

From (29-2) we see that appropriate units for the magnetic field are newtons per ampere·meter (N/A·m) as follows:

$$\frac{N}{C \cdot m/s} = \frac{N}{(C/s) \cdot m} = \frac{N}{A \cdot m}$$

In the SI system, this combination of units for the magnetic field is called the *tesla* (abbreviated T), named for N. Tesla (1856–1943). By definition,

$$1\,\text{T} \equiv \text{N}/\text{A} \cdot \text{m}$$

By laboratory standards, a magnetic field of one tesla is large. A very large electromagnet with an iron core can produce a magnetic field of only about 2 T. Magnetic fields are often given in the smaller units of the *gauss* (abbreviated G), where

$$1\,\text{T} \equiv 10^4\,\text{G}$$

The earth's magnetic field at the surface, at a latitude of 45°, has a magnitude of about 0.6 G. A typical small permanent magnet might produce a field of a few hundred gauss near its poles.

The quantity that we have been calling the magnetic field goes by other

---

* This direction for **B** agrees with the familiar behavior of a magnet in an external magnetic field. If freely pivoted, a magnet becomes aligned with **B**. For a compass in the earth's magnetic field, the so-called north pole of the compass points towards geographic north.

names — *magnetic induction field* and *magnetic induction*. For reasons to be seen in Section 29-2, **B** is also termed the *magnetic flux density*. Mostly we shall stick with the simple name *magnetic field*.*

Example 29-1. Figure 29-2(a) is a bubble-chamber photograph showing the creation of electron-positron pairs by the annihilation of highly energetic photons. The antiparticle of an electron, the positron, is like an electron but has positive electric charge of magnitude e. Charged particles, such as electrons or positrons, leave a wake of bubbles along their paths as they travel in a bubble chamber through liquid hydrogen close to its boiling point. A photon, a "particle" of electromagnetic radiation, is electrically neutral and leaves no track. Each electron-positron pair appears as an inverted V with curled ends. An external magnetic field, into the plane of the photograph in Figure 29-2(a), deflects the negative electrons and positive positrons in opposite directions. Which is which?

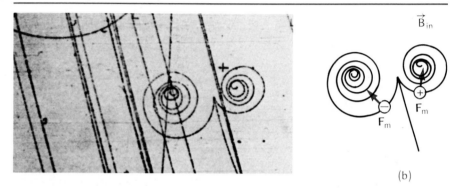

Figure 29-2. (a) Bubble-chamber photograph showing tracks of bubbles left by electron-position pairs as the particles travel in a magnetic field through liquid hydrogen. Each electron-positron pair appears as an inverted V with curled ends. The electron and positron are oppositely charged and therefore deflected in opposite directions. As each electron and positron loses kinetic energy, its radius of curvature decreases. (b) Applying the relation $\mathbf{F}_m = q\mathbf{v} \times \mathbf{B}$ to the paths of the two particles moving in an inwardly directed magnetic field shows that the negatively charged electron is on the left and the positively charged positron is on the right.

Figure 29-2(b) shows the directions of the magnetic forces on the two oppositely charged particles of one pair. The electron is deflected to the left, the positron to the right.

29-2 Magnetic-Field Lines and Magnetic Flux

Electric-field lines are used to show the direction and magnitude of **E**. Similarly, magnetic-field lines can show the direction and magnitude of **B**.† A magnetic field is strong where the magnetic lines are crowded and weak where

---

* It is not proper to call **B** the magnetic field *intensity*. That term is reserved for a related magnetic vector quantity, usually symbolized by **H**, with a physical significance and units distinct from **B**.
† We say magnetic-*field* lines, never magnetic "lines of force." After all, the magnetic force on a moving charged particle is never parallel to the lines representing **B**.

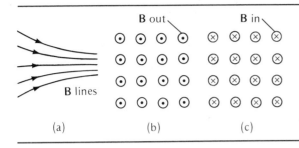

Figure 29-3. (a) Magnetic field lines. Field lines (b) out of the paper and (c) into the paper.

they are far apart. A constant, or uniform, magnetic field is represented by uniformly spaced, parallel, straight lines.

When magnetic-field lines are perpendicular to the plane of the paper, we use the following representation:

- Symbol ⊙ to show a magnetic field out of the paper.
- Symbol ⊗ to show a magnetic field into the paper.

These symbols are chosen because they remind us, respectively, of an arrow point emerging from the paper and the feathers on the tail of an arrow going into the paper. See Figure 29-3.

The electric flux $d\phi_E$ over an infinitesimal surface element dS is defined as

$$d\phi_E \equiv \mathbf{E} \cdot d\mathbf{S} \qquad (24\text{-}4)$$

Similarly, the magnetic flux $d\phi_B$ through an infinitesimal surface element dS is related to the magnetic field $\mathbf{B}$ through that small patch of area by

$$d\phi_B \equiv \mathbf{B} \cdot d\mathbf{S} = B \cos \theta \, dS \qquad (29\text{-}5)$$

where $d\mathbf{S}$ is a vector perpendicular to the surface and θ is the angle between $\mathbf{B}$ and $d\mathbf{S}$. To find the magnetic flux through surface area dS, we simply multiply dS by the component $B \cos \theta$ of the magnetic field perpendicular to the surface element. See Figure 29-4. When the $\mathbf{B}$ lines lie in the surface and $\theta = 90°$, the magnetic flux through dS is zero.

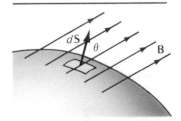

Figure 29-4. Magnetic flux through a surface element dS.

The total magnetic flux ϕ_B through a finite surface area is just the algebraic sum of the infinitesimal contributions

$$\phi_B = \int \mathbf{B} \cdot d\mathbf{S} \qquad (29\text{-}6)$$

where the integration is carried out over the surface area through which the magnetic flux is to be computed.

Suppose that magnetic field $\mathbf{B}$ is uniform over a flat area A oriented at right angles to $\mathbf{B}$. Then the flux through A is simply

$$\phi_B = BA$$

and the magnetic field B can be expressed as

$$B = \frac{\phi_B}{A} \qquad (29\text{-}7)$$

From (29-7), magnetic field B is the *magnetic flux density*, the magnetic flux divided by the area through which the magnetic-field lines penetrate at right angles.

The relations among magnetic field $\mathbf{B}$, magnetic flux ϕ_B, and the magnetic-field lines are exactly like the corresponding relations among electric field, electric flux, and electric-field lines:

- The number of magnetic-field *lines per unit transverse area* is a measure of the *magnitude of the magnetic field* **B**.
- The total *number of magnetic-field lines* through the chosen transverse area is a measure of magnetic *flux* $\phi_B = BA$.

In the SI system, the unit for *magnetic flux*—a measure of the *total number* of magnetic lines crossing a transverse area—is the *weber* (abbreviated Wb), named for W. Weber (1804–1890). A corresponding unit for the magnetic flux density B is then Wb/m². By definition, one weber per square meter is one tesla:

$$1 \text{ Wb/m}^2 \equiv 1 \text{ T}$$

Example 29-2. Cosmic radiation consists of highly energetic, positively charged particles (mostly protons) that rain upon the earth in all directions from outer space. How does the magnetic field of the earth affect protons approaching it (*a*) toward the North Pole or the South Pole and (*b*) at the equator?

The earth's magnetic field lines are shown in Figure 29-5. Protons approaching the earth at the poles travel along magnetic field lines and are consequently undeflected [Figure 29-5(a)]. On the other hand, protons approaching the earth at the equator cross magnetic field lines at right angles. These magnetic field lines go from geographic south to geographic north. Using (29-1), we see that the particles are acted on by a magnetic force that deflects them toward the east [Figure 29-5(b)].

The incoming cosmic-ray particles have a large range of energies. Some are highly energetic (energies up to 10^{18} MeV), and the earth's magnetic field is feeble (of the order of 0.6 G near the surface). Yet, this feeble magnetic field is able to influence the motion of these particles because it extends far out into space. Low-energy particles arriving at the equator are deflected so strongly that they miss hitting the earth's atmosphere. As a consequence, the intensity of the cosmic radiation is greater at the North and South poles than near the equator (the so-called latitude effect), proving that the incoming particles are electrically *charged*. Moreover, those particles arriving at the equator come preferentially from the west (the so-called east-west effect), proving that the primary particles are *positively* charged.

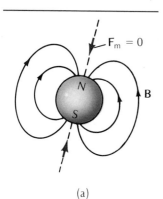

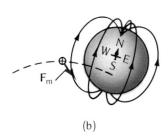

Figure 29-5. (a) Magnetic field of the earth; charged particles approaching along the magnetic axis are undeflected. (b) Positively charged particles approaching the earth at the equator are deflected toward the east.

Example 29-3. A uniform magnetic field of 0.10 T makes an angle of 30° with the plane of a square that is 20 cm on a side. See Figure 29-6. What is the magnetic flux through the square?

The angle between **B** and the *normal* to the plane is $\theta = 60°$, so that the flux through

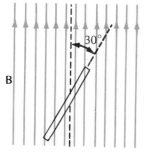
(a)

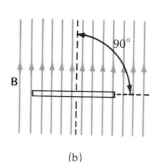
(b)

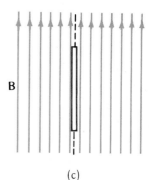
(c)

Figure 29-6.

the square loop is, from (29-5),

$$\phi_B = BS \cos\theta = (0.10 \text{ T})(0.20 \text{ m})^2 \cos 60° = 2.0 \times 10^{-3} \text{ Wb/m}^2$$

Suppose that the square is turned so that **B** becomes perpendicular to the plane. Then the magnetic flux and the number of magnetic field lines passing through the square are doubled [Figure 29-6(b)]. On the other hand, turning the square so that **B** lies in the plane of the square gives $\phi_B = 0$; no field lines then penetrate the square [Figure 29-6(c)].

29-3 Charged Particle in a Uniform Magnetic Field

The magnetic force is different from such other fundamental forces as the gravitational force or the electric force; $\mathbf{F}_m$ is *velocity-dependent*. It is therefore not possible to associate a scalar potential energy with the magnetic interaction. Furthermore, the magnetic force can do no work; the speed and kinetic energy of a charged particle are constant as it travels in a constant **B**. This follows directly from the direction of $\mathbf{F}_m$ relative to **v**; the force $\mathbf{F}_m$ is always perpendicular to **v**. Since no component of $\mathbf{F}_m$ acts along the direction of the particle's motion, $\mathbf{F}_m$ does no work.

What then is the effect of $\mathbf{F}_m$ on a moving charged particle? Since $\mathbf{F}_m$ is always at right angles to **v**, the particle is *deflected* by this sideways force. The direction of **v** changes—but not the speed.

What is the general path of a charged particle in a uniform magnetic field? We first consider two simple special cases and then the general case:

- **v**∥**B**. As (29-1) shows—and the definition of **B** requires—a particle moving initially *parallel* (or antiparallel) to **B** coasts at *constant velocity*. See Figure 29-7(a).
- **v** ⊥ **B**. The magnitude of $\mathbf{F}_m$ is constantly equal to qvB; the direction of $\mathbf{F}_m$ remains perpendicular to **v** as the velocity changes direction. These are just the conditions for *uniform circular motion*. The force $\mathbf{F}_m$ points always to the center of the circle, and the particle's speed is constant. The charged particle encircles magnetic-field lines in a plane transverse to **B**. See Figure 29-7(b).
- **v** neither ∥**B** nor ⊥**B**. The velocity component $v_\parallel = v \cos\theta$ parallel to **B** remains constant in magnitude and direction. The perpendicular velocity com-

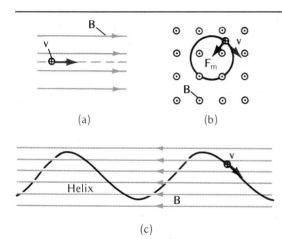

Figure 29-7. A charged particle moving in a uniform field: (a) with **v** parallel to **B**, the path is a straight line; (b) with **v** perpendicular to **B**, the path is a circle; (c) for other angles between **v** and **B**, the path is a helix.

ponent $v_\perp = v \sin\theta$ also remains constant in magnitude but changes direction continuously. The resulting motion? Simply the superposition of drift at constant speed $v_\parallel$ along **B** and circling at constant speed $v_\perp$ around **B**. The path is a *helix* whose symmetry axis coincides with the magnetic field. See Figure 29-7(c).

Electrically charged particles follow paths that are wrapped around **B** lines even when the magnetic field varies (slowly) with position. See Figure 29-8.

Consider a particle of mass m, charge q, and speed v circling at radius r in a plane perpendicular to a uniform magnetic field **B**. From Newton's second law, we have

$$\Sigma F = ma$$

$$qv_\perp B = \frac{mv_\perp^2}{r}$$

or

$$mv_\perp = qrB \tag{29-8}$$

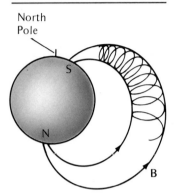

Figure 29-8. Charged particles trapped in the earth's Van Allen belt. Note that the south magnetic pole is near the geographic North Pole.

We see that the magnitude of the particle's linear momentum is directly proportional to the magnitude of B and to r, the radius of curvature of the circular arc. This gives a basis for measuring a charged particle's momentum: measure B, measure r, know q, and apply (29-8). The momentum magnitudes of the electron and positron in the bubble-chamber photograph of Figure 29-2 are readily found in this fashion.

Any charged particle projected into a uniform magnetic field travels at constant speed in a helix wrapped around the magnetic field lines. Suppose that numerous charged particles of one type, such as electrons, which differ in initial velocity and energy, are injected into the same constant magnetic field. Each electron goes in some helical path (or in special cases, a circle or a straight line). It is a remarkable fact that *all* such electrons, despite differences in their energies, speeds, and paths, will complete one loop in precisely the *same time*! All cycling particles will have the *same frequency*. Let us prove it.

The perpendicular velocity component $v_\perp$ can be written as $v_\perp = \omega r$, where ω is the circling particle's angular speed and r the radius of the helix. Using this relation in (29-8) gives

$$\omega = \frac{q}{m}B \tag{29-9}$$

The ordinary frequency $f = \omega/2\pi$ of the circling particle and its period T are then given by

$$f = \frac{1}{T} = \frac{q}{2\pi m}B \tag{29-10}$$

The characteristic frequency given in (29-10) is called the *cyclotron frequency* (for reasons to be evident in Example 29-5). The cyclotron frequency depends on the charge-to-mass ratio q/m but does not involve the particle speed $v_\perp$ or radius r.

Example 29-4. A proton (mass, 1.67×10^{-27} kg) is observed in a circular arc of 30-cm radius when it travels transverse to a magnetic field of 1.5 T. What is the proton's (a) momentum? (b) cyclotron frequency?

(a) From (29-8), we have

$$p = mv = qrB$$
$$= (1.60 \times 10^{-19} \text{ C})(0.30 \text{ m})(1.5 \text{ T})$$
$$= 7.2 \times 10^{-20} \text{ kg} \cdot \text{m/s}$$

(b) From (29-10), we have

$$f = \frac{qB}{2\pi m}$$
$$= \frac{(1.60 \times 10^{-19} \text{ C})(1.5 \text{ T})}{2\pi(1.67 \times 10^{-27} \text{ kg})}$$
$$= 2.3 \times 10^7 \text{ s}^{-1} = 23 \text{ MHz}$$

Example 29-5 The Cyclotron. The cyclotron, a machine for accelerating charged particles to relatively high kinetic energies, is based on the fact that the cyclotron frequency of a charged particle circling magnetic-field lines does not depend on the radius of the orbit or the particle's speed. The cyclotron was invented in 1932 by E. O. Lawrence (1901–1958) and M. S. Livingston (b. 1905), and it is the simplest of a whole class of particle accelerators based on the cyclotron principle.

The essential parts of a cyclotron are shown in Figure 29-9. Positive particles such as protons, deuterons (nuclei of heavy hydrogen atoms, or deuterium), or still more massive ions are injected into the evacuated central region at central point C between two flat D-shaped hollow metal conductors. An alternating high-frequency voltage at the particle's cyclotron frequency is applied to the dees and a uniform magnetic field is applied transverse to the dees. The charged particle is then subject to two forces:

• A continuous deflecting magnetic force lying in the plane of the dees, always pointing toward central point C.

• When the charged particle is in the gap between the dees, an accelerating electric force from the instantaneous potential difference between the dees. When the particle is within either dee, it is electrically shielded inside a hollow conductor and feels no electric force.

The particle is accelerated across the gap between the dees. Actually, the particle is injected so that it arrives at the gap when the sinusoidally varying potential difference reaches its maximum value. After the particle has traveled a half-circle and is at the gap again, one half-cycle has elapsed and the particle is again accelerated across the gap. The process continues, with the particle accelerated at each gap crossing. The circulating charged particles remain synchronized with the alternating voltage because the time for completing a half-circle does not depend on the particle's speed or orbital radius.

The particles are deflected by ejector plate E at the outer edge of the dee and strike target T.

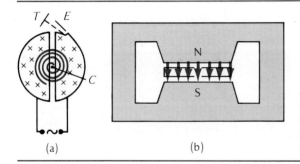

Figure 29-9. (a) Top view of cyclotron dees; a uniform magnetic field is directed into the paper. (b) Side view of cyclotron dees in magnetic field.

Suppose that a certain cyclotron with dees of 30-cm radius accelerates protons in a magnetic field of 1.5 T. What is the maximum kinetic energy of accelerated protons? We found in Example 29-4 that the cyclotron frequency of such protons is 23 MHz and their momentum is 7.2×10^{-20} kg·m/s. Therefore, the corresponding final kinetic energy is

$$K = \frac{p^2}{2m} = \frac{(7.2 \times 10^{-20} \text{ kg·m/s})^2}{2(1.67 \times 10^{-27} \text{ kg})} = 1.55 \times 10^{-12} \text{ J}$$

Electron volts are more appropriate as units for specifying the energy of such particles and we have that

$$K = 1.55 \times 10^{-12} \text{ J} \left(\frac{1 \text{ MeV}}{1.60 \times 10^{-13} \text{ J}} \right) = 9.7 \text{ MeV}$$

29-4 Charged Particles in Uniform B and E Fields

If the gravitational force is neglected, the resultant force on any charged particle is merely the vector sum of the electric force $\mathbf{F}_e = q\mathbf{E}$ and the magnetic force $\mathbf{F}_m = q\mathbf{v} \times \mathbf{B}$:

$$\mathbf{F} = \mathbf{F}_e + \mathbf{F}_m = q\mathbf{E} + q\mathbf{v} \times \mathbf{B} = q(\mathbf{E} + \mathbf{v} \times \mathbf{B}) \quad (29\text{-}11)$$

Equation (29-11) is usually referred to as the *Lorentz force relation*, after H. E. Lorentz (1853–1928), who made important contributions to the theory of electromagnetism. The Lorentz relation can be regarded as giving the basic definitions of **E** and **B**—the magnetic force is that part of the electromagnetic interaction that depends on the particle's velocity; the electric force is that part of the interaction that does not. If we know a particle's charge and initial velocity, then from knowledge of **E** and **B** at each point in space, we can predict in detail the particle's future motion.

Here we consider only the special situations in which both **E** and **B** are uniform and constant with time. We know already that with just a single field:

- *Uniform E only*, the particle has a *constant acceleration* and moves in a *parabola* whose symmetry axis is along the **E** direction (Section 24-2). See Figure 29-10(a).
- *Uniform B only*, the particle has a *constant speed* and moves in a *helix* whose symmetry axis is along the direction of **B**. See Figure 29-10(b).

How should uniform **E** and uniform **B** be arranged so that a charged particle goes through both fields undeflected? The only effect of a magnetic force is to deflect a charged particle, so that the sideways magnetic deflection must be cancelled by an oppositely directed electric deflection. The required arrangement is shown in Figure 29-11, where **v** is perpendicular to both **E** and **B** and the directions of **E** and **B** are also mutually perpendicular. The forces $\mathbf{F}_e$ and $\mathbf{F}_m$ are of equal magnitude when

$$F_e = F_m$$
$$qE = qvB$$
$$v = \frac{E}{B} \quad (29\text{-}12)$$

Notice the remarkable result. If the directions of **E** and **B** are as shown in Figure 29-11 and their relative magnitudes are adjusted to satisfy (29-12), a particle of any charge magnitude traveling at speed E/B passes through the

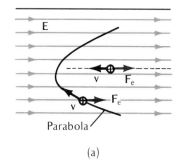

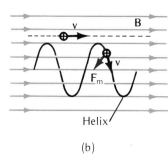

Figure 29-10. (a) In a uniform electric field, the path of a charged particle is a parabola. (b) In a uniform magnetic field, the path of a charged particle is a helix.

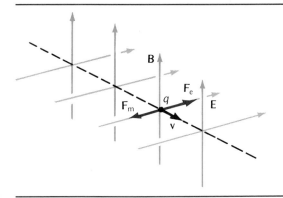

*Figure 29-11. A particle with charge q and velocity **v** moving undeflected through uniform crossed **E** and **B** fields. The electric and magnetic forces, **F**$_e$ and **F**$_m$, cancel.*

region of crossed fields undeflected. (Even the sign of q does not enter; with a negatively charged particle, the directions of *both* **F**$_e$ and **F**$_m$ in Figure 29-11 are reversed.) A device with crossed electric and magnetic fields arranged as in Figure 29-11 is appropriately known as a *velocity selector*. From a beam of polyenergetic particles, all particles save those of the design speed $v = E/B$ will be deviated from a straight path.

The relation for a velocity selector shows directly that compared with electric force, the magnetic force is usually quite feeble. Suppose, for example, that $E = 10^4$ V/m (a relatively weak electric field with a drop in potential of 100 V over 1 cm) and that $B = 1$ T (a relatively strong magnetic field). We then have from (29-12) that $v = (10^4 \text{ V/m})/1 \text{ T} = 10^4$ m/s. Only with a relatively high-speed charged particle, one moving at 10 km/s, will the magnetic force match the electric force in size.

We already have considered a *momentum selector*. It consists basically of a beam of particles directed at right angles to a uniform **B**. The particles travel in circular arcs with a linear momentum magnitude $p = mv = qrB$.

A *kinetic-energy selector* of the simplest type consists of charged particles moving along or opposite to a uniform electric field **E**. The electric potential drops by V over a distance d measured in the direction of **E**, so that

$$V = Ed \quad (25\text{-}10)$$

The kinetic energy of a charged particle traveling with or against electric-field lines changes by ΔK, where

$$\Delta K = qV = qEd$$

For example, an electron accelerated from rest across a potential difference of 1.0 kV acquires a kinetic energy of 1.0 keV. Similarly, a particle with charge e passing through a retarding potential difference of 50 V has its kinetic energy reduced by 50 eV.

In summary,

- With **B** alone, we can measure momentum mv.
- With **E** alone, we can measure kinetic energy $\tfrac{1}{2}mv^2$ (or more generally, a change in kinetic energy).
- With crossed **B** and **E**, we can measure velocity v.

Clearly, then, there are several ways of measuring the mass of a charged particle. We simply combine the separate measurements given above:

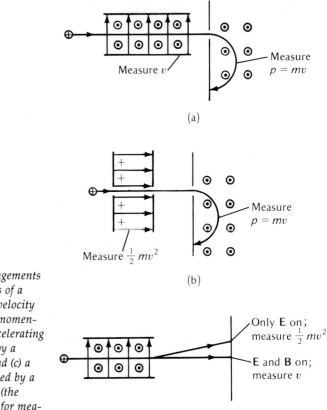

Figure 29-12. Arrangements for measuring the mass of a charged particle: (a) a velocity selector followed by a momentum selector, (b) an accelerating electric field followed by a momentum selector, and (c) a velocity selector followed by a deflecting electric field (the Thomson arrangement for measuring the electron q/m ratio).

- Measure v and mv, as shown in Figure 29-12(a).
- Measure $K = p^2/2m$, and $p = mv$, as shown in Figure 29-12(b).
- Measure v and $K = \tfrac{1}{2}mv^2$, as shown in Figure 29-12(c).

An arrangement like that shown in Figure 29-12(c) was used by J. J. Thomson in 1897 in the first measurement of the mass of the electron; indeed, Thomson is said to have discovered the electron through this experiment. Strictly, any of these procedures yields the charge-to-mass *ratio* q/m, not the mass alone. With charge measured independently, the mass can then be computed. Mass spectrometers using these procedures have been so refined that measuring atomic masses to within a few parts in 10^7 has become a routine laboratory procedure.

29-5 Magnetic Force on a Current-Carrying Conductor

To find the magnetic force on a current-carrying conductor, we first concentrate on a single charge-carrier with charge dq moving at drift velocity **v**. The magnetic force $d\mathbf{F}_m$ on dq in field **B** is

$$d\mathbf{F}_m = dq\, \mathbf{v} \times \mathbf{B}$$

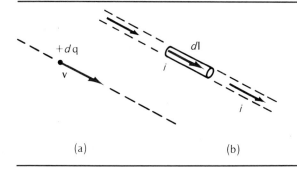

Figure 29-13. Charge $+dq$ with velocity **v** constitutes conventional current i along displacement $d\mathbf{l}$.

We can write **v** as $d\mathbf{l}/dt$, where $d\mathbf{l}$ is the particle's displacement over time interval dt. We consider **v** the drift velocity of a positive charge-carrier within a thin conductor. Then $d\mathbf{l}$ is also a length vector, pointing along the conductor in the direction of conventional current. See Figure 29-13. The quantity $dq\,\mathbf{v}$ can be written as

$$dq\,\mathbf{v} = dq\left(\frac{d\mathbf{l}}{dt}\right) = \left(\frac{dq}{dt}\right)d\mathbf{l} = i\,d\mathbf{l} \qquad (29\text{-}13)$$

where $i = dq/dt$ is the conventional current associated with charge dq.

Using (29-13) in the magnetic force relation, we then have

$$d\mathbf{F}_m = i\,d\mathbf{l} \times \mathbf{B} \qquad (29\text{-}14)$$

Current i can now represent the *total* current from *all* charge carriers in a segment of conductor of length $d\mathbf{l}$, and $d\mathbf{F}_m$ is the *total* magnetic force on this infinitesimal segment. See Figure 29-14.

The resultant magnetic force on a finite conductor of any shape is found by applying (29-14) to each infinitesimal length segment and adding the infinitesimal forces as vectors. If the conductor is straight, then the force on length **L** is

$$\mathbf{F}_m = i\mathbf{L} \times \mathbf{B} \qquad (29\text{-}15)$$

where **L** points in the direction of the current.

We can write the relation above in algebraic form as

$$F_m = iLB\sin\theta$$

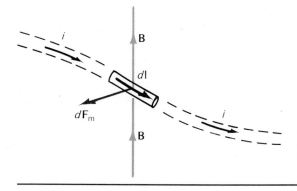

Figure 29-14. Magnetic force $d\mathbf{F}_m$ on a current element $d\mathbf{l}$ with current i in magnetic field **B**: $d\mathbf{F}_m = i\,d\mathbf{l} \times \mathbf{B}$.

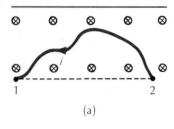

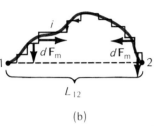

Figure 29-15. (a) A current-carrying conductor of arbitrary shape in an external magnetic field. (b) Equivalent conductor comprising segments parallel and perpendicular to the line joining the end points.

where θ is the angle between **B** and **L**. Note that $F_m = 0$ when the conductor is aligned with the external field lines.

Example 29-6. A conductor of arbitrary shape carrying current i runs from point 1 to point 2 in two dimensions, as shown in Figure 29-15(a). A constant magnetic field **B** acts into the plane of the paper. What is the resultant force on this segment of conductor?

We can approximate the continuously varying conductor as follows. We use instead a series of short, straight-line segments lying alternately parallel and perpendicular to the straight line L_{12} from point 1 to 2. The magnetic forces on these segments are, from (29-14), in the directions shown in Figure 29-15(b). We see that the total force component parallel to the line L_{12} is zero; each force to the right is matched by an equal force to the left. This leaves only the magnetic forces perpendicular to L_{12}. The resultant force perpendicular to line L_{12} is exactly the same as the magnetic force on a straight conductor running from 1 to 2. Therefore, the net magnetic force has a magnitude $F_m = iL_{12}B$; its direction is perpendicular to the line joining the end points. The arbitrarily shaped conductor is equivalent to a straight wire between its end points carrying the same current.

Now suppose that we form a current loop by imagining points 1 and 2 brought together. From the arguments above, the resultant magnetic force on the loop must be zero. A little thought will show that this conclusion applies whatever the shape of the loop — and whether the loop lies entirely in a plane or not — and that it is also independent of the loop's orientation relative to the magnetic-field lines. *The magnetic field must, however, be uniform.* Although a closed current loop is subject to no resultant magnetic force in a uniform magnetic field, it may be subject to a resultant torque, as we shall see in Section 29-7.

29-6 The Hall Effect

When a magnetic field is applied at right angles to a current-carrying conductor, there is not only a magnetic force on the conductor, but also a *transverse electric potential difference* across the conductor. This phenomenon, discovered in 1879 by E. H. Hall, is known as the *Hall effect*.

Consider Figure 29-16, where magnetic field **B** is applied at right angles to a conducting slab carrying current i. We suppose that the charge carriers composing the current are electrons; they drift with velocity $\mathbf{v}_d$ in a direction opposite to the conventional current. Then the deflecting magnetic force $\mathbf{F}_m$ on a negatively charged particle has the direction shown in Figure 29-16(a). Electrons are deflected to one edge of the slab until the outward magnetic force is balanced by an equal and inward electric force arising from excess electrons already at the edge. Excess electrons accumulate at one conductor edge, so that it acquires a net negative charge. The opposite edge, with a corresponding deficiency of electrons, acquires a net positive charge. The transverse electric potential difference between the opposite edges of the conductor is known as the *Hall potential difference* V_H.

Which side of the conductor is at the higher electric potential depends on the sign of the charge carriers. If the electrons in Figure 29-16(a) were replaced by positive charge carriers, with the direction of **B** and i kept unchanged, then the charge carriers would drift in the same direction as the current, the direction of the deflecting magnetic force $\mathbf{F}_m$ would be reversed, charges of reversed signs would accumulate on the two edges of the conductor, and the polarity of

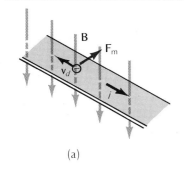

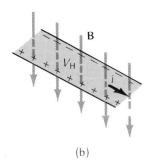

Figure 29-16. The Hall effect. (a) An electron with drift velocity $\mathbf{v}_d$ (opposite to the conventional current i) is deflected by magnetic force $\mathbf{F}_m$ when the conductor is subject to transverse magnetic field $\mathbf{B}$. (b) Opposite charges appear at the conductor edges and produce a Hall potential difference V_H.

the Hall potential difference in Figure 29-16(b) would be *reversed*. The sign of the Hall potential difference indicates the sign of the charge carriers; it can be shown that the magnitude of V_H is a measure of the concentration of charge carriers.

The sign of the observed Hall effect for ordinary metallic conductors confirms that the charge carriers are indeed negatively charged particles, electrons. Semiconducting materials, however, may exhibit either polarity. Silicon doped with arsenic is an *n-type semiconductor*; here the charge carriers, electrons, are negatively charged. Silicon doped with gallium is a *p-type semiconductor*; in such a material the effective charged carriers, known as *holes*, are positively charged.

A hole in the semiconducting material is a site at which an electron is missing because impurity atoms have been introduced. A semiconductor with electron holes remains electrically neutral. When an electron adjoining a hole shifts position to fill the hole, it creates a new hole at its former location. The electron moves in one direction; the hole shifts in the opposite direction, thereby effecting a current in the direction in which the holes move. Electron holes in a sea of electrons behave like bubbles in liquid. Bubbles are holes in a sea of liquid; as a bubble rises, liquid descends to fill the space occupied by the hole, and the situation can be described by assigning a negative mass to the bubble.

29-7 Magnetic Torque on a Current Loop

To find the resultant magnetic torque that acts on a loop of current, first we concentrate on a rigid rectangular loop with sides W and L carrying current i.

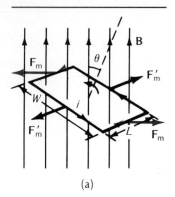

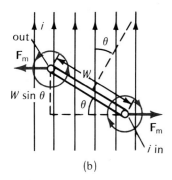

Figure 29-17. (a) Magnetic forces on a rectangular current-carrying conducting loop in a magnetic field. (b) A side view of part (a).

Consider Figure 29-17(a), where external field **B** makes an angle θ with the normal to the rectangular loop. We first find the directions of the magnetic forces on each of the four sides. As Figure 29-17(a) shows, we have two pairs of oppositely directed magnetic forces of the same magnitude. The resultant *force* on the loop is *zero*. The forces on the sides of length W are collinear; they produce no torque. The equal and opposite forces on the sides of length L are not collinear; these forces produce a torque on the loop that aligns its normal with the direction of **B**. Let us find the torque's magnitude.

From (29-15), the magnitude of the magnetic force on a side of length L is

$$F_m = iLB$$

For convenience, we compute torques relative to the lower of the two sides of length L. Then, using the definition of torque magnitude (12-4), we have

$$\tau = r_\perp F = (W \sin \theta)(iLB)$$

We can write this relation in simpler form by recognizing LW as the area A of the loop:

$$\tau = iAB \sin \theta \qquad (29\text{-}16)$$

For the situation shown in Figure 29-17(b), the torque is counterclockwise. From the vector properties of torque [(12-5), $\tau = \mathbf{r} \times \mathbf{F}$], we see that the direction of torque in Figure 29-17(b) is out of the paper. We can incorporate the vector character of torque by writing (29-16) in vector form as

$$\tau = i\mathbf{S} \times \mathbf{B} \qquad (29\text{-}17)$$

Here **S** is a vector whose magnitude represents the loop's area A. The direction of **S** is perpendicular to the plane of the loop and related to the sense of current around the loop by a right-hand rule. See Figure 29-18, which shows that the direction of τ is such as to align **S** with external field **B**.

Equation (29-17) is actually a general relation; it gives the magnetic torque on a plane loop of current of any shape. We can simply imagine the conducting wire to be approximated by small straight segments at right angles. Of course, if a loop has N identical turns (all with current in the same sense) rather than just a single turn, the torque in (29-16) and (29-17) is increased by the factor N.

The magnetic torque on a current-carrying conductor immersed in a magnetic field is the basis of many important devices. One is the *galvanometer*, a sensitive current-measuring device used in some ammeters and voltmeters. A galvanometer has a multiturn coil typically in the magnetic field of a perma-

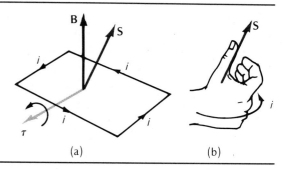

Figure 29-18. A current loop subject to a magnetic torque τ in magnetic field **B**. (a) Vector **S** represents the normal to the plane of the loop with current i. (b) The direction of **S** is related to the sense of current i through a right-hand rule.

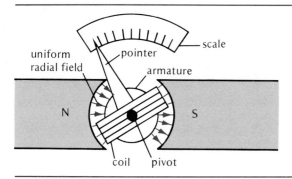

Figure 29-19. The coil of a galvanometer is subject to a magnetic torque proportional to the current in the coil.

nent magnet, as shown in Figure 29-19. The magnet is so shaped that the magnetic field at the location of the coil is not uniform but always at right angles to the plane of the coil; in effect, angle θ of (29-16) and Figure 29-19 is always 90°. The galvanometer coil is also subject to an elastic linear restoring torque from a spring ($\tau = -\kappa\phi$), which balances out the magnetic torque. Therefore, in equilibrium, the angular deflection ϕ of the coil and a needle attached to it is directly proportional to the current i through the galvanometer coil.

The magnetic torque on a current-carrying loop is also the fundamental principle behind the operation of an electric motor.

29-8   Magnetic Dipole Moment

Equation (29-17), for the magnetic torque on a current loop in a magnetic field, is like the relation for the electric torque on electric dipole moment **p** in electric field **E**:

$$\tau = \mathbf{p} \times \mathbf{E} \tag{24-9}$$

This suggests that we attribute a *magnetic dipole moment,* symbolized by vector $\boldsymbol{\mu}$, to a current loop, where by definition,

$$\boldsymbol{\mu} \equiv i\mathbf{S} \tag{29-18}$$

See Figure 29-20. The direction of $\boldsymbol{\mu}$ is related to the sense of conventional

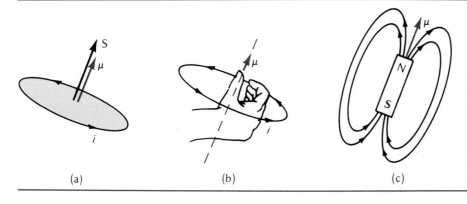

Figure 29-20. Magnetic dipoles. (a) Magnetic dipole moment $\boldsymbol{\mu}$ is aligned with vector **S** representing the enclosed area. (b) The direction of $\boldsymbol{\mu}$ is related to the sense of current i through a right-hand rule. (c) Magnetic dipole for a permanent magnet.

current in the loop by a right-hand rule. With the right-hand fingers curled in the sense of the current, the right thumb points in the direction of μ. Using (29-18) in (29-17), we get

$$\tau = \mu \times \mathbf{B} \qquad (29\text{-}19)$$

The torque is a maximum for μ at right angles to $\mathbf{B}$ and zero for a dipole aligned with the external field. (Although the torque is zero for μ opposite to $\mathbf{B}$, this orientation is highly unstable.)

Just as an electric dipole in a *uniform* field is subject to no resultant force, a magnetic dipole in a uniform magnetic field is subject to no resultant magnetic force. Both electric and magnetic dipoles are, however, subject to resultant forces in nonuniform fields. Dipoles are attracted to the region in which the field is strongest.

The electric potential energy of an electric dipole in an external electric field is given by:

$$U_E = -\mathbf{p} \cdot \mathbf{E} \qquad (24\text{-}11)$$

The corresponding relation for the magnetic potential energy U_B of a magnetic dipole μ in field $\mathbf{B}$ is

$$U_B = -\mu \cdot \mathbf{B} \qquad (29\text{-}20)$$

Note that zero potential energy corresponds to μ perpendicular to $\mathbf{B}$. Potential energy UB is maximum ($+\mu B$) when the dipole is aligned with the field and the angle between μ and $\mathbf{B}$ is zero; U_B is minimum ($-\mu B$) when the dipole points opposite to $\mathbf{B}$.

The behavior of a rigid current-carrying loop in a magnetic field reminds us of the behavior of a magnet in a magnetic field. A magnet (such as a compass needle) not aligned with a magnetic field is subject to a magnetic torque; it takes work to turn a magnet away from alignment with the field. Indeed, one may attribute a magnetic dipole moment to any object—a current-carrying loop or a magnet—that follows (29-19) and (29-20) for the torque and potential energy in an external magnetic field. See Figure 29-20(c). Although the term *dipole* may seem especially appropriate for a magnet with its magnetic effects concentrated at its two ends, we must emphasize that single magnetic poles, or magnetic monopoles, do not exist in nature (Section 30-4).

Example 29-7. A freely pivoted coil is initially aligned with an external magnetic field of 1.5 T. The coil has 100 turns and a 10.0-cm radius, and carries 1.0 A. (a) How much work must be done on the coil to turn it 180°? (b) What minimum force applied directly to the coil will keep $\mathbf{B}$ lines lying in the plane of the loop?

(a) From (29-20), we see that the coil's potential energy goes from $-\mu B$ to $+\mu B$ as it is turned from $\theta = 0$ to 180°. The total work done is $2\mu B$. The magnetic moment of the coil is [(29-18)] $\mu = Ni\pi r^2$. Therefore,

$$\text{Work done} = 2\mu B = 2(Ni\pi r^2)B$$
$$= 2(100)(1.0\ \text{A})\pi(0.10\ \text{m})^2(1.5\ \text{T}) = 9.4\ \text{J}$$

(b) The torque is maximum and the force minimum with a force applied at the outer edge of the loop and perpendicular to $\mathbf{B}$ so that the moment arm $r_\perp$ equals the loop's radius. From (29-19), we then have, with an angle of 90° between μ and $\mathbf{B}$,

$$F_{\min} = \frac{\mu B}{r} = \frac{\frac{1}{2}(\text{work from part } a)}{r} = \frac{\frac{1}{2}(9.4\ \text{J})}{0.10\ \text{m}} = 47\ \text{N}$$

Example 29-8. A fixed length of conducting wire is to be used to construct a coil. What form of coil—a single turn of large radius or many turns of small radius—will produce a maximum torque when the coil is placed in the earth's magnetic field?

For maximum magnetic torque, $\tau = \mu \times \mathbf{B}$, the coil's magnetic moment $\mu = NiS$ must be a maximum. For conducting wire of length L, the number of turns is $N = L/2\pi r$, where r is the coil radius. The coil area is $S = \pi r^2$. Therefore,

$$\mu = NiS = \left(\frac{L}{2\pi r}\right) i(L\pi r^2) = \left(\frac{L^2 r}{2\pi}\right) i$$

The coil current i is determined only by the potential difference applied across the conductor ends. With $\mu \propto r$, we have maximum magnetic moment and therefore maximum torque for a *single* turn.

Summary

Definitions

The observed *magnetic force* $\mathbf{F}_m$ on charge q with velocity $\mathbf{v}$ defines the magnetic field $\mathbf{B}$:

$$\mathbf{F}_m = q\mathbf{v} \times \mathbf{B} \qquad (29\text{-}1)$$

For a charged particle moving at right angles to the direction for which $\mathbf{F}_m = 0$, the magnetic force has its maximum magnitude $F_{m(max)}$, and the magnitude of $\mathbf{B}$ is defined as

$$B \equiv \frac{F_{m,(max)}}{qv_\perp}$$

The *Lorentz force* on charge q and velocity $\mathbf{v}$ in electric field $\mathbf{E}$ and magnetic field $\mathbf{B}$ is

$$\mathbf{F} = q\mathbf{E} + q\mathbf{v} \times \mathbf{B} \qquad (29\text{-}11)$$

Magnetic flux $d\phi_B$ of magnetic field $\mathbf{B}$ over surface element $d\mathbf{S}$:

$$d\phi_B \equiv \mathbf{B} \cdot d\mathbf{S} \qquad (29\text{-}5)$$

Magnetic flux through a surface is a measure of the net number of magnetic-field lines through that surface. Magnetic field, or magnetic-flux density, is a measure of the net number of magnetic-field lines per unit transverse area.

The magnetic dipole moment μ of a loop of current i around the periphery of surface $\mathbf{S}$ is

$$\mu \equiv i\mathbf{S} \qquad (29\text{-}18)$$

The direction of μ is related to the sense in which current goes around the edge of $\mathbf{S}$ by a right-hand rule.

Units

Magnetic field (also magnetic flux density, magnetic induction) $\mathbf{B}$

1 tesla $\equiv$ 1 newton/ampere·meter
1 T $\equiv$ 1 NA^{-1}m^{-1}
1 T = 10^4 gauss = 10^4 G

Magnetic flux

Weber $\equiv$ Tesla·meter^2
1 Wb $\equiv$ 1 T·m^2

Important Results

Properties of a particle of charge q and velocity $\mathbf{v}$ in a uniform magnetic field $\mathbf{B}$:

- Magnetic force $\mathbf{F}_m$ does *no* work on q.
- Path for $\mathbf{v} \| \mathbf{B}$: straight line.
- Path for $\mathbf{v} \perp \mathbf{B}$: circle.
- Path in general: helix.
- The charged particle circles magnetic-field lines at the *cyclotron frequency* f, where

$$f = (q/2\pi m)B \qquad (29\text{-}10)$$

- The magnitude of the particle's momentum in a plane perpendicular to $\mathbf{B}$ is

$$mv_\perp = qrB \qquad (29\text{-}8)$$

Velocity selector: the speed of any charged particle that can travel undeviated through crossed $\mathbf{E}$ and $\mathbf{B}$ fields:

$$v = \frac{E}{B} \qquad (29\text{-}12)$$

The magnetic force $d\mathbf{F}_m$ on an element $d\mathbf{l}$ of a conductor carrying current i in magnetic field $\mathbf{B}$ is

$$d\mathbf{F}_m = i\, d\mathbf{l} \times \mathbf{B} \qquad (29\text{-}14)$$

where $d\mathbf{l}$ is a length element of the current-carrying conductor in the direction of conventional current.

Hall effect: the appearance of a transverse electric potential difference, whose sign indicates the sign of charge carriers, when a magnetic field is applied transversely to a current-carrying conductor.

The magnetic torque τ and potential energy U_B of a magnetic dipole moment μ in magnetic field $\mathbf{B}$:

$$\tau = \mu \times \mathbf{B} \qquad (29\text{-}19)$$

$$U_B = -\mu \cdot \mathbf{B} \qquad (29\text{-}20)$$

Problems and Questions

Section 29-1 Magnetic Field Defined

· **29-1 Q** In the equation $\mathbf{F} = q\mathbf{v} \times \mathbf{B}$,
(A) force and velocity vectors can have any angle between them.
(B) force and magnetic-field vectors can have any angle between them.
(C) velocity and magnetic-field vectors can have any angle between them.

· **29-2 Q** Charged particles are influenced by a uniform magnetic field
(A) under all circumstances.
(B) only if they have a component of velocity parallel to the magnetic field.
(C) only if they have a component of velocity perpendicular to the field.
(D) only if they are moving in a current loop.
(E) only if they are negatively charged.

Section 29-2 Magnetic Field Lines and Magnetic Flux

· **29-3 P** A circular plane loop of radius 2 cm is oriented so that its plane makes an angle of 30° with a uniform magnetic field of 0.30 T. What is the magnetic flux through the loop?

· **29-4 Q** A magnetic field of 40 G acts parallel to the x axis in a certain region. A 20 cm × 20 cm square of cardboard is oriented with its plane at an angle of 30° to the x axis (Figure 29-21). What is the magnetic flux through the cardboard?

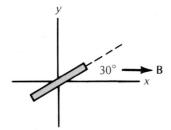

Figure 29-21. Question 29-4.

(A) Zero, since cardboard is not a conductor.
(B) Zero, but not because cardboard is not a conductor.
(C) 1.6×10^{-4} T·m²
(D) 1.4×10^{-4} T·m²
(E) 8.0×10^{-5} T·m²

: **29-5 P** At a point on the earth's surface at a latitude of 50° N, the earth's magnetic field has a magnitude of 0.4 G and is inclined downward below the horizontal at an angle of 70°. A compass at this site points due north geographically. What is the magnetic flux at this location through a loop of area 20 cm² when the loop is oriented (a) horizontally? (b) vertically in an east-west plane? (c) vertically in a north-south plane?

: **29-6 P** A magnetic field points in the $+x$ direction. It does not vary with z, but does vary in the y direction according to $B_x(y) = K/y$. Determine the magnetic flux through a loop of wire in the shape of a square of side 4 cm oriented as shown in Figure 29-22, given $K = 0.001$ T cm.

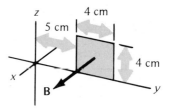

Figure 29-22. Problem 29-6.

: **29-7 P** A hemispherical shell of radius R is oriented so that the maximum flux due to a uniform magnetic field $\mathbf{B}$ passes through it (Figure 29-23). Show by direct integration over the spherical surface that the flux passing through that surface is indeed $\pi R^2 B$.

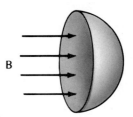

Figure 29-23. Problem 29-7.

: **29-8 P** A loop of area 10 cm² is oriented so that the components of a unit vector perpendicular to the loop are $(\frac{1}{3}, \frac{1}{3}, \frac{1}{3})$. It is placed in a magnetic field with x, y, and z components (0.010 T, 0.010 T, 0.005 T). What is the magnetic flux through the loop? (Hint: try vector methods using the scalar product.)

Section 29-3 Charged Particle in a Uniform Magnetic Field

· **29-9 P** Find the direction of the magnetic force acting on the charged particle moving as shown in each of the situations in Figure 29-24.

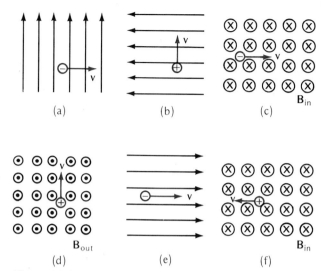

(a) (b) (c) (d) (e) (f)

Figure 29-24. Problem 29-9.

· **29-10 Q** An electron moving in the uniform magnetic field indicated in Figure 29-25 will experience a force in the direction
(A) toward A.
(B) toward B.
(C) toward C.
(D) toward D.
(E) out of the paper.
(F) into the paper.

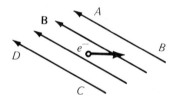

Figure 29-25. Question 29-10.

· **29-11 Q** In a mass spectrograph, a positively charged ion traveling in a horizontal path is injected into a region where a uniform vertical magnetic field is present. In such a situation, the magnetic field will
(A) increase the kinetic energy of the particle.
(B) decrease the kinetic energy of the particle.
(C) not change the kinetic energy of the particle.
(D) exert a force parallel to the magnetic field lines.
(E) exert a force parallel to the velocity of the particle.

· **29-12 P** A positively charged deuterium nucleus of momentum 2×10^{-19} kg·m/s enters a region of uniform magnetic field. The magnetic-field lines are perpendicular to the deuteron's momentum vector. For what magnitude of magnetic field will the particle move in a circle of 1.8-m radius?

· **29-13 Q** Which of the following is the most accurate statement concerning the operation of a cyclotron?
(A) The speed of the particles remains constant.
(B) The energy of the particles is increased by the magnetic field.
(C) The energy of the particles is increased by the electric field.
(D) The energy of the particles is increased by both the electric and magnetic fields.
(E) The particles travel in circular orbits of fixed radius.

: **29-14 P** A beam of 10-keV electrons moving along the positive x axis is deflected by a magentic field of 0.002 T directed along the positive z axis. (*a*) Toward what axis are the electrons deflected? (*b*) What is the radius of curvature of the path?

: **29-15 P** (*a*) What is the orbit radius of 100-eV proton that moves perpendicular to a magnetic field of 1 G? (*b*) What is its frequency of revolution?

: **29-16 Q** An electron moves in a circular orbit of radius r in a uniform magnetic field **B**. The acceleration of the electron is thus
(A) zero.
(B) eBr.
(C) e^2B^2r/m^2.
(D) eBr/m.
(E) impossible to determine without more information.

: **29-17 P** Consider a 100-eV electron moving perpendicular to the earth's magnetic field of 0.00010 T. (*a*) What is the frequency of revolution? (*b*) What is the radius of the orbit?

: **29-18 P** A particle detected in a bubble chamber is found to have the same charge-to-mass ratio as an electron. It bends to its right in a circle of radius 2 mm while traveling horizontally in a vertical magnetic field of 0.01 T directed upward. (*a*) Is the particle an electron or a positron (a positive electron)? (*b*) What is the particle's speed? (*c*) What is the particle's kinetic energy, expressed in electron volts?

: **29-19 P** At what angle with a uniform magnetic field direction must a charged particle be projected into the field so that the distance it moves parallel to field lines equals the distance it moves along a circular arc perpendicular to the field?

: **29-20 Q** In a synchrotron accelerator, charged particles are accelerated at a fixed radius. For a given accelerator magnetic field and for a given type of particle, how will the final energy depend on R, the radius of the orbit?
(A) E is proportional to R.
(B) E is proportional to $R^{3/2}$.
(C) E is proportional to R^2.
(D) E is proportional to R^{-1}.
(E) E is independent of R.

: **29-21 P** In a mass spectrometer, a beam of chloride ions, Cl$^-$, each with a kinetic energy of 1 keV, enter perpendicularly a magnetic field of 0.30 T. After the beam bends through 180°, the ions are detected when they strike a photographic plate. The beam consists of two isotopes of chlorine, ^{35}Cl (mass 35 u) and ^{37}Cl (mass 37 u, where 1 u = 1.67×10^{-27} kg). What is the separation distance between these two components of the beam when they strike the detector plate?

29-22 P A proton with velocity components $v_x = 2 \times 10^6$ m/s, $v_y = 0$, $v_z = 2 \times 10^6$ m/s passes through the origin at $t = 0$ into a region where a constant magnetic field $B_z = 0.010$ T. (*a*) Sketch the projection of the path of the particle in the *xy* plane. (*b*) What is the maximum value of *x* and of *y* for the particle's motion? (*c*) Where does the particle first pass the *z* axis again after leaving the origin?

29-23 P A particle velocity selector can be made in the following way. A beam of particles, all with the same charge and mass but varying velocities, passes through a small hole in a metal plate. A uniform magnetic field **B** is directed perpendicular to the plate. The particle velocities all make angle θ with the field. See Figure 29-26. A second plate is placed a distance *d* from the first. It also contains small hole, in line with the hole in the first plate. Show that only particles with velocity $v = qBd/2\pi m \cos\theta$ will pass through the second hole.

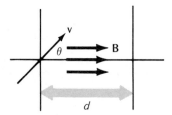

Figure 29-26. Problem 23-23.

29-24 P Several particles emitted in a cosmic-ray shower are detected in a bubble chamber, yielding the tracks sketched in Figure 29-27. The tracks are shown actual size, and they lie in the plane of the paper. A magnetic field was directed perpendicularly into the paper when the tracks were photographed. Two of the tracks were due to electrons. (*a*) Which tracks were made by electrons? (*b*) What is

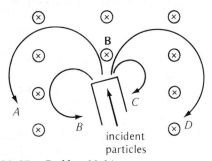

Figure 29-27. Problem 29-24.

the value of E_f/E_s, where E_f and E_s are the energies of the fast and slow electron respectively?

Section 29-4 Charged Particles in Uniform B and E Fields

29-25 Q An electron is shot into a region of constant non-zero magnetic field and zero electric field. Its path will most likely be
(A) a straight line.
(B) a circle.
(C) a parabola.
(D) a hyperbola.
(E) an ellipse.
(F) a helix.
(G) a sine curve.

29-26 Q An electron is not deflected in moving through a certain region of space. We can be certain that
(A) no magnetic field is there.
(B) no electric field is there.
(C) neither a magnetic nor an elecric field is there.
(D) the velocity of the electron remains constant.
(E) none of the above is true.

29-27 Q In a mass spectrometer, the purpose of the magnetic field is
(A) to cause the particles to travel in a circular path.
(B) to increase the speed of the particles.
(C) to compensate for the electrical force acting on the particles.
(D) to decelerate the particles at the detector.
(E) to ionize the particles so that they can then be accelerated.

29-28 P A charged particle has velocity components $v_x = 1 \times 10^5$ m/s and $v_z = 2 \times 10^5$ m/s, and is subject to a uniform mangetic field $B_z = 0.020$ T. Determine the magnitude and direction of an electric field that will produce no deflection of the particle.

29-29 P A beam of 20-keV electrons passes between two plates separated by 2 cm. A uniform magnetic field of 200 G acts into the plane of the paper. What polarity and magnitude of voltage must be applied to the plates in Figure 29-28 so that the beam is undeflected?

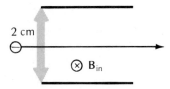

Figure 29-28. Problem 29-29.

29-30 P A particle of mass *m* and charge *q* is released from rest at one plate of a parallel-plate capacitor across which an accelerating potential *V* is applied. A uniform magnetic field B is into the paper. (a) Show that there will be no current between the plates if $V < qB^2d^2/2m$. (b) With the electron just grazing the upper plate, how far from its starting point will the electron hit the lower plate? (See Figure 29-29).

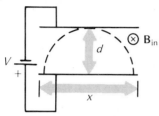

Figure 29-30. Problem 29-34.

Section 29-5 Magnetic Force on a Current-Carrying Conductor

· **29-31 P** A long, straight wire carrying a current of 3.0 A makes an angle of 60° with a uniform magnetic field of 0.040 T. What force, magnitude and direction, is exerted on a 3-m-long segment of the wire?

: **29-32 Q** A straight wire, of length L, carries a current. (The vector **L** is directed in the direction of the conventional current.) In calculating the force **F** such a wire will experience when placed in a magnetic field **B**, we recognize that for given vectors **B** and **L**,
(A) the direction of **F** will depend on the sign of the charge carriers in the wire.
(B) the angle between **F** and **B** can have any value.
(C) the angle between **F** and **L** can have any vaue.
(D) the total net force on the segment of wire of length L will be zero.
(E) the angle between **F** and **L** will always be 90°.

: **29-33 P** A long, straight transmission line carries a current of 100 A. It makes an angle of 50° with the earth's magnetic field, with magnitude 0.80×10^{-4} T. What is the magnetic force per unit length on the wire?

: **29-34 P** A conducting rod of mass m slides on two parallel fixed conducting bars separated by d and inclined at angle θ above horizontal. See Figure 29-30. A constant magnetic field B acts vertically upward. What magnitude of current must exist in the conductor if once started, the rod is to slide up the bars with constant speed?

Figure 29-29. Problem 29-30.

: **29-35 P** (a) What current density would be needed to cause an aluminum conductor (density, 2.7 gm/cm³) to "float" in the earth's magnetic field at the equator, where the field strength is about 10^{-4} T? (b) How should the conductor be oriented for best results, and in what direction should the current flow? (c) Estimate whether this would be a practical undertaking. At what rate would thermal energy be generated in the aluminum? Is it likely to melt?

: **29-36 P** The horizontal arm of the beam balance shown in Figure 29-31 carries current I. A uniform magnetic field **B** acts perpendicularly out of the paper. (a) Determine the magnetic field B needed to keep the scale in balance. Express your answer in terms of m_1, m_2, g, x_1, x_2, and I. (b) What is the force on the pivot when B has the value found above?

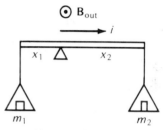

Figure 29-31. Problem 29-36.

: **29-37 P** A circular loop of wire of radius R carries current I. A uniform magnetic field **B** is directed perpendicular to the plane of the loop. What is the tension in the wire?

: **29-38 P** A wire 8 m long carries a current of 2 A. It lies on a horizontal table, 3m × 4m, as shown in Figure 29-32. A uniform magnetic field of 0.050 T acts parallel to the long edge of the table. What is the net magnetic force on the wire?

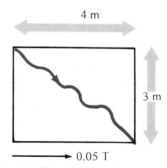

Figure 29-32. Problem 29-38.

Section 29-7 Magnetic Torque on a Current Loop

· **29-39 P** A rectangular loop of wire of 10 turns has its long edge parallel to the z axis, and the plane of the loop makes an angle of 30° with the y axis. See Figure 29-33. A uniform field **B** directed along the y axis is present. A cur-

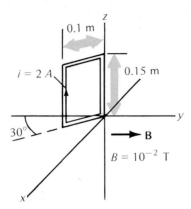

Figure 29-33. Problem 29-39.

rent of 2 A exists in the coil. (*a*) What is the net force on the coil? (*b*) What is the net torque on the coil?

· **29-40 P** A circular coil of diameter 2.4 m and 150 turns is wrapped around a satellite to provide attitude control. This is done by passing a current through the coil and allowing the resulting torque due to the earth's magnetic field to rotate the spacecraft. What torque could be produced by a current of 1 A at an elevation of 550 km, where the earth's field is 8.0×10^{-5} T? Although small, the torque can produce appreciable rotation if allowed to act long enough, inasmuch as no friction acts on the satellite.

· **29-41 P** A square 2 cm × 2 cm coil of 100 turns carries 2 A. The plane makes an angle of 60° with a uniform magnetic field of 0.050 T. What is the torque on the coil?

: **29-42 P** A circular loop of wire with current I experiences a torque τ when placed in a uniform magnetic field **B**. Suppose that this same wire is wrapped into a coil of N small loops, oriented in the same way with the magnetic field. What torque now acts on the coil?

: **29-43 P** A circular coil of diameter 6 cm and 500 turns carrying a current of 10 mA experiences a maximum torque of 0.028 m·N when rotated in a uniform magnetic field. What is the minimum value of the magnetic field?

: **29-44 P** A thin disk carries a uniform surface-charge density σ on one side. It is rotated with angular velocity ω about a diameter. A uniform magnetic field **B** is present and makes an angle θ with the axis of rotation. What is the magnitude of the torque on the disk?

: **29-45 P** A thin rod of length L has charge Q uniformly distributed along its length. It is rotated with angular velocity ω about an axis passing through its center and perpendicular to the rod. A uniform magnetic field **B** acts at an angle θ to the axis of rotation. What torque acts on the rod?

Section 29-8 Magnetic Dipole Moment

· **29-46 Q** When placed in a uniform magnetic field, a current-carrying coil of wire will
(A) experience a net force perpendicular to the magnetic field.
(B) experience a net force in the direction of the magnetic field.
(C) tend to align itself so that it encompasses the maximum number of magnetic-field lines.
(D) experience a torque $\mu \cdot \mathbf{B}$.
(E) experience no torque and no net force.

: **29-47 P** A charged particle moving in a circle constitutes a loop of electric current, and a magnetic dipole moment can be associated with it. For example, in a simple atomic model, the electron in a hydrogen atom circles the nucleus, and the atom can have a magnetic dipole moment from the orbiting electron. Suppose an electron with mass m and charge magnitude e is in a circular orbit and has orbital angular momentum of magnitude L. Show that the magnetic dipole moment of the orbiting electron is given by $-(e/2m)\mathbf{L}$. The quantity $e/2m$ is known as the *magnetogyric ratio*. It turns out that the ratio of the orbital magnetic dipole moment to orbital angular momentum has the same value for any orbital path.

Supplementary Problems

29-48 P Clouds can accumulate significant amounts of electrostatic charge (this is what gives rise to lightning). As the cloud drifts through the earth's magnetic field, forces can act, affecting the shape of the cloud. Using the following crude model, estimate the magnitude of the forces that would arise. Suppose the cloud has a charge $+1$ C on its upper half and -1 C on its lower half (the order of magnitude of the charge released in a lightning bolt). Suppose the cloud is drifting westward at the equator with a speed of 10 m/s at a point at which the earth's field is horizontal and directed north-south and has magnitude 1 G. What force acts to separate the two halves of the cloud?

29-49 P What is the magnetic force on an arc 40 mm long that carries a current of 500 A if it is acted on by a perpendicular magnetic field of 0.10 T? Such forces on arcs find many applications, such as in a rail gun or in a high-current circuit-breaker, where a magnetic field is used to "blow away" the arc if the current becomes too great.

Sources of the Magnetic Field

30

30-1 The Oersted Effect
30-2 The Magnetic Force between Current-Carrying Conductors
30-3 The Magnetic Field from a Current Element; the Biot-Savart Relation
30-4 Gauss's Law for Magnetism
30-5 Ampère's Law
30-6 The Solenoid
30-7 Magnetic Materials
    Summary

In Chapter 29 we dealt with one part of the magnetic interaction: a moving charged particle acted on by a magnetic force when it is in a magnetic field. The other part of the magnetic interaction involves a moving charged particle creating a magnetic field.

30-1 The Oersted Effect

An important scientific discovery is often the result of long and careful experiment. Sometimes it comes as a lucky accident. But how about a fundamental finding made with a demonstration experiment during a physics lecture?

It actually happened at the University of Copenhagen in 1820. Until then, electricity and magnetism were considered entirely separate phenomena; after all, a charged object does not attract or repel a magnet. But H. C. Oersted (1777–1851) believed that electricity and magnetism were related. Oersted's experiment (he said later that he did not have time to check it before the lecture) was simple enough; he sent an electric current through a conducting wire and looked to see whether a compass placed near the conductor was affected. The compass did indeed respond (strictly, it twitched a bit because of

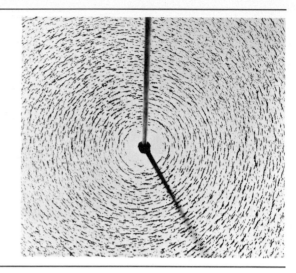

Figure 30-1. The large current in a long, straight conductor causes iron filings to arrange themselves in circular rings.

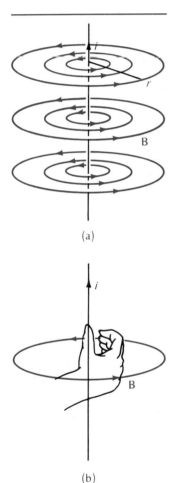

Figure 30-2. (a) A long, straight conductor with current i produces circular magnetic-field loops. (b) The right-hand rule for relating the direction of conventional current i and the sense of the associated **B** loops.

the low-current sources available then, and the people at Oersted's lecture were unimpressed). But repetitions of the experiment soon confirmed what has since that time been known as *the Oersted effect—a magnetic field is created by an electric current, by electric charges in motion.*

Figure 30-1 shows the effect on iron filings of the magnetic field from a relatively large current in a straight conductor. The magnetic field **B** produced by current i in a long straight conductor is shown in more detail in Figure 30-2(a). The **B** lines consist of circles of radius r concentric with the conductor and lying in planes transverse to the line of the conductor. The sense of the **B** loops is related to the direction of conventional current i by the right-hand rule shown in Figure 30-2(b): with the outstretched right thumb pointing in the direction of i, the curled right-hand fingers give the sense of the **B** loops. Reversing the current direction reverses the direction of **B** at each location.

How does the magnitude of **B** vary with the perpendicular distance r from the long straight conductor, where r is also the radius of the magnetic-field loop? As J. B. Biot (1774–1862) and F. Savart (1791–1841) first showed shortly after Oersted's discovery, $B \propto 1/r$; the magnetic field at any point is inversely proportional to the distance r from the conductor. Furthermore, the magnitude of **B** is directly proportional to current i, so that $B \propto i/r$. For reasons to become evident later in this chapter, it is customary to write the magnitude of **B** from a long, straight conductor as

$$B = k_m \frac{2i}{r} \tag{30-1}$$

The quantity $k_m \equiv \mu_0/4\pi$ is a magnetic interaction constant whose role in magnetism is analogous to what the constant $k_e = 1/4\pi\epsilon_0$ does in electricity. Suffice it to say at this point that k_m is *assigned* the numerical value of *exactly*

$$k_m \equiv \frac{\mu_0}{4\pi} \equiv 10^{-7} \frac{\text{T}\cdot\text{m}}{\text{A}} \tag{30-2}$$

Constant μ_0 itself is officially called the *permeability of free space*. The units assigned to $k_m = \mu_0/4\pi$, tesla meters per ampere, assure that B will have the units teslas. [Check that with (30-1).] Why the factor 4π? It is introduced here, so that it will not appear later in the mathematical formulation of Ampère's law (Section 30-5), just as $1/4\pi\epsilon_0$ appears in Coulomb's law but only ϵ_0 in Gauss's law for electricity.

We can see from (30-1) that it takes a fairly large current to produce a magnetic field of even moderate size. What is the magnetic field 2 cm from a straight conductor carrying 10 A? From (30-1), we find that $B = 10^{-4}$ T = 1 G, roughly the magnitude of the earth's magnetic field near the surface.

An *infinitely* long, *perfectly* straight conductor can never be constructed, if only because we must always have a complete loop, a return conductor, to maintain a steady current. Nevertheless, (30-1) can be applied for a straight conductor of finite length. We must simply be sure that the point at which B is computed is close to the straight conductor and relatively far from the locations where the conductor deviates from a straight line.

Example 30-1. A straight conductor lies along the x axis and carries a conventional current in the positive x direction. What is the direction of the magnetic force on an electron fired as shown in Figure 30-3: (a) in the first quadrant and in the $+x$ direction; and (b) in the first quadrant and in the $-y$ direction?

From the right-hand rule for the magnetic field from a straight conductor, we find that $\mathbf{B}$ is out of the paper (in the $+z$ direction) in the first quadrant. We apply the magnetic-force relation, $\mathbf{F}_m = q\mathbf{v} \times \mathbf{B}$, recognizing that an electron has negative charge, to find the results shown in Figure 30-3: (a) $\mathbf{F}_m$ along $+y$; and (b) $\mathbf{F}_m$ along $+x$.

Example 30-2. See Figure 30-4(a), where each of two parallel straight conductors separated by a horizontal distance d carries a current i into the paper. What is the magnetic field at point P, a distance d from each of the two conductors?

The two conductors and point P are at the corners of an equilateral triangle of side d. As shown in Figure 30-4(b), the magnetic field $\mathbf{B}_1$ produced at P by conductor 1 is perpendicular to the line from 1 to P and therefore at an angle of 30° below the horizontal. In similar fashion, $\mathbf{B}_2$ at P from conductor 2 is, as shown in Figure 30-4(c),

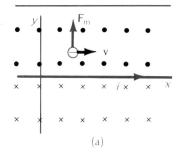

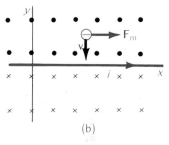

Figure 30-3. Electron with velocity $\mathbf{v}$ subject to magnetic force $\mathbf{F}_m$ when traveling through the outwardly directed magnetic field $\mathbf{B}$.

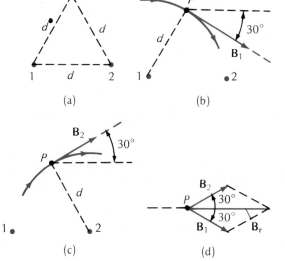

Figure 30-4. (a) Magnetic field at point P from conductors 1 and 2. (b) Conductor 1 alone produces magnetic field $\mathbf{B}_1$. (c) Conductor 2 alone produces magnetic field $\mathbf{B}_2$. (d) Resultant field $\mathbf{B}_r$ is the vector sum of $\mathbf{B}_1$ and $\mathbf{B}_2$.

30° above the horizontal. Since P is equally distant from 1 and 2, then $B_2 = B_1 = k_m (2i/d)$, from (30-1). The resultant magnetic field $\mathbf{B}_r$ at P is shown in Figure 30-4(d). The vertical components of $\mathbf{B}_1$ and $\mathbf{B}_2$ cancel, and their vector sum along the horizontal has the magnitude

$$B_r = 2B_1 \cos 30° = 2k_m \left(\frac{2i}{d}\right)\left(\frac{\sqrt{3}}{2}\right)$$

$$= 2\sqrt{3}\, k_m \frac{i}{d}$$

30-2 The Magnetic Force between Current-Carrying Conductors

Consider two long, straight, parallel, current-carrying conductors. They interact magnetically because each conductor produces a magnetic field at the site of the other. This is sometimes referred to as *the fundamental* magnetic interaction, since both conductors are electrically neutral and there is no electric force between them. We wish to find magnetic force per unit length on either conductor in terms of the currents i_1 and i_2 in the two conductors and their separation distance d.

Conductor 1 with current i_1 produces magnetic field $\mathbf{B}_1$. At the location of conductor 2, a distance d away, we have from (30-1) that

$$B_1 = k_m \frac{2i_1}{d}$$

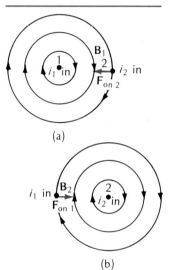

Figure 30-5. (a) Straight conductor 1 with current i into the paper produces magnetic field $\mathbf{B}$ at the location of conductor 2, with current i_2 also into the paper. The magnetic force on conductor 2 is $\mathbf{F}_{on\,2}$. (b) The converse of part (a) with current $\mathbf{i}_2$ producing magnetic field $\mathbf{B}_2$ and force $\mathbf{F}_{on\,1}$.

The field lines for $\mathbf{B}_1$ are shown in Figure 30-5(a). To find the direction of the magnetic force $\mathbf{F}_{on\,2}$ on conductor 2, we apply the rule for the magnetic force on a current element [(29-14)], $d\mathbf{F}_m = i\, d\mathbf{l} \times \mathbf{B}$. In Figure 30-5(a), with current i_2 in the same direction as i_1, we find that $\mathbf{F}_{on\,2}$ is to the left. The force magnitude on a length L of conductor 2 is

$$F_{on\,2} = B_1 i_2 L = k_m \frac{2 i_1 i_2 L}{d}$$

where we have used the result for B_1 given above.

In exactly analogous fashion, we find that the magnetic force $\mathbf{F}_{on\,1}$ on conductor 1 is to the right, as shown in Figure 30-5(b), and that it has the magnitude

$$F_{on\,1} = B_2 i_1 L = k_m \frac{2 i_2 i_1 L}{d} = F_{on\,2}$$

Two conductors with currents in the same direction attract one another magnetically. On the other hand, for currents in opposite directions, it is easy to see that the two conductors repel one another. Roughly speaking—like currents attract, unlike currents repel.

The force magnitude per unit length on either conductor is given by

$$\frac{F_{on\,1}}{L} = \frac{F_{on\,2}}{L} = k_m \frac{2 i_1 i_2}{d} \tag{30-3}$$

It is directly proportional to each of the currents, i_1 and i_2, and is inversely proportional to their separation distance d.

Suppose that each of two straight conductors carries a current of exactly 1 A and that they are separated by exactly 1 m. The force per unit length is, then, from (30-3),

$$\frac{F}{L} = k_m \frac{2i_1 i_2}{d} = \left(10^{-7} \frac{\text{T} \cdot \text{m}}{\text{A}}\right) \frac{(2)(1\text{A})^2}{(1 \text{ m})} = 2 \times 10^{-7} \text{ N/m}$$

where we have used $1 \text{ T} = 1 \text{ N/A} \cdot \text{m}$. The force per meter of conductor is exactly 2×10^{-7} N.

Now recall that the magnetic constant $k_m \equiv \mu_0/4\pi$ is assigned the value of exactly 10^{-7} T·m/A. Actually, the relation for the magnetic force between two parallel straight conductors defines the ampere as the basic electric unit for current in the SI system of units. At last we have the definition of the ampere (and therefore all other electromagnetic units that depend on the ampere):

The equal currents in two parallel straight conductors separated by 1 m in a vacuum are each, *by definition, exactly one ampere* when the magnetic force per meter on either conductor is precisely 2×10^{-7} N.

30-3 The Magnetic Field from a Current Element; the Biot-Savart Relation

Although its magnetic field is relatively simple, an *infinitely long, straight,* current-carrying conductor is a special configuration. The basic current element is one of infinitesimal length $d\mathbf{l}$ carrying current i, where vector $d\mathbf{l}$ points in the direction of the conventional current. (The infinitesimal current element does for currents and their associated magnetic fields what point electric charges do in electrostatics and their associated electric fields.)

Figure 30-6(a) shows the magnetic field originating from an infinitesimal current element. The magnetic field consists of concentric circular loops lying in planes transverse to $d\mathbf{l}$; the magnitude of **B** falls off with distance from the element in all directions. The direction of **B** is related to the direction of the current by the right-hand rule.

The magnitude of the magnetic field $d\mathbf{B}$ at any point P in Figure 30-6(b) is given by

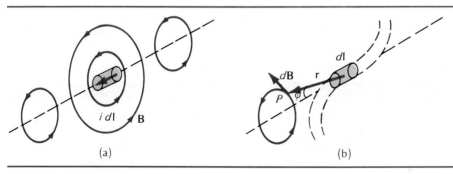

Figure 30-6. (a) Magnetic field loops from an infinitesimal current element. (b) The magnetic field $d\mathbf{B}$ at point P from current i through an infinitesimal element of conductor $d\mathbf{l}$. Radius vector **r** gives the location of P relative to $d\mathbf{l}$, and ϕ is the angle between **r** and $d\mathbf{l}$.

And the Beat Goes On

His routine never varied. During his lecture tour in America, the English novelist Charles Dickens would, upon arriving at an inn for the night's stay, whip out the compass he always carried, find north, and have his bed set along that line. The reason? The salubrious effect of having the human body aligned with the earth's magnetic field. After all, the 19th century argument went, if a compass is under least stress when it is aligned with a magnetic field, surely the human body must also enjoy minimum stress when it is aligned with a magnetic field?

Not only has magnetism confused novelists, the magnetic interaction has also been perplexing to scientists. Well, the magnetic force *is* different, even peculiar: it involves particle velocities, a couple of fairly complicated cross products, and energy conservation does not apply to it. The gravitational and the electric forces were never like this. Would it not be far better—if physics is truly to be simple at its most basic level—if there simply were no magnetic force?

Even that can be accomplished. We can, so to speak, "turn off" the magnetic force on a charged particle by a simple strategem: we merely ride with the particle as an observer in the reference frame in which the particle is always at rest. Then, since the charged particle is at rest, the magnetic force on it must be zero. It's quite extraordinary. We've made the magnetic field disappear simply by changing reference frame. But the particle's motion cannot be affected by whether we see it in motion (with a magnetic force acting on it) or travel along with it (and have no magnetic force). The resultant force on the particle must still be the same even though the magnetic force may have been turned off. So we are compelled to reach this conclusion: what had been a

$$dB = k_m \frac{i \, dl \, \sin \phi}{r^2} \tag{30-4}$$

where **r** is a radius vector from dl to P, and ϕ is the angle between dl and **r**. Note that the magnetic field falls off inversely with the square of the distance r from the current element; the *magnetic force* between a pair of infinitesimal current elements is *inverse-square*. (It is no surprise, then, that B from a long conductor varies with $1/r$, in the same fashion as E from a long, uniformly charged wire.) Note the effect of the $\sin \phi$ factor in (30-4); other things being equal, the magnetic field dB is a maximum in the transverse plane containing the current element ($\phi = 90°$) but falls to zero ahead of ($\phi = 0$) or behind ($\phi = 180°$) the current element.

The direction of $d\mathbf{B}$ shown in Figure 30-6(b) can be built into a vector relation, using the cross product (Section 12-2) as follows:

$$d\mathbf{B} = k_m \frac{i \, d\mathbf{l} \times \mathbf{r}}{r^3} \tag{30-5}$$

Satisfy yourself that (30-5) gives the direction of $d\mathbf{B}$ shown in Figure 30-6(b) in terms of the directions of $d\mathbf{l}$ and **r**. The additional factor of r in the denominator—r^3 in (30-5) as against r^2 in (30-4)—is needed to compensate for the additional magnitude of r introduced into the numerator of (30-5) by vector **r**.

It is easy to verify that the magnetic field consists of circular loops centered on the current element. Imagine that all quantities on the right side of (30-5) are fixed except for the direction **r**, which we imagine as rotating about $d\mathbf{l}$ at a constant angle ϕ. Vector $d\mathbf{B}$, unchanged in magnitude, then turns through a circle in a transverse plane.

magnetic field (with the particle in motion) has been transformed into an electric field (with the particle at rest). It turns out that an electric field can similarly be transformed into a magnetic field.

To pursue in more detail these fascinating aspects of electromagnetism would take us far beyond the intent of this text. But we should know that there is more to the story. The magnetic field and the magnetic force *do* look simple when electromagnetic effects are re-examined from the point of view of relativity theory.* To identify just one rather amazing result, recall what we found for electrically neutral conductors: they interacted *only* by the magnetic force (Section 30–2). Relativity gives a different way of looking at this. Suppose that you ride with a conduction electron in one conductor; a second conductor, with current in the same direction, is nearby. Conduction electrons are at rest in the second conductor, positive ions are in motion. Because of the relativistic space-contraction effect, the positive ions are somewhat bunched together, and the conductor as a whole carries a *net positive* charge. *That* is why conductors with like current attract. In this sense, the magnetic force is not a distinctive type of force but really just another aspect of the electric force.

So relativity gives the final insight into basic electromagnetism? Not so. Quantum theory (Chapter 41) shows electromagnetic radiation is quantized into particle-like photons. So that's the last word? Not so. Recent developments in quantum field theory show that the electromagnetic force is very closely related to the so-called "weak interaction" that shows up in the radioactive decay of certain unstable nuclei. Indeed, the two forces are separate manifestations of an even more primordial electro-weak force.

*For an introductory, intermediate-level treatment of classical electromagnetism, that is relativistic right from the start, see *Electricity and Magnetism*, Second Edition (New York: McGraw-Hill, 1985) by Nobel laureate E. M. Purcell.

Equation (30-5) for the magnetic field of an infinitesimal current element is usually referred to as the *Biot-Savart law*. How do we know it is right? There is, after all, no such thing as a single tiny piece of current unconnected to the rest of the world; a current element must always be part of a complete loop of current. Equation (30-5) is correct because it always works; that is, when all the $d\mathbf{B}$ contributions, each given by (30-5), from the infinitesimal elements that make up an actual current loop are added (as vectors), the computed resultant magnetic field at each location always agrees with experiment.

Example 30-3 B at Center of Circular Loop. The loop has radius r and carries current i, as shown in Figure 30-7. We suppose that the two lead wires that carry current to and from the loop are placed side by side. Because we then have nearly coincident equal currents in opposite directions, the magnetic effects of the lead conductors cancel, and we are left with the magnetic field of the ring of current alone. Element $d\mathbf{l}$ is on the circumference of the loop and radius vector $\mathbf{r}$ points to the center of the loop. For the counterclockwise current of Figure 30-7, the magnetic field $d\mathbf{B}$ produced by every element $d\mathbf{l}$ along the circumference is out of the plane of the paper and perpendicular to it. Therefore, we can sum the $d\mathbf{B}$ contributions algebraically. With angle ϕ between $d\mathbf{l}$ and $\mathbf{r}$ equal to $90°$, we have, from (30-4), for the magnitude of $\mathbf{B}$ at the center of the loop,

$$B = \int dB = k_m \frac{i}{r^2} \int dl$$

But $\int dl = 2\pi r$, the distance around the circumference, so that the final result is

$$B = k_m \frac{2\pi i}{r} \qquad (30\text{-}6)$$

For example, with a current of $i = 1$ A in a loop with $r = 1$ cm $= 10^{-2}$ m, Equation

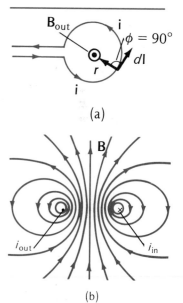

Figure 30-7. (a) The magnetic field **B** at the center of a circular current loop of radius r with current i. (b) Magnetic field lines in a plane transverse to a circular conducting loop and containing its symmetry axis.

(30-6) gives for the field at the center, $B = 0.6 \times 10^{-4}$ T $= 0.6$ G. For a coil of N identical circular loops, B at the center (and all other locations) is increased by a factor N.

Computing **B** from a circular loop at locations other than its center is far more complicated. The results for a plane transverse to the plane of the loop and containing the symmetry axis is shown in Figure 30-7(b). Close to the conducting wire, the field lines are nearly circular, since more distant current elements of the loop contribute little to the field there. The field lines are symmetrical with respect to the loop's symmetry axis. The direction of **B** at the center can be found by applying the right-hand rule to any current element. There is another way, however, to relate the direction of **B** at the center to the sense of current in the loop. Let the *curled fingers* of the right hand give the sense of the *current;* then the right *thumb* points in the direction of the *field* at the loop's center.

Example 30-4 B from a Long, Straight Conductor. Here we confirm that (30-1) for the magnetic field from an infinitely long, straight, current-carrying conductor follows from the Biot-Savart relation, (30-5), for an infinitesimal current element.

See Figure 30-8, where current i flows along the positive y axis. At point P, a distance R from the conductor, the field $d\mathbf{B}$ produced by the element dy is into the paper. The angle between current element $d\mathbf{l}$ and radius vector $\mathbf{r}$ to point P is ϕ. From (30-4), we have

$$dB = k_m \frac{i \, dy \, \sin \phi}{r^2} \quad (30\text{-}7)$$

Every current element of the straight conductor produces a field into the paper at P, so that we merely integrate the equation above algebraically over the entire length of the conductor to find the total field. Quantities y, r, and ϕ are not independent, however; we must first write (30-7) in terms of a single variable, this variable chosen as the one that will be the simplest to integrate. That variable is angle α, the complement of ϕ.

From the geometry of Figure 30-8, we see that

$$y = R \tan \alpha$$

so that

$$dy = R \sec^2 \alpha \, d\alpha$$

We also see that

$$r = R \sec \alpha$$

and

$$\sin \phi = \cos \alpha$$

Using these relations in (30-7), we get

$$dB = k_m \frac{i(R \sec^2 \alpha \, d\alpha)(\cos \alpha)}{(R \sec \alpha)^2} = k_m \frac{i \cos \alpha \, d\alpha}{R}$$

For a conductor of infinite length, we integrate α from $-90°$ to $90°$. (To find the field of a straight conductor of finite length, we should have to choose different limits.) Therefore,

$$B = \int dB = \frac{k_m i}{R} \int_{-\pi/2}^{\pi/2} \cos \alpha \, d\alpha = k_m \frac{2i}{R}$$

To conform to our earlier usage, we now replace R by r for the distance from the conductor, and have finally

$$B = k_m \frac{2i}{r} \quad (30\text{-}1)$$

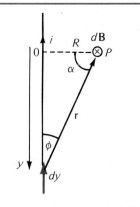

Figure 30-8. Magnetic field from a long straight current-carrying conductor.

30-4 Gauss's Law for Magnetism

First recall Gauss's law for electricity (Section 24-4):

$$\phi_E = \oint \mathbf{E} \cdot d\mathbf{S} = \frac{q}{\epsilon_0} \quad (24\text{-}5)$$

The total electric flux ϕ_E out of any closed surface is proportional to q, the net charge enclosed by the surface. Gauss's law says essentially that for electric charges at rest, *electric-field lines* **E**:

- Originate from positive charges.
- Terminate on negative charges.
- Are continuous between charges.

Therefore if we know how the electric field lines are arranged over any arbitrary closed Gaussian surface, we can immediately deduce the net charge inside.

Gauss's law for magnetism is similar. It deals with the total magnetic flux $\phi_B = \oint \mathbf{B} \cdot d\mathbf{S}$ through an imaginary closed Gaussian surface of any shape by adding the contributions $\mathbf{B} \cdot d\mathbf{S}$ of magnetic field **B** over a small, outwardly directed surface element $d\mathbf{S}$. See Figure 30-9. Positive magnetic flux corresponds to **B** lines out of the surface, negative magnetic flux to inward **B** lines. If the total magnetic flux over a closed surface is zero, equal numbers of **B** lines enter and leave the surface. This is precisely what is found in every case.* Experimental findings are consistent with the statement:

$$\phi_B = \oint \mathbf{B} \cdot d\mathbf{S} = 0 \quad (30\text{-}8)$$

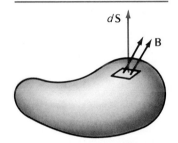

Figure 30-9. *Magnetic field* **B** *through patch of surface represented by vector d***S**.

This is *Gauss's law for magnetism*. It says, in effect:

- **B** lines always form closed loops.
- **B** lines do not originate from and terminate on magnetic "charges" (or what are known more formally as *magnetic monopoles*). The reason is simply that isolated magnetic poles do not exist in nature.† Every **B** line ends on its own tail.

Gauss's law for the magnetic fields produced by current-carrying conductors follows directly from the Biot-Savart relation (Section 30-3). Every infinitesimal current element produces circular magnetic-field lines; the magnetic flux with circular loops is clearly zero over a closed surface of any shape. Then any finite collection of current elements must also produce continuous magnetic field loops, so that $\oint \mathbf{B} \cdot d\mathbf{S} = 0$ in all situations.

What about magnets (magnetic dipoles), where **B** lines appear to emanate from the north pole and terminate at the south pole, the two ends of the magnet? See Figure 30-10. Here the magnetic field lines pass through the interior of the magnet and again form closed loops.

---

* Well, almost every case. See the next footnote.
† In very subtle experiments, B. Cabrera of Stanford University in 1982 found evidence for a single magnetic monopole, most likely one surviving from the primordial fireball at the creation of the universe.

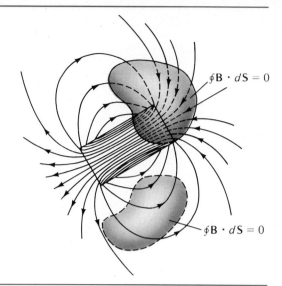

Figure 30-10. Gauss's law for magnetism: The net magnetic flux through any closed surface is zero.

30-5 Ampère's Law

Recall the two ways of relating electric field to electric charge:

• **Coulomb's law.** Each point charge dq creates an electric field $d\mathbf{E}$ at the location given by radius vector $\mathbf{r}$, where $d\mathbf{E} = k\, dq\, \mathbf{r}/r^3$ (Section 23-4).

• **Gauss's law of electricity.** The electric flux $\oint \mathbf{E} \cdot d\mathbf{S}$ over any closed surface is proportional to the net charge q within, where $\oint \mathbf{E} \cdot d\mathbf{S} = q/\epsilon_0$ (Section 24-4).

Coulomb's law is in *differential* form, Gauss's law in *integral* form. There are also differential and integral forms for relating magnetic field to electric current (see Table 30-1):

• **The Biot-Savart law.** Each current element $i\, d\mathbf{l}$ creates a magnetic field $d\mathbf{B}$ at the location given by radius vector $\mathbf{r}$, where $d\mathbf{B} = k_m i\, d\mathbf{l} \times \mathbf{r}/r^3$.

• **Ampère's law**, which relates the magnetic field around the edge of any imaginary sheet to the net current passing through the sheet.

Table 30-1. Fields and their Sources

	ELECTRICITY	MAGNETISM
Differential form	Electric field, point charge from Coulomb's law $$d\mathbf{E} = k\frac{dq\, \mathbf{r}}{r^3}$$	Magnetic field, infinitesimal current element from Biot-Savart law $$d\mathbf{B} = k_m \frac{i\, d\mathbf{l} \times \mathbf{r}}{r^3}$$
Integral form	Gauss's law for electricity $$\oint \mathbf{E} \cdot d\mathbf{S} = \frac{q}{\epsilon_0}$$	Ampère's law $$\oint \mathbf{B} \cdot d\mathbf{l} = \mu_0 i$$

More specifically, Ampère's law can be expressed by the relation

$$\oint \mathbf{B} \cdot d\mathbf{l} = \mu_0 i \qquad (30\text{-}9)$$

Here $d\mathbf{l}$ is an infinitesimal displacement vector lying along the outer edge of a sheet of arbitrary shape, and $\mathbf{B}$ the local magnetic field at that location. See Figure 30-11. Contributions of the component of $\mathbf{B}$ along $d\mathbf{l}$ are added, going all the way around the loop along the outer edge of sheet, and the complete line integral of the magnetic field around the loop $\oint \mathbf{B} \cdot d\mathbf{l}$ is thereby computed. It turns out, as we shall see, that this line integral is always equal to $\mu_0 i$, where i is the net current through the sheet. Figure 30-11 implies a convention, corresponding to the right-hand rule, relating directions; when a loop is traversed in the *counterclockwise* sense (with the interior of the loop on the left) and the *line integral* is *positive*, current coming *out* of the sheet is *positive*.

Like Gauss's law, Ampère's law is a very general principle relating a field (here $\mathbf{B}$) to its source (here i). And like Gauss's law, Ampère's law is easy to apply in concrete situations only if there is a high degree of geometrical symmetry.

Let us see how Ampère's law works in one familiar simple situation — a long, straight current-carrying conductor. Consider Figure 30-12, where as we know, the magnitude of $\mathbf{B}$ at a distance r from the conductor is

$$B = k_m \left(\frac{2i}{r}\right) = \left(\frac{\mu_0}{4\pi}\right)\left(\frac{2i}{r}\right) \qquad (30\text{-}1)$$

(Here we have replaced k_m by its equivalent, $\mu_0/4\pi$.) The relation above can also be written as

$$B(2\pi r) = \mu_0 i$$

Here we have chosen the arbitrarily shaped sheet in Ampère's law to consist of a flat circular plate of radius r, concentric with and perpendicular to the conductor, as shown in Figure 30-13. Then, at each point along the sheet's outer circular edge, $\mathbf{B}$ is constant in magnitude and parallel to $d\mathbf{l}$. Therefore, the line integral around the chosen closed loop, $\oint \mathbf{B} \cdot d\mathbf{l}$, is here just equal to B multiplied by the circumference $2\pi r$. Clearly, the relation above is just what Ampère's law gives.

Now we shall see that we get the very same result for the long straight conductor for any loop enclosing a sheet of *any* shape. Consider the loop of Figure 30-13(a), where the path consists of two circular arcs coinciding with field lines and two radial lines at right angles to $\mathbf{B}$. For radius r_1, the field has magnitude B_1; at r_2 it is B_2. But (30-1) shows that the field falls off inversely with distance r from the conductor, so that $B_1 r_1 = B_2 r_2$. Along the radial lines, $d\mathbf{l}$ is perpendicular to $\mathbf{B}_1$, so that there is no contribution to the line integral. The decrease from B_1 to B_2 in going from r_1 to a larger r_2 is exactly matched by the increase in arc length. For the loop of Figure 30-13(a), with current i enclosed, we again have

$$\oint \mathbf{B} \cdot d\mathbf{l} = \mu_0 i$$

$$B(2\pi r) = \mu_0 i$$

Suppose the loop does not enclose the conductor; this is shown in Figure 30-13(b). Again there is no contribution to the line integral along the radial

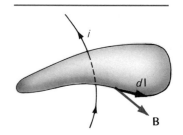

Figure 30-11. Magnetic field $\mathbf{B}$ at infinitesimal displacement vector $d\mathbf{l}$ along the edge of a sheet through which current i passes.

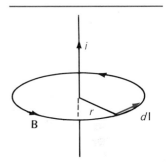

Figure 30-12. Magnetic field $\mathbf{B}$ at a distance r from a long, straight conductor with current i.

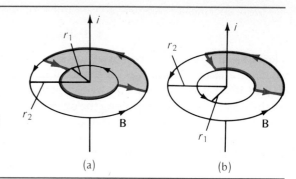

Figure 30-13. Magnetic field **B** from a long straight conductor for a loop (a) enclosing and (b) not enclosing current i.

lines. Once again *in magnitude*, $B_1 r_1 = B_2 r_2$; contributions along the two circular arcs have the same magnitude. But their signs differ. At r_2 the path is traversed in the *same* direction as **B**; at r_1, the directions of $d\mathbf{l}$ and **B** are *opposite*. Consequently, the total line integral is now zero,

$$\oint \mathbf{B} \cdot d\mathbf{l} = 0$$

which corresponds to no current passing through the loop.

To deal with more complicated loops, we simply replace the arbitrary path by small circular arcs and radial segments. See Figure 30-14(a). Circular sections subtending the same angle give the same contribution; radial segments all make zero contribution. In every case, the line integral around a closed loop is equal to the current threading the loop multiplied by the constant μ_0. Note further that if the direction of the current is reversed, and therefore also the direction of **B** at each location, a factor of -1 is thereby introduced. Even a sheet of arbitrary shape, one that does not lie in a plane perpendicular to the straight conductor, introduces no complications: **B** is entirely in planes transverse to the conductor, so that a path element parallel to the conductor does not contribute to the line integral.

Thus far we have proved that Ampère's law applies for a single long, straight conductor. Suppose that we have two or more infinitely long, straight conductors. This introduces nothing new; magnetic fields add as vectors, so that if Ampère's law holds for one straight conductor, it also applies for a collection of straight conductors. We must, however, take the current i to be the net current through the sheet whose edge constitutes the loop for the line integral, using different signs for current out of and into the sheet. One way to

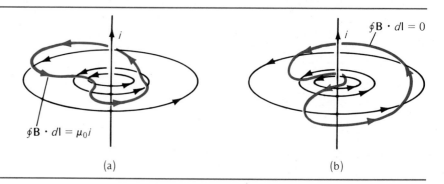

Figure 30-14. Ampère's law applied to a general loop shape (a) enclosing the conductor and (b) not enclosing the conductor.

prove that Ampère's law holds for conductors of any shape, not merely long, straight ones, is to show that any current distribution and its associated magnetic field can be produced by equivalent straight conductors.*

Although Ampère's law is altogether general, it can be used to calculate the magnetic field only when the geometry of **B** can be deduced in advance on the basis of symmetry. Then we know what sort of path to choose. (This is reminiscent of Gauss's law for electricity: If the geometry of **E** is known in advance on the basis of symmetry, we know how best to choose the Gaussian surface.)

One final remark on Ampère's law. As written in (30-9), it is not complete. In Section 34-2 we shall see that a changing electric flux, as well as a current, can contribute to the line integral of the magnetic field.

Example 30-5 Cylindrical Coaxial Conductors. A coaxial conductor consists of a solid inner circular cylinder of radius r_1, carrying current I, and a concentric outer conductor of inner and outer radii r_2 and r_3 and carrying current I in the opposite direction. See Figure 30-15. The current in both conductors is uniformly distributed over their cross sections. What is the magnetic field (a) outside the outer conductor? (b) between the two conductors? (c) within the inner conductor?

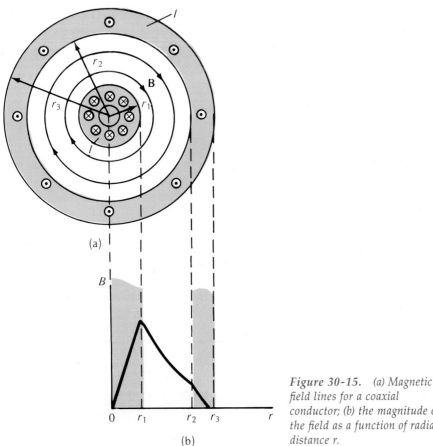

Figure 30-15. (a) Magnetic field lines for a coaxial conductor; (b) the magnitude of the field as a function of radial distance r.

* See F. Reines, *American Journal of Physics* **39**:838 (1971).

First, we know on the basis of symmetry alone that the magnetic field must, at any place where it is not zero, consist of circular loops concentric with the conductors. This means that in applying Ampère's law we shall always choose a concentric circular loop.

(a) Outside both conductors the net current, $I - I$, within the loop is *zero*. Consequently, the field there must, from Ampère's law, also be zero.

For $r > r_3$, we get $B = 0$.

(b) For a circular path in the region between the two conductors, only the current I *within the loop* contributes to the field. (The current of the outer conductor has no effect whatsoever on the region inside; it merely cancels the field of the inner conductor for $r > r_3$.) From Ampère's law we have, then,

$$\oint \mathbf{B} \cdot d\mathbf{l} = \mu_0 i$$

$$B(2\pi r) = \mu_0 I$$

For $r_1 < r < r_2$,

$$B = \frac{\mu_0 I}{2\pi r}$$

(c) Our circular loop of radius r now lies within the inner conductor. Only a fraction of the total current I is *within* this loop. With current distributed uniformly over the cross section, the fraction of I inside of r is the ratio of the circular areas $\pi r^2 / \pi r_1^2$, so that the current within r is $(r/r_1)^2 I$. Ampère's law then gives

$$B(2\pi r) = \mu_0 \left(\frac{r}{r_1}\right)^2 I$$

For $0 < r < r_1$,

$$B = \frac{\mu_0 r I}{2\pi r_1^2}$$

The magnetic field falls off linearly with r. The field within the outer conductor can be computed in similar fashion.

The variation of B with r is shown in Figure 30-15(b).

30-6 The Solenoid

A solenoid consists of a tightly wound helix of conducting wire. It is, in effect, a series of adjacent circular conducting loops arranged into a cylindrical shell. Near the center, the magnetic field within the solenoid is essentially uniform, as shown in Figure 30-16. The external magnetic field has a configuration like that of a long permanent cylindrical magnet.

We can find the magnitude of **B** at the center of an infinitely long solenoid — or at least one whose length is much greater than its diameter — from

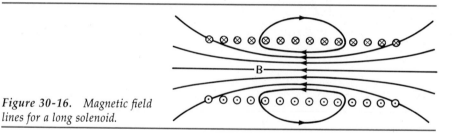

Figure 30-16. Magnetic field lines for a long solenoid.

Ampère's law. Each loop carries current i_1, and the number of turns per unit length is n. Consider, then, the rectangular closed path shown dashed in Figure 30-17. Just outside the solenoid, **B** must be essentially zero; the field lines passing through the interior are very thinly spread on the exterior. There is no contribution to the line integral of **B** along portion A. The field inside, along line B, must by symmetry be parallel to the solenoid axis; if it were not, this would imply a noninfinite solenoid length. For this reason, **B** is perpendicular to the path along the transverse portions C and D, and these parts of the closed path also do not contribute to the line integral. The only contribution comes from the inside segment of length L. The number of turns within length L is nL, and the total current enclosed is nLi_1. Therefore, from Ampère's law we have

$$\oint \mathbf{B} \cdot d\mathbf{l} = \mu_0 i$$

$$BL = \mu_0 n L i_1$$

$$B = \mu_0 n i_1 \qquad (30\text{-}10)$$

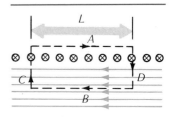

Figure 30-17. Ampère's law applied to a solenoid.

The magnetic field at any interior point near the center of a long solenoid is uniform, independent of the transverse distance from the axis; the magnitude of **B** depends only on the number of turns per unit length and i_1, the current through each.

The solenoid is a device for combining the magnetic effects of many loops to produce a strong, uniform magnetic field. As the Biot-Savart relation or Ampère's law shows, B is always $\propto i$; large currents create big fields. Since the resistance of a superconductor is zero, the very large currents that can be

Figure 30-18. The tunnel of the main accelerator at Fermilab. The upper ring of magnets (in a square casing) is the 400 GeV accelerator. The lower ring is superconducting magnets for the Tevatron. In the Tevatron, protons accelerated to a kinetic energy of 1 TeV (10^{12} eV) collide head-on with antiprotons of the same energy. The Tevatron employs the world's largest collection of superconducting magnets, more than a thousand of them. Each magnet, kept at a temperature below 4.6 K by liquid helium, produces a magnetic field of 4.5 T.

achieved in superconducting materials are now used increasingly to produce large magnetic fields. See Figure 30-18.

30-7 Magnetic Materials

Figure 30-19(a) shows the magnetic field of a solenoid, Figure 30-19(b) the magnetic field of a magnet with same dimensions. The two fields are the same. We may, in fact, think of the magnet's field as coming from currents. Such currents would encircle the magnet's external cylindrical surface.* Actually,

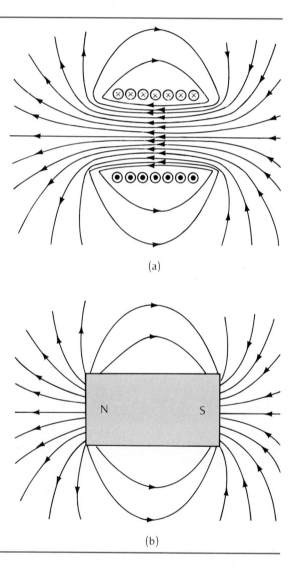

Figure 30-19. Magnetic field produced by (a) a solenoid and (b) a permanent magnet with the same dimensions.

---

* This insight—that the external magnetic field of a magnet can be attributed to circulating currents—was first given by André M. Ampère (1775–1836). He developed essentially all the fundamental relations between currents and magnetic fields within several weeks of learning of Oersted's discovery that an electric current creates a magnetic field.

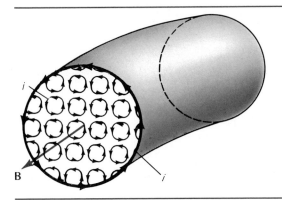

Figure 30-20. Cross section of core material within toroid, showing current loops.

separate loops of current distributed over the entire cross section of a magnet are equivalent to a single current loop around the periphery; at any interior location, the opposite currents from adjoining loops cancel, so that we are left with only the current (sometimes called the *Ampèrian current*) around the outside. See Figure 30-20.

What exactly are the atomic currents that produce magnetism in iron? Because atomic properties are understood adequately only on the basis of the quantum theory, what follows can be only qualitative and approximate. Certainly atoms contain electrically charged particles and microscopic currents; we have electrons moving in orbits about nuclei. But in ordinary materials, electron orbits are so paired that their magnetic effects cancel; for each electron going clockwise, there is another going counterclockwise in a similar orbit. The origin of the magnetism of magnetic materials is *electron spin*. There is no complete classical analog to electron spin, a strictly quantum-mechanical effect, but we may visualize it as follows. An electron regarded as a sphere of negative charge spins perpetually at a constant rotation rate about an internal rotation axis. See Figure 30-21. The circulating charge produces an electron-spin magnetic moment μ as shown (because an electron has negative charge, its spin magnetic moment is opposite in direction to its spin angular momentum). The special — and from the point of classical electromagnetism, peculiar — property of electron spin is this: the electron-spin magnetic moment has the same magnitude (1.61×10^{-23} J/T) for all electrons, and this magnetic moment (and the associated spin rate) is unaffected by an external magnetic field. In this sense, a spinning electron is the only example of a perfect permanent magnet.

In nonmagnetic materials, not only are the electron orbital magnetic moments paired off to yield no net magnetic effects. So too are electron spins, with pairs of electron spins in antialignment to yield zero electron-spin magnetization. In feebly magnetic, or *paramagnetic*, materials (for example, copper, manganese, chromium) there are unpaired electron spins, however. In zero external field, the magnetic moments of these unpaired spins produce no net magnetization, because the spin magnetic moments of the magnetic ions are oriented at random through the material. But when an external magnetic field is turned on, more electron-spin magnetic moments align themselves with the field than against the field, so that there is net magnetization along the direction of the external field. The alignment of electron-spin magnetic moments by

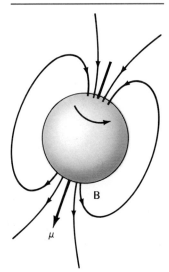

Figure 30-21. Magnetic field surrounding a spinning charged sphere, a model for electron spin.

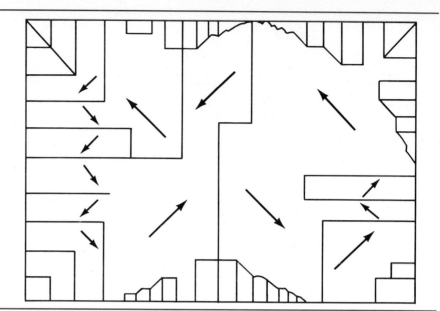

Figure 30-22. Ferromagnetic domains in a single crystal of nickel. The direction of the magnetic field in each domain is shown with an arrow.

the external field competes with the disorganizing effect of thermal motion within the paramagnetic material; the atomic vibrations tend to produce random spin orientations and no net magnetization. Therefore as the temperature is lowered and the thermal vibrations diminish, the magnetization of a paramagnetic material typically increases.

Ferromagnetic materials—such as iron, nickel, and cobalt—are strongly magnetic in the sense that the electron spins from many adjoining ferromagnetic ions are strongly coupled together by a quantum-mechanical force for which there is no classical analog, known as an *exchange force*. As a consequence large groups of atoms, or *magnetic domains,* are formed in which many electron spins are aligned together to form a fairly large magnetic moment for the domain. A typical domain has a size of about 10^{-7} m (about 10^3 atomic diameters). In an unmagnetized ferromagnetic material, the domains are oriented at random so as to give no net magnetization for the material as a whole. See Figure 30-22.

Now suppose that an external magnetic field is turned on. The boundaries of the domains change, so that those domains whose magnetic moment is pointed along the direction of the external field grow in size at the expense of other domains. In addition, the orientation of the magnetic moment within any one domain many change direction toward alignment with the external field. In fact, a very large fraction of the electron-spin magnetic moments may become aligned with the external field, and the material acquires a net magnetic moment. When the external magnetic field is turned off, the magnetic material may retain a smaller magnetic moment, and a "permanent" magnet is produced. Ferromagnetism is complicated by the fact that the properties of ferromagnetic materials depend not only on external magnetic fields but also on the past magnetic (and thermal) history of the material.

Summary

Definitions

The magnetic interaction constant k_m and permeability of the vacuum μ_0 have values that are defined to be exactly

$$k_m \equiv \frac{\mu_0}{4\pi} \equiv 10^{-7}\ \text{T}\cdot\text{m/A}$$

The *ampere* as the SI unit for current and the basis for all other electromagnetic SI units: that current in each of two infinitely long, parallel, straight conductors that produces a magnetic force on one meter of conductor of exactly 2×10^{-7} N [(30-3)].

Fundamental Principles

The Oersted effect: electric charges in motion, or currents, create magnetic-field loops. The directions are related by the right-hand rule.

Biot-Savart relation for the magnetic field $d\mathbf{B}$ at radius vector $\mathbf{r}$ from an infinitesimal element of current i in a displacement $d\mathbf{l}$:

$$d\mathbf{B} = k_m \frac{i\ d\mathbf{l} \times \mathbf{r}}{r^3} \qquad (30\text{-}5)$$

Gauss's law for magnetism:

$$\phi_B = \oint \mathbf{B} \cdot d\mathbf{S} = 0 \qquad (30\text{-}8)$$

The net magnetic flux ϕ_B out of any closed surface is zero.

Equivalently, **B** lines form closed loops; magnetic monopoles do not exist.

Ampere's law:

$$\oint \mathbf{B} \cdot d\mathbf{l} = \mu_0 i \qquad (30\text{-}9)$$

where $d\mathbf{l}$ is a displacement along the edge of a sheet of arbitrary shape and i is the net current through the loop formed by the edge of the sheet.

Important Results

Magnetic field B at distance r from an infinitely long, straight conductor with current i:

$$B = k_m \left(\frac{2i}{r}\right) \qquad (30\text{-}1)$$

Magnetic force per unit length between two parallel conductors with current i_1, and i_2 and separated by distance d:

$$\frac{F}{L} = k_m \left(\frac{2 i_1 i_2}{d}\right) \qquad (30\text{-}3)$$

"Like currents" (same direction) attract; "unlike currents" (opposite directions) repel.

Magnetic properties of magnetic materials have their origin in electron spin.

Problems and Questions

Section 30-1   The Oersted Effect

- **30-1 Q** What is the character of the magnetic-field lines produced by an infinitely long, straight wire that carries a current?
  (A) They spiral in toward the wire in planes perpendicular to the wire.
  (B) They spiral outward from the wire in planes perpendicular to the wire.
  (C) They are radial lines perpendicular to the wire.
  (D) They are concentric circles in planes perpendicular to the wire.

- **30-2 Q** Current is established along the x axis toward positive x in a long, straight wire. An electron passes through the point $y = 0$, $z = a$ moving toward positive y parallel to the y axis. The force acting on the electron will be
  (A) directed toward the $+z$ direction.
  (B) directed toward the $-z$ direction.
  (C) directed toward the $+x$ direction.
  (D) parallel to the y axis.
  (E) zero.

- **30-3 Q** A long, straight wire is oriented perpendicular to the plane of the paper. In the plane of the paper is a uni-

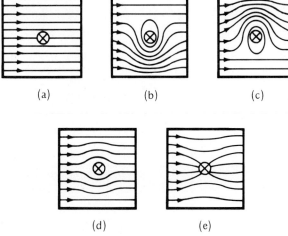

Figure 30-23.   Question 30-3.

form magnetic field directed from left to right. Which of the diagrams in Figure 30-23 best describes the magnetic field lines resulting from this arrangement when current in the wire is directed into the paper?

· **30-4 Q** Current exists in the irregular loop shown in Figure 30-24. The loop lies in the plane of the paper. Thus the magnetic field at point P is
(A) directed toward A.
(B) directed toward B.
(C) directed toward C.
(D) directed toward D.
(E) directed out of the paper.
(F) directed into the paper.
(G) zero.
(H) oriented in a direction which can't be determined without more information.

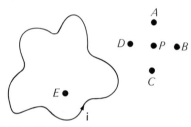

Figure 30-24. Questions 30-4, 30-5.

· **30-5 Q** An irregular loop of wire carrying a current lies in the plane of the paper (Figure 30-24). Suppose that the loop is then distorted into some other shape while still lying in the plane of the paper. Point E is still within the loop. Which of the following is a true statement concerning this situation?
(A) The magnetic field at E will not change in magnitude.
(B) The magnetic field at E will not change in direction.
(C) It is possible that the magnetic field at E may be zero.
(D) The magnetic field at point E will always lie in the plane of the paper.
(E) None of the above is true.

: **30-6 P** A current of 2 A exists along the z axis in the positive z direction, and a current of 3 A exists along the line $y = 0.02$ m in the xy plane in the positive x direction. What is the magnetic field at the point (0, 0.01 m, 0)?

: **30-7 Q** Consider a square loop of current in which a

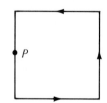

Figure 30-25. Question 30-7.

sinusoidally alternating current exists, as shown in Figure 30-25. Which of the following is an accurate statement?
(A) The loop will experience a torque that will tend to make it rotate.
(B) A magnetic force to the left will act on point P.
(C) A magnetic force to the right will act on point P.
(D) A magnetic force out of the paper will act on point P.
(E) A magnetic force oscillating in direction will act on point P.
(F) The magnetic force on point P will be zero.

: **30-8 P** In a simple model of the hydrogen atom, the electron orbits the proton as nucleus in a circle of radius 5.3×10^{-11} m at a speed of 2.2×10^6 m/s. What is the magnitude of the magnetic field produced at the site of the proton by the orbiting electron?

Section 30-2 The Magnetic Force between Current-Carrying Conductors

· **30-9 Q** Two long parallel straight wires carry currents of 2 A and 4 A as indicated in Figure 30-26. Thus the right-hand wire will experience a force that is
(A) directed to the right.
(B) directed to the left.
(C) zero.
(D) directed out of the paper.
(E) directed into the paper.
(F) different in magnitude from the force exerted on the left-hand wire.

Figure 30-26. Question 30-9.

· **30-10 P** Each of two long, straight, parallel wires carries a current of 2 A. The wires are separated by 4 cm. What force is experienced by a 3-cm segment of one of the wires?

: **30-11 P** The ampere is defined so that it is the current that exists in each of two parallel wires separated by 1 m if one meter of wire is to experience a magnetic force of 2×10^{-7} N/m drawing them together. What linear electric-charge density would each of two such wires have to carry to produce the same attractive force?

: **30-12 Q** Two concentric circular loops of flexible wire lie on a table. There is a sudden pulse of current in the outer wire, and then a second pulse of current in the opposite direction. How will the inner loop of wire be affected?

(A) It will first contract slightly and then expand slightly.
(B) It will first expand slightly and then contract slightly.
(C) It will contract with each surge of current.
(D) It will expand with each surge of current.
(E) It will not be affected by the outside loop.

: **30-13 Q** Sketch the resultant magnetic field for two parallel, straight current-carrying conductors with currents (a) in the same direction and (b) in opposite directions. The result seen here, one that is altogether general, is that the direction of the magnetic force on a current-carrying conductor is in the direction from strong-to-weak resultant magnetic field.

: **30-14 P** Four long straight wires each carry a current of I. They are arranged at the corners of a square of side d (see Figure 30-27). What is the magnitude and direction of the force per unit length acting on one of the wires? The current is in the same direction in all wires.

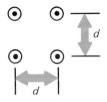

Figure 30-27. Problem 30-14.

: **30-15 P** Three long, straight wires run parallel to each other and carry currents of I, two with currents in the same direction and the third with its current in the opposite direction. Each wire passes through a corner of an equilateral triangle of side d. Determine the magnitude and direction of the magnetic force per unit length on the wires.

: **30-16 P** Two protons move parallel to each other with velocity **v**. Show that the ratio of the magnetic force $\mathbf{F}_m$ to the electric force $\mathbf{F}_e$ on either proton is $F_m/F_e = v^2\epsilon_0\mu_0 = (v/c)^2$.

Section 30-3 The Magnetic Field from a Current Element; the Biot-Savart Relation

· **30-17 P** What is the magnetic field at the center of a loop of wire, of radius 10 cm, carrying a current of 1 A?

: **30-18 P** Two semicircles of wire are joined to form a loop carrying 4 A (Figure 30-28). What is the magnetic field at point C, the center of the concentric semicircles? The inner semicircle has radius 2 cm; and the outer radius, 4 cm.

Figure 30-28. Problem 30-18.

: **30-19 P** Determine **B** at the center C of the semicircular loop carrying current I shown in Figure 30-29.

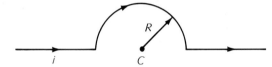

Figure 30-29. Problem 30-19.

: **30-20 P** Current I exists in the wire drawn in Figure 30-30. Determine the magnetic field at the center C of the arc.

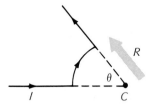

Figure 30-30. Problem 30-20.

: **30-21 P** Derive an expression for the magnetic field at a point P on the axis of a circular current loop of radius R carrying current i, in the case for which the distance z from the center of the loop to P is much greater than R.

: **30-22 Q** Two beams of protons are fired in the same direction along parallel paths. What happens to the motion of the two beams when the initial velocity of one of the beams is increased?
(A) The motion will be unaffected by this change.
(B) The beams will tend to draw together more.
(C) The beams will repel each other other more.
(D) A spiral motion of the beams will be induced.
(E) The speed of particles in the second beam will also increase.

: **30-23 P** Using the following methods, determine the magnetic field at the center of a square loop of side $2a$ with current I. Use the law of Biot-Savart to (*a*) calculate **B** due to one side. Multiply by 4 to get the desired result. (*b*) Fit a circular loop just outside the square and find **B** at its center. Then fit a circular loop just inside and again find **B**. Average these values to get an approximate value for the field due to the square. (*c*) What is the ratio of the values found in (*a*) and (*b*)? Does the approximation improve for larger loops?

: **30-24 P** Charge Q is uniformly distributed over the surface of a circular disk of radius R. The disk is rotated with constant angular velocity ω about its symmetry axis. (*a*) Determine the magnetic field at the center of the disk. (*b*) Where is the magnetic field greatest? (Give a qualitative argument.)

Section 30-4 Gauss's Law for Magnetism

30-25 Q Which of the following are accurate statements?
(A) Magnetic-field lines would "sprout" out of magnetic monopoles in the same way that static electric-field lines emanate from positive electric charges.
(B) The magnitude of a magnetic field is proportional to the spacing between its magnetic-field lines; that is, the larger the separation between lines, the stronger the magnetic field.
(C) Magnetic-field lines are analogous to whirlpool flow lines in a fluid, where the "paddlewheels" stirring up the magnetic fields are electric currents.
(D) A proton moving in a magnetic field experiences a magnetic force directed tangent to a magnetic-field line.
(E) Magnetic fields exert forces only on magnetic objects, not on electrically charged particles.

Section 30-5 Ampère's Law

30-26 Q It is important, in using the equation $\oint \mathbf{B} \cdot d\mathbf{l} = \mu_0 i$ to find the field due to a long, straight wire, to recognize that
(A) the integral is evaluated on a path along the wire.
(B) the integral is evaluated on a circular path in a plane perpendicular to the wire.
(C) $d\mathbf{l}$ is a small vector pointing from the wire to the point at which the field is to be evaluated.
(D) $d\mathbf{l}$ is a small vector directed along the wire.
(E) $\mathbf{B}$ will not have constant magnitude along the path C.

30-27 Q The equation $\oint \mathbf{B} \cdot d\mathbf{l} = \mu_0 i$ is
(A) based on experimental observation.
(B) one in which $d\mathbf{l}$ is a small vector pointing along a wire carrying current i.
(C) a result of Newton's second law of motion.
(D) a consequence of Gauss's law, $\int \mathbf{E} \cdot d\mathbf{S} = q/\epsilon_0$
(E) a consequence of $\mathcal{E} = -d\phi_B/dt$

30-28 P Current exists with uniform density across the cross-sectional area of a long, straight wire of radius R. The magnetic field at the surface of the wire is B_0. What is the value of the magnetic field inside the wire at a point a distance $\frac{1}{2}R$ from the axis?

30-29 P A uniform current I exists in a cylindrical pipe with inner radius R_1 and outer radius R_2. The current density is uniform over the cross section of the pipe. Determine the magnetic field everywhere.

30-30 P The current density in a long straight cylindrical conductor of radius R is given by $j = j_0 e^{-r/r_0}$ at a distance r from the cylinder axis. Find the magnetic field as a function of r.

30-31 P What is the magnetic field just outside an infinite plane sheet that carries a uniform surface current density $\mathbf{j}$?

30-32 P Uniform, constant surface current densities are established parallel to each other in two thin sheets separated by a distance d. Determine the magnetic field at an arbitrary point between the two sheets and outside the two sheets when (a) the current densities are in the same direction; (b) the current densities are in opposite directions (antiparallel).

Section 30-6 The Solenoid

30-33 Q Consider the magnetic field set up in space by a coil of N turns carrying current i. No other sources of magnetic field are present. If the number of turns is doubled, the magnetic field will
(A) be unchanged.
(B) increase in some regions and decrease in others, so that the total magnetic flux will be doubled.
(C) double everywhere.
(D) increase everywhere, but not by the same factor at every point.
(E) None of the above.

30-34 Q The easiest way to calculate the magnetic field within a long solenoid of N turns is to
(A) note that inductance depends only on geometric factors.
(B) use $B = \mu_0 i / 2\pi r$.
(C) use $\mathcal{E} = -d\phi/dt$.
(D) use $\oint \mathbf{B} \cdot d\mathbf{l} = \mu_0 i$.
(E) use $d\mathbf{B} = (\mu_0/4\pi) i d\mathbf{l} \times \mathbf{r}/r^3$.
(F) use $d\mathbf{F} = i\, d\mathbf{l} \times \mathbf{B}$.

30-35 P A solenoid that is 80 cm long and 2 cm in diameter is wound with 5000 turns of wire. What is the field near the center of the solenoid when it carries a current of 2 A?

30-36 Q An electron is fired along the axis of a current-carrying solenoid. It is aimed at point A on the fluorescent screen shown in Figure 30-31. It strikes the screen at one of the indicated points. Which point does it strike?

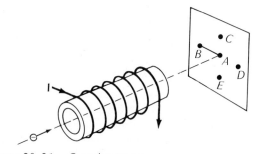

Figure 30-31. Question 30-36.

30-37 Q The pattern of magnetic field lines shown in Figure 30-32 is set up by a little black box. Such a magnetic field might result if the black box contained
(A) a toroid (a solenoid in the shape of a doughnut) carrying a current.

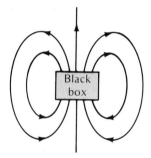

Figure 30-32.   Question 30-37.

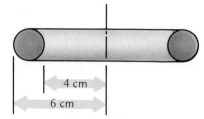

Figure 30-33.   Problem 30-41.

(B) a single small loop of wire carrying a current.
(C) a current-carrying wire wound into a sphere like a ball of yarn.
(D) a small bar magnet.
(E) none of the above.
(F) More than one of the above could be correct.

: 30-38 P A toroid (a solenoid bent into a circle to form a doughnut) has inner radius r_1 and outer radius r_2. Show that the magnetic field inside the toroid depends only on r, the distance from the center, and not on the shape of the cross section of the toroid (for example, the cross section may be circular, elliptical, or rectangular).

: 30-39 P A toroid of circular cross section has an inner radius of 4 cm and an outer radius of 6 cm (Figure 30-33). It is wound with 400 turns of wire. What is the magnetic field at the center of the toroid (that is, at a radius of 5 cm) when a current of 2 A exists in it?

: 30-40 P Show that the magnetic field at the end of a very long solenoid is just half the value at the center. (*Hint:* Think of an infinitely long solenoid as consisting of two long solenoids placed end to end.)

Supplementary Problem
30-41 P The very high currents that can exist with charged particles in motion in a plasma can give rise to the effect in magnetohydrodynamics known as the *pinch effect.* Parallel currents in the same direction attract one another. In similar fashion, current over the cross section of a beam of particles can produce inwardly directed magnetic forces over the outer surface of the beam. Such a magnetic pressure can pinch the beam of charged particles. (*a*) Show that the magnetic pressure varies as the square of the current in the beam. (*b*) Show that the pressure can reach 1 atmosphere when the current in a beam of 1 cm radius reaches about 25,000 A. (For 10^6 A, the pressure is 1600 atm.)

31 Electromagnetic Induction

31-1 Induced Currents and EMF's
31-2 EMF in a Moving Conductor
31-3 Lenz's Law
31-4 Faraday's Law and the Induced Electric Field
31-5 Eddy Currents
31-6 Diamagnetism
Summary

Basic features of the interactions between electrically charged particles are these:

• Part of the force does not depend on the particle velocities. This is the *electric*, or Coulomb, force.

• A velocity-dependent force also acts between charged particles. This is the *magnetic* force.

We might be inclined to think that so far as basic electromagnetism is concerned, this is the whole story. It is not! Electromagnetism is concerned with electrically charged particles and their interactions, of course, but equally important are electric and magnetic fields and their interrelation. Electric and magnetic fields are created by electric charges at rest and in motion. But in addition, we shall see that:

• A changing magnetic field can create an electric field. (That is the fundamental phenomenon of electromagnetic induction to be treated in this chapter.)

• A changing electric field can create a magnetic field. (That topic is treated in chapter 34.)

Beyond its role in classical electromagnetic theory, electromagnetic induction also has enormous practical importance. Such common devices as inductors, transformers, alternating-current generators—and indeed, a large fraction of all applied electric power—are based directly in the induction effect.

31-1 Induced Currents and EMF's

The phenomenon of *electromagnetic induction* was discovered in England in 1831 by Michael Faraday (1791–1867), regarded as the greatest experimental physicist of all time. The same effects were discovered independently in the United States about the same time by Joseph Henry (1797–1878), the principal scientific advisor to Alexander Graham Bell in the development of the telephone.

Figure 31-1 shows a variety of experimental situations in which electromagnetic induction is observed. Each involves a loop of conductor with a galvanometer that can register the existence of an *induced current* in the loop; we call this the detector loop. Some *changing* condition induces the current:

• Closing the switch in a nearby primary loop containing a battery, Figure 31-1(a).
• Moving an entire current-carrying primary loop toward a detector loop, Figure 31-1(b).
• Moving a permanent magnet toward a detector loop, Figure 31-1(c).
• Reorienting a detector loop when it is close to the current-carrying primary loop, Figure 31-1(d).
• Deforming a detector loop when it is close to the current-carrying primary loop, Figure 31-1(e).

The general qualitative experimental observations are these (the current direction is discussed in Section 31-3):

• An induced current exists only while the change shown in Figure 31-1 is occurring (while the switch is being closed, while the magnet is moving, and so on).
• Only relative motion matters. If, for example, in Figure 31-1(c) the magnet remains at rest with the detecting loop moved left, instead of the magnet's being moved right with the loop at rest, a current is again induced. On the other hand, with both magnet and loop at rest or with both moving at the same velocity, there is no induced current.
• The more rapid the change or motion, the greater the magnitude of induced current.
• If the change is reversed (switch opened, rather than closed; motion to left rather than right; and so on) the direction of the current is also reversed.

The effects are peculiar, at least in terms of what we have learned thus far of electromagnetism. The detector loop has no battery in it. The loop is unconnected to any other circuit and sits in apparently empty space. Yet an emf drives charges around the conductor to produce an induced current.

What common feature did Faraday identify in these varied circumstances? What was the basic underlying change associated with every induced current? Faraday was an early champion of the field concept. For him electric-field lines and magnetic-field lines (he always called them lines of force) were a palpable

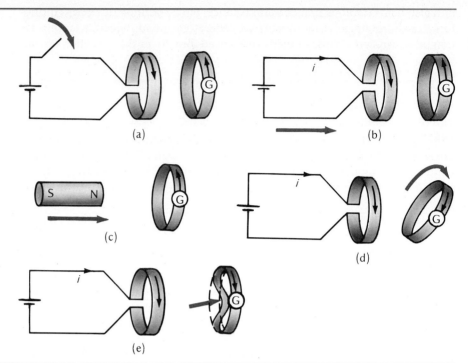

Figure 31-1. Experimental situations for observing electromagnetic induction. (a) Closing a switch in a primary loop with a battery. (b) Moving the entire current-carrying primary loop. (c) Moving a permanent magnet. (d) Reorienting the detector loop. (e) Deforming the detector loop.

presence. In Faraday's mind's eye, every charged object bristled with electric-field lines, every current-carrying conductor was wreathed with magnetic-field loops. Faraday could perceive in the situations portrayed in Figure 31-1 what cannot be shown in any single static picture:

- In graphical terms, the *number of magnetic-field lines* threading through the detector loop *changes with time*.
- Or in equivalent mathematical terms, the *magnetic flux* through the detector loop *changes with time*.

In Figure 31-1(a), for example, closing the switch and establishing a current in the primary loop produces a growing magnetic field, not only in the region of the primary loop, but also through the detector loop. Similarly, the motion of a current-carrying primary loop [Figure 31-1(b)] or the motion of a magnet [Figure 31-1(c)] changes the number of magnetic-field lines through the detector loop.

Recall that the magnetic flux (Section 29-2) $d\phi_B$ through some small patch of surface element $d\mathbf{S}$ at a location where the magnetic field is $\mathbf{B}$ is defined as

$$d\phi_B = \mathbf{B} \cdot d\mathbf{S} = BS \cos\theta = B_\perp \, dS \tag{31-1}$$

where θ is the angle between $\mathbf{B}$ and $d\mathbf{S}$, and $B_\perp$ is the component of $\mathbf{B}$ perpendicular to the surface. To find the net magnetic flux over some finite loop, one simply sums the contributions from infinitesimal surface elements. We can think of the magnitude of $\mathbf{B}$ as equal to the number of magnetic-field lines per unit transverse area; the net magnetic flux ϕ_B through a loop is then equal to the net number of magnetic-field lines threading the chosen loop.

A magnetic flux change through an element of surface $d\mathbf{S}$ can then arise from any of the following circumstances, individually or in combination:

- A change in the *magnitude* of the *magnetic field* **B** [Figure 31-1(a), (b), (c)].
- A change in the *magnitude* of *area* $d\mathbf{S}$ of the loop [Figure 31-1(e)].
- A change in the angle θ between **B** and $d\mathbf{S}$, that is, in their *relative orientation* [Figure 31-1(d)].

Although induced current is observed directly in a conducting loop, the more fundamental induced quantity is the induced emf $\mathcal{E}$ that drives charged particles around the conducting loop to create the induced current. Experiment shows that all electromagnetic induction effects can be summarized in what is also called *Faraday's law*. This fundamental principle relates the emf $\mathcal{E}$ induced around any loop to the time rate of change of the magnetic flux through that loop as follows:

$$\mathcal{E} = -\frac{d\phi_B}{dt} \qquad (31\text{-}2)$$

First, let us check that the units in this relation are all right. With $d\phi_B/dt$ given in the units webers per second, does $\mathcal{E}$ have the units volts? The unit conversions go as follows:

$$\text{Wb/s} = \text{T} \cdot \text{m}^2/\text{s} = (\text{N}/\text{A} \cdot \text{m})\text{m}^2/\text{s} = \text{N} \cdot \text{m}/\text{A} \cdot \text{s} = \text{J}/\text{C} = \text{V}$$

Implications of Faraday's law of electromagnetic induction are these:

- Whether an emf is induced around some loop depends only on the change with time in the net magnetic *flux* through that loop. The magnetic field is not the crucial quantity. The magnetic field at all locations on the periphery of a loop may be constant, even constantly zero, but if the net flux within the loop changes, an emf is created (see Example 31-2).
- An induced emf, however, cannot be localized, like a battery; an emf is associated with the entire loop and is a property thereof.
- Suppose that the conducting loop has N turns. An emf is induced in each of the N coincident loops, and the net emf is larger than that from a single loop by a factor N. In that situation, Faraday's law may be written as

$$\text{Loop of } N \text{ identical turns:} \quad \mathcal{E} = -\frac{d(N\phi_B)}{dt} \qquad (31\text{-}3)$$

where ϕ_B is the flux change through a single loop and $\mathcal{E}$ is the net emf. The product $N\phi_B$ is called the *flux linkage*.

The minus sign in (31-2) relates to the direction or sense of the emf. This is treated in Section 31-3.

Example 31-1. A uniform magnetic field of 0.20 T is applied transverse to a flexible circular loop of a conductor 10 cm in radius. See Figure 31-2(a). The opposite points along a diameter of the loop are pulled outward so that 0.20 s later, the loop lies along a straight line, as shown in Figure 31-2(b). What is the average emf induced in the loop over the 0.20-s interval?

In this example the magnetic flux changes by ϕ_B because the area of loop is reduced from πr^2 to zero in a time interval Δt. The time average of the emf is then

$$\overline{\mathcal{E}} = -\frac{\Delta \phi_B}{\Delta t} = -\frac{B(\pi r^2)}{\Delta t} = -\frac{(0.20 \text{ T})\pi(0.10 \text{ m})^2}{0.20 \text{ s}} = -31 \text{ mV}$$

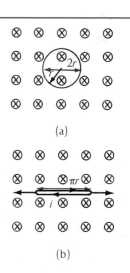

Figure 31-2. (a) A circular conducting loop in a magnetic field is pulled along opposite points along a diameter. (b) The loop lies along a straight line.

Example 31-2. A very long current-carrying solenoid passes through a square conducting loop. See Figure 31-3. The solenoid has a 1.0-cm radius, 20 turns/cm, and the current through it changes at the rate of 100 A/s. What is the magnitude of the emf induced in this square loop?

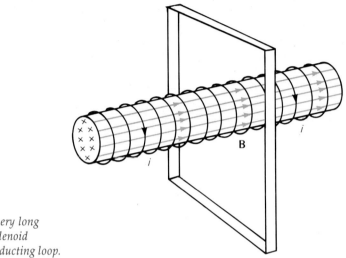

Figure 31-3. A very long current-carrying solenoid within a square conducting loop.

Inside the solenoid (Section 30-6) the magnetic field is uniform and lies along the solenoid axis; in magnitude [(30-10)], $B = \mu_0 ni$, where n is the number of turns per unit length and i the current in each turn. We assume that the solenoid's length is so great compared with the size of the square loop that the only magnetic field through the square loop is that confined to the interior of the solenoid. The magnetic field at all points of the square loop is *zero* and remains so.

The magnetic flux through a cross section of the solenoid is

$$\phi_B = B_\perp A = (\mu_0 ni)(\pi r^2)$$

where r is the solenoid radius. This is also the total magnetic flux through the entire square loop, since the only magnetic-field lines penetrating the square are the lines inside the solenoid. The emf induced in the square loop is then

$$\mathcal{E} = -\frac{d\phi_B}{dt} = -\frac{d}{dt}(\mu_0 ni\pi r^2) = -\mu_0 \pi n r^2 \frac{di}{dt}$$

$$= (4\pi \times 10^{-7} \text{ Wb/A} \cdot \text{m})\pi(20 \times 10^2/\text{m})(10^{-2}\text{ m})^2(100 \text{ A/s})$$

$$= -7.9 \times 10^{-5} \text{ V} = -79\mu\text{V}$$

(The minus sign relates to the sense of the emf.)

Note especially what the induced emf does not depend on: the dimensions of the loop, or even its shape; where the solenoid goes through the square. Even the angle between the plane of the loop and the axis of the solenoid does not enter. If the solenoid axis is not perpendicular to the rectangle, the area intercepted by the solenoid on the plane of the square is increased by a factor $(1/\cos \theta)$, but the magnetic flux and the number of magnetic-field lines is unchanged because of the appearance also of factor $\cos \theta$ in the relation for magnetic flux. All that matters in determining the emf is how rapidly the flux penetrating the loop changes with time.

Example 31-3 Electric Generator. A coil of N turns, each of area A, is rotated at constant angular speed ω in a uniform magnetic field of magnitude B. What is the emf induced in the coil?

The arrangement here is that of a simple electric generator. The magnetic flux through each turn of the coil is given by

$$\phi_B = AB \cos \theta$$

where θ is the angle between the normal to the plane of the coil and the direction of **B**; see Figure 31-4(a). With the coil rotated at constant angular speed ω, we can write $\theta = \omega t$, and the instantaneous flux is

$$\phi_B = AB \cos \omega t$$

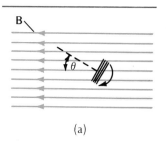

We have N identical conductor turns connected in series, so that the total emf is

$$\mathcal{E} = \frac{-d(N\phi_B)}{dt}$$
$$= \omega NAB \sin \omega t = \mathcal{E}_0 \sin \omega t$$

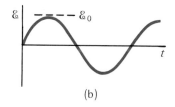

The output of the coil is a sinusoidally varying alternating emf (ac), as shown in Figure 31-4(b).

It is interesting to note how the coil is oriented when the instantaneous emf is a maximum and when it is zero:

- Maximum $\mathcal{E}$ corresponds to $\theta = \omega t = 90°$. At this instant, the magnetic flux throughout the coil is zero; $\phi_B = AB \cos \omega t = 0$. No magnetic field lines pass through the coil. But a small change in θ from 90° allows the lines to penetrate the coil, and the time rate of change of ϕ_B is a maximum.
- Zero $\mathcal{E}$ corresponds to $\theta = \omega t = 0°$. At this instant the magnetic flux has its maximum value; $\theta_B = AB \cos \omega t = AB$. Field lines pass perpendicularly through the coil but their number is unchanged for a small change in θ from 0°.

In the electric generator described here, mechanical power supplied by an external source turning the coils is transformed into electric power. Conversely, one can operate the arrangement in Figure 31-4 as an electric motor by supplying a sinusoidal electric current to the coil; then a magnetic torque acts on the coil, and electric power is converted to mechanical power. An electric generator operated in reverse is a motor.

Figure 31-4. (a) A coil rotating at constant angular speed ω in a uniform magnetic field. (b) The sinusoidally varying emf induced in the coil.

31-2 EMF in a Moving Conductor

Here we investigate one example in which a current is induced in a moving conductor, from two points of view:

- Faraday's law of electromagnetic induction.
- The detailed forces acting on conduction electrons.

Consider Figure 31-5, where a rectangular conducting loop is moving into a uniform transverse magnetic field **B** directed into the paper. For simplicity, we assume that the magnetic field drops abruptly to zero at its left boundary. The loop's height is l; its velocity to the right is v. After time t, the right end of the loop has advanced a distance vt into the field, so that an area lvt is within the field. The instantaneous magnetic flux through the loop is

$$\phi_B = Blvt$$

so that the emf induced in the loop is

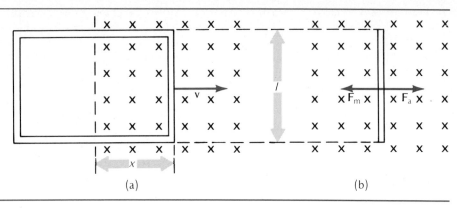

Figure 31-5. *(a) Rectangular conducting loop of height l moving at velocity **v** into a uniform magnetic field **B**. (b) Forces acting on the right-hand end of the conducting loop: retarding magnetic force **F**$_m$ and applied force **F**$_a$.*

$$\mathcal{E} = -\frac{d\phi_B}{dt} = -Blv \tag{31-4}$$

The minus sign relates to the sense of the induced emf, which is, as we shall see, counterclockwise. The induced current exists so long as the loop is in motion and the flux through it is changing. After the left end of the loop has entered the magnetic field and the loop is then entirely within the region of **B**, flux ϕ_B is constant and therefore $\mathcal{E} = 0$. As the loop's right end leaves the region of magnetic field, there is again an induced current, but in the reversed sense.

Now we examine the forces on conduction electrons. First we consider, not the entire loop in Figure 31-5, but only its right end, a rod of length l, shown in Figure 31-6. With the rod moved to the right at velocity v, so too is every conduction electron within it. The magnetic force on such a moving free electron has the direction, from $\mathbf{F}_m = q\mathbf{v} \times \mathbf{B}$, vertically downward.* As a consequence, free electrons are forced to the lower end of the moving rod; it acquires a net negative charge, so that the upper end of the rod is positively charged. A magnetic force also acts on the positive charges within the rod, but these charges remain locked in place. The charge separation continues until the downward magnetic force on a free electron is balanced by an equal upward electric force from electrons that are already at the rod's lower end. An electric field is created within the rod by opposite charges at the two ends, and an electric potential difference is established between the rod ends. This potential difference exists only so long as the rod is in motion and a magnetic force drives free electrons as they are pulled through the magnetic field. If the rod is brought to rest, the potential difference drops to zero; if the rod changes direction of motion, the potential difference reverses polarity.

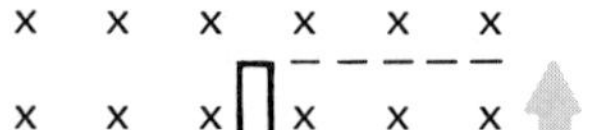

Back to Figure 31-5. The entire rectangular conducting loop has only its right end moving through **B**. Although charges are separated on the end within **B**, there is no charge separation on the left end, which is outside of **B**. Excess electrons no longer accumulate on the lower end of the right end. The electrons are driven around the loop in the clockwise sense.

With current in the right end of the loop as it is moved through the magnetic field, a magnetic force acts on this current-carrying conductor. From $d\mathbf{F}_m = i\,d\mathbf{l} \times \mathbf{B}$, we see that the magnetic force on the right end is in the direction

Figure 31-6. *(a) Rod of length l moving at velocity **v** through a magnetic field **B**. (b) The forces on a conduction electron moving to the right: a downward magnetic force **F**$_m$ and an upward electric force **F**$_e$.*

---

*Strictly, the electron moves *obliquely* downward in the rod, not just to the right. To simplify the analysis, we neglect the downward velocity component.

How Does Physics Advance?

A parable:

The room is dark. Objects don't fit, don't make sense. At first everything seems chaotic. But then the investigator finds two pieces that fit together, then another pair. Finally, after great effort, everything is in place. What had seemed to be disorderly parts now make up useful furniture; light illuminates the room brilliantly. As the investigator makes one more tour around the room, congratulating himself on his skill in bringing order out of chaos, he notices an imperfection. On the floor he sees a dark line—no, it is actually a rectangle that on further inspection turns out to be a trapdoor. Descending the stairs, the investigator finds that the room is dark. Objects don't fit, don't make sense.

Well, that's one way. Physics can progress by gradually, painstakingly fitting things together to form a coherent picture and then finding a defect that leads to a whole new and possibly quite different view of nature. That sort of thing happened when quantum theory was first introduced in 1900 by Max Planck to repair a defect in the theory of radiation from a black body; quantum physics gave the first sensible picture of atomic structure and led to an entirely new and different way of doing physics.

There are other ways also. Serendipity, the lucky accident in which an investigator stumbles upon a new effect he or she had not been looking for, usually happens to those thorough investigators who deserve it. For example, X-rays were discovered in 1895 by Wilhelm C. Röntgen, the first recipient of a Nobel prize in physics, when he noticed some materials glowing in the dark and also a mysterious darkening of photographic plates. It did not take long for this discovery to be put to practical use; X-ray pictures were being taken of bones within weeks.

Other major discoveries have come from someone's asking, "What if?" Albert Einstein, working alone in relative obscurity in 1905 kept asking himself, as he had ever since he first studied electromagnetism as a teen-ager, "What would an electromagnetic wave look like if you could ride along with the electric and magnetic fields?" The answer came in the special theory of relativity.

---

opposite the motion of the loop. Now if the loop is to be moved at constant speed v, the retarding force $F_m = Bil$ must be balanced by an external agent applying force $\mathbf{F}_a$ to the right with the magnitude $F_a = F_m = Bil$. The external agent does work dW as the loop advances a distance dx, where

$$dW = F_a \, dx = Bil \, dx$$

We can write current as $i = dq/dt$, so that

$$dW = B \frac{dq}{dt} l \, dx = Bl \frac{dx}{dt} dq$$

where $v = dx/dt$. The work per unit charge done by the applied force must equal the emf in the circuit, so that we get

$$\mathcal{E} = \frac{dW}{dq} = Bvl \tag{31-4}$$

which is the same result as that found earlier by applying Faraday's law for electromagnetic induction.

The agent does work continuously on the conducting loop; its kinetic energy does not change, since it moves at constant speed. Where does the energy go? The power P_a delivered to the loop by the agent is

$$P_a = F_a v$$
$$P_a = Bilv \tag{31-5}$$

Let R represent the electric resistance of the entire conducting loop. Then we can write

$$i = \frac{\mathcal{E}}{R} = \frac{Bvl}{R}$$

This result, when substituted in (31-5), gives finally

$$P_a = i^2 R$$

We see that the rate at which the agent does work is just equal to the rate at which thermal energy is dissipated in the loop.

Note a curious circumstance; we got the relation $\mathcal{E} = Blv$ simply by considering the magnetic force on conduction electrons. We did not need to use the electromagnetic induction principle at all. Does this mean the induction effects are not really new and fundamental? Not at all. To see that electromagnetic induction is a distinctive phenomenon, we need merely to take another look at the conductor loop moving through the magnetic field from the point of view of another observer, one riding with the conductor.

In this reference frame, the conduction electrons are at rest. They are subject to no magnetic force driving them through the conductor. Yet the observer traveling with loop does find an induced current and therefore an induced emf produced by changing magnetic flux.*

Example 31-4. An airplane flies horizontally at 270 m/s over the location in northern Canada where the earth's magnetic field is vertically downward and has a magnitude of 0.58 G. (This location is above the earth's magnetic pole at latitude 76.1°N and longitude 100°W.) (a) What, if any, is the electric potential difference between the wing tips, which are separated by 40 m? (b) How might the potential difference be observed?

(a) The situation is like that in Figure 31-6, with a conductor moving transversely through a magnetic field. The charge separation on the airplane's wings produces a potential difference V between the tips given, from (31-4), by

$$V = Blv = (0.58 \times 10^{-4} \text{ T})(40 \text{ m})(270 \text{ m/s}) = 0.63 \text{ V}$$

The left wing tip is positive, and the right wing tip is negative.

(b) Suppose that one tries to measure the electric potential difference with an ordinary voltmeter that rides with the airplane and has two conducting lead wires connected to the wing tips. No current goes through the voltmeter, and it reads zero. We can see this from two angles:

- The voltmeter with its lead wires also constitutes a conductor extending between the wing tips, and therefore the same electric potential difference exists across it as across the airplane wings.
- The voltmeter with leads and the airplane wings constitute a conducting loop. The magnetic flux through this loop does not change, and there can be no induced current.

Suppose that somehow the voltmeter remained at rest on the ground with its leads still connected to the airplane wing tips. Then the area of the conducting loop would change with time, as would the magnetic flux through it, and an emf would be induced in the loop. The voltmeter would register this emf.

* The traveling observer finds, in effect, the magnetic field turned off and an electric field turned on; or more fundamentally, in the new reference frame a magnetic field has been transformed into an electric field.

This example illustrates a special consideration that must always be kept in mind in dealing with moving conductors and induced emfs; an emf or an electric potential difference is specified meaningfully only if, at the same time, we specify the reference frame relative to which it is measured.

Example 31-5. As shown in Figure 31-7, a conducting rod of mass m slides to the right at initial speed v_0 over parallel conducting tracks separated by distance l and connected at their left ends across a resistance R. A uniform magnetic field **B** is applied perpendicular to the plane of the tracks. (a) Find the instantaneous velocity of the rod as a function of time. (b) What is the total thermal energy dissipated in the resistor?

(a) The induced emf $\mathcal{E} = Blv$ from (31-4) produces an induced current $i = \mathcal{E}/R = Blv/R$. The retarding magnetic force on the rod is then

$$F = Bil = B\left(\frac{Blv}{R}\right)l = \frac{B^2l^2v}{R}$$

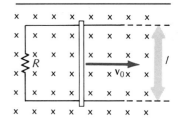

Figure 31-7.

From Newton's second law, we have

$$\Sigma F = ma$$

$$-\left(\frac{B^2l^2}{R}\right)v = m\frac{dv}{dt}$$

The minus sign indicates that the magnetic force retards the motion. The equation above is simplified when we put

$$k \equiv \frac{B^2l^2}{mR}$$

so that it can be written as

$$-\int_0^t k\,dt = \int_{v_0}^v \frac{dv}{v}$$

Evaluating the definite integral yields

$$v = v_0 e^{-kt}$$

The speed decreases exponentially with time.

(b) Energy is dissipated in resistor R at the rate

$$i^2R = \left(\frac{\mathcal{E}}{R}\right)^2 R = \frac{(Blv)^2}{R} = \frac{B^2l^2v_0^2}{R}e^{-2kt}$$

and the total energy dissipated is

$$\int_0^\infty i^2R\,dt = \frac{B^2l^2v_0^2}{R}\int_0^\infty e^{-2kt}dt = \frac{B^2l^2v_0^2}{2kR}$$

Substituting the value for k given above yields finally

$$\int_0^\infty i^2R\,dt = \tfrac{1}{2}mv_0^2$$

The total energy dissipated is just equal to the rod's initial kinetic energy.

31-3 Lenz's Law

For any loop of conducting material, there are always two possible directions, or senses, for the electric current. But when a current is induced by a changing magnetic flux, the current goes in one direction only. Which is it? The basis for finding the direction of the induced current, or the induced emf, is Lenz's law, named for H. F. E. Lenz (1804–1865).

Consider again the situation of Figure 31-5, where a conducting loop is pulled into a magnetic field. As we have seen, an external agent had to apply a force to balance the retarding magnetic force, and the work done by the agent appeared as thermal energy dissipated in the loop's resistance. The current in the loop of Figure 31-5 was found to be counterclockwise.

Suppose, for the sake of argument, that it went the other way, clockwise current in the loop of Figure 31-5. Then the magnetic force would be in the direction of the loop's motion, and an agent would not be required to drag the loop through the magnetic field. Once given a little push to get it started, the conductor would be accelerated by the magnetic force. Consequently, the current and the magnetic force would grow; the conductor would go still faster; and all the while thermal energy would be dissipated at an ever increasing rate. Clearly, this is impossible. It violates the principle of energy conservation. In this, as in all other examples of induced currents, the *direction of the current is always such as to preclude a violation of energy conservation.* That is one way of stating Lenz's law.

Here is another. *The direction of the induced current is always such as to oppose the change (in magnetic flux) that produces it.* For the situation in Figure 31-5, with a conducting loop pulled into a magnetic field, the opposition to magnetic flux change is manifest as follows. The magnetic field arising from the induced current (*out* of the paper for counterclockwise induced current) opposes the external magnetic field (*into* the paper) and thereby "tries" to maintain the magnetic flux through the loop unchanged. Opposition is also manifest in the retarding force.

In every example of induced emf, Lenz's law implies that:

• When a force acts on a conductor, it is a *retarding* force, one that opposes the relative motion.

• The magnetic field produced by an induced current is always in such a direction as to *oppose* the change in magnetic flux.

Another situation illustrating Lenz's law is shown in Figure 31-8(a). Here one current-carrying loop is being moved toward a second, fixed loop, in which a current is then induced. The magnetic flux through the fixed loop on the right increases with time to the *right*. Consequently, the situation in the fixed loop must be as follows. The magnetic field produced by the induced current must be in such a direction that it annuls (or tends to annul) the magnetic field producing the increasing flux to the right through the fixed loop.

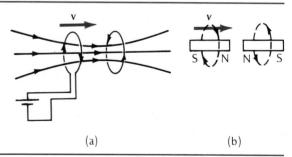

Figure 31-8. An example of Lenz's law. (a) The left-hand current-carrying conductor moves to the right, inducing a current in the fixed right-hand coil in the opposite sense. (b) Magnets equivalent to the coils of part (a).

That is, the induced current must itself produce a magnetic field to the *left*. Clearly then, from the right-hand rule relating the current to the magnetic field, the current in the fixed loop must be in the sense shown in Figure 31-8(a).

Here is another way of looking at it. Imagine a current-carrying loop to be equivalent to a magnet (see Figure 30-19). Then, we see from Figure 31-8(b) that the equivalent magnets associated with the two loops repel each other. Actually, it is the magnetic force between the two current-carrying conducting loops that is responsible for the repulsion between them. We have already found that the two currents are in opposite senses. This is in accord with the general result that two parallel current-carrying conductors repel one another when the currents are in opposite directions (Section 30-2).

For practice in applying Lenz's law, check that the directions of induced current shown in Figure 31-1 are correct.

The minus sign that appears in the mathematical statement of Faraday's law, $\mathcal{E} = -d\phi_B/dt$, is a symbolic representation of Lenz's law. We discuss it in Section 31-4.

Example 31-6. Magnetic field **B** is applied along the normal of the conducting loop, shown in Figure 31-9; **B** is decreasing with time. What is the sense of the current induced in the loop?

Because **B** is decreasing in magnitude, the change in this field Δ**B** is opposite in direction to **B**. From Lenz's law, the field **B**$_{\text{ind }i}$ produced by the induced current must be opposite to Δ**B**. (Said a little differently—**B**$_{\text{ind }i}$ points in the direction of **B** to make up for its decreased magnitude.) The right-hand rule applied to field **B**$_{\text{ind }i}$ and the induced current i that produces it then show the current to be in the direction displayed in Figure 31-9.

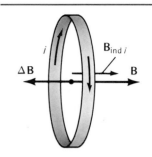

Figure 31-9. Magnetic field **B** decreases and changes by Δ**B**. The magnetic field **B**$_{\text{ind }i}$ produced by induced current i is opposite to Δ**B**.

31-4 Faraday's Law and the Induced Electric Field

According to the electromagnetic induction principle, a current is induced in a loop of conductor whenever the loop is exposed to a changing magnetic flux. But a current driven by an emf means, basically, that charged particles are driven by an electric field. Does a changing magnetic flux create an electric field in space, even when no electric charges are present? The answer: yes. Indeed, the fundamental electromagnetic induction effect is this: *A changing magnetic flux creates an electric field.*

First let us relate the induced emf $\mathcal{E}$ to the associated electric field **E**. The work dW done on charge q by electric field **E** as the charge is carried around a closed loop is

$$W = \oint \mathbf{F}_e \cdot d\mathbf{l} = \oint q\mathbf{E} \cdot d\mathbf{l}$$

But emf $\mathcal{E}$ is defined as work done per unit charge in taking it around the loop, so that

$$\mathcal{E} = \frac{W}{q} = \oint \mathbf{E} \cdot d\mathbf{l} \qquad (31\text{-}5)$$

Using (31-5) in (31-2), we get the fundamental form of Faraday's law:

$$\oint \mathbf{E} \cdot d\mathbf{l} = -\frac{d\phi_B}{dt} \qquad (31\text{-}6)$$

The loop chosen for computing the line integral $\oint \mathbf{E} \cdot d\mathbf{l}$ of the electric field is arbitrary, but the chosen loop must also be that through which magnetic flux change $d\phi_B/dt$ is computed.

We must distinguish carefully two general types of electric field: $\mathbf{E}$ originating directly from electric changes (Coulomb's law), and $\mathbf{E}$ created by a changing magnetic flux (Faraday's law):

- $\mathbf{E}$ *from static charges* is a *conservative* electric field. The line integral $\oint \mathbf{E} \cdot d\mathbf{l}$ is always zero. That is to say, if one transports an electric charge around a closed loop in a conservative electric field, no net work is done on the charge. Consequently, we can associate an electric potential with a conservative $\mathbf{E}$. The electric field lines originate from positive charges, terminate on negative charges, and are continuous in between.

- $\mathbf{E}$ *from changing* ϕ_B is a *nonconservative* electric field. Electric field is still defined as force per unit charge, but the line integral $\oint \mathbf{E} \cdot d\mathbf{l}$ around a closed loop is not zero. Instead, it is equal to the rate of magnetic flux change through the loop. A scalar electric potential cannot be associated with a nonconservative electric field; the work done per unit charge around a closed loop is not zero but equal to the emf around that loop. Moreover, since the electric-field lines do not originate from and terminate upon electric charges but are continuous nevertheless (Gauss's law for electricity), the *electric-field lines* from a changing magnetic flux form *closed loops*.

Example 31-7. The magnetic field produced by a long solenoid is uniform and confined to the circular region inside the windings. A cross section is shown in Figure 31-10(a), where $\mathbf{B}$ is directed into the paper. Suppose that the magnitude of $\mathbf{B}$ increases with time. Find the electric field induced by the changing magnetic field.*

By symmetry we know that the closed $\mathbf{E}$ loops consist of circles concentrically surrounding the magnetic field and lying in a plane perpendicular to the solenoid symmetry axis. To find the sense of $\mathbf{E}$ loops, we can imagine a circular conducting wire coinciding with an $\mathbf{E}$ loop and determine the direction of the induced current (hence, the direction of $\mathbf{E}$) from Lenz's law. Since ϕ_B is increasing into the paper, the field produced by the induced $\mathbf{E}$ must be out of the paper. From the right-hand rule connecting current and magnetic-field directions, this means that the $\mathbf{E}$ loops must be counterclockwise, as shown.

To find how the magnitude of $\mathbf{E}$ depends on the distance r from the center of the magnetic field, we consider separately two regions: first outside the boundary of $\mathbf{B}$ ($r > R$) and then inside ($r < R$).

We evaluate the line integral in Faraday's law around a circular loop of radius r:

$$\oint \mathbf{E} \cdot d\mathbf{l} = -\frac{d\phi_B}{dt}$$

$$E(2\pi r) = -\left(\frac{d\phi_B}{dt}\right)_R$$

Here $(d\phi_B/dt)_R$ represents the total time rate of flux change within R, the region to which the field is confined. We have, then,

$$\text{For } r > R, \quad E = -\left(\frac{1}{2\pi r}\right)\left(\frac{d\phi_B}{dt}\right)_R$$

The significance of the minus sign is given (finally!) at the end of this section.

---

* This example is very similar in mathematical form to Example 30-5 for finding the magnetic field from a current-carrying cylindrical conductor by applying Ampère's law.

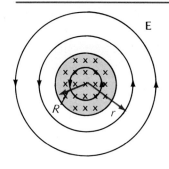

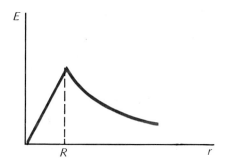

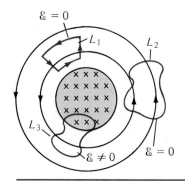

Figure 31-10. (a) A uniform magnetic field is into the paper over a region of radius R, and the field increases with time. The induced electric-field lines consist of counterclockwise circular loops. (b) Magnitude of **E** as a function of r. (c) Some loops for computing the induced electric field.

For any point outside the region of changing magnetic field, the electric field falls off inversely with distance r. Note a curious circumstance. We have a nonzero **E** in a region where **B** remains zero. What matters is not whether there is a changing magnetic field *at* the location of the loop, but whether the total magnetic *flux* changes *within* the loop. Put an electric charge in such an electric field and it is accelerated.

Now we go inside, with $r < R$. Only a fraction of the total flux penetrates a circle of radius r. Indeed, the fraction of flux within r equals the area ratio, $\pi r^2/\pi R^2$, since **B** is uniform within R. The rate of flux change within r is then $(r/R)^2 (d\phi_B/dt)_R$, and Faraday's law gives

$$\oint \mathbf{E} \cdot d\mathbf{l} = -\frac{d\phi_B}{dt}$$

$$E(2\pi r) = -\left(\frac{r}{R}\right)^2 \left(\frac{d\phi_B}{dt}\right)_R$$

$$\text{For } r < R, \quad E = -\frac{r}{2\pi R^2}\left(\frac{d\phi_B}{dt}\right)_R$$

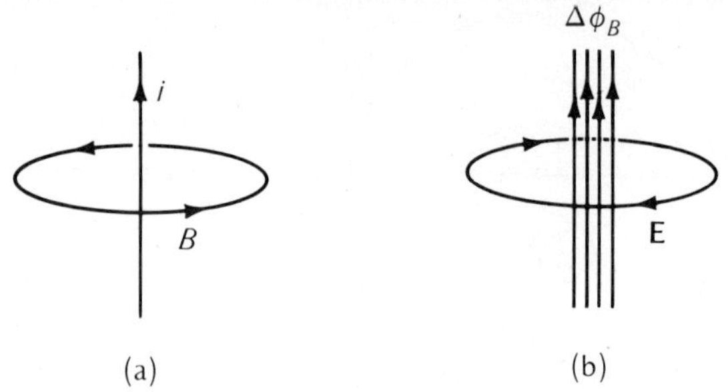

Figure 31-11. (a) The right-hand rule relating magnetic field **B** to conduction current i. (b) The left-hand rule relating electric field **E** to magnetic flux change $\Delta\phi_B$.

Within R, the electric field is directly proportional to r. The magnitude of the circumferential induced electric field is shown in Figure 31-10(b) as a function of r.

Now consider still other closed loops, different from circles, as shown in Figure 31-10(c). Loop L_1 has two circular arcs and two radial segments. The electric field **E** falls off inversely with r, and the line integral around L_1 is zero, because the contribution to $\oint \mathbf{E} \cdot d\mathbf{l}$ is positive for one circular arc and negative but of the same magnitude for the other circular arc. Similarly, loop L_2, through which there is no change in magnetic flux, has no emf around it. Loop L_3, with ϕ_B changing through at least some of it, has a nonzero emf.

The reason for the minus sign in Faraday's law is shown in Figure 31-11. Recall first the right-hand rule for relating the directions of an electric current to the magnetic-field loops that surround it; see Figure 31-11(a). The relation between **B** and i is given, in mathematical terms, by Ampère's law, $\oint \mathbf{B} \cdot d\mathbf{l} = \mu_0 i$, (30-9). No minus sign appears in this equation. The relative "directions" of the induced electric field **E** and magnetic-flux change $\Delta\phi_B$ are shown in Figure 31-11(b) [also shown in Figure 31-10(a)]. The right-hand rule does not apply here; if the thumb gives the direction of $\Delta\phi_B$, the curled right-hand fingers are opposite to the direction of the induced electric-field lines. This is indicated formally by the minus sign in Faraday's law.

31-5 Eddy Currents

Consider Figure 31-12(a), where a permanent magnet is moved toward a circular conducting loop. A current is induced in the loop, and its sense is, from Lenz's law, counterclockwise. Now consider a similar situation in which the north pole of a permanent magnet is moved toward a conducting sheet. As Figure 31-12(b) shows, there are now many concentric induced current loops, in each of which the current is counterclockwise.

These currents, induced by a changing magnetic flux in an extended conductor, are called *eddy currents* because they resemble the whirling eddies in a liquid.

Different forms of eddy currents are produced when the magnet moves parallel to the conducting sheet, as shown in Figure 31-13. We now have two sets of loops, one clockwise and the other counterclockwise. Their associated equivalent magnets oppose the motion of the magnet creating them.

Eddy currents are created not merely by moving a magnet near a conductor, but whenever the magnetic flux changes over a conducting sheet or bulk conductor. For every eddy current:

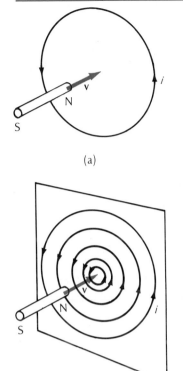

Figure 31-12. (a) Magnet moving toward a conducting loop induces a current. (b) Magnet moving toward a conducting sheet induces eddy currents.

- The sense of the current is always such as to oppose the magnetic flux change; a resistive force always acts on a moving magnet or conductor.
- Energy is dissipated.

The resistive force arising from an eddy current may be used, for example, in the magnetic dumping of an analytical balance. The energy losses from eddy currents may be highly undesirable in electrical devices, for example, in the conducting magnetic materials used as cores of transformers. The eddy-current losses can be reduced by cutting slots in conducting materials, or equivalently, by using laminations with high-resistance outer surfaces.

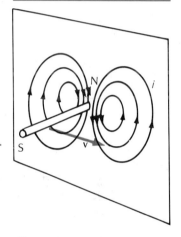

Figure 31-13. Magnet moving transverse to a conducting sheet induces two sets of oppositely directed eddy currents.

31-6 Diamagnetism

We have already seen (Section 30-7) that when a paramagnetic or a ferromagnetic material is immersed in an external magnetic field, such materials enhance the magnetic field because electron spins are reoriented by the applied field. A material is *diamagnetic* however, if its presence reduces the magnitude of an applied magnetic field.

Lenz's law tells us that any material must show some degree of diamagnetism. There are, after all, mobile electric charges in any type of material. When the external magnetic flux over a diamagnetic material goes from zero to some final value, the motion of the charged particles must be such as to *oppose* this flux change and *reduce* the applied magnetic field. As a consequence, the net magnetic field is less than the value it would have reached in the absence of material with mobile charged particles.

On a simple classical atomic model, the electrons in a material can be thought of as orbiting their parent nuclei; each such orbiting electron constitutes a current loop, and it produces a magnetic moment (Section 29-8). In the absence of an external magnetic field, the electron orbits in any material are so paired off that the magnetic moments cancel. For every electron orbiting clockwise in one plane, there is another orbiting counterclockwise in the same plane. The magnetic moments from orbital motion are, in the absence of an external field, oppositely directed and of the same magnitude. They produce no net magnetic effect.

Now suppose that an external field **B** is turned on. For simplicity, we suppose that **B** is perpendicular to the planes of electron orbits. Changing the magnetic field from zero to **B** has two effects:

- The changing magnetic flux through the current loop (the circular electron orbit) induces an emf; the induced electric field is tangent to the orbit, so that the electron's linear speed v and angular speed ω are changed.
- A magnetic force acts on the orbiting electron, in addition to the electrostatic force from the force center; this magnetic force is radially inward or outward.

The effects are shown in Figure 31-14. We see that the electron speed for the left orbit (with its magnetic moment μ aligned with **B**) is reduced, as is also its angular speed from ω_0 to $\omega_0 - \Delta\omega$. The effect for the orbit on the right (with μ opposite B) is just the reverse; here the electron speed is increased, and so is the angular velocity to $\omega_0 + \Delta\omega_0$. A curious result that we simply state without detailed proof: although the linear speeds, angular speeds, and net radial forces change as **B** changes, the orbital radii are essentially unchanged. As Figure 31-14(b) shows, with **B** turned on, the orbital magnetic moments no longer cancel. Both magnetic moments change, and the net induced magnetization opposes the external field. This is diamagnetism.

All materials exhibit diamagnetism, although the effect may be masked by stronger para- and ferromagnetic effects.

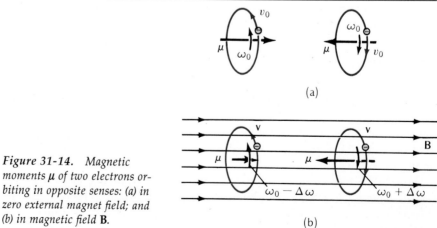

Figure 31-14. Magnetic moments μ of two electrons orbiting in opposite senses: (a) in zero external magnet field; and (b) in magnetic field **B**.

Summary

Definitions

Flux linkage $\equiv N\phi_B$ for a coil of N identical turns with magnetic flux ϕ_B through each.

Eddy current: loops of current induced in an extended conducting material by a changing magnetic flux.

Diamagnetism: the phenomenon whereby introducing a material into an external magnetic field reduces the magnitude of the net field.

Fundamental Principles

Law of electromagnetic induction (also known as Faraday's law):

$$\mathcal{E} = -\frac{d\phi_B}{dt} \qquad (31\text{-}2)$$

where $\mathcal{E}$ is the emf induced around a loop by net magnetic flux changing over that loop at the rate $d\phi_B/dt$.

In more fundamental form, the electromagnetic induction law is written as

$$\oint \mathbf{E} \cdot d\mathbf{l} = -\frac{d\phi_B}{dt} \qquad (31\text{-}6)$$

Here $\mathbf{E}$ is the nonconservative induced electric field. The line integral $\oint \mathbf{E} \cdot d\mathbf{l}$ is taken around the same loop as that through which the rate of change of magnetic flux $d\phi_B/dt$ is computed.

Lenz's law: the direction of the induced current (or induced emf, or induced electric field) is always such as to oppose the change in magnetic flux that produces it. The minus signs in the relations above symbolize Lenz's law; in effect, $\mathbf{E}$ and $\Delta\phi_B$ are related by a *left*-hand rule.

Problems and Questions

Section 31-1   Induced Currents and EMF's

· **31-1 Q** A coil of wire is placed in the field of an electromagnet so that magnetic flux passes through the coil. In what ways could you induce an emf in the coil?
(A) Rotate the coil.
(B) Pull the coil out of the electromagnet.
(C) Change the shape of the coil.
(D) Change the current in the electromagnet.
(E) None of the above will work.

· **31-2 P** A small coil of 4 cm² area and 10 turns is jerked out of a magnetic field of 0.040 T in 20 ms. What is the induced emf?

· **31-3 P** A coil has 80 turns and an area of 2800 cm². It rotates in a magnetic field of 0.75 T. At what frequency should it be rotated to produce a peak emf of 150 V?

· **31-4 Q** The number of magnetic field lines passing through a surface is a measure of the
(A) magnetic moment of the current flowing around the boundary of the surface.
(B) magnetic flux through the surface.
(C) induced emf along a path bounding the surface.
(D) magnetic field at the surface.

· **31-5 P** An anemometer used to measure wind velocity is constructed of four light cups mounted on a shaft to which is attached a coil. See Figure 31-15. When the shaft rotates, the coil also rotates in the field of a small permanent magnet that provides a magnetic field **B** perpendicular to the axis of rotation. Experimentally it is found that the cups rotate with a velocity that is 80 percent of the wind velocity. With this device, what emf would you expect to generate in

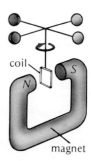

Figure 31-15. Problem 31-5.

a 60-km/h wind for the following design parameters: area of coil = 4 cm²; 500 turns; cup arm radius = 15 cm; magnetic field = 300 G.

· **31-6 Q** A beam of electrons is shot through a gap in a toroidal coil, as shown in Figure 31-16. When there is no

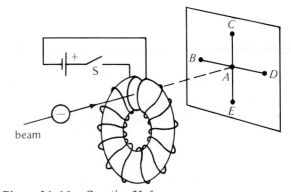

Figure 31-16. Question 31-6.

current in the coil, the beam strikes the screen at point A. When the current is turned on, the beam strikes the screen at one of the indicated points. Which is it? (A is a possible answer also.)

: **31-7 Q** A bar magnet is suspended as a pendulum, as shown in Figure 31-17. A coil of N turns lies on the table below the point of suspension. The pendulum is pulled to one side and released. It swings back and forth with frequency f. Which of the following is the most accurate statement?
(A) An emf oscillating at frequency f will be induced in the coil.
(B) An emf oscillating at frequency $2f$ will be induced in the coil.
(C) Whether or not an emf will be induced in the coil depends on whether or not the switch is closed.
(D) The magnitude of the induced emf will depend on the value of the resistance R.
(E) The induced emf will be independent of the number of turns in the coil.

Figure 31-17. Question 31-7.

: **31-8 P** A coil of wire is hung as a pendulum with the plane of the coil vertical in an east-west plane. The pendulum is pulled to the south and released from rest at $t = 0$. It subsequently swings back and forth in a north-south direction. Carefully sketch the current induced in the coil as a function of time after it is released. Let T be the period of the pendulum. Take the current to be positive if it flows counterclockwise when viewed from the north.

: **31-9 P** What is the amplitude of the emf produced by a generator coil of area 0.1 m² and 500 turns that rotates at 60 Hz in a magnetic field of 0.1 T?

: **31-10 P** A circular coil of 6-cm diameter has 200 turns. The coil resistance is 100 Ω. At what rate must a magnetic field normal to the plane of the coil change to produce Joule heating in the coil at the rate of 2.0 mW?

: **31-11 P** A large coil and a small coil lie in the same plane, with the small coil at the center of the large one. The large coil has 4000 turns and a radius of 40 cm, and the small coil has 2000 turns and a radius of 2 cm. What is the induced emf in the small coil when the current in the large coil is changed at the rate of 8000 A/s? Assume that the field due to the large coil is uniform over the area of the small coil.

: **31-12 P** An air-core solenoid has 2000 turns, a length of 60 cm, and a diameter of 1.8 cm. What emf is induced in it if the current in the windings increases from 2 A to 8 A in 3 s at a steady rate?

: **31-13 P** A flip coil consists of a circular coil of 2.0-cm diameter and 100 turns. It is situated in a magnetic field of 0.050 T with the field at an angle of 60° with the plane of the coil. What emf will be induced in the coil if it is pulled out of the field in 10 ms?

: **31-14 P** The perpendicular magnetic field through a single loop of area 2.0×10^{-4} m² changes with time as shown in Figure 31-18. The loop has resistance 0.0030 Ω. Make graphs as a function of time of (a) the emf induced in the loop; (b) the current in the loop; (c) the rate at which energy is dissipated in the loop.

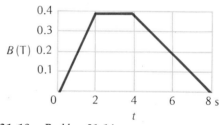

Figure 31-18. Problem 31-14.

: **31-15 P** A coil of 200 turns and resistance 2.0 Ω is wrapped around a solenoid with 10,000 turns per meter and of radius 0.20 m. What current is produced in the coil when the current in the solenoid is changed at the rate of 2.0 A/s?

: **31-16 P** A rectangular loop of wire of resistance R is placed next to a long straight wire carrying current, as shown in Figure 31-19. The current in the long wire changes at a rate di/dt. What is the current in the loop?

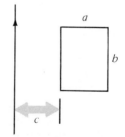

Figure 31-19. Problem 31-16.

: **31-17 P** Current i exists along the z axis. A circular loop of radius R has its center at a point x along the x axis. The plane of the loop lies in the xz plane. What emf is induced in the loop when the current changes at a rate di/dt?

31-18 P Current i exists in a long straight wire placed along the z axis. A circular loop of radius a has its plane parallel to the xz plane. The loop moves away from the wire along the x axis with constant velocity v. At $t = 0$, the center of the loop is at position x_0. Determine the emf induced in the loop as a function of time.

31-19 P A small search coil of 250 turns and radius 1 cm has a resistance of 40 Ω. In order to measure the earth's magnetic field, the coil is suddenly rotated through 180°. The coil is connected to a ballistic galvanometer, and a total charge of 3.2×10^{-7} C is found to flow through the coil because of the induced emf. Determine the magnitude of the earth's magnetic field. Assume that the coil was initially oriented for maximum flux through it.

31-20 P A coil of N turns and area A is suspended as a pendulum of length L. It is displaced through an angle θ_0 at $t = 0$, as shown in Figure 31-20. A uniform horizontal magnetic field $\mathbf{B}$ acts in the plane in which the pendulum swings. Determine the induced emf in the coil as a function of time.

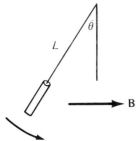

Figure 31-20. Problem 31-20.

31-21 P Shown in Figure 31-21 are three identical light bulbs lighted by a battery. The shaded circle represents the end view of a solenoid in which the magnetic flux can be increased at a constant rate to produce an induced emf. What rate of change of magnetic flux, expressed in terms of $\mathcal{E}_0$, is required to shut off lamp No. 1? What is the sense of the change in flux required?

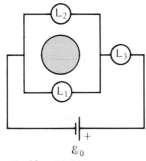

Figure 31-21. Problem 31-21.

31-22 P The planar circuits shown in Figure 31-22 are divided into two areas, A_1 and A_2, as indicated. A magnetic

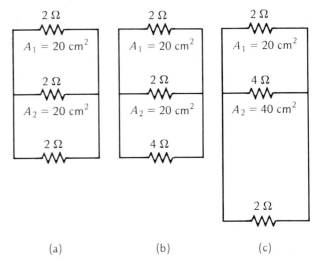

Figure 31-22. Problem 31-22.

field directed out of the paper is increasing at a constant rate of 0.10 T/s. Determine, for each case, the magnitude and direction of the induced current in each resistor.

31-23 P A coil of 400 turns and area 4.0 cm² is connected to a so-called "ballistic" galvanometer of resistance 24 Ω. The maximum angular displacement of the galvanometer coil is proportional to the charge that flows through it. The flux through the coil changes suddenly, and a charge of 0.8 C flows through the galvanometer. What is the change in flux through the coil?

31-24 P Three long straight wires pass through the corners of an equilateral triangle of side a. The plane of the triangle is perpendicular to the wires. A sinusoidal current of amplitude I_0 and angular frequency ω passes through each wire, but there is a phase difference of 120° between the current in each wire; that is,

$$I_1 = I_0 \sin \omega t \qquad I_2 = I_0 \sin\left(\omega t + \frac{2\pi}{3}\right)$$

$$I_3 = I_0 \sin\left(\omega t + \frac{4\pi}{3}\right)$$

(a) Show that the magnetic field at the center of the triangle can be represented by a vector of constant magnitude rotating with angular frequency ω. (b) Determine the emf induced in a small coil of N turns and area A placed at the center of the triangle, with its plane parallel to the wires. (c) What emf will be induced if the coil is rotated clockwise with frequency ω_1? (d) What emf will be induced if the coil is rotated counterclockwise with frequency ω_1?

Section 31-2 EMF in a Moving Conductor
• **31-25 Q** A uniform magnetic field directed into the plane of the paper acts in the shaded region of Figure 31-23(a). A circular loop of wire is moved with constant

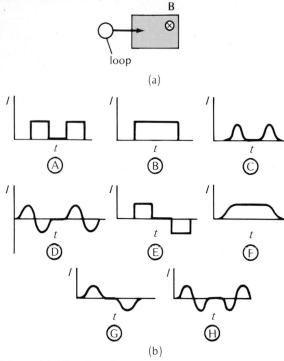

Figure 31-23. Question 31-25.

velocity through the magnetic-field region and out the other side, starting at $t = 0$ at the position where it is drawn. Which of the graphs A–H in Figure 31-22(b) depict the induced current in the loop as a function of time? A counterclockwise current is taken as positive.

· **31-26 P** An electromagnetic flowmeter works like this. A conducting fluid (such as blood) moves with velocity v through a magnetic field **B** oriented perpendicular to the direction of flow. Suppose that the fluid flows through a rectangular tube of cross section ab, with b parallel to the magnetic field. Write an expression for the fluid velocity as a function of the voltage between electrodes attached to opposite sides of the tube.

· **31-27 P** A square wire loop of side 2.0 cm is moved at a constant velocity of 6.0 cm/s along the x axis. In the region from $x = 8.0$ cm to $x = 14.0$ cm, a uniform magnetic field of 0.010 T acts perpendicular to the plane of the loop. Graph the induced current in the loop as a function of time, starting and ending when the loop is outside the magnetic field. The loop lies in the xz plane with one side parallel to the x axis. It moves in the $+x$ direction, and the magnetic field acts in the $+y$ direction.

: **31-28 P** The conducting blades of a helicopter rotor are 5.0 m long and rotate at 640 rpm in normal flight. The earth's magnetic field has a horizontal component of 0.60 G and a vertical component of 0.30 G. What potential difference is induced between the tip of a blade and the shaft on which it rotates?

: **31-29 P** A rectangular loop of width a and height b is released from rest with its plane vertical. See Figure 31-24. As it falls, its lower end enters a uniform magnetic field **B** oriented normal to the plane of the loop. Dimension b is great enough that the loop can reach a constant terminal falling velocity (with the top end of the loop remaining out of the magnetic field). What is this terminal velocity? The mass of the loop is m and its resistance is R.

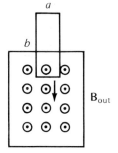

Figure 31-24. Problem 31-29.

: **31-30 Q** A bar magnet is placed inside a coil, as shown in Figure 31-25. In the following, "work" refers to any work done because the bar is magnetic. If the bar is suddenly pulled out of the coil,
(A) no work will be done, independent of whether or not the switch is closed.
(B) more work will be done if the switch is open.
(C) more work will be done if the switch is closed.
(D) equal (nonzero) amounts of work will be done whether or not the switch is open.
(E) whether the work done is positive or negative depends on whether the magnet is pulled out of the right end of the coil or the left end.

Figure 31-25. Question 31-30.

: **31-31 P** A prototype model of a "mass driver" to be used to launch objects into orbit around the moon is sketched in Figure 31-26. A battery and a resistor are connected to two copper rails and a bar, which have negligible electric resistance. The bar and projectile have a mass of 5.0 gm and slide without friction on the rails. A magnetic field of 0.20 T is directed perpendicularly into the paper. (a) What is the initial acceleration of the bar and projectile when the switch is closed? (b) What is the maximum velocity reached by the projectile, if the length of the rails is not a limitation?

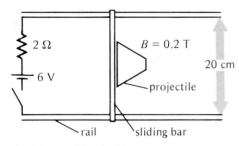

Figure 31-26.   Problem 31-31.

: **31-32 P** A uniform magnetic field of 0.050 T acts vertically upward. A conducting rod 60 cm long is rotated clockwise (as viewed from above) at 100 Hz about one end in a horizontal plane. (a) Which end of the rod is positive as a result of the induced emf? (b) What is the potential difference between the ends of the rod?

: **31-33 P** A rectangular frame is rotated at angular frequency ω about the axis XX^1 shown in Figure 31-27. The two sides of the frame have resistances R_1 and R_2. A uniform magnetic field **B** acts normal to the paper. Find the ratio of the power dissipated in the two arms.

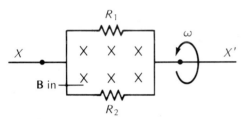

Figure 31-27.   Problem 31-33.

Section 31-3   Lenz's Law

· **31-34 Q** A magnet is placed above a horizontal conducting ring, as sketched in Figure 31-28. If the magnet is moved upward at constant speed,
(A) no current will exist in the ring.
(B) a clockwise current, as viewed from above, will exist in the ring.
(C) a counterclockwise current, as viewed from above, will exist in the ring.

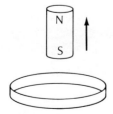

Figure 31.28.   Question 31-34.

· **31-35 Q** Verify the directions of the induced currents in Figure 31-1.

· **31-36 Q** All the effects associated with electromagnetic induction (induced currents and emf's and magnetic forces) always act in the sense of opposing the change that caused them. This is because
(A) of the law of conservation of energy.
(B) of the law of conservation of momentum.
(C) magnetism and electricity are two different aspects of the same phenomenon.
(D) opposites attract and likes repel.
(E) magnetic fields decrease as the distance to the source increases.

· **31-37 Q** Two wire loops are oriented with their axes parallel, as shown in Figure 31-29. If a current is suddenly established in one of the loops in the direction shown, what force acts on the second loop?
(A) No net force acts on the loop.
(B) The loop experiences a force to the right.
(C) The loop experiences a force to the left.

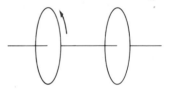

Figure 31.29.   Question 31-37.

Section 31-4   Faraday's Law and the Induced Electric Field

· **31-38 Q** A copper ring and a wooden ring of the same dimensions are placed so that there is the same changing magnetic flux through each. How do the induced electric fields in each compare?
(A) The induced electric fields are the same in both.
(B) The induced fields are greatest in the copper.
(C) The induced fields are greatest in the wood.

· **31-39 Q** Two coils are wound on a wooden dowel as shown in Figure 31-30. Immediately after the switch is closed, the current in the resistor will be
(A) from right to left.
(B) from left to right.

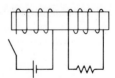

Figure 31.30.   Question 31-39.

(C) zero, since wood is nonmagnetic.
(D) zero, but not because wood is nonmagnetic.

· **31-40 P** The magnetic field in a solenoid is uniform and increases at a constant rate of 50 G/s. The radius of the solenoid is 0.20 m. What is the electric field in the plane normal to the magnetic field at a distance from the axis of the solenoid of (a) 0.10 m? (b) 0.20 m? (c) 0.40 m?

: **31-41 Q** When a bar magnet is dropped lengthwise down a long, vertical copper pipe in the absence of air friction, it will
(A) fall with constant acceleration less than g.
(B) fall with acceleration g.
(C) alternately speed up and slow down.
(D) eventually reach a constant terminal velocity.

: **31-42 Q** Use the answers (A) left, (B) right, or (C) no current for the following questions: (a) Immediately after the switch S in Figure 31-31 is closed, in what direction is the current in resistor R? (b) Quite a long time after the switch is closed, what is the direction of the current in the resistor? (c) After the switch has been closed for a long time, it is then opened. In what direction is the current is resistor R?

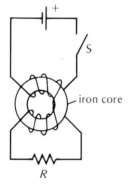

Figure 31.31.   Question 31-42.

Section 31-5   Eddy Currents

: **31-43 P** A plane conducting object falls through a transverse magnetic field. Show that the power dissipated in eddy currents is proportional to the square of the speed of the falling object.

: **31-44 P** A metal disk of radius a, thickness d, and resistivity ρ experiences a magnetic field $B_0 \sin \omega t$ parallel to its axis. Determine (a) the total current in the disk; and (b) the average power dissipated in the disk.

: **31-45 P** A metal disk of radius a, thickness t, and resistivity ρ is oriented normal to a uniform magnetic field **B**. The disk is rotated about its axis with constant angular velocity ω. What is (a) the current in the disk? (b) the potential difference between the center of the disk and a point on the perimeter?

: **31-46 P** An electromagnetic brake consists of a conducting disk, of thickness d and resistivity ρ, rotating with angular velocity ω about its axis. A magnetic field **B** is applied normal to the disk only over a small area a^2 at a distance x from the axis. The induced current experiences both a force and a magnetic torque that slows the disk. Show that this torque has the approximate magnitude of $(\omega/\rho)B^2x^2a^2d$.

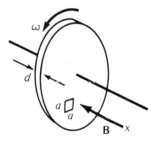

Figure 31-32.   Problem 31-46.

Supplementary Problem

31-47 P Very large magnetic fields can be obtained with a technique called flux compression. Consider Figure 31-33, with a thin metal tube positioned inside a solenoid where the field is B_0. Between the tube end and the solenoid is a layer of high explosive (HE), and outside the solenoid is thick armor plate to strengthen the solenoid and keep it from blowing apart when the explosive is detonated. When the explosive is set off, the tube collapses. The flux through it is drastically reduced in a short time, thereby inducing huge currents in the tube wall. This current creates a very large **B** field.

(a) Show that if the tube shrinks very rapidly, the magnetic field reaches a value $B \cong B_0(R_0/R)^2$ when the tube radius is reduced from R_0 to R. (b) What surface current is needed in the tube (approximately) to produce a field of the order of 10^9 T? (c) What **B** can be obtained if $B_0 = 10$ T, $R_0 = 6$ cm, and $R = 3$ mm?

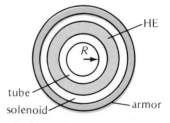

Figure 31-33.   Problem 31-47.

Inductance and Electric Oscillations

32

32-1 Self-Inductance Defined
32-2 The *LR* Circuit
32-3 Energy of an Inductor
32-4 Energy of the Magnetic Field
32-5 Electrical Free Oscillations
32-6 Electrical-Mechanical Analogs
32-7 Mutual Inductance (Optional)
Summary

Inductance is the property of a circuit element that derives from electromagnetic induction. We shall treat an inductor in combination with one other circuit element, first with a resistor in an *LR* circuit and then with a capacitor in an *LC* oscillator circuit. Close analogies can be identified between electric-circuit elements and their mechanical counterparts.

32-1 Self-Inductance Defined

Consider Figure 32-1(a), where current i goes through a coil with several turns. The current is driven by some external energy source, such as a battery with emf $\mathcal{E}_b$. Suppose that the current is changing—more specifically, that it is increasing. The time rate of change of current, di/dt, is positive. This would happen just after a switch connecting a battery to the coil is closed. As the current i through the conducting wires changes, so too does the magnetic flux through turns of the coil, so that an emf is induced in the coil. The *induced emf* here opposes the emf of the battery and is sometimes called a *back emf*.

An emf is also induced if the current decreases, and $di/dt < 0$, See Figure 32-1(b). We know that the induced emf is always in the direction opposing the *change* that produces it, so that here the induced emf is in the same direction as

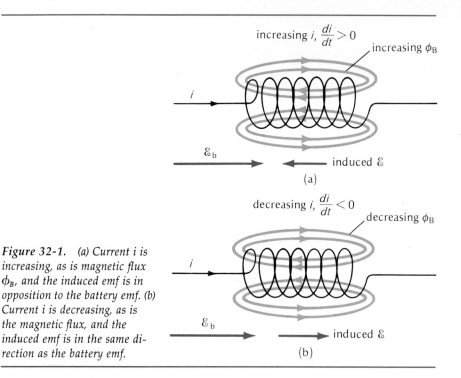

Figure 32-1. (a) Current i is increasing, as is magnetic flux ϕ_B, and the induced emf is in opposition to the battery emf. (b) Current i is decreasing, as is the magnetic flux, and the induced emf is in the same direction as the battery emf.

the applied emf $\mathcal{E}_b$. The coil "tries," so to speak, to maintain the current unchanged. A constant current of any magnitude through the coil produces no induced emf. Self-induction depends on a changing magnetic flux and therefore, for conductors fixed in position, on changing a current.

Self-induction occurs in every electric circuit, not merely in a multiple-turn coil. Every circuit with a continuous current must have at least one loop, so that when the current changes, the magnetic flux through this single loop will change and induce an emf in the circuit. For example, merely closing the switch on a simple circuit with a battery and a resistor will change the magnetic flux through the loop and thereby induce an emf, possibly very feeble. It is easy to see that using N turns enhances the self-induction effect by a factor N^2. First, the magnetic flux through each turn is increased by factor N; additionally, an emf is induced in each of N turns.

Any device showing the self-induction effect, such as a coil of conducting wire, is called an *inductor*. It is represented in a circuit diagram by the symbol —⟋⟋⟋⟋— .

It is useful to compare general characteristics of an inductor with the corresponding properties of a capacitor:

A *capacitor*:

- Stores separated charge of opposite sign.
- Stores potential energy associated with separated electric charges; alternatively, we can say that a capacitor stores energy in the electric field in the region between oppositely charged plates.

As we shall see, an *inductor*:

- Acts to maintain a constant electric current.

- Stores energy associated with electric charges in motion; alternatively, we can say that an inductor stores energy in the magnetic field in the region surrounding the current-carrying conductor.

An inductor is characterized by its *self-inductance L*; unless there is need to avoid confusion with mutual inductance (Section 32-7), L is usually called simply the inductance.

In defining inductance, we first recognize that the magnitude of the magnetic field B produced by any current element at any point in space is directly proportional to the current magnitude i (Ampère's law). Further, the magnetic flux ϕ_B is proportional to B, which is in turn proportional to i. It follows that $\phi_B \propto i$. For an inductor of N identical turns, we write

$$N\phi_B \equiv Li \tag{32-1}$$

where ϕ_B is the flux enclosed by one loop of the inductor and L is a proportionality constant that depends on the size and shape of the conductor. The product $N\phi$ is referred to as the *flux linkage*.

From Faraday's law, (31-3), we have the induced emf given by

$$\mathcal{E} = -\frac{d}{dt}(N\phi) \tag{32-2}$$

Substituting (32-1) in (32-2), we get

$$\mathcal{E} = -L\frac{di}{dt} \tag{32-3}$$

This relation is another equivalent definition of L, where

$$L \equiv -\frac{\mathcal{E}}{di/dt} \tag{32-4}$$

Inductance may be defined in two ways:

- In terms of the magnetic flux and its relation to current, $L \equiv N\phi_B/i$, (32-1). This form is most useful in computing the self-inductance of a particular conductor arrangement.
- In terms of the emf produced by a change in current, $L \equiv -\mathcal{E}/(di/dt)$, (32-4). This form is most useful in describing the behavior of an inductor in an electric circuit or measuring inductance experimentally.

From (32-4) we see that inductance has the SI units volts per ampere per second. A special name, the *henry* (abbreviated H) is given to this combination of units:*

$$\text{Henry} \equiv \frac{\text{volt}}{\text{ampere per second}}$$

$$1\ \text{H} \equiv \frac{\text{V}\cdot\text{s}}{\text{A}}$$

---

* The unit for inductance honors the American physicist Joseph Henry (1797–1878), who discovered electromagnetic induction independently of Faraday. Henry was Professor of Physics at Princeton University, and later he was appointed the first director of the Smithsonian Institution in Washington, D.C.

Thus, an inductor producing an emf of 1 V when the current through it is changed at the rate of 1 A/s has an inductance of 1 H. Related units are the millihenry (1 mH = 10^{-3} H) and the microhenry (1 μH = 10^{-6} H). Air-filled laboratory inductors of moderate size have inductances typically of the order of several millihenries. With cores of magnetic material (Section 30-7), the inductance may rise to several henries. As examples below show, any inductor's inductance is the product of μ_0, some characteristic dimension, and a dimensionless constant relating to the geometry of the magnetic field.

Example 32-1. Every coil has both inductance and resistance. When the current through a certain coil is 0.50 A and increasing at the rate of 100 A/s, the potential difference across the coil terminals is 6.0 V. When the current through the coil is again 0.50 A and in the same direction, but decreasing at the rate of 100 A/s, the potential difference across the coil terminals is only 4.0 V. What are the inductance and the resistance of the coil?

The two situations may be described by the relations

$$V_1 = iR + L\frac{di}{dt}$$

$$V_2 = iR - L\frac{di}{dt}$$

where V_1 is the larger and V_2 the smaller potential difference.

We have then that

$$V_1 + V_2 = 2iR$$

$$R = \frac{V_1 + V_2}{2i} = \frac{(6.0 + 4.0) \text{ V}}{2(0.50 \text{ A})} = 10 \text{ }\Omega$$

$$V_1 - V_2 = 2L\frac{di}{dt}$$

$$L = \frac{V_1 - V_2}{2di/dt} = \frac{(6.0 - 4.0) \text{ V}}{2(100 \text{ A}\cdot\text{s}^{-1})} = 10 \text{ mH}$$

Example 32-2. A toroidal inductor consists basically of a long solenoid bent into a circle so that its ends meet and its external shape is that of a toroid (a doughnut). See Figure 32-2. What is the inductance for a toroidal inductor of n turns per unit length with central circumference l and transverse area A? Assume that the radius of any turn is small compared with the radius of the toroid.

The toroid is one of a very few inductor configurations in which the geometry of the magnetic field is simple enough for the magnetic flux, and therefore the inductance, to be readily computed. Here all the magnetic field is confined within the turns of the toroid. Moreover, since the turn radius is small compared with the toroid radius, the magnetic field can be taken to be uniform through any turn.

The magnetic field at any point near the center of a long solenoid, and therefore any interior point for the toroid, is

$$B = \mu_0 ui \qquad (30\text{-}10), (32\text{-}5)$$

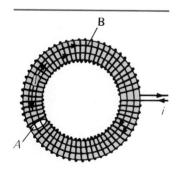

Figure 32-2. A toroidal solenoid with current i, magnetic field B, and cross-sectional area A.

The turns per unit length n can be written as $n = N/l$, where N is the total number of turns and l is the circumference of the toroid.

From the definition of inductance, (32-1), we have

$$L = \frac{N\phi_B}{i} = \frac{NBA}{i} = \frac{nlBA}{i}$$

and using the relation for B above, we get

$$L = \mu_0 n^2 (Al)$$

or

$$L = \mu_0 n^2 V \qquad (32\text{-}6)$$

where $V \equiv Al$ is the total volume within the windings. Note that as we anticipated earlier, the inductance varies with the *square* of the number of turns.

Suppose, for example, that a toroidal inductor has 2000 turns, a turn radius $r = 2.0$ cm, and a toroid mean radius $R = 20$ cm. Then $A = \pi r^2$, $l = 2\pi R$, and $n = N/2\pi R$; and the relation above yields

$$L = \mu_0 n^2 Al = \mu_0 \left(\frac{N}{2\pi R}\right)^2 (\pi r^2)(2\pi R)$$

$$= \frac{\mu_0 N^2 r^2}{2R} = \frac{(4\pi \times 10^{-7}\ \text{Wb/A}\cdot\text{m})(2000)^2 (2.0 \times 10^{-2}\ \text{m})^2}{2(20 \times 10^{-2}\ \text{m})}$$

$$= 5.0\ \text{mH}$$

The inductance would be increased substantially by winding the turns on a doughnut of strongly magnetic material, such as iron.

Example 32-3. An air-filled coaxial cable consists of concentric cylindrical conductors of radius r_1 and r_2 with equal currents, each of magnitude i, in opposite directions in the inner and outer conductors. What is the inductance per unit length of the coax cable?

As shown in Example 30-5, a magnetic field exists only in the region between the inner and outer conductors. The magnetic-field loops are circular and lie in planes transverse to the symmetry axis of the cylindrical conductors. Between r_1 and r_2, the magnitude of the magnetic field is

$$B = \frac{\mu_0 i}{2\pi r}$$

(The magnetic field is the same as that from a single, infinitely long conducting wire at the symmetry axis carrying current i.) See Figure 32-3.

In applying the definition for inductance,

$$L = \frac{N\phi_B}{i}$$

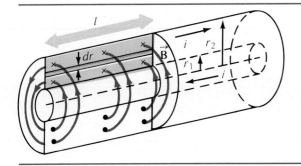

Figure 32-3. *A coaxial conductor with inner radius r_1 and outer radius r_2. The magnetic flux is computed over the shaded area.*

we choose the surface for computing magnetic flux ϕ_B to be a rectangle of length l along the symmetry axis, and of width $r_2 - r_1$. All **B** lines in a length l along the axis are perpendicular to this rectangle and pass through it. For an infinitesimal radial displacement dr, the magnetic flux is

$$d\phi_B = Bldr = l\frac{\mu_0 i}{2\pi r} dr$$

Therefore

$$\phi_B = \int_{r_1}^{r_2} d\phi_B = \frac{\mu_0 il}{2\pi} \int_{r_1}^{r_2} \frac{dr}{r} = \frac{\mu_0 il}{2\pi} \ln \frac{r_2}{r_1}$$

Here $N = 1$, so that the definition above gives, for the inductance per unit length.

$$\frac{L}{l} = \frac{\phi_B}{il} = \frac{\mu_0}{2\pi} \ln \frac{r_2}{r_1}$$

For example, any air-filled coaxial cable with $r_2 = 2r_1$ has an inductance per unit length of

$$\frac{L}{l} = \frac{(4\pi \times 10^{-7} \text{ Wb/A} \cdot \text{m}) \ln 2}{2\pi} = 13.9 \ \mu\text{H/m}$$

As shown in Example 26-2, the capacitance per unit length of coax cable was of a similar form: $C/l = 2\pi\epsilon_0 / \ln (r_2/r_1)$.

32-2 The *LR* Circuit

When a charged capacitor C is connected across a resistor R, the charge on either plate does not fall to zero instantaneously. Rather, the charge Q decays exponentially with a characteristic time constant RC, the time required for the charge to fall to $1/e$ of its initial value (Section 27-5).

The same sort of behavior is found when a current-carrying inductor is connected with a resistor. The inductor exhibits inertia as a consequence of Lenz's law: When the current through an inductor changes, the inductor opposes the change, and the induced emf is always such as to tend to offset the change in current.

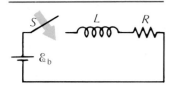

Figure 32-4. An LR circuit, with inductor L, resistor R, battery emf $\mathcal{E}_b$, and switch S in series.

Consider the circuit in Figure 32-4, where an inductor L and a resistor R are in series with a switch and a battery of emf $\mathcal{E}_b$ (we neglect its internal resistance). The current i in the circuit does not reach its steady-state value instantaneously, because as the current grows from an initial zero value, the back emf $-L\, di/dt$ from the inductor opposes the emf of the battery $\mathcal{E}_b$, and therefore, the buildup of current.

Applying the loop theorem (Kirchhoff's second circuit rule) we have

$$\Sigma \mathcal{E} = \Sigma V$$

$$\mathcal{E}_b - L\frac{di}{dt} = iR$$

(Note that the term $-L\, di/dt$ is put on the left side of the equation above; it is an emf, not a potential drop. This is equivalent to placing a term $+L\, di/dt$ on the right side of the equation and then counting it as an equivalent potential drop.)

The equation above may be written as

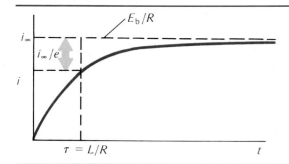

Figure 32-5. Current i in the LR circuit of Figure 32-4 plotted as a function of the time t after the switch is closed. The time constant $\tau = L/R$ represents the time it takes for the current to come within $1/e$ of its steady-state final value $i_\infty = \mathcal{E}_b/R$.

$$\mathcal{E}_b = L\frac{di}{dt} + iR \qquad (32\text{-}7)$$

It is not difficult to solve the differential equation for the current i as a function of time t, and we shall merely state that the solution is

$$i = \frac{\mathcal{E}_b}{R}\left[1 - \exp\left(-\frac{t}{L/R}\right)\right] \qquad (32\text{-}8)$$

That (32-8) is indeed the solution is easily verified by substituting it in (32-7).

Figure 32-5 is a plot of (32-8). The current grows steadily with time, and approaches its steady value $i_\infty = \mathcal{E}_b/R$. The quantity L/R is the characteristic time constant τ of the LR circuit where

$$\tau \equiv \frac{L}{R} \qquad (32\text{-}9)$$

We see from (32-8) that the current differs from its final value i_∞ by $1/e$, or 36.8 percent, after a time interval $\tau = L/R$ elapses following the initial closing of the switch. For example, with $L = 1$ mH, $R = 1$ MΩ, we have

$$\tau = \frac{L}{R} = \frac{10^{-3}\text{ H}}{10^6\ \Omega} = \frac{10^{-3}\text{ V}\cdot\text{s}/\text{A}}{10^6\text{ V}/\text{A}} = 1 \times 10^{-9}\text{ s} = 1\text{ ns}$$

Example 32-4. Both the inductance and the internal resistance of a very large induction coil are unknown and are to be determined indirectly. An observer finds that when the inductor alone is connected to a source maintained a constant voltage, the current through the inductor rises to half its final value in 0.60 s. When a 4.0-Ω resistor is placed in series with inductor, it takes only 0.30 s for the current to reach half its new final value. What are the inductor's inductance L and resistance R?

We see from (32-8) that the current reaches half its final value at time $t_{1/2}$, where

$$i = \frac{1}{2}i_\infty = i_\infty\left[1 - \exp\left(-\frac{t_{1/2}}{L/R}\right)\right]$$

This equation simplifies to

$$\exp\left(\frac{t_{1/2}}{L/R}\right) = 2$$

or

$$t_{1/2} = \frac{L}{R}\ln 2 = 0.693\frac{L}{R}$$

The time it takes to reach one-half the final current is 69.3 percent of the characteristic time constant, L/R.

With the inductor alone, we have in the equation above

$$0.60 \text{ s} = 0.693 \frac{L}{R}$$

and with the additional 4.0-Ω resistor in series,

$$0.30 \text{ s} = 0.693 \frac{L}{R + 4.0 \text{ }\Omega}$$

Solving for L and R in the two simultaneous equations above yields

$$L = 3.5 \text{ H}$$
$$R = 4.0 \text{ }\Omega$$

32-3 Energy of an Inductor

To find the energy associated with an inductor through which a current is passing, we consider again the circuit of Figure 32-4, in which a battery is connected to a circuit containing an inductor and a resistor. The loop theorem yielded (32-7),

$$\mathcal{E} = L \frac{di}{dt} + iR$$

With this equation multiplied by the instantaneous current i, we get

$$\mathcal{E}i = Li \frac{di}{dt} + i^2 R$$

The terms in this equation are interpreted as follows. The left-hand term $\mathcal{E}i$ represents the rate at which the energy source of emf $\mathcal{E}$ delivers energy to the circuit elements; it is the input power. The term $i^2 R$ is the rate at which electric energy is being dissipated into thermal energy in the circuit's resistance; it is, so to speak, the output power. This leaves $Li\,(di/dt)$. This term is positive, since both the current i and its time rate of change di/dt are positive. The term $Li\,(di/dt)$ must therefore represent electric power delivered to the circuit but not yet dissipated. It is the rate at which energy is supplied to and stored in the inductor. Labeling the instantaneous power into the inductor P_L and the energy associated with the inductor U_L, we can then write

$$P_L = Li \frac{di}{dt}$$

and

$$U_L = \int_0^t P_L \, dt = \int_0^t Li \frac{di}{dt} \, dt = \int_0^i Li \, di$$

$$U_L = \tfrac{1}{2} L i^2 \tag{32-10}$$

The energy stored in an inductor with current i is proportional to the inductance and to the square of the current. Note the similarity to the relation for the electric energy U_C stored in a charged capacitor of capacitance C with charge of magnitude Q on each plate:

$$U_C = \frac{1}{2}\frac{Q^2}{C} = \frac{1}{2}CV^2 \qquad (26\text{-}9)$$

32-4 Energy of the Magnetic Field

A charged capacitor establishes an electric field between its plates, and we may speak of the capacitor's energy as residing in the electric field between the plates (Section 26-5). Similarly, the energy of a current-carrying inductor may be considered to reside in its magnetic field. We wish to compute the energy density u_B or magnetic energy per unit volume, of magnetic field B.

The energy U stored in the magnetic field of an inductor carrying current i is

$$U = \tfrac{1}{2}Li^2 \qquad (32\text{-}10)$$

We take the conductor to be a toroid; the magnetic field is then confined entirely to the region within the windings. The inductance of a toroid is given by (32-6), and the current i by (32-5). Then the equation above becomes

$$U = \frac{1}{2}(\mu_0 n^2 V)\left(\frac{B}{\mu_0 n}\right)^2 = \frac{1}{2}\frac{B^2}{\mu_0}V$$

where V is the volume of the inductor. The magnetic-energy density $u_B = U/V$ is then

$$u_B = \frac{B^2}{2\mu_0} \qquad (32\text{-}11)$$

The energy density of the magnetic field is proportional to the *square* of the magnetic field.

Although derived for the special case of a toroidal inductor, (32-11) holds in general. The relation for the magnetic energy density is like that for the energy density u_E of an electric field:

$$u_E = \tfrac{1}{2}\epsilon_0 E^2 \qquad (26\text{-}10)$$

32-5 Electrical Free Oscillations

Here we treat the oscillations of a simple circuit consisting of an initially charged capacitor connected across an inductor. The more general case of an electric oscillator with resistance also driven by a sinusoidally varying emf of arbitrary frequency is treated in Chapter 33, on ac circuits.

Consider the circuit of Figure 32-6(a), where a capacitor C, initially carrying a charge of magnitude Q_m on each plate, is connected across an inductor L by closing a switch. For simplicity we assume that the circuit contains no resistance.

We can see on a qualitative basis that the circuit will oscillate. Immediately after the switch is closed, charges start to leave the capacitor plates, thereby creating a current; this current changes with time, and an emf is also set up in

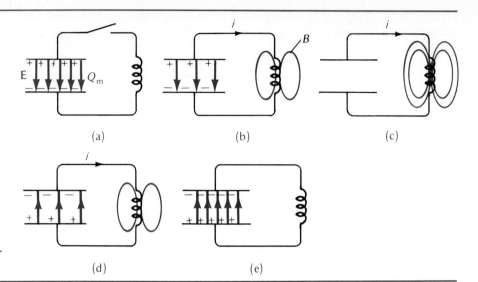

Figure 32-6. Stages in the oscillation of an LC circuit over one-half cycle.

the inductor [Figure 32-6(b)]. When the charge on each capacitor plate has reached zero [Figure 32-5(c)], the current continues because of the inertia of the inductor. The inductor's emf now opposes a decrease in current, and charges accumulate on the capacitor plates in the reverse sense (Figure 32-5(d)); the current now has decreased in magnitude as the first charges to arrive on the capacitor plates repel other charges arriving later. Still later, the current falls to zero and the capacitor is again fully charged, but with opposite polarity and again with charges of magnitude Q_m on each plate [Figure 32-5(e)]. At this point, the electric oscillator has completed exactly *one-half* of a cycle. The process is then repeated but in the opposite sense (for the sign of charge and the direction of current), until the capacitor reaches its initial charge state [Figure 32-5(a)].

Oscillations in the electric charge and electric current continue. No energy is dissipated; the circuit has no resistance. The oscillations can also be characterized by the continuous alternation of energy stored in the electric field of the charged capacitor and energy stored in the magnetic field of the current-carrying inductor. The two circuit elements play different roles:

• The capacitor C stores energy in its electric field when charged, but tends to lose its charge and be restored to its equilibrium state of electric neutrality.

• The inductor L stores magnetic energy when carrying a current, and it displays electrical inertia in that its self-induced emf tends to maintain the charges in motion.

Now we analyze the electric oscillator with mathematics. Applying Kirchhoff's second rule (energy conservation) to the circuit loop of Figure 32-6, we have

$$\Sigma \mathcal{E} = \Sigma V$$

$$-L\frac{di}{dt} = \frac{q}{C}$$

The inductor's emf is $-L\,di/dt$; the electric potential difference across the capacitor is q/C with q as the charge on one capacitor plate at any instant. By definition, $i = dq/dt$; substituting this in the equation above, we have

$$-L\frac{d^2q}{dt^2} = \frac{q}{C}$$

$$\frac{d^2q}{dt^2} = -\frac{1}{LC}q$$

We make the substitution

$$\omega_0^2 \equiv \frac{1}{LC} \quad (32\text{-}12)$$

Then the equation above can be written as

$$\frac{d^2q}{dt^2} + \omega_0^2 q = 0 \quad (32\text{-}13)$$

This is a particularly important linear second-order differential equation, that for the simple harmonic oscillator [(11-8)]. Its solution is

$$q = Q_m \cos \omega_0 t \quad (32\text{-}14)$$

as can easily be verified by substituting (32-14) in (32-13). The charge Q_m is the maximum on either capacitor plate; it is also, in this problem, the initial charge

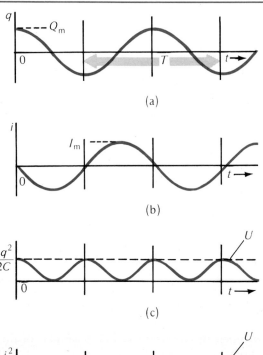

Figure 32-7. Time variation for an electric oscillator of (a) charge on a capacitor plate, (b) current, (c) capacitor energy, and (d) inductor energy.

($q = Q_m$ at $t = 0$). The zero subscript on ω_0 is a reminder that the circuit contains no resistance.

The charge varies sinusoidally with time, as shown in Figure 32-7(a). The angular frequency ω_0 of the free oscillations is given from (32-12) by

$$\omega_0 = \frac{1}{\sqrt{LC}}$$

The (ordinary) frequency $f = \omega_0/2\pi$ and its reciprocal, the period of oscillation T, are given by

$$f = \frac{1}{T} = \frac{1}{2\pi\sqrt{LC}} \tag{32-15}$$

Instantaneous current $i = dq/dt$ also oscillates sinusoidally. We see this by taking the time derivative of (32-14):

$$i = -\omega_0 Q_m \sin \omega_0 t = -I_m \sin \omega_0 t \tag{32-16}$$

where $I_m = \omega_0 Q_m$ is the maximum value of i.

Comparing (32-16) and (32-14) [and Figures 32-6(a) and (b)], we see that the charge (here varying directly with the cosine) and the current (here varying directly with the sine) are 90° out of phase. That is, when the capacitor is fully charged and the energy resides entirely in the capacitor's electric field, the current through the inductor and the magnetic field associated with it are zero. Conversely, when the capacitor is discharged, the inductor has all the energy.

The circuit's total energy U consists of the capacitor's energy $U_C = q^2/2C$ and the inductor's energy $U_L = Li^2/2$:

$$U = U_C + U_L = \frac{q^2}{2C} + \frac{Li^2}{2} = \frac{Q_m^2 \cos^2 \omega_0 t}{2C} + \frac{LI_m^2 \sin^2 \omega_0 t}{2} \tag{32-17}$$

To see that this total energy remains constant with time, we use $I_m = \omega_0 Q_m$ from (32-16) and $\omega_0^2 = 1/LC$ from (32-12) in LI_m^2. We have then

$$LI_m^2 = L\omega_0^2 Q_m^2 = \frac{Q_m^2}{C}$$

so that (32-17) becomes

$$U = \frac{Q_m^2}{2C}(\sin^2 \omega_0 t + \cos^2 \omega_0 t) = \frac{Q_m^2}{2C} \tag{32-18}$$

The energies of the capacitor and inductor vary sinusoidally with time (at *twice* the frequency of q and i), as shown in Figures 35-7(c) and (d); their sum is constant.

Any actual circuit has some resistance, even if only the resistance of the inductor's windings. With resistance included, we have a damped harmonic electrical oscillator. The energy then decreases with time exponentially (in the fashion of Figure 11-13 for a damped mechanical oscillator) because of the ever-present i^2R loss in the resistor.

Example 32-5. An electric oscillator is a part of every radio receiver. One tunes to a station typically by adjusting a variable capacitance with a tuning knob so that the radio oscillator's frequency is brought to the frequency of the incoming radio wave.

The oscillator of a certain radio receiver has an inductor of 1.0 mH and is tuned to an AM station at 710 kHz. (a) What is the capacitance in this oscillator? (b) In what sense, clockwise or counterclockwise, must the knob attached to the variable capacitor in Figure 32-8 be turned to tune to a station at 880 kHz?

(a) From (32-13), we have

$$f = \frac{1}{2\pi\sqrt{LC}}$$

or

$$C = \frac{1}{4\pi^2 f^2 L} = \frac{1}{4\pi^2 (710 \times 10^3 \text{ s}^{-1})^2 (1.0 \times 10^{-3} \text{ H})} = 50 \text{ pF}$$

(b) Tuning to a higher frequency implies, from the equation above, that the capacitance must decrease. For a parallel-plate capacitor of adjustable plate area [$C = \epsilon_0 A/d$, (26-2)], the area of the overlapping adjacent capacitor plates must be reduced to decrease C. The knob must be tuned clockwise.

As (32-15) shows, the frequency of an electric oscillator increases as the magnitudes of L and C decrease. It follows that if one is to construct an electric oscillator of very high frequency, the capacitor and inductor must be small, not merely in the magnitudes of C and L, but in the actual dimensions of these circuit elements.

Figure 32-9 shows the evolution of an ordinary LC circuit, with obvious, lumped capacitance and inductance elements, into two varieties of high-frequency oscillators, in which these circuit elements are less easily recognized. In Figure 32-9(a), first the inductance is reduced drastically by replacing the coil with a single conducting wire; then the capacitance also is reduced drastically by shrinking the area of the capacitor plates. What remains is a single conducting loop broken by a gap at one point. This is indeed an electric oscillator, and if the loop's size is of the order of 1 m with a gap of perhaps 1 cm, it oscillates at a frequency of tens of megahertz, a frequency lying in the radiofrequency region of the electromagnetic spectrum.

An oscillator of just this type was used in the historic experiments of Heinrich Hertz (1857–1894), who first demonstrated the existence of electromagnetic radiation in 1887. Hertz used two such oscillators, both resonant at the same frequency. A spark at the gap served to identify the oscillations. Electromagnetic radiation was detected by observing that when the second oscillator was moved relatively far from the first oscillator, the sparks across its

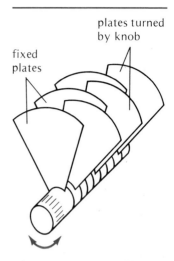

Figure 32-8. A variable parallel-plate capacitor. Turning the knob clockwise reduces the area that overlaps between adjacent capacitor plates.

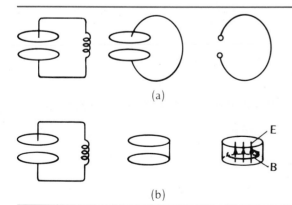

Figure 32-9. Evolution of an ordinary electric oscillator into (a) a high-frequency oscillator consisting of a single conducting loop with a gap and (b) a microwave cavity oscillator consisting of a hollow right circular cylindrical conductor.

gap persisted, although this effect could not be attributed to the direct action of the electric and magnetic fields of the first oscillator on the second.

Figure 32-9(b) shows the evolution of a simple *LC* circuit into a different type of high-frequency oscillator. Here the inductance is reduced first by connecting the capacitor plates with a single straight conducting wire. Then the inductance is reduced still further by connecting additional wires in parallel with the first. (Inductors in parallel follow the same rule as resistors.) Indeed, one constructs an entire cylindrical surface between the two capacitor plates, and thereby forms a hollow, closed right circular cylinder of conducting material. Superficially at least this certainly does not look like an *LC* oscillator of the ordinary variety. It is, in fact, one simple type of a *microwave* oscillator. For dimensions of the order of a few centimeters, the free oscillations occur at microwave frequencies of the order of a few gigahertz (or electromagnetic waves having wavelengths of a few centimeters). The oscillating electric field, as well as the oscillating magnetic field, is confined entirely within the closed cylinder. Here it becomes more useful to describe the electric oscillations in terms, not of the current through the circuit or of the potential difference across various pairs of points, but of the electric and magnetic fields within the closed resonating cylinder.

32-6 Electrical-Mechanical Analogs

An electric oscillator reminds us in many respects—the sinusoidal variation in time, the constant total energy, to name just two—of an ordinary mechanical oscillator. There is indeed an exact analogy between mechanical quantities and properties and their counterparts in electromagnetism and electric circuit elements.

Consider, for example, a mechanical oscillator with mass m attached to spring of force constant k. We get the equation of motion from Newton's second law:

$$\Sigma \mathbf{F} = m \mathbf{a}$$

$$-kx = m \frac{d^2x}{dt^2}$$

or

$$\frac{d^2x}{dt^2} = -\frac{k}{m} x$$

This equation has *exactly* the same form as the equation for an electric oscillator:

$$\frac{d^2q}{dt^2} = -\frac{1}{LC} q$$

One equation can be transformed into the other simply by making the substitutions listed in Table 32-1.

The correspondence is more than formal. Like mass m, which is a measure of the inertial tendency of a particle to maintain constant velocity v, inductance L is a measure of an inductor's tendency to maintain a constant current i.

Table 32-1

MECHANICAL QUANTITY			ELECTRIC (OR MAGNETIC) QUANTITY
Displacement		$x \quad q$	Electric charge
Velocity		$v = \dfrac{dx}{dt} \quad \dfrac{dq}{dt} = i$	Electric current
Acceleration	$a = \dfrac{dv}{dt} = \dfrac{d^2x}{dt^2}$	$\dfrac{d^2q}{dt^2} = \dfrac{di}{dt}$	Rate of current change
Mass		$m \quad L$	Inductance
Force		$F = ma \quad L\dfrac{di}{dt} = -\mathcal{E}$	EMF, potential difference
Stiffness		$k \quad \dfrac{1}{C}$	Reciprocal of capacitance
Force of spring		$F = -kx \quad \dfrac{q}{C} = V$	Potential difference, capacitor
Potential energy of spring		$\dfrac{1}{2}kx^2 \quad \dfrac{1}{2}\left(\dfrac{1}{C}\right)q^2$	Potential energy of capacitor
Kinetic energy		$\tfrac{1}{2}mv^2 \quad \tfrac{1}{2}Li^2$	Energy of inductor
Mechanical (linear) resistive force		$F_r = -rv \quad iR = V$	Potential difference across electrical resistor
Constant weight		$W = mg \quad \mathcal{E}_b$	Constant emf battery
Rate of energy dissipation		$P = rv^2 \quad i^2R = P$	Rate of energy dissipation

Similarly, a spring's restoring constant k corresponds to what might be called the capacitor's restoring constant $1/C = V/q$. (Be careful about the symbols, however. Both an inductor and a helical spring are typically represented in a diagram by the same sort of symbol. But a capacitor is analogous to a spring, and an inductor to mass.)

Note, further, that the electric potential difference q/C corresponds to the elastic restoring force $-kx$, and the induced emf $L(di/dt)$ to the force $m(dv/dt)$. The parallel to the mechanical dissipative force $F = -rv$, proportional to the particle's velocity, is the electrical resistance R, with potential difference $V = iR$. Similarly, the potential energy $\tfrac{1}{2}kx^2$ of the stretched spring corresponds to the electric energy $q^2/2C$ of the charged capacitor, and the kinetic energy $\tfrac{1}{2}mv^2$ of the moving particle to the magnetic energy $Li^2/2$ of the inductor.

The analogs are complete, in both mathematical and physical behavior, between mechanical and electrical "circuit" elements. It may be useful to analyze a complicated mechanical system by constructing its electrical analog with ordinary electric circuit elements. Then the measured charges, currents, and potential differences give the corresponding particle displacements, velocities, and forces.

Example 32-6. An object falls from rest under the influence of gravity and it is also subject to linear resistive force (proportional to the object's velocity). How does the velocity of the falling object change with time?

This problem was dealt with in Section 6-4 as an example of Newton's laws. Now we can write down the solution immediately, simply by recognizing its electric analog: an LR circuit, with an inductor and resistor in series connected to a battery of constant

CHAPTER 32 Inductance and Electric Oscillations

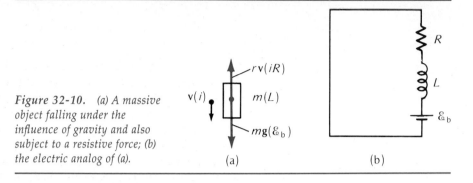

Figure 32-10. (a) A massive object falling under the influence of gravity and also subject to a resistive force; (b) the electric analog of (a).

emf $\mathcal{E}_b$. See Figure 32-10. Using the translation dictionary in Table 32-1 and (32-8) for the current as a function of time in an LR circuit, we have at once

$$v = \frac{mg}{r}\left[1 - \exp\left(-\frac{t}{m/r}\right)\right]$$

The time constant m/r now represents the time elapsing until the falling object comes within $(1/e)$th of its terminal velocity mg/r.

32-7 Mutual Inductance (Optional)

Self-inductance L relates the emf induced in an inductance coil to the rate at which current changes in that same coil. *Mutual* inductance, on the other hand, has to do with the emf induced in one inductor by a current change in another inductor, not a part of the same circuit. The term *transformer* (Section 33-7) is also used, especially in connection with ac circuits, to denote a mutual inductor.

Consider Figure 32-11, where we have two inductors at rest, each a part of separate circuits. The coils are, however, linked by magnetic flux: an electric

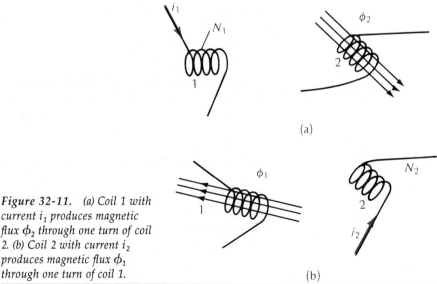

Figure 32-11. (a) Coil 1 with current i_1 produces magnetic flux ϕ_2 through one turn of coil 2. (b) Coil 2 with current i_2 produces magnetic flux ϕ_1 through one turn of coil 1.

current in either inductor produces a magnetic flux through the other one. Current i_1, in coil 1 with N_1 identical turns, produces a magnetic flux ϕ_2 through one turn at coil 2. Similarly, current i_2 in coil 2 of N_2 identical turns produces a flux ϕ_1 through one turn of coil 1. The emf $\mathcal{E}_2$ induced in coil 2 when the flux ϕ_2 changes with time is, from Faraday's law,

$$\mathcal{E}_2 = -N_2 \frac{d\phi_2}{dt} \tag{32-19}$$

Magnetic flux ϕ_2 is proportional to the current i_1 that creates it; and the *mutual inductance* M_{21} is defined, in analogy with self-inductance [(32-1)], by

$$N_2 \phi_2 \equiv M_{21} i_1 \tag{32-20}$$

We see that M_{21} is a quantitative measure of how the flux linkage $N_2 \phi_2$ in coil 2 is influenced by current i_1 in coil 1. When we take the time derivative of (32-20) and substitute the result in (32-19), we get

$$\mathcal{E}_2 = -M_{21} \frac{di_1}{dt} \tag{32-21}$$

a relation analogous to (32-3) for self-inductance. For any pair of inductors fixed in position, the mutual inductance for the pair is a constant whose value depends on the geometry of the arrangement. Like self-inductance, mutual inductance has units of henries.

Mutual inductance can be considered defined either by (32-20) or (32-21). The first equation is most useful for computing the value for M_{21} from a knowledge of the flux linking the two inductors. The second equation is most useful for describing the behavior of a mutual inductor in coupled circuits.

In arriving at (32-20) and (32-21), we began with changing current i_1 producing a magnetic flux ϕ_2 and emf $\mathcal{E}_2$. The reciprocal arrangement is that in which changing current i_2 produces a magnetic flux ϕ_1 and emf $\mathcal{E}_1$. The equations corresponding to (32-20) and (32-21) are, with subscripts 1 and 2 interchanged,

$$N_1 \phi_1 \equiv M_{12} i_2 \tag{32-22}$$

and

$$\mathcal{E}_1 = -M_{12} \frac{di_2}{dt} \tag{32-23}$$

We assert without proof a remarkable result. The two mutual inductances, M_{21} and M_{12}, are exactly equal for every mutual inductor. We can then use a single symbol M for either mutual inductance:

$$M \equiv M_{12} = M_{21} \tag{32-24}$$

Summary

Definitions

Self inductance L for a coil of N identical turns with magnetic flux ϕ_B through each turn produced by current i:

$$N\phi_B \equiv Li \tag{32-1}$$

When the current through an inductor changes at the rate di/dt, a back emf $\mathcal{E}$ is produced:

$$\mathcal{E} = -L \frac{di}{dt} \tag{32-3}$$

Units

Henry, the unit for inductance:

$$1\,H \equiv 1\,V \cdot s/A$$

Important Results

Time constant τ for an LR circuit:

$$\tau = \frac{L}{R} \quad (32\text{-}9)$$

Energy U_L of an inductor of inductance L carrying current i:

$$U_L = \tfrac{1}{2}Li^2 \quad (32\text{-}10)$$

Energy density (energy for unit volume) u_B of magnetic field **B**:

$$u_B = \frac{B^2}{2\mu_0} \quad (32\text{-}11)$$

LC electric oscillator: the charge (on a capacitor plate) and current oscillate sinusoidally with time at frequency f and period T, where

$$f = \frac{1}{T} = \frac{1}{2\pi\sqrt{LC}} \quad (32\text{-}15)$$

Electric-mechanical analogs (see Table 12-1): electric-circuit elements are exactly parallel, in mathematical form and physical properties, to mechanical quantities.

Problems and Questions

Section 32-1 Self-Inductance Defined

· **32-1 P** An emf of 4.0 V is induced in a certain coil when the current in the coil changes at a rate of 4000 A/s. What is the self-inductance of the coil?

· **32-2 P** A current of 100 A has been established in a large electromagnet with an inductance of 20 H. Suppose that the circuit is broken by accident and the current falls to zero in 1 ms. What emf would be induced across the magnet?

· **32-3 P** A 5-H inductor carries a current of 2 A. How can an emf of 200 V be made to appear across the inductor?

· **32-4 Q** Show that μ_0 has units of henries per meter.

· **32-5 Q** The inductance of a coil depends on
(A) the current in it.
(B) the flux in it.
(C) the emf applied to it.
(D) geometrical factors such as size, shape, and number of turns.
(E) none of the above.

· **32-6 P** What is the inductance of a solenoid of length 50 cm and diameter 2.0 cm wound with 10,000 turns of wire?

· **32-7 P** An inductor consists of a closely wound, long solenoid. Its inductance is 200 mH. What is the total magnetic flux through the middle cross section of the coil when a current of 5 A exists in the solenoid?

: **32-8 Q** Sometimes when you first turn on a large electric motor in your garage, you find that the house lights dim momentarily. Why is this?
(A) A large current is used briefly to charge the capacitor attached to the motor.
(B) The large back emf generated by the motor reduces the voltage available to the house.
(C) Before the motor armature reaches full speed, it does not generate much back emf and as a result a large current is drawn in the wires coming into the house.
(D) This is a transient effect associated with the fact that the current in the motor is ac.
(E) A large torque is needed to cause the armature to start rotating; but once it reaches full speed, essentially very little torque needs to be applied.

: **32-9 P** Two concentric circular loops lie in a plane. They have radius R_1 and R_2, where $R_1 \gg R_2$. What emf is induced in the small loop when the current in the large loop changes at the rate di/dt?

: **32-10 P** Determine the inductance of a toroid of rectangular cross section $a \times b$ of N turns, with inner radius R and outer radius $R + a$. See Figure 32-12.

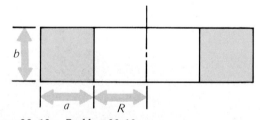

Figure 32-12.   Problem 32-10.

: **32-11 P** Two long, straight, parallel wires of radius r are separated by a distance d. Show that if they each carry current I in opposite directions, their self-inductance per meter length is, with the flux within the wires themselves neglected,

$$L = \frac{\mu_0}{\pi} \ln\left(\frac{d-r}{r}\right)$$

Section 32-2 The LR Circuit

· 32-12 P A coil with a inductance of 2.5 H is suddenly connected to a 24-V battery. The inductor has a resistance of 12 Ω. Graph current as a function of time.

· 32-13 P How long would it take for the voltage across the resistor in an LR circuit to drop to 1 percent of its initial value if $L = 1$ mH and $R = 1000$ Ω?

· 32-14 P An inductor is connected to a 6-V battery with an internal resistance of 0.5 Ω. The inductor has 4.2-Ω resistance. The current is found to grow to 95 percent of its final value in 15 ms. What is the inductance of the coil?

· 32-15 P Show that equation (32-8) *is* a solution to (32-7).

: 32-16 P After how many time constants will the current in an LR circuit be within 1 percent of its final value?

: 32-17 Q You have momentarily forgotten the relation for the time constant in an RL circuit. (Is it L/R, or LR, or R/L?) Show what the correct relation is by using dimensional analysis.

: 32-18 Q An inductance L and a resistance R are connected in a circuit as shown in the text in Figure 32-4. When the switch is closed,
(A) the current in the resistor will initially be greater than the current in the inductance.
(B) the current in the inductance will initially be greater than the current in the resistance.
(C) the current in the battery will decay exponentially.
(D) the current wll rise slowly at first, and then more rapidly.
(E) the time required to reach the final current in the resistor is proportional to R/L.
(F) None of the above is true.

: 32-19 P A coil with an inductance L and a resistance R is connected to a battery with emf $\mathcal{E}$. How long after the switch is closed will (*a*) the current reach 99.9 percent of its steady-state value? (*b*) the energy stored in the inductor reach 99.9 percent of its steady-state value?

: 32-20 P A short time after the switch is closed in an LR circuit, the current is 2 mA. Ten milliseconds later the current is 6 mA. What is the time constant of the circuit?

: 32-21 P A solenoid of inductance L is connected to a power supply that applies a potential difference V_0. (*a*) Suppose the switch to the supply is opened in a short time t. Show that the voltage that appears across the solenoid is $V = (\tau/t)V_0$, where $\tau = L/R =$ the time constant for the charging circuit. (*b*) Suppose that when the switch is opened, an arc 2 cm long jumps across the two poles of the switch. The breakdown strength of air is 30×10^6 V/m. Estimate the sparking time t.

: 32-22 P For the circuit shown in Figure 32-4, (*a*) show that at $t = 0$, $di/dt = i_f/\tau$, where $\tau = L/R$ and i_f is the final value of the current. (*b*) Find what the ratio is of voltage drop across inductor to voltage drop across resistor when $i = \frac{1}{2}i_f$.

: 32-23 Q A large inductance (such as a big electromagnet) is connected to a battery as shown in Figure 32-13. Which of the following is the most accurate statement? Assume that the circuit has reached equilibrium before the switch is opened or closed.
(A) Just before the switch is closed the potential difference between its contacts is $\frac{1}{2}\mathcal{E}$.
(B) Just after the switch is opened the potential difference across it is $\frac{1}{2}\mathcal{E}$.
(C) Just after the switch is opened, the potential difference across it can never be greater than $\mathcal{E}$.
(D) Just after the switch is opened, the potential difference across it can be much greater than $\mathcal{E}$.
(E) Corrosion of the switch contacts due to arcing is more likely to occur in closing than in opening the switch.
(F) None of the above is true.

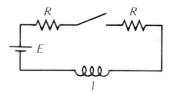

Figure 32-13. Question 32-23.

: 32-24 Q Suppose that an electromagnet with a very large inductance and a fairly low resistance are connected in series to a 100-V dc power supply, as shown in Figure 32-4. If the switch were suddenly opened, the potential difference between the contacts of the swtich
(A) would be zero initially.
(B) would be less than 100 V.
(C) would be equal to 100 V.
(D) could be much greater than 100 V.
(E) None of the above is true.

: 32-25 P For the circuit shown in Figure 32-14, find what the values of i_1 and i_2 are (*a*) immediately after the switch is closed; (*b*) long after the switch is closed; (*c*) just after the switch is then opened; (*d*) long after the switch has been opened. ($L = 3.0$ H, $R_1 = 5$ Ω, $R_2 = 12$ Ω, $R_3 = 20$ Ω, $\mathcal{E} = 60$ V)

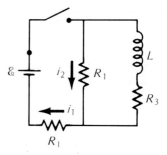

Figure 32-14. Problem 32-25.

Section 32-3 Energy of an Inductor

32-26 The current in a circuit with inductance L and resistance R grows to within $1/e$ of its final value in a time interval L/R. What is the time interval for the energy of the inductor to grow to within $1/e$ of its final value?

: **32-27 P** For the circuit shown in Figure 32-4, $\mathcal{E}_b = 12$ V, $R = 3\ \Omega$ and $L = 6$ mH. Six milliseconds after the switch is closed, what is (a) the power being dissipated in the resistor? (b) the power being supplied by the battery? (c) the rate at which energy is being stored in the magnetic field of the inductor?

· **32-28 P** A current of 1 A has been established in an inductor of 1.0 H. What capacitance would store the same amount of energy when charged to a voltage of 1 V?

· **32-29 P** A solenoid of 4-cm radius is 80 cm long and has 5000 turns. (a) What is the self-inductance of the coil? (b) What magnetic energy is stored in the coil for a current of 1 A?

: **32-30 P** Show that the equivalent self-inductance for two inductors in series is $L_s = L_1 + L_2$, and that the equivalent self-inductance for two inductors in parallel is L_p, where $1/L_p = 1/L_1 + 1/L_2$. Assume that there is negligible magnetic coupling between the two inductors (zero mutual inductance).

: **32-31 P** (a) Two solenoids of inductance 2 mH and 6 mH are connected in series. They are well separated. What is the total magnetic energy stored in them for a current in each of 1 A? (b) The two coils are connected in parallel and a total current of 1 A enters the junction with the two inductors. The inductors are still well separated. What is the total energy stored in the inductors?

: **32-32 P** Two coils of inductance 2.0 H and 4.0 H are connected in parallel. They are well separated. What current flows in each when a total magnetic energy of 6 J is stored in them?

Section 32-4 Energy of the Magnetic Field

· **32-33 Q** The current through a conducting loop is doubled. The total energy of the magnetic field associated with the circuit is thus changed by a factor of
(A) $\tfrac{1}{4}$.
(B) $\tfrac{1}{2}$.
(C) 2.
(D) 4.
(E) The change cannot be determined without more information about the shape and size of the loop.

· **32-34 P** A magnetic field of 0.10 T is set up in a certain region of space. What electric field would have the same energy density?

· **32-35 P** A long, straight wire carries current i. What is the magnetic-field-energy density at a distance r from the wire?

· **32-36 P** Very large magnetic fields can be obtained by discharging a capacitor through a low-inductance coil. The capacitor leads must have very low inductance themselves. Such "fast" capacitors cost about $3 per joule of stored energy capacity. (a) Estimate the cost of a capacitor that could store as much energy as contained in a 200-T field in a volume of one liter. (b) Assuming electricity costs $0.10 per kW·h, what does the electricity for charging the capacitor cost?

: **32-37 P** A flat circular coil of N turns and radius R carries current i. What is the magnetic-field energy density at the center of the coil?

: **32-38 P** The switch in the circuit of Figure 32-4 is closed at $t = 0$. Plot as a function of time (a) the current in the circuit; (b) the voltage across the inductance; (c) the energy stored in the magnetic field of the inductance.

: **32-39 P** Energy density and pressure have the same dimensions, and it can be shown that a magnetic pressure equal to the magnetic energy density $B^2/2\mu_0$ can be associated with magnetic field B. More specifically, a surface with magnetic field B on one side and zero field on the other is subject to magnetic pressure $B^2/2\mu_0$. For what magnetic field is the magnetic pressure equal to one atmosphere?

: **32-40 P** A long, straight wire carries current i uniformly distributed over its circular cross section. Find the magnetic energy stored per unit length within the conducting wire and show that it is independent of the wire radius.

32-5 Electrical Free Oscillations

· **32-41 P** An inductor with an inductance of 10 mH is available for use in an electric oscillator that is to resonate over the range of the AM radio dial from 530 kHz to 1600 kHz. A suitable variable capacitor is to be chosen. What is its (a) minimum and (b) maximum capacitance?

· **32-42 P** An electric oscillator resonates at frequency f_0. Then the original air-filled capacitor is filled with a dielectric having a dielectric constant κ. What is the new resonance frequency?

: **32-43 P** An inductor of fixed inductance and two identical variable capacitors are available to construct an electric oscillator. Each capacitor's capacitance can be changed continuously by a factor 2. The capacitors can be connected in series, or in parallel, or used singly. Over what range of frequencies can the oscillator be used?

: **32-44 Q** Suppose that you have temporarily forgotten how the resonance frequency of an LC circuit depends on L and C. Find the functional dependence of frequency on L and C by using dimensional analysis.

: **32-45 P** A certain electric oscillator consists of a parallel-plate capacitor and long cylindrical solenoid. (a) What happens to the resonance frequency if every dimension is reduced by a factor 2? (b) What then happens to the total resistance of the circuit?

Section 32-6 Electrical-Mechanical Analogs

: **32-46 Q** The analogy between an LC circuit and a harmonic oscillator is helpful in gaining insight into the system. In this analogy, the energy stored in the inductor is like
(A) the mass of the harmonic oscillator.
(B) the spring constant of the oscillator.
(C) the potential energy of the oscillator.
(D) the velocity of the oscillating mass.
(E) none of the above.

: **32-47 Q** As Example 32-6 shows, the mechanical analog of an LR circuit is a massive object moving through a resistive medium. What is the mechanical analog of an RC circuit? How might one achieve such a mechanical device in practice?

Supplementary Problem

32-48 P A magnetic rail gun is represented schematically in Figure 32-15. A power supply charges a large capacitor bank, which is then discharged by closing the switch S. Current in the right-hand loop creates magnetic forces, which push the movable armature to the right. In guns being studied at the Los Alamos and Livermore Laboratories, a plasma arc is the moving armature, and it is contained in the bore of the gun in such a way that it is able to push a projectile along in front of it. The velocity obtainable by a conventional gun is limited by the rate at which a hot gas can expand, but there is no such limitation here. It is possible to get velocities well in excess of 10 km/s. These devices are being investigated for many applications, including high-pressure research, industrial manufacturing processes, impact thermonuclear fusion, and space propulsion. These guns could be used to propel material into space for use in manufacturing. Such a device, called a mass driver, is planned for use on the moon. It would hurl material out to a manufacturing space station.

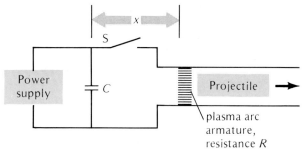

Figure 32-15.   Problem 32-48.

To see what is involved in the design of such a system, consider a rail gun with the following characteristics:

$$L_1 = \text{inductance per meter of rail length} = 4 \times 10^{-7} \text{ H/m}$$

$$I = 2.5 \times 10^6 \text{ A}$$

$$m = \text{projectile mass} = 4 \text{ kg}$$

$$\text{Energy efficiency of device} = 11\% = \frac{E_{k,\text{projectile}}}{E_{\text{stored,capacitor}}}$$

(a) Calculate the force on the projectile in terms of L_1 and I. (Hint: Consider what happens to the energy provided by the capacitor. It goes into magnetic energy, kinetic energy, and heat.) (b) What length of rails would be needed to accelerate the projectile to 9.5 km/s, a typical velocity needed for an orbit of the earth? (c) How much energy must the power supply provide for each launch? (d) What is the average power delivered by the capacitor bank?

33 AC Circuits

33-1 Some Preliminaries
33-2 Series *RLC* Circuit
33-3 Phasors
33-4 Series *RLC* Circuit with Phasors
33-5 RMS Values for AC Current and Voltage
33-6 Power in AC Circuits
33-7 The Transformer
Summary

Some electric devices run on direct current (dc)—on batteries, or other sources of constant emf. Nearly all other electric devices operate on alternating current (ac). A particularly important ac circuit is one involving a resistor, an inductor, and a capacitor, all in series and driven by a sinusoidally varying emf of variable frequency.

33-1 Some Preliminaries

A constant danger in physics is that the basic ideas, which are usually straightforward, can become obscured in mathematical analysis—sometimes messy, and always potentially distracting. This is particularly true of ac circuits. Here we deal with sinusoidally varying currents and a circuit with all three basic circuit elements—resistance R, capacitance C, and inductance L. It is easy to understand these elements taken singly; their combined behavior is far more complicated.

After we have made some basic definitions, we shall consider separately the qualitative behavior of resistor, capacitor, and inductor for a current that changes with time.

We have already discussed a simple form of ac generator (Example 31-3). It consists of a coil of wire rotated at a constant angular speed ω in a uniform magnetic field perpendicular to the rotation axis of the coil. The generator produces a sinusoidally varying emf $\mathcal{E}$, which has the same frequency as the rotating coil:

$$\mathcal{E} = \mathcal{E}_m \sin(\omega t + \phi)$$

where the phase constant ϕ determines the value of the emf at the arbitarily chosen zero of time:

$$\mathcal{E}\ (\text{at } t = 0) = \mathcal{E}_m \sin \phi$$

Figure 33-1 illustrates the variation in time of a constant dc source of emf and a sinusoidal ac source. It is customary to represent the dc source by the symbol ⊣⊢ and the ac source by ∼. The special importance of the *sinusoidal* variation in time of the emf is that first, most ordinary ac sources, such as household current, are of this type, but more fundamentally, *any* variation in emf with time can be regarded as the superposition of strictly sinusoidal variations (Optional Section 11-6).

We now consider in turn the qualitative behavior of R, C, and L for a constant emf and also a time-varying emf.

The potential drop V across a resistance R carrying current is $V = iR$. This relation holds for dc certainly. It also applies at each instant for ac, provided only that the frequency of the current is not too high. To see this, we recall first that the origin of resistance in any conductor is basically the collision of conduction electrons with the remaining atomic lattice and with other conduction electrons. The average time between collisions in a typical conductor at room temperature is of the order of 10^{-14} s. Therefore, a change in current with time, and hence a change within a conductor in the electric field driving conduction electrons, will appear to a conduction electron to occur so slowly that the current is effectively constant, so long as the frequency of the changing current is small compared with 10^{14} Hz (roughly the frequency of infrared light). Consequently, for all ordinary ac frequencies, the behavior of a resistor is the simplest possible; at each instant, $V = iR$. As the current changes with time, the potential drop across R keeps pace. In more formal language, i and V are exactly in phase for a resistor.

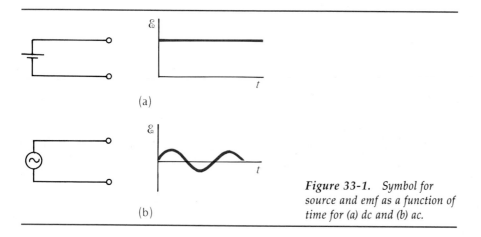

Figure 33-1. *Symbol for source and emf as a function of time for (a) dc and (b) ac.*

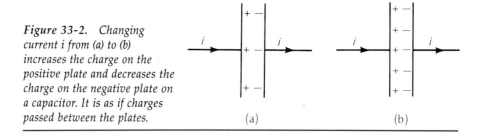

Figure 33-2. Changing current i from (a) to (b) increases the charge on the positive plate and decreases the charge on the negative plate on a capacitor. It is as if charges passed between the plates.

Now consider an inductor. If the current through an inductor is constant so that $di/dt = 0$, the inductor produces no induced emf, and $\mathcal{E} = -L\, di/dt = 0$; an inductor does not oppose a constant direct current. An inductor opposes a changing current, however; the higher the frequency of the varying current, the greater the rate of change of current and therefore the larger the opposing emf. In short, an inductor passes a *constant* current without opposition, but impedes the flow of charge through it increasingly as the frequency of the current increases.

How does a capacitor behave in a circuit with varying current? Strictly, there can never be any real current through a capacitor since no charged particles actually move from one charged plate to the other, oppositely charged plate through the space between them. With a capacitor maintained at a constant potential difference, the charge magnitude on each plate is also fixed. Then, not only is there no current between the plates; the current is also zero in the lead wires connecting the capacitor to other parts of an electric circuit. At constant voltage, a capacitor is an open circuit.

When the potential difference across a capacitor changes with time, however, there is actually an *equivalent* current through C. We can see this in Figure 33-2, where the potential difference across the capacitor plates increases from (a) to (b). With voltage V increased, the charge magnitude q on each plate must also increase by the same factor ($q = CV$). But if the positive plate becomes more positive, a current in the conductor connected to this plate must have brought additional positive charges to this plate; at the same time, an equal current removes positive charges from the negatively charged plate and makes it more negative. In short, when the potential drop across a capacitor is changing, the same instantaneous current exists in both lead wires connected to the capacitor—it is as if charges passed between the plates—and we can regard this current as effectively "passing through" the capacitor. Whereas a capacitor impedes completely constant direct current, it passes an alternating current.

Table 33-1

	DC (OR LOW FREQUENCIES)	AC (OR HIGH FREQUENCIES)
R	Passes	Passes
L	Passes	Impedes
C	Impedes	Passes

The qualitative features of R, L, and C circuit elements for dc and ac are summarized in Table 33-1. Note especially the reciprocal behavior of an inductor and a capacitor:

- L passes low frequencies but impedes high frequencies.
- C passes high frequencies but impedes low frequencies.

Example 33-1. A sinusoidal current superimposed on constant current is applied to the input terminals of the two circuit arrangements shown in Figure 33-3: (a) a low-pass filter and (b) a high-pass filter. Show on the basis of the qualitative behavior of an inductor and a capacitor that (a) the low-pass filter "passes" the low-frequency, or dc, signal to the output terminals but impedes the ac signal, whereas (b) the high-pass filter passes the ac signal but impedes the dc signal.

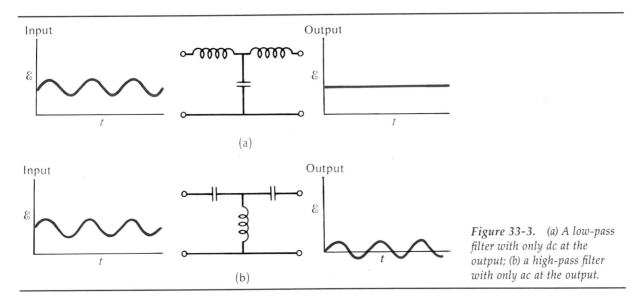

Figure 33-3. (a) A low-pass filter with only dc at the output; (b) a high-pass filter with only ac at the output.

(a) The inductors act as "choke coils"; the ac signal is impeded by them from reaching the output terminals. The ac signal passes readily through the "by-pass" capacitor, however, and is returned to the input terminals. The signal at the output terminal is more nearly a constant emf.

(b) In the high-pass filter, the circuit elements are reversed. Now dc is blocked from reaching the output terminals by the capacitors, but dc passes readily through the inductor. The output signal is more nearly a purely sinusoidal emf.

33-2 Series RLC Circuit

The situation: An ac emf of variable frequency is applied to a circuit of R, L, and C all in series. The question: What is the magnitude of the current and its phase relative to the sinusoidal applied emf?

Before beginning the formal analysis, let us consider qualitatively what the general result must be. We know that a simple LC circuit (Section 32-5) oscillates at its characteristic angular frequency $\omega_0 = 1/\sqrt{LC}$, (32-10), with an

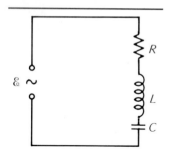

Figure 33-4. A series RLC circuit driven by a sinusoidal emf of variable frequency.

undamped current amplitude. A more realistic electric oscillator, one with at least some small resistance, would oscillate at essentially the same frequency. Because of energy lost in the resistor, the oscillations would die out with time in the same fashion as a damped mechanical oscillator (Optional Section 11-7). Finally, an RLC circuit driven by an external emf would respond appreciably — that is, oscillate at resonance with large current amplitude — only when the driving frequency matches the characteristic oscillation frequency of the damped electric oscillator (Optional Section 11-8).

The series RLC circuit is shown in Figure 33-4. Applying Kirchhoff's loop rule (energy conservation) to this circuit, we have

$$\Sigma \mathcal{E} = \Sigma V$$

$$\mathcal{E} - L\frac{di}{dt} = iR + \frac{q}{C}$$

where $\mathcal{E}$ is the instantaneous applied emf; $-L\,di/dt$, the emf induced in the inductor; iR, the potential drop across the resistor; and q/C, the potential drop across the capacitor.

We shift the term $-L\,di/dt$ to the right side of the equation above and it becomes

$$\mathcal{E} = L\frac{di}{dt} + iR + \frac{q}{C} \qquad (33\text{-}1)$$

The instantaneous current i is the *same* at each instant through all circuit elements (Kirchhoff's junction rule, charge conservation). We take the current to be of the form

$$i = I_m \sin \omega t \qquad (33\text{-}2)$$

where I_m is the maximum instantaneous value of current and ω is the angular frequency of both the current and the applied sinusoidal emf.

The rate of change of current is then

$$\frac{di}{dt} = \omega I_m \cos \omega t \qquad (33\text{-}3)$$

The charge q on the positively charged capacitor plate is related to the current in the circuit by

$$q = \int i\,dt = \int I_m \sin \omega t\,dt$$

$$q = -\frac{I_m}{\omega} \cos \omega t \qquad (33\text{-}4)$$

So much for the physics; what follows is strictly mathematics. We substitute (33-2), (33-3), and (33-4) in (33-1), to get

$$\mathcal{E} = \omega L I_m \cos \omega t + R I_m \sin \omega t - \frac{1}{\omega C} I_m \cos \omega t$$

$$\mathcal{E} = I_m \left[R \sin \omega t + \left(\omega L - \frac{1}{\omega C} \right) \cos \omega t \right] \qquad (33\text{-}5)$$

The following general definitions allow this equation to be written more compactly:

$$\text{Reactance} = X \equiv X_L - X_C \tag{33-6a}$$

where

$$\text{Inductive reactance} = X_L \equiv \omega L \tag{33-6b}$$

and

$$\text{Capacitive reactance} = X_C \equiv \frac{1}{\omega C} \tag{33-6c}$$

Equation (33-5) then can be written as

$$\mathcal{E} = I_m(R \sin \omega t + X \cos \omega t) \tag{33-7}$$

We see that the term with R varies directly with the sine of ωt, whereas the term with X varies directly with the cosine of ωt. We wish to manipulate the equation so that it contains a single sinusoidal function.

Consider the trigonometric identity:

$$\sin(A + B) = \sin A \cos B + \cos A \sin B$$

We set $A = \omega t$ and $B = \phi$, so that this equation becomes

$$\sin(\omega t + \phi) = \sin \omega t \cos \phi + \cos \omega t \sin \phi \tag{33-8}$$

For reasons soon to be apparent, it is a good idea to define quantities Z and ϕ as follows:

$$\text{Impedance} = Z \equiv \sqrt{R^2 + X^2} = \sqrt{R^2 + \left(\omega L - \frac{1}{\omega C}\right)^2} \tag{33-9}$$

$$\text{Tangent of phase angle} = \tan \phi \equiv \frac{X}{R} = \frac{\omega L - 1/\omega C}{R} \tag{33-10}$$

where we have used (33-6a), (b), and (c).

These definitions correspond to the simple geometrical construction shown in Figure 33-5, which is a helpful mnemonic for keeping straight the relations among Z, ϕ, X, and R. (This geometrical construction acquires additional significance in the discussion of Section 33-4).

From Figure 33-5 we see that

$$R = Z \cos \phi \quad \text{and} \quad X = Z \sin \phi$$

By these relations, (33-7) can then be written as

$$\mathcal{E} = I_m Z(\sin \omega t \cos \phi + \cos \omega t \sin \phi)$$

which simplifies, with the use of (33-8), to

$$\mathcal{E} = I_m Z \sin(\omega t + \phi)$$

We write this instantaneous applied emf as

$$\mathcal{E} = \mathcal{E}_m \sin(\omega t + \phi) \tag{33-11}$$

where $\mathcal{E}_m$ is the maximum value of $\mathcal{E}$, and ϕ is the phase angle by which the sinusoidal emf leads the sinusoidal current. Comparing the two equations immediately above then gives for the *magnitude of the maximum value of instantaneous current*

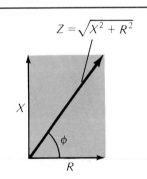

Figure 33-5. Geometrical construction serving as a mnemonic for the relations among Z, ϕ, X, and R.

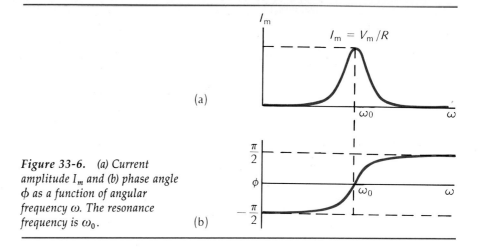

Figure 33-6. (a) Current amplitude I_m and (b) phase angle ϕ as a function of angular frequency ω. The resonance frequency is ω_0.

$$I_m = \frac{\mathcal{E}_m}{Z} = \frac{\mathcal{E}}{\sqrt{R^2 + (\omega L - 1/\omega C)^2}} \quad (33\text{-}12)$$

The *phase ϕ of current i relative to the applied emf $\mathcal{E}$* is, as we write (33-10) once more,

$$\tan \phi = \frac{\omega L - 1/\omega C}{R} \quad (33\text{-}10)$$

That ends the mathematical analysis.* Now let us see what the results mean.

• Impedance Z plays a role analogous to that of resistance in a dc circuit. Impedance Z and reactance X have the units ohms. From (33-12), we see that the *maximum* value $\mathcal{E}_m$ of the applied emf is related to the *maximum* value I_m of the current by $\mathcal{E}_m = Z I_m$ (reminiscent of $\mathcal{E} = RI$). Note especially that we cannot write $\mathcal{E} = Zi$, since $\mathcal{E} = \mathcal{E}_m \cos(\omega t + \phi)$ and $i = I_m \sin \omega t$ do not reach their maximum values simultaneously—there is the phase difference ϕ between $\mathcal{E}$ and i. The impedance is frequency-dependent. From (33-9), we see that Z grows increasingly large at high frequencies, where the inductive reactance [(33-6b)] ωL is large. Impedance Z also becomes large at low frequencies, where the capacitive reactance [(33-6c)], $1/\omega C$, inversely proportional to angular frequency, dominates. Resistance is the only frequency-independent part of impedance.

• From (33-12), the amplitude I_m of the current varies with ω, as shown in Figure 33-6(a). The peak current $I_m = \mathcal{E}_m/R$ occurs at that angular frequency ω_0, for which the impedance has its minimum value. From (33-9), we see that minimum Z corresponds to the condition

$$\omega_0 L - \frac{1}{\omega_0 C} = 0$$

---

* Another way to get the same final results: use the electrical-mechanical analogs (Table 32-1) in the solution for a driven, damped mechanical oscillator, (11-21). The solution given above is the *steady-state* solution; in addition, there is a transient solution, applicable, however, only for a short period after the emf is first applied.

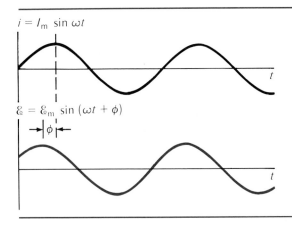

Figure 33-7. *Instantaneous current i and emf & as a function of time showing that & leads i by phase angle φ.*

or

$$\omega_0 = \frac{1}{\sqrt{LC}} \tag{33-13}$$

The resonant frequency ω_0 of the driven *RLC* circuit is the same as for a simple *LC* circuit [(32-10)]; resonance corresponds to the circuit's being driven at its natural resonance frequency, $\omega = \omega_0$.

- At the resonant angular frequency ω_0, the inductive and capacitive reactances have the same magnitude, and cancel, so that the impedance consists of resistance only: Z (for $\omega = \omega_0$) $= R$.
- The resonance is also manifest in the variation of the phase angle ϕ with ω. See Figure 33-6(b), which is a plot of (33-10).

First, let us be clear on the meaning of ϕ, by considering Figure 33-7, where both *i* and & are plotted as a function of time.* We see that the applied emf & *leads* the current *i* by phase angle ϕ because & reaches a peak, for example, *before i* does. (For a more specific example, take $t = 0$. Then *i* is zero, whereas & has already reached a positive value.) Of course, if & leads *i* by ϕ, then *i* lags & by ϕ.†

As Figure 33-6(b) shows, resonance, $\omega = \omega_0$, corresponds to $\phi = 0$. At this one frequency, the current and applied emf are exactly *in phase*; both *i* and & reach peak values at the same instant. For higher frequencies, $\omega > \omega_0$, at which the inductive reactance dominates, (33-10) shows that ϕ is positive. Angle ϕ reaches its maximum value, $\pi/2$, at infinitely high frequency; the emf then *leads* the current by $\pi/2$. At frequencies below resonance, $\omega < \omega_0$, the capacitive reactance dominates, and (33-10) shows that ϕ is negative. Angle ϕ reaches its minimum value, $-\pi/2$ rad, at the lowest frequencies; the emf then *lags* the current by $\pi/2$.

Still further aspects of an *RLC* series resonant circuit remain to be treated:

---

* What is shown plotted in Figure 33-7 can be seen directly on the screen of a dual-trace oscilloscope. The applied emf controls one trace; the potential drop across the resistor *iR*, which is proportional to *i*, controls the second trace.
† Be very careful about the phase angle, especially what is leading or lagging what. In some treatments of ac circuits, the phase angle is so defined that it represents the lag of & behind *i* rather than, as here, the lead of & ahead of *i*.

the relative phases of voltage drops across circuit elements; rms values for current and voltage; and power considerations (Sections 33-4 through 33-6).

Example 33-2. An overhead electric transmission line has a series inductive reactance 0.5 Ω/km at the frequency of 60 Hz. What is the inductance per unit length of the transmission line?

From (33-6b),

$$X_L = \omega L = 2\pi f L$$

$$L = \frac{X_L}{2\pi f} = \frac{0.5 \ \Omega/\text{km}}{2\pi(60 \ \text{s}^{-1})} = 1.3 \ \text{mH/km}$$

Example 33-3. What is the capacitive reactance of a 500-pF capacitor in a high-fidelity audio amplifier at the extremes of the audible range, (a) 20 Hz and (b) 20 kHz?

(a) From (33-6c),

$$X_C = \frac{1}{\omega C} = \frac{1}{2\pi f C} = \frac{1}{2\pi(20 \ \text{s}^{-1})(500 \times 10^{-12} \ \text{F})} = 16 \ \text{M}\Omega$$

(b) With the frequency increased by a factor 10^3, the capacitive reactance is reduced by a factor 10^3 to 16 kΩ.

Example 33-4. An *RLC* series circuit is connected across the terminals of a sinusoidal emf with an amplitude of 170 V and a frequency of 60 Hz. The circuit elements are a resistor of 50 Ω, a large capacitor of 27 μF, and a large inductor of 133 mH. (a) What is the total impedance of the circuit? (b) What is the phase angle by which the instantaneous applied emf leads the instantaneous current? (c) What is the peak value of the current? (d) Is the *RLC* circuit being driven above its resonance frequency, or below?

(a) We first compute the angular frequency

$$\omega = 2\pi f = 2\pi(60 \ \text{s}^{-1}) = 377 \ \text{s}^{-1}$$

The inductive reactance is

$$X_L = \omega L = (377 \ \text{s}^{-1})(133 \times 10^{-3} \ \text{H}) = 50 \ \Omega$$

The capacitive reactance is

$$X_C = \frac{1}{\omega C} = \frac{1}{(377 \ \text{s}^{-1})(27 \times 10^{-6} \ \text{F})} = 100 \ \Omega$$

With $R = 50 \ \Omega$, we then have for the impedance, from (33-9),

$$Z = \sqrt{R^2 + (X_L - X_C)^2} = \sqrt{(50 \ \Omega)^2 + (100 \ \Omega - 50 \ \Omega)^2} = 71 \ \Omega$$

(b) From (33-10),

$$\phi = \tan^{-1}\left(\frac{X_L - X_C}{R}\right) = \tan^{-1}\left(\frac{50 \ \Omega}{50 \ \Omega}\right) = \frac{\pi}{4} \ \text{rad}$$

(c) The maximum value of current is, from (33-12),

$$I_m = \frac{\mathcal{E}_m}{Z} = \frac{170 \ \text{V}}{71 \ \Omega} = 2.4 \ \text{A}$$

(d) From (33-13),

$$f_0 = \frac{\omega_0}{2\pi} = \frac{1}{2\pi\sqrt{LC}} = \frac{1}{2\pi\sqrt{(0.133 \ \text{H})(27 \times 10^{-6} \ \text{F})}} = 84 \ \text{Hz}$$

The circuit is driven at 60 Hz, which is *below* its resonance frequency. At the resonance frequency with a voltage amplitude still equal to 170 V, we get $Z = R$, and the peak current rises to $I_m = \mathcal{E}_m/R = 170\text{ V}/50\text{ }\Omega = 3.4\text{ A}$.

33-3 Phasors

Phase relations between voltage and current in an ac circuit are made especially easy to visualize and compute through the use of phasors. Phasors are, in fact, useful in any physical situation involving two or more sinusoidal oscillations at the same frequency but with a difference in phase. (Later we shall use phasors in optics.)

Recall how a rotating vector—a *phasor*—can be used to represent a sinusoidal oscillation (Section 11-1 and Figure 11-2). Consider, for example, the alternating voltage

$$\mathcal{E} = \mathcal{E}_m \sin(\omega t + \phi)$$

It corresponds in Figure 33-8(a) to a vector of length $\mathcal{E}_m$ rotating counterclockwise at the constant angular speed ω. At $t = 0$, the vector makes an angle ϕ with the positive x axis. Figure 33-8(b), a plot of $\mathcal{E}$ as a function of time t, is derived from Figure 33-8(a) simply by taking the component of the vector along the y axis.

The special advantage of phasors is apparent when we have two or more vectors such as **A** and **B** in Figure 33-9(a), rotating together at the *same* angular frequency ω in the same sense. The angle between **A** and **B**, the relative phase difference between the two oscillations, remains fixed with time. Suppose that vector **B** lags vector **A** in phase by the angle ϕ. Vector **A** lies along the x axis at time $t = 0$, so that the equation describing the y component of vector **A** is

$$y_A = A \sin \omega t$$

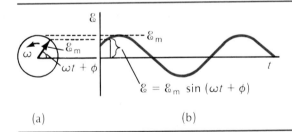

Figure 33-8. (a) Rotating phasor $\mathcal{E}_m$ and (b) sinusoidal $\mathcal{E}$ as a function of time.

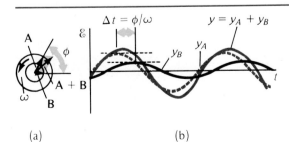

Figure 33-9. (a) Rotating vectors **A**, **B**, and **A + B**; (b) the corresponding displacements as a function of time.

The corresponding equation for the y component of **B** is then

$$y_B = B \sin(\omega t - \phi)$$

Figure 33-9(b) shows both y_A and y_B plotted as a function of time. Here again y_B lags behind y_A by the angle ϕ; y_B reaches a peak later than y_A does. The phase lag is shown by the circumstance that the peak of the curve for y_B is shifted, relative to y_A, to the *right* (to *later* times) by Δt. The phase shift measured in terms of the time interval Δt between the two curves can be related to the phase difference ϕ and the common period T and angular frequency ω of the oscillation by

$$\frac{\Delta t}{T} = \frac{\phi}{2\pi}$$

$$\Delta t = \frac{\phi}{2\pi} T = \frac{\phi}{\omega} \tag{33-14}$$

Suppose that we are to find the sum of two sinusoidal functions with the same frequency ω but differing in phase by ϕ. The resultant y would be written as

$$y = y_A + y_B = A \sin \omega t + B \sin(\omega t - \phi)$$

Computing an actual value for y or finding it graphically from Figure 33-9(b) can be tedious, since the two oscillations are not in phase.

There is an easy way to find the sum. We simply recognize that the projection of the vector sum of two vectors is equal to the sum of the projections of the two vectors. Thus, we can immediately find the resultant oscillations of **A** and **B** by adding these as vectors and then taking the component along the y axis of their vector sum. The resultant oscillation is at the same frequency as the component oscillations; it differs from them both in amplitude and in relative phase, as shown in Figure 33-8. The three vectors **A**, **B**, and **A** + **B** remain locked together as they all rotate at the same rate ω. The amplitude of each oscillation is merely the length of the respective vectors, **A**, **B**, or **A** + **B**. In general, the amplitude of the resultant oscillation is not the sum of the amplitudes of the component oscillations.

Example 33-5. Use phasors to find the sum of the sinusoidal oscillations $y_A = A \cos \omega t$ and $y_B = B \sin \omega t$.

Phasors for y_A and y_B are shown in Figure 33-10(a). Vector **A** leads vector **B** by 90°;

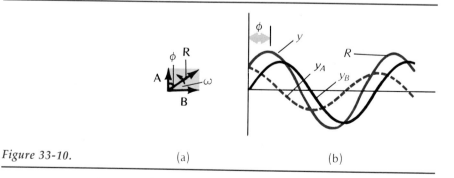

Figure 33-10.   (a)   (b)

A reaches a maximum one-quarter-cycle before vector **B**. To find the resultant oscillation, we merely add vectors **A** and **B** to obtain **R**, and then take the y component of this phasor to obtain $y = y_A + y_B$. From Figure 33-9(a), we see that **R** lags vector **A** by angle ϕ, where

$$\tan \phi = \frac{B}{A}$$

and the amplitude of **R** is

$$R = \sqrt{A^2 + B^2}$$

Therefore, the equation for the resultant oscillation is

$$y = R \cos(\omega t - \phi)$$

Figure 33-9(b) shows y_A, y_B, and y as functions of time.

33-4 Series *RLC* Circuit with Phasors

The phase relationships make ac circuits somewhat complicated. Such oscillating quantities as emf and current do not reach their maximum values simultaneously. But phasors, as we have seen, simplify the representation of oscillating quantities, and they are especially helpful in displaying the phase relationships in ac circuits. We resume our consideration of the *RLC* series resonance circuit driven by an external sinusoidal emf, but now with the aid of phasors and with special attention to the potential drops, or voltages, across the several circuit elements.

We again write the applied emf $\mathcal{E}$ with a maximum value $\mathcal{E}_m$ as

$$\mathcal{E} = \mathcal{E}_m \sin(\omega t + \phi) \qquad (33\text{-}15a)$$

From the general relation $\Sigma \mathcal{E} = \Sigma V$, we see that V is the instantaneous potential difference across all three circuit elements (R, L, and C); V_m is its maximum value, and we can write, using (33-15a),

$$V = V_m \sin(\omega t + \phi) \qquad (33\text{-}15b)$$

The current through all three elements is still written as

$$i = I_m \sin \omega t \qquad (33\text{-}2)$$

Plots of i and of V as functions of time are shown respectively in Figures

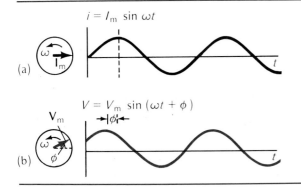

Figure 33-11. Representations of ac current i and ac voltage V.

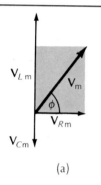

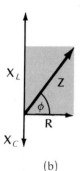

Figure 33-13. (a) Relative orientations for voltage phasors of the three circuit elements, together with their "vector sum" V_m, the instantaneous voltage across all three circuit elements. (b) Impedance vector diagram, derived from part (a) by dividing each vector there by the maximum current amplitude I_m.

Figure 33-12. (a) Phasors for maximum voltages across the L, R, and C series circuit elements. (b) The respective instantaneous voltages as a function of time.

33-11(a) and 33-11(b); the phasors also show the same information. Again, V leads i by phase angle ϕ.

The instantaneous potential drops across the three circuit elements are, by (33-2),

$$V_R = iR = RI_m \sin \omega t \tag{33-16a}$$

$$V_L = L\frac{di}{dt} = \omega L I_m \cos \omega t = X_L I_m \sin\left(\omega t + \frac{\pi}{2}\right) \tag{33-16b}$$

$$V_C = \frac{q}{C} = -\frac{1}{\omega C} I_m \cos \omega t = X_C I_m \sin\left(\omega t - \frac{\pi}{2}\right) \tag{33-16c}$$

We used (33-6) in the above.

The three voltages are plotted as a function of time in Figure 33-12(b). The corresponding phasors in Figure 33-12(a) display the same information, especially the phase relationships, in far simpler fashion. Here the magnitudes of the three phasor vectors are V_{Rm}, V_{Lm}, and V_{Cm}, where from (33-16) we have

$$V_{Rm} = RI_m \tag{33-17a}$$

$$V_{Lm} = X_L I_m \tag{33-17b}$$

$$V_{Cm} = X_C I_m \tag{33-17c}$$

All three voltages V_R, V_L, and V_C, oscillate at the same angular frequency, so that all three phasors in Figure 33-12(a) rotate at the same rate. The three phasors are locked together, with V_L always leading V_R by $\pi/2$ and V_C always lagging V_R by $\pi/2$. For our purposes, all that really matters is the relative orientation of the three phasors, not their rotation as such.

The phasors of Figure 33-12(a) are shown again in Figure 33-13(a), where an additional vector $\mathbf{V}_m$, equal to their "vector sum,"

$$\mathbf{V}_m = \mathbf{V}_{Rm} + \mathbf{V}_{Lm} + \mathbf{V}_{Cm} \tag{33-18}$$

is also shown. The vector $\mathbf{V}_m$ is the phasor for the voltage across all three circuit elements [(33-15b)]. In *magnitude only* this maximum voltage is

$$V_m = ZI_m \tag{33-19}$$

where Z is the impedance of the series RLC circuit.

Phase angle ϕ in Figure 33-13(a) has exactly the same meaning as given earlier: ϕ is the angle by which the total voltage $\mathbf{V}_m$ leads the voltage across the

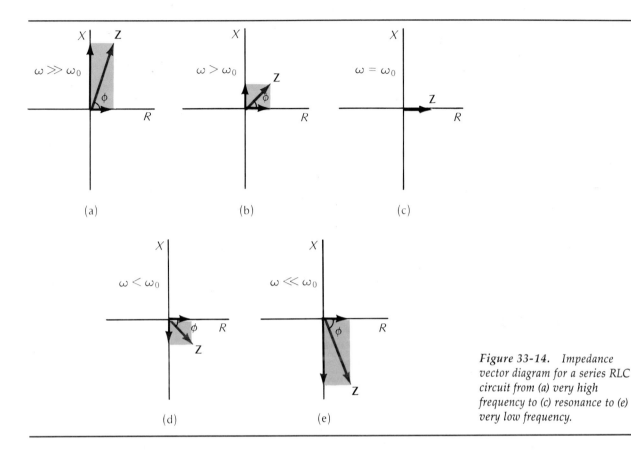

Figure 33-14. Impedance vector diagram for a series RLC circuit from (a) very high frequency to (c) resonance to (e) very low frequency.

resistor V_{Rm}; this is equivalent to saying that ϕ is the angle by which the applied emf $\mathcal{E} = V$ leads the current i (proportional to iR).

All four of the phasor vectors in Figure 33-13(a) have a magnitude proportional to I_m. See (33-17) and (33-19). It is therefore still simpler to remove this common factor and redraw the vectors as shown in Figure 33-13(b). Here we have an *impedance vector diagram*, identical to Figure 33-5. Although the displayed quantities are resistance, inductive and capacitive reactance, and total impedance, it must be recognized that what the diagram really portrays are the *phase relationships among the voltages across the various circuit elements*. (In no way does Figure 33-13(b) say, for example, that inductance is "perpendicular" to resistance.)

An impedance diagram summarizes very neatly a lot of information on the properties of an ac circuit. Consider Figure 33-14, which shows how the magnitude and direction (really, the phase) of the impedance vector changes as the frequency of the applied $\mathcal{E}$ is reduced from an initial high value and the circuit goes through resonance. In Figure 33-14(a) at high ω, the inductance reactance $X_L = \omega L$ is large, and ϕ is close to $\pi/2$. In Figure 33-14(b) at a lower frequency, X_L is smaller and so is ϕ. Figure 33-14(c) corresponds to the resonance peak with $\omega = \omega_0$; here impedance has its minimum value, $Z = R$, and $\phi = 0$. At still lower frequencies, Figure 33-14(d) and (e), the capacitance reactance has grown larger, the impedance is also larger, and phase angle ϕ is negative and approaches $-\pi/2$.

Example 33-6. An RLC series circuit is driven by a sinusoidal emf with maximum value 170 V at 60 Hz. The circuit parameters are $R = 50\ \Omega$, $C = 27\ \mu\text{F}$, and $L = 133$ mH. (These values are the same as those given in Example 33-4). Find (a) the peak voltage drop across each element separately and (b) the peak voltage across the inductor and resistor together in series.

(a) In Example 33-4, we found that

$$X_L = 50\ \Omega$$
$$X_C = 100\ \Omega$$
$$I_m = 2.4\ \text{A}$$

From (33-17), we have for the maximum voltage drops

$$V_{Rm} = RI_m = (50\ \Omega)(2.4\ \text{A}) = 120\ \text{V}$$
$$V_{Lm} = X_L I_m = (50\ \Omega)(2.4\ \text{A}) = 120\ \text{V}$$
$$V_{Cm} = X_C I_m = (100\ \Omega)(2.4\ \text{A}) = 240\ \text{V}$$

Note that these values are quite different from the peak value 170 V of the applied emf. The voltages combine as *vectors* (phasors), as shown in Figure 33-12a.*

(b) The peak voltage across L and R in series with L is given by

$$Z_{R,L} I_m = \sqrt{R^2 + X_L^2}\ I_m = \sqrt{(50\ \Omega)^2 + (50\ \Omega)^2}\ (2.4\ \text{A}) = 170\ \text{V}$$

Note the curious result; the peak voltage across all three circuit elements is the same as that across L and R by themselves.

---

* The fact that in this example $V_{Rm} = V_{Lm}$ and $V_{Cm} = 2V_{Rm}$ is strictly accidental; it is a consequence of the particular parameters chosen.

33-5 RMS Values for AC Current and Voltage

An ordinary ac voltmeter is connected to a household outlet. It reads a steady 120 V. The alternating voltage has a frequency 60 Hz, so that the reading cannot be the rapidly changing instantaneous voltage. After all, an alternating voltage makes equal excursions to positive and negative values, so that the time average value of the sinusoidally oscillating voltage is zero. What kind of average does 120 V represent?

We show below why ordinary ac voltmeters and ammeters are calibrated to read effective, or rms (root-mean-square) values, where

$$I_{\text{rms}} \equiv \frac{I_m}{\sqrt{2}} \tag{33-20a}$$

$$V_{\text{rms}} \equiv \frac{V_m}{\sqrt{2}} \tag{33-20b}$$

Here I_m and V_m are, as before, the maximum values of the instantaneous current and voltage, respectively. For 120 V, 60 Hz, household voltage, a voltmeter reads $V_{\text{rms}} = 120$ V, so that $V_m = \sqrt{2}\ V_{\text{rms}} = \sqrt{2}\ (120\ \text{V}) = 170$ V. (The peak-to-peak value from positive to negative extreme is 340 V.)

Effective values of current and voltage are so defined that familiar relations for dc circuits work also for ac circuits. Consider power dissipated in a resistor. It is given by $i^2 R$ for constant current i through a resistor R. A relation of the same form, $I_{\text{rms}}^2 R$, is to hold for ac.

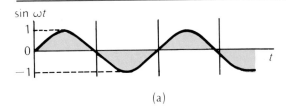

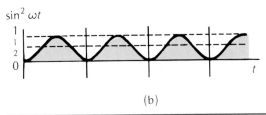

Figure 33-15. The time average value (a) of $\sin \omega t$ is zero and (b) of $\sin^2 \omega t$ is ½.

The *instantaneous* power p dissipated in a resistor through which ac current i passes is

$$p = i^2 R$$
$$= (I_m \sin \omega t)^2 R$$
$$= I_m^2 R \sin^2 \omega t$$

To find the time average $\bar{p}$ of this power, we must find the time average of $\sin^2 \omega t$. The result follows immediately from Figure 33-15, where (a) $\sin \omega t$ and (b) $\sin^2 \omega t$ are plotted as a function of time t. Clearly, the time average of $\sin \omega t$ is zero in Figure 33-15(a), because over one cycle, there are equal areas above and below the horizontal zero line. In Figure 33-15(b), the areas are all positive, and the time average of $\sin^2 \omega t$ is equal to ½, since every area above the ½ line can be fitted exactly into a corresponding empty space below the ½ line. We have then that

$$\bar{p} = \tfrac{1}{2} I_m^2 R$$
$$= \left(\frac{I_m}{\sqrt{2}}\right)^2 R$$

Using (33-20a), we have finally what we set out to prove:

$$\bar{p} = I_{rms}^2 R$$

The term "root-mean-square" is appropriate because we have computed the square *root* of the time *average* value of the *square* of the oscillating current. The proof that voltage follows the same rule, $V_{rms} = V_m/\sqrt{2}$, is given in Section 33-6.

33-6 Power in AC Circuits

The power p delivered to any circuit is given by

$$p = iV \qquad (27\text{-}6)$$

where V is the voltage across and i is the current through the load. Here p is the instantaneous power, i the instantaneous current [(33-2)], and V the instantaneous voltage [(33-15b)] across all three elements in a series RLC circuit. The equation above can be written in more detail as

$$p = iV = (I_m \sin \omega t)[V_m \sin(\omega t + \phi)]$$

From the trigonometric identity in (33-8), this equation can also be written

$$p = I_m V_m \cos\phi \sin^2 \omega t + I_m V_m \sin\phi \sin \omega t \cos \omega t$$

We are interested in the time average value $\bar{p}$ of the power delivered to the circuit. The time average of the second term in the equation above is zero because $\sin \omega t \cos \omega t = \tfrac{1}{2} \sin 2\omega t$; the time average of $\sin^2 \omega t$ in the first term is $\tfrac{1}{2}$ (both results proved in Section 33-5). Therefore, the time-average power is

$$\bar{p} = \tfrac{1}{2} I_m V_m \cos \phi$$

$$\bar{p} = \left(\frac{I_m}{\sqrt{2}}\right)\left(\frac{V_m}{\sqrt{2}}\right) \cos \phi$$

$$\bar{p} = I_{\text{rms}} V_{\text{rms}} \cos \phi \qquad (33\text{-}21)$$

where we have used (33-20). Equation (33-21) is just like the relation for power in a dc circuit except for the additional factor, $\cos \phi$, called the *power factor*, that enters in an ac circuit.

Equation (33-21) can be written in other useful forms by introducing the circuit's impedance Z. From (33-12), we have

$$V_m = Z I_m$$

which can also be written

$$V_{\text{rms}} = Z I_{\text{rms}} \qquad (33\text{-}22)$$

Using (33-22) in (33-21), we get

$$\bar{p} = I_{\text{rms}}^2 Z \cos \phi \qquad (33\text{-}23)$$

From Figure 33-5, we have that

$$R = Z \cos \phi$$

so that (33-23) can also be written as

$$\bar{p} = I_{\text{rms}}^2 R \qquad (33\text{-}24)$$

Equations (33-21), (33-23), and (33-24) are alternative forms for the same basic relation giving the time-average power delivered to a circuit with R, L, and C in series. The implications of these relations are as follows:

- On a time average, *no power is delivered to a purely reactive load*, that is, to a circuit with inductance or capacitance, or both, but no resistance. Suppose that $R = 0$. Suppose further that $C = 0$, so that the load then consists of an inductor L only. Then $\phi = \pi/2$ [(33-10)] and $Z = \omega L$ [(33-9)]. Voltage leads current by $\pi/2$, and the power factor $\cos \phi$ is zero. From any of the equations (33-21), (33-23), or (33-24), we have that $\bar{p} = 0$. We can see in more detail that the time average power is zero from Figure 33-16. As Figure 33-16(d) shows, power alternately goes into and comes out of the inductor as the inductor

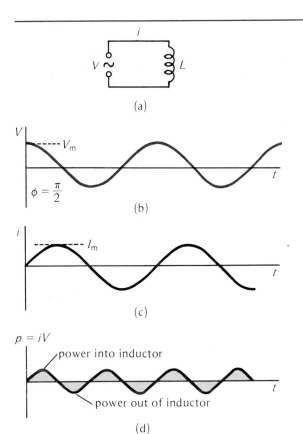

Figure 33-16. (a) Circuit with pure inductive reactance. As a function of time, (b) voltage, (c) current, and (d) power to the inductive load.

temporarily gains and loses energy. The inductor merely reacts (hence, *reactance*) to the applied oscillating voltage; it does not extract energy permanently. The same sort of behavior is exhibited with a single capacitor as load. In that situation, $\phi = -\pi/2$. Once again the power factor and average power to the load are zero.

- On a time average, *power is delivered only to the resistance in the circuit.* Equation (33-24), $\bar{p} = I_{rms}^2 R$, shows this directly. A power factor different from zero implies that phase angle ϕ is not as large as $\pi/2$ nor as small as $-\pi/2$. Said differently, the impedance vector **Z** has some nonzero "component" along the resistance axis [Figure 33-13(b)]. On the average, more power goes into the load than comes out; see Figure 33-17.

- Resonance shows up in the power delivered to an RLC circuit as a function of the driving frequency. Using (33-12) in (33-24), we have, with $\mathcal{E}_m = V_m$,

$$\bar{p} = I_{rms}^2 R = \frac{V_{rms}^2 R}{Z^2}$$

$$\bar{p} = \frac{V_{rms}^2 R}{R^2 + (\omega L - 1/\omega C)^2} \tag{33-25}$$

The *resonance* frequency, $\omega = \omega_0$, with $\omega_0 L = 1/\omega_0 C$, corresponds to *maximum* power to the load, $\bar{p}_{max} = V_{rms}^2/R$. The power dissipated in the resistor is less at lower and higher frequencies, as shown in Figure 33-18.

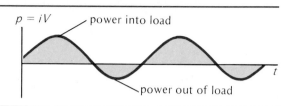

Figure 33-17. Instantaneous power as a function of time to a load with some resistance.

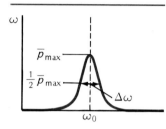

Figure 33-18. Average power dissipated in the resistor of a series RLC circuit as a function of frequency. The full width in angular frequency at half of the maximum power is $\Delta\omega$.

• Both the height and width of the resonance peak in Figure 33-18 are controlled by the magnitude of the resistance. At the resonance peak, $\bar{p}_{max} = V_{rms}^2/R$, so that the power absorbed at the resonance frequency decreases as R is increased, with all other parameters unchanged. At the same time, an increase in R broadens the width of the resonance peak. As (33-25) shows, the average power has half its peak value when $R = \omega L - 1/\omega C$. If the half-power points are designated by $\omega = \omega_0 \pm \tfrac{1}{2}\Delta\omega$, so that the full width in angular frequency at half maximum power is $\Delta\omega$, then it follows from the relation $R = (\omega_0 \pm \tfrac{1}{2}\Delta\omega)L - [1/(\omega_0 \pm \tfrac{1}{2}\Delta\omega)C]$ after a little algebra that

$$\frac{\Delta\omega}{\omega_0} = \frac{R}{\omega_0 L} \tag{33-26}$$

As (33-26) shows, the smaller the resistance, the narrower the peak.

The ratio of the inductive reactance $\omega_0 L$ at the resonance peak to the resistance R is defined as the Q (for "quality") of the circuit:

$$Q \equiv \frac{\omega_0 L}{R} \tag{33-27}$$

Comparing (33-26) and (33-27), we see that the dimensionless Q factor gives a direct measure of the sharpness of the resonance peak, and the capacity of the resonance circuit for differentiating the resonance frequency ω_0 from other nearby frequencies. Since the higher the Q, the narrower the peak.

Example 33-7. Design an RLC series resonance circuit with the following properties: resonance at 200 kHz, a Q of 100, and power dissipation at the resonance peak of 2.0 μW, with an ac source of 10 mV rms.

The average power dissipated at the resonance peak is, from (33-25),

$$\bar{p}_{max} = \frac{V_{rms}^2}{R}$$

so that

$$R = \frac{V_{rms}^2}{\bar{p}_{max}} = \frac{(10 \times 10^{-3}\ V)}{2.0 \times 10^{-6}\ W} = 50\ \Omega$$

The inductance can then be computed from (33-27) as

$$L = \frac{QR}{\omega_0} = \frac{(100)(50\ \Omega)}{2\pi(200 \times 10^3\ s^{-1})} = 4.0\ mH$$

Finally, the capacitance is, from (33-13),

$$C = \frac{1}{L\omega_0^2} = \frac{1}{(4.0 \times 10^{-3}\ H)[2\pi(200 \times 10^3\ s^{-1})]^2} = 158\ pF$$

33-7 The Transformer

As electric energy is transmitted from a generating station to such a simple household electric device as a doorbell, the magnitude of the alternating voltage changes several times. The turbo generator output, typically several kilovolts, is transformed into very high voltage (up to 765 kV rms) for long-distance transmission over the countryside along lines suspended from tall towers. The ac coming down the street along lines on utility poles may be at several to tens of kilovolts. Ordinary household current is 120 V and the doorbell operates on 6 V. The several voltage transformations are readily accomplished for ac by the circuit element known as a transformer and symbolized in circuit diagrams by ▆▆ .* Indeed, the principal advantage of alternating current over direct is the relative ease of changing voltage amplitude with transformers.

A transformer allows the ac voltage amplitude to be changed with little energy dissipation, so that the voltage amplitude can be made most appropriate for the specific application. To minimize i^2R losses in long-distance transmission lines, the current i must be very low; with fixed power $p = iV$, the voltage V must therefore be made as large as possible. On the other hand, safety considerations dictate low voltages for household applications.

A transformer, a circuit device operating on the basis of electromagnetic induction, consists of two multiturn coils typically wound around a laminated core of soft iron. See Figure 33-19. The input coil, known as the primary, has N_1 turns; the output coil, the secondary, has N_2 turns. When a current exists in, say, the primary coil, a magnetic field is created in the core. Ideally, the field lines lie entirely within the soft iron core. Therefore, the same number of field lines pass through every turn of primary and secondary coil, which is to say, the magnetic flux is the same through any one turn of either coil.

The basic transformer effect is that a changing emf $\mathcal{E}_1$ applied at the primary coil changes the magnetic flux through the primary and secondary coils and thereby creates a varying emf $\mathcal{E}_2$ at the secondary coil. More specifically, from Faraday's law of induction,

$$\mathcal{E}_1 = -N_1 \frac{d\phi_1}{dt}$$

and

$$\mathcal{E}_2 = -N_2 \frac{d\phi_2}{dt}$$

But the rate of magnetic flux change per turn is the same for primary and secondary, so that $d\phi_1/dt = d\phi_2/dt$. The two equations above then yield

$$\frac{\mathcal{E}_2}{\mathcal{E}_1} = \frac{N_2}{N_1}$$

The input and output voltages are, it can be shown for an ideal, lossless

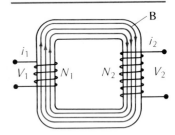

Figure 33-19. *Transformer with equal magnetic flux through any turn of input or output coils. At the input, N_1 turns, current i_1, and voltage V_1 with corresponding terms for the output with subscript 2.*

---

* The transformer is one important type of mutual inductor (Optional Section 32-7).

transformer, in the same ratio as the corresponding emf's, so that the relation above can be written as

$$\frac{V_2}{V_1} = \frac{N_2}{N_1} \qquad (33\text{-}28)$$

where V_1 and V_2 can represent either the peak or rms values at the primary and secondary.

If there are more secondary than primary turns, with $N_2 > N_1$, then the output voltage V_2 exceeds the input voltage V_1, and we have a *step-up transformer*. With $N_2 < N_1$, and therefore $V_2 < V_1$, the device is a *step-down transformer*.

Transformers can be made so highly efficient that the output power differs from the input power by less than 1 percent. Eddy currents within the core are drastically reduced by using laminated sheets; the iron oxide that forms on the surfaces of the laminations has a very high resistance. A more detailed analysis than we give here shows that the power factor $\cos \phi$ (Section 33-6) can be close to 1. Then, if no losses take place in the transformer, the input power $i_1 V_1$ is related to the output power $i_2 V_2$ for a lossless transformer by

$$i_1 V_1 = i_2 V_2$$

where i_1 and i_2 are the currents in the primary and secondary circuits. Using this result in (33-26), we then have

$$\frac{V_2}{V_1} = \frac{i_1}{i_2} = \frac{N_2}{N_1} \qquad (33\text{-}29)$$

The current ratio equals the inverse turns ratio. To the degree that the output voltage is raised, the output current is reduced correspondingly; the converse is also true.

Example 33-8. A doorbell operates on 6 V rms at 0.5 A rms. What current would be required in a high-"tension" transmission line operating at 765 kV rms to run the doorbell?

Apart from small resistive losses, the power supplied by the transmission line is nearly the same as power supplied to the doorbell, so that the current is

$$\frac{0.5 \text{ A}}{765 \times 10^3 \text{ V}/6 \text{ V}} = 3.9 \; \mu\text{A rms}.$$

Summary

All results apply to an ac circuit with R, L, and C in series. The angular frequency of current and voltage is $\omega = 2\pi f$.

Definitions

$$\text{Inductive reactance} = X_L \equiv \omega L \qquad (33\text{-}6b)$$

An inductor passes dc, impedes ac.

$$\text{Capacitive reactance} = X_C \equiv \frac{1}{\omega C} \qquad (33\text{-}6c)$$

A capacitor passes ac, impedes dc.

$$\text{Impedance} = Z \equiv \sqrt{R^2 + (X_L - X_C)^2} \qquad (33\text{-}9)$$

Impedance is the ac analog of resistance.

Phase angle by which applied emf leads current:

$$\phi \equiv \tan^{-1}\left(\frac{X_L - X_C}{R}\right) \qquad (33\text{-}10)$$

where $\mathscr{E} = \mathscr{E}_m \sin(\omega t + \phi)$ and $i = I_m \sin \omega t$.

Phasor: rotating vector whose projection along a diameter is used to represent a sinusoidal variation with time.

$$Q \text{ value} \equiv \frac{\omega_0 L}{R} \qquad (33\text{-}27)$$

Units

Reactance X and impedance Z are in ohms.

Important Results

Nomenclature:
Subscript m denotes maximum, or peak, value.
Subscript rms denotes root-mean-square, or effective, value.
Current-voltage-impedance relation:

$$I_m = \frac{\mathcal{E}_m}{Z} \qquad (33\text{-}12)$$

or $\quad V_m = Z I_m \qquad (33\text{-}19)$

or $\quad V_{rms} = Z I_{rms}$

RMS versus peak values:

$$I_{rms} = \frac{I_m}{\sqrt{2}} \qquad (33\text{-}20a)$$

$$V_{rms} = \frac{V_m}{\sqrt{2}} \qquad (33\text{-}20b)$$

Time-average power $\bar{p}$ dissipated in the load:

$$\bar{p} = I_{rms} V_{rms} \cos \phi \qquad (33\text{-}21)$$
$$= I_{rms}^2 Z \cos \phi \qquad (33\text{-}23)$$
$$= I_{rms}^2 R \qquad (32\text{-}24)$$

where $\cos \phi$ is called the *power factor*.

Phase relations among the voltage drops across the several circuit elements are summarized in an impedance diagram, shown in Figure 33-12(b). The *resonance peak* in a series *RLC* circuit is characterized, for an applied emf of constant voltage amplitude but variable frequency ω, by:

- $\omega = \omega_0 = \dfrac{1}{\sqrt{LC}} \qquad (33\text{-}13)$
- Maximum current
- Maximum power
- Purely resistive load, $Z = R$
- Instantaneous current in phase with instantaneous applied emf; $\phi = 0$.

Full width $\Delta \omega$ in angular frequency for half maximum power:

$$\frac{\Delta \omega}{\omega_0} = \frac{R}{\omega_0 L} = \frac{1}{Q} \qquad (33\text{-}26)$$

Transformer with primary coil of N_1 turns, current i_1, voltage V_1, and secondary coil of N_2 turns, current i_2, and voltage V_2:

$$\frac{V_2}{V_1} = \frac{i_1}{i_2} = \frac{N_2}{N_1} \qquad (33\text{-}29)$$

Problems and Questions

Section 33-1 Some Preliminaries

· **33-1 P** Commercial power in the United States is at a frequency of 60 Hz. To what angular frequency does this correspond?

· **33-2 P** A sinusoidal voltage of amplitude 10 V is applied with a frequency of 400 Hz. (a) What is the period of the voltage? (b) What is the maximum instantaneous rate of change of the voltage?

· **33-3 Q** For which of the following would ac or dc voltages be equally acceptable? (None or more than one may be correct.)
(A) Incandescent light bulb.
(B) Electric toaster.
(C) Electric stove.
(D) Neon sign transformer.
(E) Battery charger.
(F) Chrome plating a metal object.

: **33-4 P** A 20-μF capacitor is charged to 100 V and connected to a resistanceless inductance of 40 mH. What is the maximum current in the circuit after the switch is closed?

: **33-5 P** An 8.0 μF capacitor charged to 40 V is connected across a 2.0 mH inductor. (a) What is the maximum value of the current in the inductor? (b) Sketch the voltage across the capacitor, V_C, the voltage across the inductor, V_L, and their sum $V_C + V_L$ as a function of time. (c) What is the total energy of the oscillator at any time?

· **33-6 P** A sinusoidal potential difference is applied to an *RLC* circuit. An experimenter observes on an oscilloscope the potential difference across the resistor in the circuit and finds that successive positive peaks occur at intervals of 1.0 ms. What is the angular frequency of the applied emf?

Section 33-2 Series *RLC* Circuit

· **33-7 Q** When the switch of the circuit in Figure 33-20 is closed, the potential difference across the resistor R is found to vary as shown. From this we should deduce that the black box probably contains
(A) a capacitor.
(B) an inductor.
(C) a neon lamp.

732 CHAPTER 33 AC Circuits

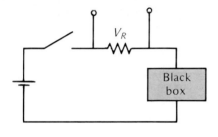

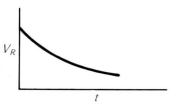

Figure 33-20. Question 33-7.

(D) a battery.
(E) a resistor.
(F) a resonant circuit.

· **33-8 Q** The circuit sketched in Figure 33-21 is used in a radio receiver. Which of the following best describes it?
(A) Since the current in the circuit is not in phase with the voltage across either the inductor or the capacitor, no energy will be dissipated in them and thus the detector will have maximum possible efficiency.
(B) This circuit is a demodulator, which enables us to separate the desired audio signal from the high-frequency carrier signal.
(C) This is a so-called flip-flop circuit, which inverts the polarity of the detected signal on alternate half-cycles.
(D) One adjusts the capacitance until the natural resonant frequency of the circuit matches the modulation frequency of the signal to be detected, thereby causing very large currents to flow in the resistance R.
(E) One adjusts the capacitance until the natural resonant frequency of the circuit matches the carrier frequency of the signal to be detected, thereby causing very large currents to flow in the resistance R.

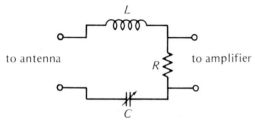

Figure 33-21. Question 33-8.

· **33-9 P** At a microwave frequency of 10 GHz, (*a*) what capacitance has a reactance of 100 Ω? (*b*) what inductance has a reactance of 100 Ω?

· **33-10 P** At what frequency will the phase angle be 30° for the circuit in Figure 33-22?

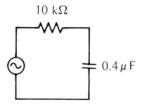

Figure 33-22. Problem 33-10.

· **33-11 P** What capacitance must be used with an inductance of 200 μH if the circuit is to resonate at 2.0 MHz?

: **33-12 P** The circuit shown in Figure 33-23 can be used to attenuate a narrow band of frequencies. Here R is the resistance of the inductor. Make a schematic sketch of V_{out} versus frequency and determine the ratio of the minimum value to the value far from resonance.

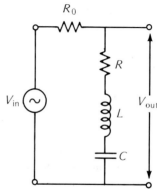

Figure 33-23. Problem 33-12.

: **33-13 P** Shown in Figure 33-24 is a high-pass filter, which attenuates low-frequency signals. (*a*) What is the voltage gain, V_{out}/V_{in}, for the filter? (*b*) Make a qualitative sketch of $\log(V_{out}/V_{in})$ versus angular frequency. (*c*) Show that the voltage gain is $\frac{1}{2}$ at the break-point frequency, $\omega_B = 1/RC$.

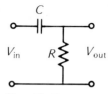

Figure 33-24. Problem 33-13.

: **33-14 P** In Figure 33-25 is diagrammed a low-pass filter, which attenuates high-frequency signals. (*a*) What is the voltage gain, V_{out}/V_{in}, of the circuit? (*b*) Make a qualitative sketch of $\log(V_{out}/V_{in})$ versus $\log \omega$. (*c*) Show that the voltage gain is $1/\sqrt{2}$ when $\omega = \omega_B = 1/RC$. This is called the break point of the curve found in (*b*).

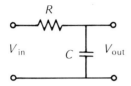

Figure 33-25.   Problem 33-14.

: **33-15 Q** The reason that inductive reactance varies proportionately with frequency is that
(A) as the frequency increases, the current decreases because electron motion is not able to respond to the driving force.
(B) more rapid changes in current induce larger back emf's.
(C) the potential difference across an inductance is proportional to the current through it.
(D) the more rapidly the current through an inductance changes, the more rapidly the voltage across it changes.
(E) capacitive reactance varies inversely with frequency, and the product $X_L X_C = 1/LC$ is independent of frequency.

: **33-16 P** A series RLC circuit has $R = 10\ \Omega$, $L = 1.0$ H, and $C = 20\ \mu F$. (a) For what angular frequency of applied emf will the current be a maximum? (b) At what angular frequencies will the current be half maximum?

: **33-17 P** An initially charged capacitor is connected across an inductor at time $t = 0$. The magnetic energy stored in the circuit first reaches a maximum at $t = 1\ \mu s$. At what time does the capacitor first return to its initial state of charge?

: **33-18 P** An emf of 20 V at a frequency of 2 kHz is applied to a series RLC circuit with $R = 10\ \Omega$, $L = 6.0$ mH, and $C = 2.4\ \mu F$. (a) For what applied frequency will a maximum current exist in the circuit? (b) Should capacitance be added in series or in parallel to the 2.4 μF to maximize the current? (c) What is this capacitance?

Section 33-4 Series RLC Circuit with Phasors

: **33-19 Q** A series RLC circuit is driven by an alternating applied potential difference. If V_R, V_C, and V_L represent the voltages across the resistor, capacitance, and inductor, then
(A) V_R, V_C, and V_L will all be in phase.
(B) V_C and V_L will always be 180° out of phase.
(C) the current will always be in phase with the applied emf.
(D) the current in the resistor will not be in phase with the current in the capacitor.
(E) equal amounts of power will be dissipated in the resistor, the capacitor, and the inductor.

: **33-20 Q** Consider a series RLC circuit driven by an alternating supplied emf. Quantities V_R, V_L, and V_C represent the potential differences across the resistance, inductance, and capacitance and i is the current in the circuit. Then
(A) V_R, V_L, and V_C will all be in phase.
(B) i will always be in phase with the applied emf.
(C) i will always be in phase with V_R.
(D) i will always be in phase with V_L.
(E) i will always be in phase with V_C.

: **33-21 P** An RLC circuit has $R = 20\ \Omega$, $C = 4$ pF, $L = 1.0$ H. Calculate (a) the impedance at resonance; and (b) the impedance at 60 kHz. (c) At what other frequency besides 60 kHz will the impedance have the same value it has at 60 kHz?

: **33-22 P** A voltage $V_0 \sin \omega t$ is applied to the input terminals of the phase shifter in Figure 33-26. (a) What is the phase of the output voltage with respect to the input voltage? (b) What is the ratio of the output voltage amplitude to the input voltage amplitude?

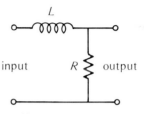

Figure 33-26.   Problem 33-22.

: **33-23 P** Draw the phasor diagram at resonance for a series RLC circuit, with $R = 10\ \Omega$, $C = 1\ \mu F$, $L = 20$ mH, to which a peak voltage of 12 V is applied.

: **33-24 P** At an instant when the applied voltage has the value zero in the circuit of Figure 33-27, what is the instantaneous voltage across each of the elements R, L, and C?

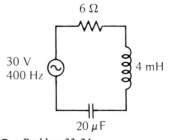

Figure 33-27.   Problem 33-24.

: **33-25 P** The impedance of a certain coil is 30 Ω at 100 Hz and 60 Ω at 500 Hz. (a) What is the inductance of the coil? (b) What is the resistance of the coil? (c) Find the phase angle between the current and the instantaneous voltage across the inductor for each of these frequencies.

: **33-26 P** The applied emf and potential differences for a series RLC circuit are shown in Figure 33-28. The resistance is 10 Ω. What is the capacitive reactance of the circuit?

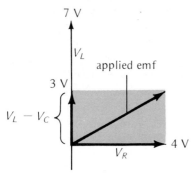

Figure 33-28. Problem 33-26.

: **33-27 P** A rudimentary tone control for a radio consists of a variable low-pass RC filter placed between the audio frequency source of ac and the load. In the circuit of Figure 33-29, the load is 600 Ω, and the filter resistance can be varied from 100 Ω to 2000 Ω. (a) At what frequency will there be equal currents in the load and through the capacitor when the filter resistance is 1000 Ω? (b) For a frequency of 400 Hz, sketch the ratio I_L/I_C as a function of filter resistance that is varied over its full range; I_L = load current and I_C = current in capacitor.

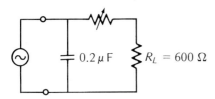

Figure 33-29. Problem 33-27.

: **33-28 P** A tunnel diode has the current-voltage characteristic shown in Figure 33-30a. It is used in the tunnel-diode oscillator circuit shown in Figure 33-30b. For the circuit to oscillate, what is the minimum value of $\mathcal{E}_0$ (in terms of i_a, V_a, and R)?

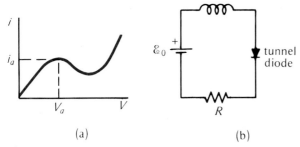

Figure 33-30. Problem 33-28.

: **33-29 P** A phasor diagram for a series RLC circuit is shown in Figure 33-31. From this picture we can see that the applied emf is
(A) 4 V.
(B) 6 V.
(C) 9 V.

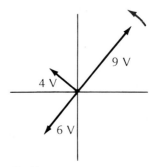

Figure 33-31. Problem 33-29.

(D) more than 9 V.
(E) oscillating at a frequency below resonance.
(F) in phase with the current in the circuit.
(G) None of the above is true.

Section 33-5 RMS Values for AC Current and Voltage

· **33-30 P** An emf with an rms value of 115 V is applied at a frequency of 50 Hz to a circuit. The voltage is a maximum at $t = 5$ ms. Write an expression giving the instantaneous emf as a function of time.

· **33-31 P** A circuit has current $i = 2 \sin(288\pi t + 3\pi/8)$ ampere. (a) What is the peak current? (b) What is the rms current? (c) What is the frequency? (d) What is the phase of the current at $t = 1.0$ s?

· **33-32 P** The current through a 2-mH inductance is $20 \sin 500t$ ampere. What is the rms voltage across the inductor?

: **33-33 Q** Suppose that you apply an alternating potential difference to an inductor and measure the rms value of the current in the inductor. You find that the current decreases as you increase the frequency. The reason for this is that
(A) capacitive reactance varies inversely with frequency, and the product $X_L X_C = L/C$ is independent of frequency.
(B) the more rapidly the current in an inductor changes, the more rapidly the potential difference across it changes.
(C) energy is conserved.
(D) the induced emf in the inductor is proportional to the rate of change of the magnetic field in the inductor.
(E) in any tuned circuit, the current decreases as you move off resonance.

: **33-34 P** An rms voltage of 12 V is applied at a frequency of 3200 Hz to a series RLC circuit with $R = 50$ Ω, $L = 2.0$ mH, and $C = 1.0$ μF. Determine the phase difference between the applied voltage and the current in the circuit. What is the rms current?

: **33-35 P** Determine the rms voltages for each of the waveforms (a) square pulse, (b) rectified sine wave, and (c) ramp (saw-tooth) voltage shown in Figure 33-32.

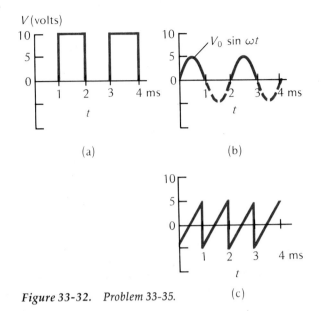

Figure 33-32. Problem 33-35.

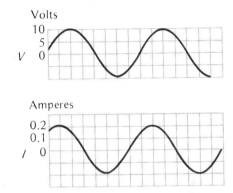

Figure 33-33. Problem 33-42.

Section 33-6 Power in AC Circuits

· **33-36 P** What is the average power dissipated in a circuit where $i = 2 \cos \omega t$ ampere and $V = 12 \cos (\omega t + 60°)$ volt?

· **33-37 P** The current in a circuit is $i = 4 \sin (600\pi t + \frac{1}{4}\pi)$ ampere and the applied emf is $V = 60 \sin (600\pi t)$ volt. What is the average power delivered to the circuit?

· **33-38 P** A pure inductor of inductance 2.4 H is connected to a 60-Hz supply, with $V_{rms} = 115$ V. Find (a) the inductive reactance; (b) the rms current; (c) the maximum power delivered to the inductor; (d) the maximum energy stored in the inductor.

: **33-39 P** A series RLC circuit has $R = 8\ \Omega$, $L = 2.0$ mH, and $C = 30\ \mu F$. A 40-V ac potential difference is applied across the three elements in the circuit. What are the current and the phase angle when the applied frequency is (a) equal to the resonant frequency? (b) half the resonant frequency? (c) three-halves the resonant frequency?

: **33-40 Q** A series RLC tuned circuit is driven by the signal from an antenna in a radio receiver. A particular station is usually "tuned in" by varying the capacitance in the circuit. The purpose of doing this is to
(A) make $1/\omega C \gg \omega L$
(B) make $1/\omega C \ll \omega L$.
(C) maximize the current in the resistor.
(D) minimize the power dissipated in the resistor.
(E) decrease the time constant of the circuit and thereby increase the response speed.
(F) increase the phase difference between the current and the applied emf.

: **33-41 P** In a series RLC circuit, what happens to I_{max}, ω_0, and $\Delta\omega$, the width of the resonance, when (a) L is increased? (b) C is increased? (c) R is increased?

: **33-42 P** A dual-trace oscilloscope shows traces for the current and the applied emf in a circuit as shown in Figure 33-33. What is the average power delivered to the circuit?

: **33-43 P** A 60-Hz rms voltage of 100 V is applied to a series RLC circuit. A current of 2.5 A exists in the circuit, and the voltages across the elements are 60 V across R, 40 V across L, and 120 V across C. (a) Find R, L, and C. (b) What is the power dissipated in R, in L, and in C?

: **33-44 P** Design an RLC circuit that has the following properties for $V_{rms} = 10$ V: $Q = 500$, $P_{max} = 2$ W, and $\omega_0 = 10^4$ rad/s.

: **33-45 P** A generator applies a peak voltage of 100 V at a frequency of 400 Hz to an RLC circuit with $R = 10\ \Omega$, $L = 60$ mH, and $C = 20\ \mu F$. As the frequency is increased, the power dissipated in the resistor will
(A) not change.
(B) continuously decrease.
(C) decrease at first and then later increase.
(D) continuously increase.
(E) increase at first and then later decrease.
(F) None of the above is correct.

: **33-46 P** For a series RLC circuit, characterize the behavior at the resonance peak for each of the following quantities in terms of the Q factor for the circuit: (a) The impedance Z. (b) The phase angle. (c) Magnitude of i. (d) Power dissipated. (e) The reactance X.

: **33-47 P** Show that the curve of current amplitude versus frequency is broader at half-maximum than is the curve of time average power dissipated versus frequency.

Section 33-7 The Transformer

: **33-48 Q** A certain transformer has thin primary wires

but thick secondary wires. Which of the following is the most probable statement concerning the transformer's characteristics?

(A) The transformer may operate on ac or dc.
(B) The output power of the transformer exceeds the input power.
(C) The primary and secondary voltages of the transformer are equal.
(D) Any device connected to the secondary operates on less than 110 V.
(E) The transformer probably steps up the voltage delivered to the apparatus.

Maxwell's Equations

34

34-1 The General Form of Ampère's Law
34-2 Maxwell's Equations
34-3 Electromagnetic Waves from Maxwell's Equations (Optional)
Summary

The fundamentals of electromagnetism tell how electric and magnetic fields are related to one another and to electric charge and current. These fundamentals are summarized by Gauss's laws for electricity and magnetism, Ampère's law, and Faraday's law.* The four mathematical relations expressing these laws are known as Maxwell's equations because of the epochal contributions around 1865 of the Scottish physicist, James Clerk Maxwell (1831–1879). More specifically,

• Maxwell showed that Ampère's law as it had been formulated up to that time was incomplete, and he suggested how to remedy the shortcoming.
• He formulated for the first time in precise and comprehensive mathematical language what are now known as Maxwell's equations.
• Maxwell predicted that electric and magnetic fields, unattached to electric charge, propagate at the speed of light through otherwise empty space as electromagnetic waves.

34-1 The General Form of Ampère's Law

According to Ampère's law, a magnetic field $\mathbf{B}$ is created by an electric current i. More specifically, the line integral $\oint \mathbf{B} \cdot d\mathbf{l}$ around any closed loop equals the

---

* Sections 24-4, 30-4, 30-5, and 31-4.

net current *i* (multiplied by constant μ_0) crossing *any surface* bounded by the loop:

$$\oint \mathbf{B} \cdot d\mathbf{l} = \mu_0 i \qquad (30\text{-}9)$$

As given above, Ampère's law recognizes only one source of a magnetic field: moving electric charges. Here we shall see that Ampère's law must be generalized to include another source of magnetic field: a changing electric field (or more precisely, a changing electric flux).

Consider the simple circuit shown in Figure 34-1. Here a capacitor is being charged by connecting it across a battery with the closing of a switch. Immediately after the switch is closed we have this transient effect: the current rises from zero to some maximum value and then falls to zero again when the capacitor has become fully charged. We are interested in that interval during which the current in the circuit is changing. Actually, to say it a little more carefully, we are interested in the time during which the current *through the conducting wire* is changing, since a real current does not exist at any time in the region between the capacitor plates. After all, no charged particles ever pass through this region.

Suppose that we apply the equation to the circular loop shown in Figure 34-2(a); the loop is centered on the conducting wire and is perpendicular to it. The simplest surface bounded by the circular loop is the flat circular plate shown in Figure 34-2(a). Elecric current penetrates this surface, and according to Ampère's law, there is then a magnetic field circling in the loop around the conductor. But recall that Ampère's law relates the line integral of the magnetic field around a chosen loop to the net current through a closed surface of any shape bounded by that loop (Section 30-4). Suppose then that we keep the

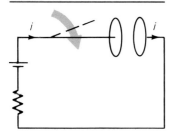

Figure 34-1. A parallel-plate capacitor being charged by switching it across a battery. Current *i* changes with time.

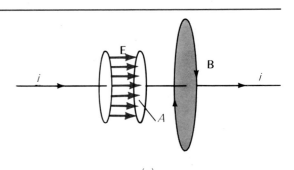

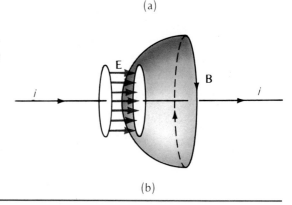

Figure 34-2. Ampère's law applied to a battery being charged. (a) The loop is circular and the surface flat. A magnetic field **B** circles the loop, and a real current passes through the surface. (b) The same circular loop as in part (a) but with a surface that passes between the capacitor plates. A magnetic field **B** still circles the loop but no real current passes through the surface.

circular loop at the location shown in Figure 34-2(a) but choose the surface to be a hemisphere that passes through the region between the capacitor plates. The real current through the hemispherical surface is now zero. But there must still be a magnetic field at the circular loop bounding this surface, just as before. In short, Ampère's law as given thus far does not work when it is applied to the situation shown in Figure 34-2(b). What must be done to fix it?

Enter Maxwell. He conjectured that when there is no real current, a changing electric flux can act as an effective current, or *displacement current*, in creating magnetic field loops.* Why a changing electric flux? Maxwell's argument was basically esthetic, one based on symmetry: if a changing magnetic flux creates electric-field loops (Faraday's law), should we not expect that a changing electric flux might create magnetic-field loops?

To allow for displacement current i_d, as well as real current i, we write Ampère's law as

$$\oint \mathbf{B} \cdot d\mathbf{l} = \mu_0 (i + i_d) \qquad (34\text{-}1)$$

For the situation shown in Figure 34-2 we must have, because of the continuity of current, that

$$i \text{ (in conductor)} = i_d \text{ (between capacitor plates)}$$

Electric flux ϕ_E is defined by

$$\phi_E \equiv \oint \mathbf{E} \cdot d\mathbf{S} \qquad (34\text{-}2)$$

where **E** is the electric field through surface element $d\mathbf{S}$.

Instantaneous current $i = dq/dt$ gives the rate at which charges pass through any cross section of the conductor; dq/dt is also the rate at which charges accumulate on a capacitor plate. The uniform electric field **E** between capacitor plates with area A is related to the charge q on one plate by

$$E = \frac{q}{\epsilon_0 A} \qquad (24\text{-}6)$$

so that $q = \epsilon_0 EA = \epsilon_0 \phi_E$ where we have identified EA as the electric flux ϕ_E through a transverse area between the capacitor plates.

The displacement current is then given by

$$i_d = i = \frac{dq}{dt} = \frac{d}{dt}(\epsilon_0 \phi_E) = \epsilon_0 \frac{d\phi_E}{dt}$$

Using this result in (34-1), we get for the generalized form of Ampère's law

$$\oint \mathbf{B} \cdot d\mathbf{l} = \mu_0 \left(i + \epsilon_0 \frac{d\phi_E}{dt} \right) = \mu_0 i + \epsilon_0 \mu_0 \frac{d\phi_E}{dt} \qquad (34\text{-}3)$$

This relation implies that magnetic field loops can be produced by a changing electric flux even in the absence of electric charges. Magnetic field loops can of course be created by either a steady current or a changing real current, but only a *changing* electric flux creates **B** loops. Although "derived" from the special

---

* The term *displacement current* originated in the nineteenth century concept of an invisible medium, or ether, pervading all space. The ether concept has been thoroughly discredited and has been discarded, but the term *displacement current* lingers on.

740 CHAPTER 34 Maxwell's Equations

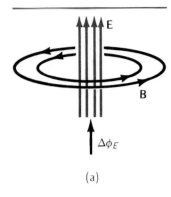

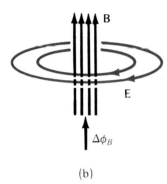

Figure 34-3. (a) **B** loops surrounding a region in which electric flux ϕ_E is changing (right-hand rule); (b) **E** loops surrounding a region in which magnetic flux ϕ_B is changing (left-hand rule).

case of a parallel-plate capacitor, the generalized form of Ampère's law, with the effects of changing electric flux included, holds in all situations.

Suppose some region of space is entirely free of electric charges and currents. Then a magnetic field can be created there only by a changing electric flux, and (34-3) reduces to

$$\oint \mathbf{B} \cdot d\mathbf{l} = \epsilon_0 \mu_0 \frac{d\phi_E}{dt} \qquad (34\text{-}4)$$

This relation is similar in form to Faraday's law, in which an electric field is created by a changing magnetic flux:

$$\oint \mathbf{E} \cdot d\mathbf{l} = -\frac{d\phi_B}{dt} \qquad (31\text{-}6), (34\text{-}5)$$

The two situations are illustrated in Figure 34-3. In Figure 34-3(a), **B** loops circle a region in which ϕ_E is changing; in Figure 34-3(b), **E** loops circle a region in which ϕ_B is changing. The sense of the **B** loops is related to the direction in which ϕ_E increases by a *right*-hand rule, just as the sense of **B** loops is related to the direction of real current by a right-hand rule. On the other hand, the **E** loops in Figure 34-3(b) surrounding a region in which ϕ_B is changing is related to the direction of ϕ_B change by a *left*-hand rule (a consequence of Lenz's law). This difference in the sense of the field loops is reflected in (34-4) and (34-5); one has a plus sign, the other a minus sign.

Because of the close parallel between (34-4) and (34-5), the conclusions we drew earlier from Faraday's law concerning the **E** field accompanying a change in ϕ_B (Section 31-4) hold equally well for the **B** field accompanying a change in ϕ_E. For example, the line integral $\oint \mathbf{B} \cdot d\mathbf{l}$ of the magnetic field depends only on the time rate of change of the total electric flux through a surface enclosed by the path about which we evaluate the line integral, not on whether there is flux at every interior point within the loop. Thus, if we choose concentric circular paths of various radii outside a region in which the electric flux is changing, the integral $\oint \mathbf{B} \cdot d\mathbf{l}$ is the same for all such closed paths. This implies that the magnitude of **B** falls off inversely with r, where r is the radius of the path and also the distance from the center of the region of changing electric flux. This result corresponds exactly to the fact that the magnetic field from a long straight wire falls off inversely with the distance from the wire.

Example 34-1. A parallel-plate capacitor with circular plates of radius R is being charged at a constant rate. Assume that the electric field is confined entirely to the region between the capacitor plates; see Figure 34-4. What is the magnitude of the magnetic field **B** at any distance r from the center of the capacitor plates?

We worked an exactly analogous problem with Faraday's law (as Example 31-7), so that we need not go through it again in detail. Comparing (34-4) with (34-5), we see that we need merely interchange **E** and **B** and introduce the additional factor $\epsilon_0 \mu_0$. The results are shown in Figure 34-4(c). The magnitude of **B** increases linearly with r for $r < R$ and decreases inversely with r for $r > R$. At $r = R$, $B = \frac{1}{2}\epsilon_0 \mu_0 R(dE/dt)$.

To get an idea of the magnitude of the magnetic field induced by the changing electric flux, suppose that the capacitor-plate radius R is 10 cm and that the electric field changes at the rate $dE/dt = 10^{10}$ V·m/s (corresponding to a time rate of change in electric potential difference across the capacitor of 10 V/μs for capacitor plates separated by 1 mm). Then the maximum magnetic field (at $r = B$) is

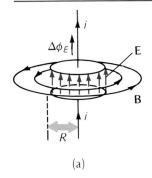

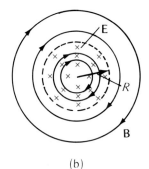

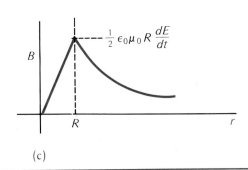

Figure 34-4. (a) Parallel-plate capacitor of radius R with a changing electric field **E** producing magnetic field **B**. (b) Magnetic field loops corresponding to (a). (c) Magnetic field magnitude B as a function of distance r from the center.

$$B = \frac{1}{2}\epsilon_0\mu_0 R \frac{dE}{dt}$$
$$= \tfrac{1}{2}(8.9 \times 10^{-12} \text{ C}^2/\text{N}\cdot\text{m}^2)(4\pi \times 10^{-7} \text{ T}\cdot\text{m/A})(0.10 \text{ m})(10^{10} \text{ V/m}\cdot\text{s})$$
$$= 5.6 \times 10^{-9} \text{ T} = 0.056 \text{ mG}$$

Even for the relatively large rate of electric-flux change in this example, the induced magnetic field is very small indeed. This contrasts with induced electric fields produced by a changing magnetic flux, where emf's of the order of volts are relatively easily obtained.

34-2 Maxwell's Equations

Faraday invented the field concept as a useful and picturesque means of visualizing electric and magnetic effects. Maxwell took the electric and mag-

The Odd Couple

He was something of a fitness freak. While he was in college at Cambridge University he used, at least for a while, to jog an hour every day. He would repeatedly run along the dormitory corridor, down a flight of steps, back along a lower corridor, and up stairs to finish the loop—starting at 2:00 AM.

James Clerk Maxwell (1831–1879) is remembered especially for fixing Ampère's law, for the equations in classical electromagnetism that bear his name, and for his prediction of electromagnetic waves at the speed of light. But he had many other accomplishments. Born to a family of very comfortable means with a country estate in Scotland, Maxwell showed extraordinary talent at an early age. He was only 15 when he submitted his first paper to the Edinburgh Royal Society. Its title: "On the Description of Oval Curves and Those Having a Plurality of Foci." His college tutor remarked that, "It appears impossible for him to think incorrectly on physical subjects."

Maxwell was only 26 when he won the prestigious Adams prize for the best paper on a scientific subject; Maxwell's was on the rings of Saturn, a topic that would concern him for many years. He was, first of all, a mathematical physicist; he deduced, strictly on theoretical grounds, the distribution of molecular speeds (henceforth to be known as the Maxwell dis-

tribution). But he also did experiments: measuring both the viscosity of gases and the standard ohm. He became director of the Cavendish Laboratory at Cambridge University at its founding, which was to become a world center for advances in atomic and nuclear physics.

One of Maxwell's delightful inventions has become known as the *Maxwell demon*. Visualize gas in thermal equilibrium in a container with a partition between the two halves. A small hole in the partition allows a molecule headed for the hole to pass from one side to the other. The demon sits near the hole with a paddle, and he keeps his eye on approaching molecules. If a fast molecule is heading toward the left half, he lets it pass through the hole, but if he sees that a slow molecule would go through the hole to the left he blocks it with his paddle. In similar fashion, he keeps fast molecules from going from the left to the right side, but allows slow ones to pass through. The upshot, of course, is that eventually the left side has mostly fast molecules and is at a higher temperature than the right side. It costs almost no energy for the demon to put the tiny paddle over the hole or

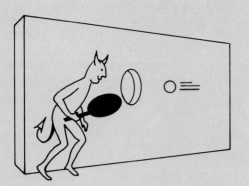

netic fields seriously, and he developed the general mathematical expressions for their properties and interrelations. These four fundamental relations, which say everything there is to say about classical electromagnetism, are known as *Maxwell's equations*.*

---

* *Classical* electromagnetism excludes quantum effects, especially the basic quantum phenomenon that electromagnetic radiation consists not of continuous fields, but of discrete particlelike photons. Photons are discussed in Chapter 41.

remove it, so that Maxwell's demon has foiled the second law of thermodynamics!

Well, not quite. For the demon to perform his feat, he needs to know whether a molecule approaching the hole is a fast one or a slow one. He needs information. Actually, information can be contrasted with ignorance in the same way that we contrast order, or low entropy, with disorder, noise, or high entropy. Indeed, information theory, a fundamental discipline that relates to the transmission of intelligence, is altogether analogous to the statistical thermodynamics that Maxwell knew so well.

Maxwell also acknowledged his profound indebtedness to Faraday for clarifying a wide variety of electromagnetic effects through his experiments and especially for Faraday's championing what has ever after been a central concept in physics—the field.

Michael Faraday (1791–1867) had humble beginnings; his father was a blacksmith. Faraday was first apprenticed to a bookseller, and within a couple of years he had two boys working for him. But his real interest was chemistry, on which he had read considerably. Faraday's big break came when he was 22. He wanted to become an assistant to Sir Humphrey Davy, chemist of the Royal Institution, so he wrote up and bound very neatly notes he had kept while attending public lectures by Davy.

In Faraday's words, "My desire to escape from trade, which I thought vicious and selfish, and to enter into the service of Science, which I imagined made its pursuers amiable and liberal, induced me at last to take the bold and simple step of writing to Sir H. Davy, expressing my wishes . . . at the same time I sent the notes I had taken of his lectures."

Faraday landed the job. At first his duties were fairly onerous—washing out test tubes, and that sort of thing. But, by looking over Davy's shoulder and diligent self-study, Faraday soon became an expert chemist himself. Indeed, he is acknowledged to have been perhaps the most talented experimentalist in chemistry and physics of all time.

Everybody knows about Faraday's discovery of and interpretation of electromagnetic induction. It is easy to miss what a remarkable accomplishment it really was. After all, Faraday had to do everything from scratch—make the batteries, the meters, and even the conducting wire. Faraday never really studied mathematics, and he was a bit suspicious about it. He was not really sure that he understood what Maxwell had done. Faraday concentrated on electric and magnetic fields; for him the fields were the most real part of electromagnetism. But he did other things of great significance. He clarified electrolysis and the notions of atomic mass and valence. He showed by experiment that a magnetic field can change the direction of polarization of a beam of light, the very first direct evidence that light and electromagnetism are closely related.

Faraday did other things besides physics and chemistry. Among his special enthusiasms were poetry, the beauties of nature, various games, acrobatics, Punch and Judy shows, children. He never lost his sense of joy and wonder, not only in the laboratory, but in everything around him.

Let us first be clear on the meanings of electric field **E** and magnetic field **B**. They are given by the relation for the total force on an electric charge q from electric and magnetic fields:

$$\mathbf{F} = q(\mathbf{E} + \mathbf{v} \times \mathbf{B}) \tag{29-11}$$

This equation defines **E** and **B**. That part of the force $q\mathbf{v} \times \mathbf{B}$ that depends on the charge's velocity **v** is the magnetic force; the remaining part $q\mathbf{E}$ is the electric force.

Table 34-1

NAME	EQUATION	EXPERIMENTAL EVIDENCE
Gauss's law for electricity **(34-6a)**	$\epsilon_0 \oint \mathbf{E} \cdot d\mathbf{S} = q$	Electric (or Coulomb) force is inverse-square; no net charge on interior of hollow charged conductor under steady-state conditions. **E** lines either originate from and terminate on charges, or form closed loops.
Gauss's law for magnetism **(34-7b)**	$\oint \mathbf{B} \cdot d\mathbf{S} = 0$	No isolated magnetic poles. **B** lines form closed loops.
Faraday's law **(34-6c)**	$\oint \mathbf{E} \cdot d\mathbf{l} = -\dfrac{d\phi_B}{dt}$	Electromagnetic induction effects.
Ampère's law **(34-6d)**	$\oint \mathbf{B} \cdot d\mathbf{l} = \mu_0 i + \epsilon_0 \mu_0 \dfrac{d\phi_E}{dt}$	Magnetic force between current-carrying conductors; electromagnetic waves.

Table 34-1 lists the four Maxwell equations for charged particles and currents in a vacuum and gives the common name and the primary experimental evidence for each.

Gauss's Law for Electricity Gauss's law for electricity says that the net electric flux through any closed surface depends only on the net charge within:

$$\epsilon_0 \oint \mathbf{E} \cdot d\mathbf{S} = q \tag{34-6a}$$

Gauss's law (Section 24-4) can be derived from Coulomb's law; indeed Gauss's law for electricity is merely an alternative way of saying that the force between point charges varies inversely with the square of the distance between them. Gauss's law is confirmed by experiments showing that the Coulomb force is inverse-square. A more precise verification of Gauss's law comes from the observation that for static conditions there is never any net charge within a conductor.

Gauss's Law for Magnetism Gauss's law for magnetism says that the net magnetic flux through any closed surface is always zero:

$$\oint \mathbf{B} \cdot d\mathbf{S} = 0 \tag{34-6b}$$

If it were not zero, one would have "magnetic charges," or single magnetic monopoles, on which magnetic field lines would originate and terminate. Gauss's law for magnetism is based on the observation that isolated magnetic poles do not exist in nature.

Faraday's Law Faraday's law of electromagnetic induction says that a changing magnetic flux generates electric-field loops:

$$\oint \mathbf{E} \cdot d\mathbf{l} = -\dfrac{d\phi_B}{dt} \tag{34-6c}$$

Faraday's law is confirmed by electromagnetic induction effects, for example, a current can be induced in a conducting loop by a nearby and separate conducting loop in which the current is changing.

Ampère's Law According to Ampère's law, a magnetic field has two origins: electric current i, from electric charges in motion, and changing electric flux (displacement current):

$$\oint \mathbf{B} \cdot d\mathbf{l} = \mu_0 i + \epsilon_0 \mu_0 \frac{d\phi_E}{dt} \qquad (34\text{-}6d)$$

The magnetic force between straight, parallel current-carrying conductors, varying inversely with the separation distance, is a direct experimental test of the magnetic field originating from moving charges. The most convincing evidence that a magnetic field is generated also by a changing electric flux comes from the observed properties of electromagnetic waves.

The equations for Ampère's law and Faraday's law are very nearly symmetrical, but not completely. We do not have in Faraday's law a term corresponding to the current i. This merely reflects the fact that since isolated magnetic poles do not exist, there can be no "magnetic current" arising from magnetic poles in motion. Since all electromagnetic phenomena can be accounted for without magnetic monopoles, we see that in this instance Nature chose economy over symmetry.

Other relations in electromagnetism are not fundamental. Ohm's law, for example, merely describes the properties of certain conducting materials.

34-3 Electromagnetic Waves from Maxwell's Equations (Optional)

One of the most notable predictions of theoretical physics of all time was that made by James Clerk Maxwell in 1865.* Maxwell predicted from the four equations of classical electromagnetism that electric and magnetic fields may exist in space far removed from electric charges and currents and that these fields propagate at the speed of light as electromagnetic waves. For empty space, with both q and i equal to zero, Maxwell's equations (34-6) reduce to the following:

$$\oint \mathbf{E} \cdot d\mathbf{S} = 0 \qquad \oint \mathbf{E} \cdot d\mathbf{l} = -\frac{d\phi_B}{dt}$$

$$\oint \mathbf{B} \cdot d\mathbf{S} = 0 \qquad \oint \mathbf{B} \cdot d\mathbf{l} = \epsilon_0 \mu_0 \frac{d\phi_E}{dt}$$

The first two equations say, in effect, that both electric and magnetic fields must form closed loops. The third equation (Faraday's law) says that a changing magnetic field (strictly, a changing magnetic flux) generates an electric field, and the fourth equation (Ampère's law) says just the reverse, that a changing electric flux generates a magnetic field.

The derivation that follows shows that electric and magnetic fields unat-

---

* The principal results of this section are summarized in Section 35-1.

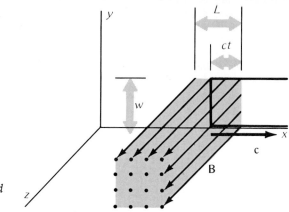

Figure 34-5. Constant magnetic field **B** along $+z$, of thickness L, advancing at speed c along $+x$.

tached to electric charges and currents do propagate through empty space at the speed of light. The strategy will be this:

- We first suppose that a pulse of **B** field can travel through space and we find that an **E** field must accompany it.
- Then we just do the reverse: imagine an **E** pulse to travel through space and find that **B** must accompany it.

Suppose then that a constant pulse of magnetic field **B** directed along z moves through otherwise empty space at speed c along the x axis. See Figure 34-5. How such a magnetic field can be separated from currents and launched into space does not concern us for the moment.

The thickness of the **B** pulse from its leading to its trailing edge is L; along the y and z axes we take **B** to be effectively infinite in extent.* Outside this region, $\mathbf{B} = 0$.

Now consider an imaginary rectangular loop lying in the xy plane. The loop has width w along the y direction and is indefinitely long along x. We suppose that the leading edge of the magnetic pulse reaches the left end of the loop at time $t = 0$. Then, after a time interval t has elapsed, the pulse will have progressed into the loop a distance ct. The magnetic flux through the loop is changing so that there must be an induced electric field whose magnitude we get from Faraday's law.

The magnetic flux penetrating the loop at time t is

$$\phi_B = BA = B(wct)$$

The rate at which the flux changes is, then,

$$\frac{d\phi_B}{dt} = Bwc$$

After a time $t = L/c$, the trailing edge of the magnetic pulse will have entered the loop, and thereafter ϕ_B will be constant. The induced electric field is in the y

---

* The **B** field cannot be strictly uniform in magnitude and truly infinite in extent because the **B** lines must form closed loops. We are here considering a limited region of space in which the **B** lines are effectively straight.

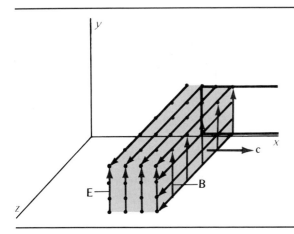

Figure 34-6. The magnetic field **B** of Figure 34-5 is accompanied by a transverse electric field **E** along $+y$.

direction. The only contribution to the line integral $\oint \mathbf{E} \cdot d\mathbf{l}$ comes from the left end of the loop:

$$\oint \mathbf{E} \cdot d\mathbf{l} = -Ew$$

Equating $\oint \mathbf{E} \cdot d\mathbf{l}$ and $-d\phi_B/dt$, as required by Faraday's law, then yields

$$Ew = Bwc$$

$$B = \frac{E}{c} \qquad (34\text{-}7)$$

The magnetic field is smaller than the electric field by the factor c.

What is the direction of **E**? We get that from Lenz's law. Imagine the loop to be conducting; then an induced current must, from Lenz's law, circulate in the clockwise sense. This means, in turn, that at the loop's left end, **E** points in the $+y$ direction. Now **E** exists only so long as the flux ϕ_B is changing. We see that the original magnetic pulse must be accompanied by an electric pulse extending over the same region of space, as shown in Figure 34-6. The **E** and **B** fields are mutually perpendicular, and both fields are perpendicular to the direction in which what we must call the electromagnetic pulse is traveling.

Let us review what we have thus far. We began with a moving **B** and found, from Faraday's law, that **E** must accompany it. Now we do just the reverse: start with an **E** pulse and find, from Ampere's law, that **B** must accompany it.

Figure 34-7(a) shows an electric pulse **E** pointed along the $+y$ direction traveling at speed c along the $+x$ direction. We consider an imaginary indefinitely long rectangular loop of width l lying in the xz plane. The distance from the leading to trailing edge of the pulse is again L. We use Ampère's law to find the magnetic field produced by the changing electric flux ϕ_E through the loop. The electric flux penetrating the loop at time t is

$$\phi_E = EA = E(lct)$$

so that the electric flux changes at the rate

$$\frac{d\phi_E}{dt} = Elc$$

This flux changes only so long as the trailing edge is outside the loop; once the

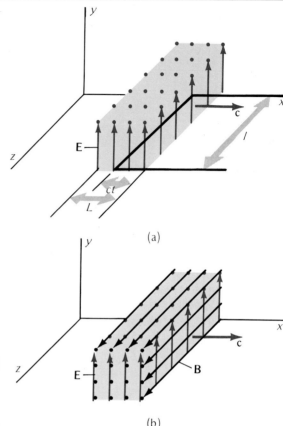

Figure 34-7. (a) Constant electric field **E** along +y, of thickness L, advancing at speed c along +x. (b) The electric field is accompanied by a transverse magnetic field along +z. Compare with Figure 34-6.

pulse is entirely within the loop and ϕ_E is constant, we no longer have a magnetic field induced.

The only contribution from the line integral $\oint \mathbf{B} \cdot d\mathbf{l}$ comes again from the left end of the loop, so that we have

$$\oint \mathbf{B} \cdot d\mathbf{l} = Bl$$

Substituting these results in Ampère's law then gives

$$\oint \mathbf{B} \cdot d\mathbf{l} = Bl = \epsilon_0 \mu_0 \frac{d\phi_E}{dt}$$

$$Bl = \epsilon_0 \mu_0 Elc$$

$$B = \epsilon_0 \mu_0 Ec \tag{34-8}$$

The direction of **B** at the loop's left end is along +z. This result follows from taking an electric flux to be equivalent to a current and then applying the right-hand rule. Again the induced **B** field is confined entirely to the region of the traveling **E** pulse. We see, moreover, by comparing Figures 34-6 and 34-7(b) that the relative directions of **E**, of **B**, and of pulse propagation are the *same*. The magnitudes must also be consistent. We use (34-7) in (34-8) and find that

$$\frac{E}{c} = \epsilon_0 \mu_0 Ec$$

$$c = \frac{1}{\sqrt{\epsilon_0 \mu_0}} \qquad (34\text{-}9)$$

The speed of the electromagnetic pulse depends solely on the two fundamental constants of electromagnetism:

$$\epsilon_0 = 8.854\ 187\ 82 \times 10^8\ C^2/m^2 \cdot N$$

$$\mu_0 \equiv 4\pi \times 10^{-7}\ N \cdot s^2/C^2$$

so that (34-9) yields*

$$c \equiv 2.997\ 924\ 58 \times 10^8\ m/s \approx 3.00 \times 10^8\ m/s$$

The speed of electromagnetic waves is the same as the speed of light. More fundamentally, light is one type of electromagnetic wave.

In the preceding derivation, it was assumed that the electric and magnetic fields change with time in a very simple way; the fields are "turned on and off" abruptly at the leading and trailing edges of a constant pulse. Far more typical is an electromagnetic wave in which the electric and magnetic fields vary continuously with time, especially monochromatic sinusoidal waves. But any more general waveform involves no basically new considerations, however, since any continuously varying field can be approximated by a succession of short constant pulses, or step functions.

* The speed of light is now *assigned* its numerical value because the meter is *defined* in terms of c. The value for ϵ_0 is *computed*, using (34-9), from c and the assigned value for μ_0.

Summary

Definitions

Electric flux: the flux $d\phi_E$ of the electric field $\mathbf{E}$ through a surface element $d\mathbf{S}$ is

$$d\phi_E \equiv \mathbf{E} \cdot d\mathbf{S} \qquad (34\text{-}2)$$

Fundamental Principles

Ampère's law in complete form is

$$\oint \mathbf{B} \cdot d\mathbf{l} = \mu_0 i + \epsilon_0 \mu_0 \frac{d\phi_E}{dt} \qquad (34\text{-}3)$$

The second term on the right implies that a changing electric flux, as well as an electric current i, creates magnetic field loops. The quantity $\epsilon_0 d\phi_E/dt$ is sometimes referred to as the displacement current.

Important Results

Maxwell's equations for classical electromagnetism: see Table 34-1.

Properties of electromagnetic waves are summarized in Section 35-1.

Problems and Questions

Section 34-1 The General Form of Ampère's Law

· **34-1 Q** Which of the following is a true statement concerning displacement current?
(A) It always has the same magnitude as a nearby conduction current.
(B) It is a magnetic current, whereas the conduction current is an electric current.
(C) Its importance is relatively greater at lower frequencies.
(D) Its importance is relatively greater at higher frequencies.
(E) It is a measure of the rate at which charge is transported across the gap in a parallel-plate capacitor.

· **34-2 Q** Which of the following has a meaning closest to that of "displacement current"?
(A) The current associated with the small displacements of bound charges in a dielectric medium.
(B) The rate of change of electric flux.
(C) The rate of change of magnetic flux.
(D) A current that is changing with time, as opposed to a "dc" current.

(E) The current associated with the motion of charged particles through a vacuum.

· **34-3 Q** Maxwell, in his equations, introduced the concept of the displacement current and the resulting term $\epsilon_0(d\phi_E/dt)$, using
(A) an intuitive guess.
(B) experimental evidence of its existence.
(C) the mathematical necessity of its inclusion to make the generalized equation for Ampère's law consistent.
(D) dimensional analysis.
(E) an attempt to explain the absence of magnetic monopoles in nature.

· **34-4 P** Show that $\epsilon_0(d\phi/dt)$ has the dimensions of amperes.

· **34-5 P** Show that the displacement current in a parallel-plate capacitor is given by the expression $i_d = C(dV/dt)$, where C is the capacitance and dV/dt the rate at which the potential difference between the plates changes with time.

· **34-6 P** What is the magnitude of the maximum displacement current in a capacitor consisting of two parallel plates, each of area 0.25 m², separated by an air gap of 1.0 mm, when a sinusoidal potential difference with amplitude $V_m = 2.5$ kV and at frequency 1.0 kHz is applied across the capacitor plates?

: **34-7 P** An alternating voltage of 10 V rms is applied at frequency f to an RC circuit with $R = 10$ Ω, $C = 20$ μF. Determine the current in the resistor and the displacement current in the capacitor for (a) $f = 60$ Hz and (b) $f = 2$ MHz.

: **34-8 P** An AC voltage of 100 V (rms) is applied at angular frequency ω to a resistor $R = 2$ Ω in series with capacitance $C = 10$ μF. The parallel-plate capacitor is air filled and has circular plates of radius 12 cm separated by 1.0 mm. Determine the magnetic and electric fields (rms values) at a distance of 10 cm from the center of the capacitor for (a) $\omega = 400$ s^{-1} and (b) $\omega = 6.0 \times 10^6$ s^{-1}.

Section 34-2 Maxwell's Equations

· **34-9 Q** Fields **E** and **B** are defined by
(A) the energy they store.
(B) potential differences and currents.
(C) the forces they exert on electric charges.
(D) vector fields.
(E) two of the fundamental constants of nature, ϵ_0 and μ_0.

· **34-10 Q** The equation $\oint \mathbf{B} \cdot d\mathbf{S} = 0$ leads us to the conclusion that
(A) the magnetic field **B** is zero.
(B) magnetic field lines have no beginnings or ends; that is, they end on themselves.
(C) magnetic field lines have constant separation; that is, they are parallel lines.
(D) the universe is comprises equal numbers of positive and negative magnetic charges.

: **34-11 Q** Are Maxwell's equations fully symmetric for **E** and **B**?
(A) Yes. They are symmetric in all respects.
(B) No, because **B** exerts a force perpendicular to **B** lines, whereas **E** exerts a force along **E** lines.
(C) No, because there is no source of **E** analogous to the displacement current as a source of **B**.
(D) No, because no magnetic monopoles have been discovered.
(E) No, because they are measured in different units.

: **34-12 Q** In the comparison with fluid flow, we can think of magnetic-field lines as analogous to
(A) the flow lines when sources and sinks are present.
(B) the flow lines when energy is not conserved in a viscous fluid.
(C) lines of constant pressure in a fluid.
(D) the flow lines describing flow in a whirlpool.

: **34-13 Q** An important difference between electric and magnetic fields is that
(A) magnetic fields do not act on electrically charged particles, whereas electric fields do.
(B) a magnetic field is created by a changing electric flux, but an electric field is not created by a changing magnetic flux.
(C) the laws describing magnetic phenomena are deduced from the experimental observation of electric phenomena, but the reverse is not true.
(D) magnetic fields due to a current of electric charge have been observed in nature, but electric fields due to a current of magnetic charge have not been found.
(E) one field stores energy, whereas the other does not.

: **34-14 Q** The law $q = \epsilon_0 \int \mathbf{E} \cdot d\mathbf{S}$ is equivalent to
(A) Ampère's law.
(B) the fact that a time-varying magnetic flux creates an electric field.
(C) the fact that the force between two charges varies inversely with the square of the distance between them.
(D) a statement that all the flux lines into a closed volume must also flow out.
(E) a statement of conservation of charge.

: **34-15 P** Suppose that magnetic monopoles were discovered and their existence well confirmed experimentally. Let q_m be the "magnetic charge" analogous to the electric charge q. (a) What units would q_m have? (b) Write new Maxwell equations incorporating the magnetic monopoles.

Electromagnetic Waves

35

- 35-1 Basic Properties of Electromagnetic Waves
- 35-2 Sinusoidal Electromagnetic Waves and the Electromagnetic Spectrum
- 35-3 Energy Density, Intensity, and Poynting Vectors
- 35-4 Electric-Dipole Oscillator
- 35-5 The Speed of Light
- 35-6 Radiation Force and Pressure, and the Linear Momentum of an Electromagnetic Wave
- 35-7 Polarization
  Summary

35-1 Basic Properties of Electromagnetic Waves

For those who skipped the derivation of the basic properties of electromagnetic waves from Maxwell's equations (Optional Section 34-3), here are some important results:

- Electromagnetic (abbreviated EM) waves can exist in otherwise empty space because a changing electric flux creates a magnetic field (Ampère's law) and a changing magnetic flux creates an electric field (Faraday's law).
- EM waves are *transverse*. At each instant and in each location in space, the instantaneous **E** and **B** fields are mutually perpendicular vectors that lie in a plane transverse to the direction of wave propagation. See Figure 35-1.
- The speed of an EM wave through a vacuum is

$$c = \frac{1}{\sqrt{\epsilon_0 \mu_0}} \equiv 2.997\,924\,58 \times 10^8 \text{ m/s} \simeq 3.00 \times 10^8 \text{ m/s} \quad (35\text{-}1)$$

- The relative magnitudes of **E** and **B** at each instant and location in space are

$$B = \frac{E}{c} \quad (35\text{-}2)$$

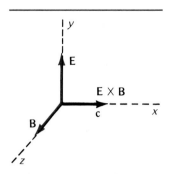

Figure 35-1. Relative directions for an electromagnetic wave of electric field **E**, magnetic field **B**, and direction of wave propagation **c**.

35-2 Sinusoidal Electromagnetic Waves and the Electromagnetic Spectrum

The simplest EM wave is a monochromatic, sinusoidal, plane wave (Section 17-5).

For a sinusoidal wave, **E** varies sinusoidally with time at each location. Furthermore, at each instant, the magnitude of **E** varies sinusoidally with position along the direction of wave propagation. Magnetic field **B** varies in similar fashion. A wave traveling along the $+x$ axis can be represented (Section 17-5) by

$$E_y = E_0 \sin k(x - ct)$$
$$B_z = B_0 \sin k(x - ct) \tag{35-3}$$

with the other rectangular components of **E** and **B** equal to zero.*

Wave number k is

$$k = \frac{2\pi}{\lambda} \tag{17-5}$$

and the wavelength λ and frequency ν are related by

$$c = \nu\lambda \tag{17-1}$$

Figure 35-2 shows the EM wave of (35-3) at one instant in two different representations:

- By **E** and **B** *vectors*. The envelopes of the tips of the vectors are sine waves.
- By **E** and **B** *field lines* whose density is a measure of the field magnitude.

All sinusoidal plane EM waves are identical in their basic properties. The complete spectrum of EM radiation is shown in Figure 35-3, with frequency and wavelength plotted on logarithmic scales. The various regions differ in:

- How the radiation is produced and detected.
- How the radiation affects materials.

The waves of lowest frequency (and longest wavelength) are radio waves; they are generated by oscillating electric currents. Short-wavelength radio waves, or microwaves, have wavelengths comparable to those of audible sound through air; they are generated by specialized electron vacuum tubes. Infrared radiation is produced by heated solids or the molecular vibrations in gases and liquids. Visible light is produced by rearrangement of the outer electrons in atoms. The very narrow range of wavelengths, between 400 nm and 700 nm (from violet to red light), to which the human eye is sensitive corresponds, in musical terminology, to slightly less than one octave (a factor of 2 in frequency) and is to be contrasted with the enormous frequency range (20 to 20,000 Hz) to which the human ear is sensitive. Ultraviolet radiation immediately adjoins the visible spectrum; it is easily absorbed by ordinary glass and many other materials transparent to visible light. X-rays have wave-

---

* The EM wave given by (35-3) is *linearly polarized* along the y axis; at each point in space, **E** is always along y.

SECTION 35-2   Sinusoidal Electromagnetic Waves and the Electromagnetic Spectrum   753

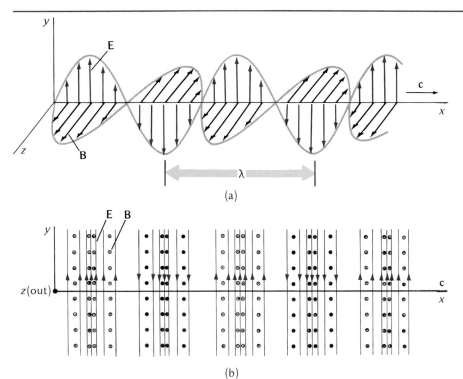

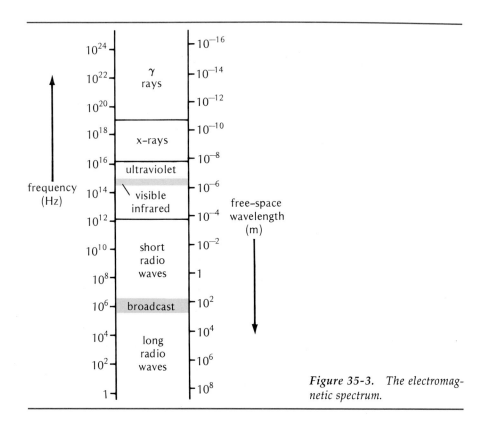

Figure 35-2. Two representations of the **E** and **B** fields of a linearly polarized, plane, monochromatic electromagnetic wave of wavelength λ: (a) **E** and **B** vectors. (b) **E** and **B** field lines.

Figure 35-3. The electromagnetic spectrum.

Table 35-1. The Visible Spectrum

COLOR	WAVELENGTH IN VACUUM (nm)
Ultra Violet (UV)	<400
Violet	400–424
Blue	424–491
Green	491–575
Yellow	575–585
Orange	585–647
Red	647–700
Infra-red (IR)	>700

lengths of the approximate size of atoms, and they originate in the rearrangement of innermost electrons of atoms. Gamma rays are the electromagnetic waves of the highest frequency and shortest wavelength; they originate in the rearrangement among the particles within the atomic nucleus.

Table 35-1 shows the visible spectrum delinated by color and wavelength.

The boundaries between the adjoining regions are not sharply defined. For example, one cannot distinguish between a short-wavelength x-ray and a long-wavelength γ ray.

35-3 Energy Density, Intensity, and Poynting Vector

The heating effect of sunlight is obvious to anyone standing outside on a bright clear day. Any EM wave represents the transport of energy through space, and in this section we derive some important results relating to the energy of an EM wave.

First, where does the energy come from? An EM wave is created by an accelerated electric charge and the energy in the wave simply equals the total work done by the agent accelerating the charge less the energy acquired by the charged particle itself. By the same token, the energy extracted from an EM wave by an absorber is the work done, primarily by the electric field **E**, on a charged particle within the absorbing material.

Energy Density Consider now the energy density, or energy per unit volume, of an EM wave. The energy density u_E of the electric field **E** is, from (26-10) in Section 26-5,

$$u_E = \tfrac{1}{2}\epsilon_0 E^2 \tag{35-4}$$

The energy density u_B of magnetic field B is, from (32-11) in Section 32-4,

$$u_B = \frac{1}{2\mu_0} B^2 \tag{35-5}$$

If **E** and **B** vary with time, so do the energy densities u_E and u_B. The relations above give the *instantaneous* energy densities at any point in space. Using (35-2) in (35-5) allows us to write u_B as

SECTION 35-3 Energy Density, Intensity, and Poynting Vector

$$u_B = \frac{1}{2\mu_0}\left(\frac{E}{c}\right)^2 = \frac{\epsilon_0\mu_0}{2\mu_0}E^2 = \frac{1}{2}\epsilon_0 E^2$$

Comparing this relation with (35-4), we see that

$$u_B = u_E \qquad (35\text{-}6)$$

The energy densities of the electric and the magnetic fields of an EM wave are equal; the energy transported is shared equally between the electric and magnetic fields.

The total instantaneous energy density u of an EM wave is

$$u = u_E + u_B = 2u_E = \epsilon_0 E^2 \qquad (35\text{-}7)$$

Equivalently, we can write $u = 2u_B = B^2/\mu_0$. The energy density is proportional to the *square* of either the electric field or the magnetic field.

For a sinusoidal EM wave, electric field **E** at any point in space varies with time as $E = E_0 \cos \omega t$, where E_0 is the electric-field amplitude and ω is the angular frequency. Then the time average of E^2 is given by

$$\overline{E^2} = E_0^2 \,\overline{\cos^2 \omega t} = \tfrac{1}{2}E_0^2$$

In the last step, we have used the fact that the time average of the square of any sinusoidal function is $\tfrac{1}{2}$ (proved in detail in Section 33-5). The time average of the total energy density of an EM wave can then be written as

$$\bar{u} = \tfrac{1}{2}\epsilon_0 E_0^2 \qquad (35\text{-}8)$$

Intensity The intensity I of any wave is defined as the energy flow per unit time, or power P, passing through a unit area oriented at right angles to the direction of wave propagation. In symbols,

$$I = \frac{P}{A} \qquad (35\text{-}9)$$

We wish to relate the intensity I of any wave to its energy density u and propagation speed c. Suppose that a wave is propagated along the axis of a cylinder. The cylinder's cross-sectional area is A and its length is L, as shown in Figure 35-4. If the cylinder's thickness L is small, the wave's energy density is constant throughout the cylinder. We can write the thickness L as $L = ct$, where t is the time for the wave to travel from the front to the back cylinder face. Over the interval t, all energy originally contained in the cylinder's volume $AL = Act$ will have passed through the area A. We can write

$$\text{Intensity} = \frac{\text{energy}}{\text{area} \times \text{time}} = \frac{(\text{energy/volume}) \times \text{volume}}{\text{area} \times \text{time}}$$

$$I = \frac{u \times Act}{At}$$

$$I = uc \qquad (35\text{-}10)$$

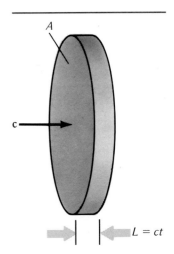

Figure 35-4. *Wave propagating at speed c through a cylinder of thickness L and cross-sectional area A.*

Equation (35-10) relates the instantaneous energy density u and propagation speed c of *any* wave to its instantaneous intensity I. For an EM wave we have, using (35-7),

$$I = \epsilon_0 c E^2 \qquad (35\text{-}11)$$

The instantaneous intensity is proportional to the *square* of the electric field. Note that neither the energy density nor the intensity of an EM wave depends on its frequency.

Poynting Vector Equation (35-11) for the intensity can be written in an interesting alternative form by using (35-1) and (35-2):

$$I = \epsilon_0 E^2 c = \epsilon_0 E(Bc)c = \frac{\epsilon_0 EB}{\epsilon_0 \mu_0} = \frac{EB}{\mu_0}$$

This relation can be written as a vector equation,

$$\mathbf{I} = \mathbf{E} \times \frac{\mathbf{B}}{\mu_0} \qquad (35\text{-}12)$$

where the intensity vector **I** points in the direction of wave propagation and of energy transport. See Figure 35-1. Vector **I** is referred to as the *Poynting vector*,* named for J. H. Poynting (1852–1914). To find the energy flow per unit time, or power P, from EM fields through a surface with a differential element $d\mathbf{S}$, we merely integrate $\int \mathbf{I} \cdot d\mathbf{S}$ over the surface

$$P = \int \mathbf{I} \cdot d\mathbf{S} = \int \left(\mathbf{E} \times \frac{\mathbf{B}}{\mu_0} \right) \cdot d\mathbf{S} \qquad (35\text{-}13)$$

Example 35-1. A radio antenna has a very modest power output of 1.0 W and broadcasts uniformly in all directions. At a location 1000 km from the source, what is (a) the intensity? (b) the energy density of the EM wave? (c) the electric field? (d) the magnetic field?

(a) The power P of the transmitting antenna is spread uniformly over a spherical area of $4\pi r^2$, so that at a distance r from the source, the intensity is

$$I = \frac{P}{A} = \frac{P}{4\pi r^2} = \frac{1.0 \text{ W}}{4\pi (1.0 \times 10^6 \text{ m})^2} = 7.9 \times 10^{-14} \text{ W/m}^2$$

(b) The energy density u is, from (35-10),

$$u = \frac{I}{c} = \frac{7.9 \times 10^{-14} \text{ W/m}^2}{3.0 \times 10^8 \text{ m/s}} = 2.7 \times 10^{-22} \text{ J/m}^3 = 2.7 \times 10^{-10} \text{ }\mu\text{J/cm}^3$$

(c) Electric field E is, from (35-7),

$$E = \sqrt{u/\epsilon_0} = \sqrt{(2.7 \times 10^{-22} \text{ J/m}^3)(8.85 \times 10^{-12} \text{ C}^2/\text{N} \cdot \text{m}^2)} = 5.5 \text{ }\mu\text{V/m}$$

It is relatively easy to detect the electric field with a modern radio receiver. Suppose that the electric field of the incoming EM wave is parallel to a receiving antenna 1 m long; then a signal of 5.5 μV is produced along it. Note that the computed value for E is the instantaneous value; if the antenna transmits a sinusoidal wave with an *average* power output of 1.0 W, then the computed value above is the rms value of the oscillating electric field.

(d) The magnetic field at the receiver location is, from (35-2),

---

* The Poynting vector can be written in still simpler form by introducing the vector **H** for *magnetic field intensity*, defined as $\mathbf{H} \equiv \mathbf{B}/\mu_0$. Then $\mathbf{I} = \mathbf{E} \times \mathbf{H}$.

$$B = \frac{E}{c} = \frac{5.5 \times 10^{-6} \text{ V/m}}{3.0 \times 10^8 \text{ m/s}} = 1.8 \times 10^{-14} \text{ T} = 1.8 \times 10^{-10} \text{ G}$$

The magnetic field of the received EM wave is indeed small. It is smaller than the earth's magnetic field by a factor of more than 10^{10}.

35-4 Electric-Dipole Oscillator

Every electric charge has electric fields attached to it, and every charged particle in motion has magnetic-field loops surrounding the line of its velocity. But an EM wave in otherwise empty space consists of **E** and **B** lines unattached to charges. How is an EM wave launched? How do the **E** and **B** lines become detached from charges?

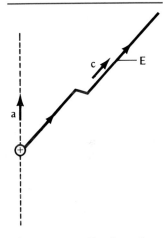

The essential requirement is *acceleration;* a charged particle must be accelerated to generate an EM wave. A positively charged particle at rest or in motion at constant velocity has in effect electric-field lines rigidly attached to it and extending straight outward. But if such a particle initially at rest is accelerated briefly and then brought to rest again, the acceleration shows up as a kink, with a transverse component, in the **E** lines, traveling outward at speed c. This is shown in Figure 35-5. The transverse **E** component is part of an outwardly expanding EM wave. Detailed analysis (not to be given here) shows not only that acceleration produces *transverse* **E** and **B** field components but also that the magnitude of these transverse components falls off with distance less rapidly than the radial electric field characteristic of Coulomb's law.

Figure 35-5. *If a charged particle is given an acceleration* **a**, *a kink, or transverse component, is produced in the electric field* **E** *that travels outward at speed c.*

The simplest generator of sinusoidal EM radiation is an *electric-dipole oscillator*. Recall that an electric dipole consists of two separated point charges of opposite sign and equal magnitude (Section 24-6). One simple form of an electric-dipole oscillator is an electric dipole in which the two charges oscillate at the same frequency but 180° out of phase, so that the electric-dipole moment oscillates sinusoidally with time. Another equivalent electric-dipole oscillator consists of two straight-line conductors to which a sinusoidally alternating voltage is applied. When the tip of one conductor is positive, the other conductor tip is negative with the same charge. See Figure 35-6.

The electric- and magnetic-field patterns from an electric-dipole oscillator,

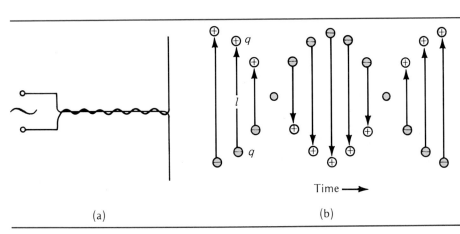

Figure 35-6. (a) Electric-dipole oscillator. (b) Equivalent oscillating electric charges.

Everything Was Big but the Particles

Its nickname, the Z^0. Its full name, the Neutral Intermediate Vector Boson. Forty years ago this fundamental exotic particle was first predicted to exist; contemporary elementary-particle theory said that a Z^0 would have, among others, these specific properties: a mass about 120 times that of a proton, and a lifetime so short (10^{-10} s) that it can not be observed directly but only by finding the oppositely charged particles into which the highly unstable Z^0 can decay (either an electron-positron pair or a muon-antimuon pair).

Is there really such a creature? A research report in 1983 said yes [*Physics Letters* **126B** 398 (7 July 1983)]. After an enormous effort, a total of five Z^0's were positively identified. This work, together with research on finding the closely related W^+ and W^- particles (see Figure 10–20), was the basis of the 1984 Nobel prize in physics.* The recipients were Carlo Rubbia of Harvard University, who masterminded the two projects, and Simon van der Meer, of CERN (the European Organization for Nuclear Research, near Geneva, Switzerland), who devised the ingenious devices for handling—steering, deflecting, accelerating, bunching—vast numbers of antiprotons traveling at nearly the speed of light for many hours at a time. The antiprotons were then directed to slam head-on into protons of the same energy. The hoped-for result of such a collision: one of the quarks within a proton would combine with an antiquark within an antiproton to create a Z^0.

It was a remarkable feat. Everything about the project was big:

- *The team of scientists*. A total of 128 physicists from a dozen research centers in Europe and the U.S., together with many more hundreds of technicians and assistants in supporting roles, participated in the project.
- *The accelerating machine*. It was the SPS Collider, a Super Proton Synchrotron accelerating ma-

*A large fraction of the November, 1983, issue of the *CERN Courier* **23** is devoted to this story. This international journal of high energy physics is available in most physics libraries.

---

for one instant, are shown in Figure 35-7. We can see several features in the figure:

- Close to the dipole, electric-field lines go from positive to negative charge; this is the Coulomb field. [See Figure 23-15(a).] At larger distances, the electric-field lines have become detached from the oscillating dipole and form closed **E** loops of the radiated EM wave. Similarly, **B** loops become detached from the oscillating charges.
- The **E** and **B** fields are mutually perpendicular at each location and also perpendicular to the radially outward direction of the radiated EM wave.

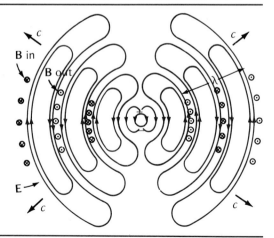

Figure 35-7. Electric and magnetic fields produced by a sinusoidally varying electric dipole.

chine modified to allow protons and antiprotons to circulate and be accelerated while traveling in opposite directions inside an evacuated (~10^{-13} atm) underground tube of 7 km circumference. The 6×10^{11} antiprotons in a pulse, each particle with a kinetic energy of 270 GeV (2.7×10^{11} eV), were made to collide head-on a few thousand times a second with a roughly equal number of protons of the same kinetic energy.

• *The detecting apparatus.* Known as the UA 1 (Underground Area #1), the detecting device was really a collection of many thousands of individual particle detectors. The largest ever constructed for an accelerating machine, the detector was located in a vast cavern 20 m underground and weighed 2000 tons, but was nevertheless moveable. It alone cost many hundreds of man-years of effort in design and construction, and $20 million.

• *The data handling system.* Each time a sufficiently energetic proton hits an equally energetic antiproton head-on, the two particles mutually annihilate one another and many newly created particles splatter off in all directions. The data from just one such event would fill a large telephone directory. All told about 10^9 such collisions were analyzed in detail, while the experiment was running, to find just the five Z^0 events. Closely articulated microprocessors and computers were essential parts of the apparatus.

• *The stakes.* A lot was riding on this experiment; the Nobel committee had handed out a few Nobel prizes in physics a few years earlier to people who said that intermediate vector bosons surely existed and would be observed in the laboratory. It would have been awkward, to say the least, if the Z's and W's had not been found. More importantly, the grand unified field theories, according to which the fundamental forces (the gravitational, the weak, the electromagnetic, and the strong) are united in one primordial interaction, would have been put in very serious jeopardy if these peculiar and highly transient particles were missing. Elementary-particle physicists now have enough confidence in the essential correctness of their ideas of what the truly fundamental particles are and how they interact with one another that they can tell in remarkable detail what happened very shortly—within even 10^{-43} s—of the initial Big Bang that marked the instant when the Universe came into being.

• At great distances from the dipole, the wave fronts, which are surfaces of constant phase, are spheres centered on the dipole. The dipole as an effective point source radiates a spherical wave.

• The frequency of the EM wave is the same as the frequency of the dipole oscillator. It takes a full period for **E** at any location to return to its initial value, and, therefore, a half period of oscillation to produce just one **E** loop.

• The wave is *linearly polarized;* that is, the transverse electric field at any location in space oscillates along a fixed line. This direction is called the *direction of polarization* of the EM wave.

• The radiation pattern is symmetric relative to the vertical electric-dipole axis (think of it here as a north-south axis). The radiated intensity is a maximum at the equator, and the intensity reaches zero in the directions outward from the north and south poles. No EM radiation is emitted along the direction of a charged particle's acceleration. See Figure 35-8. It can be shown that the intensity I of EM waves from an electric-dipole oscillator varies with distance r from the source and angle θ from the dipole axis according to $I \propto \sin^2 \theta / r^2$.

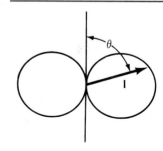

Figure 35-8. Radiation pattern from an electric-dipole oscillator. Intensity **I** in a polar plot as a function of angle θ from the line of the dipole's acceleration direction.

35-5 The Speed of Light

The fundamental constant of electromagnetism is c. It relates through (35-1), $c^2 = 1/\epsilon_0 \mu_0$, the basic constants for the electric (ϵ_0) and the magnetic (μ_0) interactions. Atomic structure is governed by the EM interaction between electrically charged particles, and c is also a fundamental constant of atomic theory. Furthermore, relativity physics is intimately connected with c; the most famous equation in physics, $E = mc^2$, contains it.

The Standard Meter and c Since the action in Paris on October 20, 1983, by the International Committee of Weights and Measures, the speed of light is, in effect, *defined* as having the magnitude 299 792 458 m/s in the SI system of units. Strictly speaking, the standard meter is now defined as the distance traveled in a vacuum by an electromagnetic wave in a time interval equal to 1/299 792 458 s.

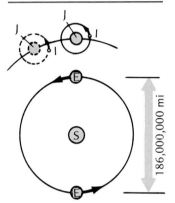

Figure 35-9. Roemer's measurement of the speed of light. The sun (S), earth (E), Jupiter (J), and one of Jupiter's moons, Io (I).

Measuring c Timing light signals is an important means of measuring distance indirectly. For example, laser pulses directed to mirrors placed on the moon during the Apollo flights make it possible to measure very small changes (to within a few centimeters) in the distance from the earth's surface to the moon's. Still greater planetary distances have been measured very precisely by timing signals from Voyager spacecraft that have reached distant planets.

Recognizing that light might not be transmitted instantaneously, with an infinite speed, was an important advance in early science, but measuring such a high speed as c requires great subtlety. Since c is now defined to have its present numerical value, the ways in which it has been measured over the last several hundred years are mostly of historical interest. Here, briefly, are some of the principal methods of measuring c:

- Use astronomical distances so that the travel time is long enough to be measured readily. This was done by Ole Roemer (1644–1710), first in 1666. He measured the time (about 1000 s) for light to traverse the diameter (186,000,000 miles) of the earth's orbit around the sun. The signals came from an eclipsing moon orbiting the slower-moving and more distant planet Jupiter. As the earth moved farther and farther from Jupiter, the disappearance of a moon behind Jupiter was found to be progressively later because of the additional distance the light signals had to cover to reach the observer on earth. See Figure 35-9.

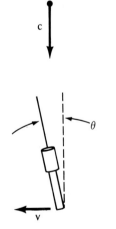

Figure 35-10. Stellar aberration: when a telescope is moving at right angles to the direction of light propagation, the telescope must be tipped at angle θ in the direction of its velocity **v**.

- *Stellar aberration* was the method used in 1725 by the astronomer James Bradley (1692–1762). This method depends on combining one very high velocity, that of light, with another high velocity, that of the earth as it orbits the sun at 3×10^4 m/s. To see a star directly overhead, one does not point a telescope straight up when the earth is moving at right angles to the line joining the earth and the distant star. The telescope must be tilted a little (about 20 seconds of arc). See Figure 35-10. Six months later, with the telescope moving in the opposite direction, the telescope must be tilted the other way. The angle of tilt, together with the known speed of the earth around the sun, yields c.

- To measure the time for light to travel several kilometers to a distant mirror (by terrestial standards) and back again was first done with rotating mechanical devices. The first determinations were made by A. H. L. Fizeau (1819–1896) in 1849; the American physicist, A. H. Michelson (1852–1931) carried the method to its greatest refinement through several decades. Basically, the rotating device chops a continuous light beam into a series of pulses and its speed of rotation serves also to measure the time interval for a round trip. See Figure 35-11.

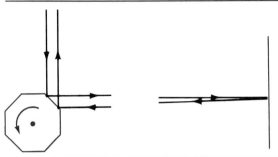

Figure 35-11. A rotating octagonal mirror used by A. A. Michelson in measuring the speed of light. If the mirror makes an eighth turn in the time it takes for a light pulse to travel to a distant mirror and back, the light beam is reflected in the same direction as if there were no rotation of the mirror.

- Until the recent definition of c, the most recent measurements of c depended on measuring the wavelength of nearly monochromatic radiation against the standard meter, defined in terms of wavelength of light, and also measuring the frequency of the same radiation against the standard second as defined in terms of a standard atomic clock. Then $c = \nu\lambda$. This method was limited, finally, by the uncertainty in the measurement of wavelength (about 4 parts in 10^9); the uncertainty in the measurement of time was far less (about 1 part in 10^{10}, which is equivalent to an uncertainty of 3 s in a millennium). With the meter defined in terms of c, uncertainties are limited to those in time intervals. Comparing the frequencies of highly monochromatic light sources over the visible and adjoining regions of the EM spectrum depends on two basic measurement effects: *nonlinear materials*, which produce harmonics, oscillations at integral multiples of the incident radiation (see Section 11-6); and *beats* (Section 18-6), in which a measurement of the beat frequency between two oscillations yields their frequency difference.

- Speed c has such a large numerical value and its direct measurement is so difficult that you can easily overlook simple indirect means for determining it. Suppose, for example, that you measure the attractive electric force between the two oppositely charged plates of a parallel-plate capacitor. Then, if you also know the charges on the plates, and the dimensions of the capacitor, you can compute the value for ϵ_0. This means you can compute the value of c from $\epsilon_0 = 1/\mu_0 c^2$ using (35-1).

Doppler Effect for Light When we say that the speed of sound through air is, say, 344 m/s, we mean that that is how fast a pressure disturbance travels relative to the medium—air—in which sound waves are propagated. An observer in motion relative to air measures a different speed. Not so for an EM wave, however. The speed of light through a vacuum is the *same* for *all* observers, quite apart from their state of motion or the motion of the source emitting the light. A basic postulate for the special theory of relativity (Chapter 40), confirmed in detail by experiment, is that c has the very same value for all observers.

One important consequence of the constancy of c is that the simple relations for the shift in frequency, the Doppler effect, as derived in Section 18-7 for mechanical waves, do not hold for light. For mechanical waves, we distinguish between the wave source in motion and the observer of waves in motion, both relative to the medium propagating the waves. But for EM waves propagated through empty space, the speed is always c, so that we cannot distinguish between source in motion and observer in motion. We can speak only of the speed of the observer relative to the source; and therefore a single relation gives the relativistic Doppler shift for EM waves. It is, as derived from the special theory of relativity,

$$f_o = f_s \frac{1 + v/c}{\sqrt{1 - (v/c)^2}} \tag{35-14}$$

where v is the relative speed of approach between source and observer, f_s is the source frequency, and f_o is the observed frequency. (The observed wavelength is always $\lambda_0 = c/f_o$.) For low speeds, with $v/c \ll 1$, the radical in the denominator of (35-14) is essentially equal to 1, and the relation for the relativistic Doppler effect reduces to the simple classical result

$$\frac{\Delta f}{f} = \pm \frac{v}{c} \tag{18-7}$$

where Δf is the frequency shift.

Equation (35-14) is confirmed directly in detail for the light emitted along the direction of their motion by high-speed atoms. This equation is also used by astrophysicists to measure the speed of stellar objects from the frequency shifts in their emitted light—a red shift to longer wavelengths for receding objects and a blue shift to shorter wavelengths for approaching objects. Indeed, Doppler-shift observations indicate that some very distant stellar objects are in motion away from us at speeds approaching c.

Example 35-2. At what speed would a star have to recede from the earth so that light emitted from the star at the violet limit of the visible spectrum would, because of the Doppler shift, be observed on earth as light at the red limit of the visible spectrum?

Wavelengths at the extremes of the visible spectrum differ by a factor of 2, so that the Doppler-shifted wavelength would be twice the emitted wavelength, or the observed frequency would be half the source frequency.

With $f_o = f_s/2$ and $v/c \equiv x$, the equation (35-14) becomes

$$\frac{1}{2} = \frac{1 + x}{\sqrt{1 - x^2}}$$

After a little algebra and the solving of a quadratic equation, we get

$$x = \frac{v}{c} = 0.6$$

For stars receding from us at speeds greater than 60 percent c, the "visible" spectrum from the high-speed emitter is shifted to the infrared region of the EM spectrum.

35-6 Radiation Force and Pressure, and the Linear Momentum of an Electromagnetic Wave

An EM wave carries energy. It is easy to show that an EM wave also has *linear momentum* in the direction of propagation and that such a wave can exert a force, or *radiation pressure*, on a material upon which it impinges.*

For simplicity, suppose that an EM wave is incident on a material that absorbs all the energy striking it, and reflects and transmits none. This implies

---

* P. N. Lebedev (1866–1912) first measured the pressure of light in 1901.

that when the electric field **E** does work on a charged particle within the material, the energy removed per unit time from the EM wave is exactly the power absorbed by the material. Note that we say the work done by the *electric* field. The magnetic force, since it always acts at right angles to a charged particle's velocity, does *no work*.

Consider the situation in Figure 35-12. Here an EM wave travels along the positive x axis; at one instant, the electric field is **E** along the positive y axis and the magnetic field **B** along the positive z axis. We are interested in the forces produced by **E** and **B** on an electron within the material. The electric force on the electron has a magnitude $F_e = eE$. This force acts in the direction opposite to that of **E** and accelerates the electron in a direction transverse to the wave propagation.

What is the effect of the magnetic field? We take the electron to be moving with speed v along the negative y axis. In general, the direction and magnitude of the magnetic force is given by $\mathbf{F}_m = q\mathbf{v} \times \mathbf{B}$. Here the magnitude of $\mathbf{F}_m$ is evB, and its direction is along the positive x axis. The EM wave produces a *force* on the electron (and therefore also on the material to which the electron is bound) *along the direction of wave motion*. In short, when an EM wave impinges on an electric charge, the **E** field accelerates the charge in the transverse direction and does work on it, while the **B** field, acting on the moving charge, produces a longitudinal force.

We wish to find relations for the radiation force and pressure, and the linear momentum of the EM field, in terms of such quantities as the intensity I and power P of the wave. We found above that the radiation force F_r is given by

$$F_r = evB$$

where all quantities are instantaneous values. Since the magnitudes of **E** and **B** are related by $B = E/c$, this equation can be rewritten as

$$F_r = \frac{v}{c} eE \qquad (35\text{-}15)$$

But eE is just the magnitude of the electric force $\mathbf{F}_e$, the force that does work. In general, the rate of doing work, or the power P, is given by

$$P = Fv$$

where F is the force doing work and v is the speed of the particle acted on. Then (35-15) can be written

$$F_r = \frac{v}{c} F_e$$

Total absorption: $\quad F_r = \dfrac{P}{c} \qquad (35\text{-}16)$

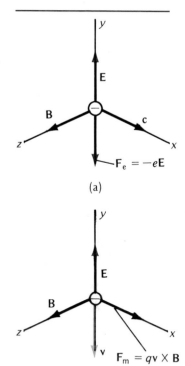

Figure 35-12. An electromagnetic wave incident upon an electron in an absorbing material. (a) An electric force $\mathbf{F}_e = -e\mathbf{E}$ acts on the electron. (b) A magnetic force $\mathbf{F}_m$, in the direction of wave propagation, acts on electron in motion with velocity **v** in magnetic field **B**.

The radiation force of an EM wave on a material that *absorbs it completely* is simply the power of the wave divided by the speed of light.

A longitudinal force P/c is exerted by an EM wave on a material that absorbs it completely. It follows that when a material *emits* radiation of power P in one direction, this emitter must recoil under the action of a recoil radiation force of magnitude P/c. We can see this most easily by noting that emission is, so to speak, absorption run backward in time. In emission, charges within the material *lose* energy and create an outgoing EM wave. With time reversal, the directions of the electric field **E**, of the electric force $\mathbf{F}_e$, and of the radiation force $\mathbf{F}_r$ all remain *the same;* but the direction of the *velocity* **v** is reversed, and so is the direction of the *magnetic field* **B**. Under time reversal—that is, with emission rather than absorption—the direction of the Poynting vector is reversed, and energy then flows away from the material rather than toward it.

What is the radiation force on a material that reflects all the radiation striking it, absorbing none? Think of reflection as taking place in two stages: absorption of the incident radiation followed by reemission in the reverse direction. Since a radiation force of magnitude P/c acts on the material both in absorption and in emission, the radiation force for complete reflection is*

$$\text{Complete reflection:} \quad F_r = \frac{2P}{c} \tag{35-17}$$

The radiation force given in (35-16) and (35-17) applies for radiation that is incident in a direction *perpendicular* to the absorbing or reflecting surface. For oblique incidence, with an angle θ between the direction of wave propagation and the normal to the plane of the absorber or reflector, the radiation force normal to the surface is obtained by multiplying F_r by the factor $\cos\theta$. For total reflection, when $\theta \neq 0$, there is no tangential force.

It is also useful to have relations for the *radiation pressure* p_r, the radiation force F_r per unit transverse area A. Since pressure p_r is, by definition, F/A, we have for complete absorption,

$$p_r = \frac{F_r}{A} = \frac{P}{cA}$$

The intensity I is given by $I = P/A$. Therefore,

$$\text{Complete absorption:} \quad p_r = \frac{I}{c}$$

$$\text{Complete reflection:} \quad p_r = 2\frac{I}{c} \tag{35-18}$$

Example 35-3. A 3-W beam of EM radiation shines on a black object and is completely absorbed by it. (a) What is the radiation force on the absorber? (b) What is the recoil force on the source emitting the beam?

(a) For complete absorption,

$$F_r = \frac{P}{c} = \frac{3\text{ W}}{3.0 \times 10^8 \text{ m/s}} = 1 \times 10^{-8}\text{ N}$$

which is a very small force indeed.

(b) The source emitting the 3-W beam, whether it is a source of light or a radio transmitter, will, as long as the emitted waves travel outward in a single direction, recoil

---

* That the radiation force for complete reflection is twice the force for complete absorption has an exact analog in mechanics. When a particle with initial momentum $+mv$ strikes and sticks to an object, the linear momentum transferred to the struck object is $+mv$; but when a particle with initial momentum $+mv$ is "reflected" from the struck object, rebounding with the same speed, the particle's final momentum is $-mv$, its momentum having been changed by $\Delta(mv) = -mv - (+mv) = -2mv$. Thus, for reflection, the struck object acquires a momentum $+2mv$, just *twice* the momentum acquired in absorption. Equivalently, the force (average) on a struck object is twice as great for reflection as for absorption.

under the action of a force of 10^{-8} N. Thus, a flashlight emitting light constitutes a very elementary form of a rocket.

For any sources of moderate intensity or power, the radiation force is negligibly small. In stellar phenomena, where very high intensities are encountered, the radiation force may equal or exceed the gravitational force, as evidenced by an exploding star, or a supernova.

If EM radiation can exert a force and transfer linear momentum to an object upon which it impinges, linear momentum must be associated with the EM field itself. It is easy to derive the expression for the momentum M of an EM wave. (We use the symbol M for linear momentum, rather than the conventional symbol p, to avoid confusion with the pressure p and the power P.) By definition, the force F is related to the momentum M by

$$F = \frac{dM}{dt}$$

Similarly, the power P is related to the energy by

$$P = \frac{d(\text{energy})}{dt}$$

From (35-16),

$$F_r = \frac{P}{c}$$

so that

$$\frac{dM}{dt} = \frac{1}{c} \cdot \frac{d(\text{energy})}{dt}$$

We use the relation for the radiation force in *absorption* since we want to count the energy transfer only *once*. Integrating yields

$$\text{EM momentum} = \frac{\text{EM energy}}{c} \qquad (35\text{-}19)$$

The magnitude of the linear momentum of an EM wave is the energy of the wave divided by c; the direction of the momentum is along the direction of energy propagation.

We have seen that we can attribute energy and linear momentum to an EM wave. We can also attribute *angular* momentum (actually, *spin* angular momentum) to circularly polarized electromagnetic waves. The magnitude of the EM angular momentum of a circularly polarized wave of frequency f is the energy of the wave divided by $\omega = 2\pi f$, and the direction of the angular momentum vector is parallel or antiparallel to the direction of wave propagation.

35-7 Polarization

Types of Polarization Imagine a plane transverse to the direction of propagation of an EM wave. We concentrate on the path traced out on this plane by

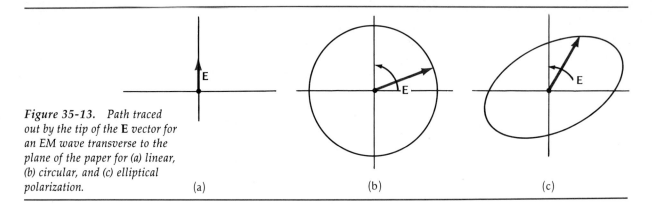

Figure 35-13. Path traced out by the tip of the **E** vector for an EM wave transverse to the plane of the paper for (a) linear, (b) circular, and (c) elliptical polarization.

the tip of the vector **E**, representing the instantaneous electric field, as shown in Figure 35-13.

• *Linearly polarized* wave. The path is a single straight line. This line gives the *direction* of polarization; the *plane* of polarization contains this line and the line representing direction of wave propagation.

• *Circularly polarized* wave. The path is a circle. It is easy to confirm that two simple harmonic oscillations of the same frequency and amplitude, but out of phase by 90°, yield uniform circular motion. Circular polarization can then be thought of as the superposition of two linearly polarized waves of the same amplitude and frequency but out of phase by 90° traveling in the same direction.

• *Elliptically polarized* wave. The path is an ellipse. Mutually perpendicular simple harmonic oscillations of the same frequency but differing in amplitude or phase by an amount different from 90° produce elliptical polarization.

Unpolarized Light Visible light from ordinary light sources (electric excitation, heating) comes from the random radiation of many individual atoms, each atom "turned on" about 10^{-8} s. The emitted light is incoherent; the phases of successive wave trains are not the same (Section 38-6). Furthermore, the polarization of the light from the various radiating atoms does not take one direction but is distributed randomly in all possible directions. When the polarization direction changes randomly and rapidly, so that one cannot follow it in time, the light is said to be *unpolarized*.

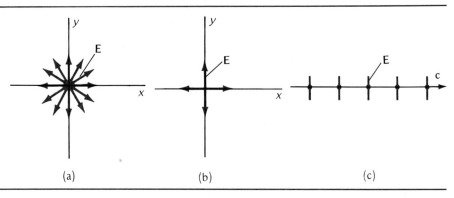

Figure 35-14. Representation of states of polarization. Unpolarized light transverse to the plane of the paper. (a) random orientations of the **E** vectors, and (b) equivalent polarizations along mutually perpendicular directions. (c) Unpolarized light propagated in the plane of the paper to the right.

Figure 33-14 shows ways of representing unpolarized light. In (a), we have linear oscillations in all transverse directions. Since each oblique oscillation can be replaced by its rectangular components, the two mutually perpendicular oscillations, which are neither continuous nor coherent, of part (b) also represent unpolarized light. In part (c), the propagation direction is in the plane of the paper rather than perpendicular to the plane.

The Law of Malus Suppose a microwave beam is linearly polarized and encounters a set of parallel conducting wires oriented along the direction of polarization of the wave. Then the electric field of the EM wave will induce currents in the conducting wire, thermal energy will be dissipated in the wires, and the polarized wave will be absorbed.

The commercial material Polaroid behaves in the same way for visible light. The material has needlelike molecules (herapathite) aligned mostly along one direction (by stretching a flexible transparent sheet as it solidifies). Suppose that the electric field of a polarized light wave is parallel to the long axis of molecules; then the polarized wave is absorbed. On the other hand, with the polarization direction perpendicular to the long axis, the wave is transmitted (mostly). The easy transmission direction is called the polarization direction.

Suppose that unpolarized light passes in turn through two Polaroid sheets, with the second one rotatable relative to the first. Light emerging from the first sheet is linearly polarized. If the second Polaroid sheet has its polarization direction at angle θ relative to the first sheet, only the parallel component of the electric field, $E \cos \theta$, will be transmitted though the second sheet; the perpendicular component will be absorbed. See Figure 35-15.

What is the intensity of the light emerging from the second sheet? Since intensity I varies with the square of the electric field amplitude, we have that

$$I = I_0 \cos^2 \theta \qquad (35\text{-}20)$$

where I_0 is the maximum intensity, corresponding to $\theta = 0$.

When $\theta = 90°$ and the polarization directions of the two sheets are at right angles, the trasmitted intensity is zero. Said differently, when the *polarizer* and the *analyzer* are at right angles, we have *extinction*. Equation (35-20) thus governs the intensity of light through *any* two devices, each of which transmits only one polarization direction; it is known as *Malus's law*, after its discoverer, E. L. Malus (1775–1812).

Polarization by Scattering The EM radiation from an electric-dipole oscillator is linearly polarized. It follows that radiation scattered from a material may also be polarized, since scattering means essentially that the electric field of an EM wave is driving a charged particle in a material into ocillation and

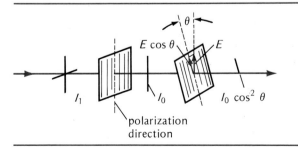

Figure 35-15. With the polarization directions of two successive polarizing materials differing by angle θ, the component of the electric field transmitted through the second material is $E \cos \theta$.

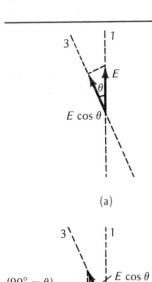

Figure 35-16.

causing it to become effectively an electric-dipole oscillator. For example, sunlight scattered by particles and molecules in the earth's atmosphere is at least partially polarized.

Example 35-4. An unpolarized beam of light passes through a single sheet of Polaroid. Compare the transmitted beam and incident beam in intensity.

If the incident light is truly unpolarized, then the two equivalent perpendicular oscillations into which the beam may be resolved, shown in Figure 35-14(b), must be of equal magnitude. We may choose the orientation of one oscillation to be parallel to the polarization direction of the Polaroid sheet. This component is completely transmitted; the other is completely absorbed. Therefore, the transmitted beam has an intensity half that of the incident beam.

If there were no absorption whatsoever for oscillations parallel to the polarization direction of a Polaroid sheet, adding a second Polaroid sheet with its polarization direction to that of the first would produce no further attenuation in the light intensity. This ideal behavior is not found, however, because partial absorption occurs even in the preferred orientations.

Example 35-5. An unpolarized light beam falls on two Polaroid sheets so oriented that no light is transmitted through the second sheet. A third Polaroid sheet is then introduced between the first two sheets. How does the intensity of transmitted light vary with the orientation of the third sheet? (The intensity is not zero for all orientations!)

The polarization directions of sheets 1 and 2 are mutually perpendicular. We take the angle between the polarization directions of sheets 1 and 3 to be θ. Then, as Figure 35-16(a) shows, if E is the magnitude of the electric field transmitted through the first sheet, the component emerging through the third (the one placed between the first and second sheets) is $E \cos \theta$. From Figure 35-16(b), we see that the component emerging through the last sheet has the magnitude $E \cos \theta \cos (90° - \theta) = E \cos \theta \sin \theta$. Thus, the intensity varies according to

$$I = I_0 \cos^2 \theta \sin^2 \theta = \tfrac{1}{4} I_0 \sin^2 2\theta$$

where I_0 is the intensity transmitted through the first sheet. Note that there are four positions at which the intensity is maximum: 45°, 135°, 225°, and 315°. Note also that interchanging sheets 1 and 3 does not change the dependence of I and θ.

Summary

Definitions

Poynting vector **I** of an EM wave:

$$\mathbf{I} = \mathbf{E} \times \frac{\mathbf{B}}{\mu_0} \qquad (35\text{-}12)$$

Vector **I** gives the direction and magnitude of the intensity, the power per unit transverse area.

Polarization state of an EM wave is given by the path traced out by the **E** vector in a plane transverse to wave propagation: linear, circular, or elliptical.

Fundamental Principles

The propagation speed c of an EM wave through a vacuum is

$$c = \frac{1}{\sqrt{\epsilon_0 \mu_0}} \approx 3.00 \times 10^8 \text{ m/s} \qquad (35\text{-}1)$$

Electromagnetic radiation is produced by accelerated electric charges.

Important Results

Relative magnitudes of the **E** and **B** fields of an EM wave:

$$B = \frac{E}{c} \qquad (35\text{-}2)$$

Instantaneous energy density u of an EM wave:

$$u = u_E + u_B = \epsilon_0 E^2 = \frac{B^2}{\mu_0} \qquad (35\text{-}7)$$

Intensity for any wave of energy density u and propagation speed c:

$$I = uc \qquad (35\text{-}10)$$

Radiation force F_r and radiation pressure p_r for total absorption at normal incidence of an EM wave:

$$F_r = \frac{P}{c} \qquad (35\text{-}16)$$

$$p_r = \frac{I}{c} \qquad (35\text{-}18)$$

where P is the power and I the intensity. For total reflection of the EM wave at normal incidence, the force and pressure are doubled.

$$\text{EM momentum} = \frac{\text{EM energy}}{c} \qquad (35\text{-}19)$$

Law of Malus for intensity I of radiation transmitted through two polarizing materials with angle θ between their polarization directions:

$$I = I_0 \cos^2 \theta \qquad (35\text{-}20)$$

Problems and Questions

Section 35-1 Basic Properties of Electromagnetic Waves

· **35-1 Q** When electromagnetic waves travel through a vacuum, they all have the same
(A) frequency.
(B) wavelength.
(C) velocity.
(D) More than one of the above.
(E) None of the above.

· **35-2 Q** Electromagnetic waves can travel
(A) only through a vacuum.
(B) only through gravitational fields.
(C) only through static electric and magnetic fields.
(D) only through gases.
(E) through all of the above.

· **35-3 Q** In a beam of sunlight that is incident on the earth, the **E** and **B** fields are
(A) 180° out of phase.
(B) 90° out of phase, with **E** leading **B**.
(C) 90° out of phase, with **B** leading **E**.
(D) in phase.
(E) varying independently with no fixed phase relation between them.

· **35-4 P** An electromagnetic wave is directed vertically upward from a point on the earth's equator. At an instant when the electric field is directed toward the north, the magnetic field will be directed.
(A) south.
(B) east.
(C) west.
(D) north.
(E) vertically upward.
(F) vertically downward.

· **35-5 Q** Which of the following statements is not true for an electromagnetic wave in free space?
(A) **E** and **B** are perpendicular to each other and to the direction of propagation.
(B) The magnitudes of **E** and **B** are in the same ratio at every instant.
(C) **E** and **B** are out of phase with each other by 90°.
(D) The velocity of the wave is always the same, independent of variations in frequency.
(E) Such waves always travel at 3×10^8 m/s.

: **35-6 Q** Which of the following statements is false?
(A) Changing electric fields create magnetic fields, and changing magnetic fields create electric fields.
(B) A Maxwell displacement current exists only when there is a magnetic field that changes with time.
(C) A Maxwell displacement current gives rise to a magnetic field, just as an ordinary current does.
(D) Without a displacement current, there can be no electromagnetic waves.

Section 35-2 Sinusoidal Electromagnetic Waves and the Electromagnetic Spectrum

· **35-7 Q** In which of the following pairs are the members in the same relation as microwaves and ultraviolet radiation?
(A) A baritone and a tenor.
(B) A whisper and a shout.
(C) An electron and a photon.
(D) A baby whale and an adult whale.
(E) Waves and particles.

· **35-8 Q** Which of the following forms of electromagnetic radiation has the longest wavelength?
(A) Gamma rays.
(B) Visible light.
(C) Microwaves.
(D) Ultraviolet rays.
(E) Infrared radiation.

· **35-9 Q** A certain light wave has a frequency of 6.38×10^{14} Hz. (a) What is its wavelength in a vacuum? (b) What color is it?

· **35-10 P** Electromagnetic waves with a frequency as low as 100 Hz can be used to communicate with a submerged submarine. Compare the wavelength of this radiation in free space with the earth's diameter.

· **35-11 P** What is the frequency of yellow light with a wavelength of 570 nm?

· **35-12 P** What is the wavelength of a radar wave of frequency 3.0 GHz?

Section 35-3 Energy Density, Intensity, and Poynting Vector

- **35-13 P** If the distance from a point source of light to a detector is doubled, the intensity of light at the detector will
  (A) be unchanged.
  (B) double.
  (C) increase by a factor of 4.
  (D) be half as great.
  (E) be one-quarter as great.

- **35-14 P** The intensity of sunlight at the top of the earth's atmosphere is 1.4 kW/m². Determine the magnitude of the electric and magnetic fields of this radiation.

: **35-15 P** What is the energy density in the beam from a 2-mW laser if the area of the beam is 2×10^{-4} m²?

: **35-16 P** A 10-W point source of radiation radiates uniformly in all directions. For a distance 4 m from the source, determine (a) the electric field; (b) the magnetic field; (c) the intensity.

: **35-17 P** The intensity of the sun's radiation at the top of the earth's atmosphere is approximately 1400 W/m². If the sun is considered a sphere of radius 7.0×10^8 m, what is (a) the intensity of sunlight at the surface of the sun? (b) the value of the electric field in the sunlight at the surface of the sun? (c) the value of the magnetic field in the sunlight at the surface of the sun?

Section 35-6 Radiation Force and Pressure, and the Linear Momentum of an Electromagnetic Wave

- **35-18 P** What is the momentum of a laser pulse of 10 MW and with a pulse duration of 1.0 μs?

- **35-19 P** A 10 kW light source is turned on for 10 s and it produces a beam with a plane wavefront. What are (a) the energy and (b) the momentum of the beam? (c) The recoil force on the light source while it is turned on? (d) The distance between the leading and trailing edges of the beam?

- **35-20 P** A continuously operating high-intensity laser can have a power output of 1 kW. Suppose that such a laser is used as a photon rocket on a device with a total mass of only 2 kg. If the rocket starts from rest in interstellar space and operates continuously for one year, what is then the rocket's speed?

- **35-21 P** According to the quantum theory, electromagnetic radiation consists of particle-like photons, where the energy of a photon of frequency ν is given by hν, where h is a constant. Show that the momentum of a photon is h/λ, where λ is the photon wavelength.

: **35-22 P** The intensity of electromagnetic radiation from the sun just outside the earth's atmosphere is 1.4 kW/m². (a) What is the sun's power output? (b) What is the intensity of the sun's radiation at the surface of the sun? (c) What is the radiation pressure (in atm) at the surface of the sun?

: **35-23 P** Dye lasers can generate ultra-short pulses. A pulse of 0.65 fs corresponds to just 8 full cycles of oscillation. (a) What is the wavelength of the radiation? (b) Over what distance along the propagation direction does the pulse extend? (c) What is the total energy per pulse if the power of the pulse is 1.0 GW?

: **35-24 P** A beam of electromagnetic radiation of power P is incident normally on a surface. One-third of the incident radiation is absorbed, and the rest is reflected. What is the radiation force on the surface?

: **35-25 P** A beam of electromagnetic radiation of intensity I is incident at a angle of 60° relative to the normal upon a surface that aborbs 25 per cent of the radiation and reflects the rest. What is the radiation pressure on the surface?

: **35-26 P** A 10 kW beam of electromagnetic radiation shines normally on an absorbing surface of a 1.0-kg object. Ten percent of the incident radiation is reflected. (a) What is the total energy absorbed by the object over a period of 1.0 hour? (b) Assuming the object to be initially at rest and subject only to the radiation force, what is the final kinetic energy of the object?

: **35-27 P** (a) What is the momentum of a typical nitrogen molecule (N_2) in air at 300 K? (b) What is the energy of a pulse of light with the same momentum? (c) If the pulse duration is 1.0 μs, what is the average power of the pulse?

: **35-28 P** a horizontal electromagnetic beam with an intensity of 10 MW/m² shines on a perfectly absorbing rectangular sheet of 1.0 m² area that hangs freely and rotates about a horizontal axis at the top edge. The sheet's mass is 0.050 kg. At what angle with respect to the vertical will the sheet remain in equilibrium?

Supplementary Problems

35-29 P A *solar sail vehicle* can be propelled by radiation pressure. (a) Suppose that a solar sail vehicle is in equilibrium because the only gravitational force on it, that from the sun, is exactly balanced by the sun's radiation force on the sail. Show that if the orientation of the sail is not changed, it will be in equilibrium at *any* distance from the sun. (b) Suppose that a sail can be constructed whose total area is 1 km². What is the maximum allowable mass of the solar-sail vehicle? (c) What is the total volume of solid, assuming it be aluminum (2.7×10^3 kg/m³). (d) If one-fourth of the total mass were used for the capsule and the rest for the sail, what would be the sail's average thickness? (At the earth, the sun has I = 1.4 kW/m².)

35-30 P A simple way to tell whether a laser beam can effectively eat away at a material it strikes is to see, first of all, whether the electric field in the electromagnetic field is comparable in magnitude to the interatomic field in an ordinary material. (a) A relatively strong interatomic field would be that at a distance of 10 nm from a proton. What is its magnitude? (*b*) What is the average energy density of an electromagnetic wave with an electric field of this magnitude? (*c*) What is the intensity of the beam? (*d*) Assume that this intensity is produced in a laser pulse that is focused to a cross-sectional area of 0.10 mm². What is the power of the pulse? (*e*) Assume the pulse duration to be 1.0 μs. What is the energy per pulse? (*f*) Assume that each atom in the material upon which the laser pulse shines is bound to its neighbors with an energy of 2.0 eV and that adjacent atoms are separated by 0.2 nm. Assume further that only about one-fifth of the energy of the pulse goes into eating away material. How deep a hole is produced by one pulse? (*g*) How many pulses are required to make the hole 0.1 mm deep?

35-31 P A sufficiently intense laser beam shining on a material may produce, because of *nonlinear* effects, coherent radiation at twice the frequency of the illuminating radiation. In the simplest model of atomic structure, an electron is bound to its parent atom by a linear restoring force. Then, when the electron is driven into simple harmonic motion by a monochromatic electromagnetic field, the electron acts as a kind of subatomic antenna, radiating waves of the same frequency. If the driving electric field is sufficiently strong, however, the equivalent force on the electron is nonlinear, the oscillations are not simple harmonic, and the electron generates, not merely radiation of the same frequency as the driving radiation, but also harmonics at integral multiples of the fundamental frequency (Section 11-6).

To see what characteristics are required of laser radiation capable of generating harmonics, we suppose that the electron oscillations are surely nonlinear when the magnitude of the electric field in a laser beam is comparable in magnitude to an interatomic field, say, the electric field at a distance of 1.0 nm from a proton. (*a*) What is that electric field magnitude? (*b*) What are the energy density and intensity of a laser pulse with the requisite electric field magnitude? (*c*) Suppose that the laser beam can be focused to a cross sectional area of 0.1 mm². What is the required laser power output? (A continuously operating carbon-dioxide laser can have a power output of the order of a kW.)

36 Ray Optics

36-1 Ray Optics and Wave Optics
36-2 The Reciprocity Principle
36-3 Rules of Reflection and Refraction
36-4 Reflection
36-5 Index of Refraction
36-6 Refraction
36-7 Total Internal Reflection
      Summary

In this chapter we deal with waves, primarily electromagnetic waves of light, traveling in two and three dimensions and encountering the boundaries between media under the special condition in which the wavelength is small compared with the size of obstacles or apertures. Then the only phenomena occurring at the interfaces—reflection and refraction—can be understood, and the progress of a wave can be charted, by a simple geometrical procedure, ray tracing. In our examination of ray optics, or geometrical optics, we exclude such distinctive wave effects as interference and diffraction.

36-1 Ray Optics and Wave Optics

An opaque object with a sharp boundary casts a sharp shadow when illuminated with visible light from a small source. Light travels strictly along a straight line; and it is not obvious, certainly not to the casual observer, that light is actually a wave phenomenon. (Early physicists, including even Isaac Newton, thought that light consisted of particles, or corpuscles.) Consider first the circumstances under which we can assume light, or any other wave disturbance, to follow the "paths" given by the rays associated with wave fronts (Section 18-4) and ignore distinctive wave effects.

Figure 36-1(a) shows waves of various wavelengths λ impinging on an opaque object with an aperture of width d. When $\lambda > d$, the waves spread outward from the aperture in all directions; the wave fronts are circular. For shorter wavelengths, the spreading, or diffraction, of the waves beyond the limits of the geometrical shadow is less pronounced. Finally, when $\lambda \ll d$, the wave fronts remain straight lines and the wave disturbance lies strictly within the limits of the "shadow" of the opening, so that any ray entering the opening continues through without a change in direction.

A similar behavior is seen for waves encountering an isolated opaque object, as shown in Figure 36-1(b). With the wavelength relatively large compared with d, the rays are bent and the wave fronts are curved as a result of diffraction by the object. For $\lambda \ll d$, however, the object casts a sharp shadow. Diffraction effects are then negligibly small, and we can draw any ray as undeviated. In the remainder of this chapter, we assume that the condition $\lambda \ll d$ is always met. (Diffraction effects, which arise when this condition is not satisfied, are dealt with in Chapter 39.)

Under what conditions, then, may we treat visible light by the procedures of ray (or geometrical) optics rather than wave (or physical) optics? A typical wavelength for visible light is somewhat less than 10^{-3} mm. Therefore, ray optics works all right so long as we deal with objects or openings of ordinary size, that is, much larger than 10^{-3} mm ($= 1\ \mu$m $= 10^3$ nm). The requirement for ray optics applies equally well to other wave types. For example, an audible sound wave through air with a frequency of 1000 Hz and a wavelength of 34 cm, or an electromagnetic microwave with a wavelength of several centimeters, can be traced by geometrical "optics" only when such waves strike objects whose characteristic dimensions are much larger than a few centimeters.

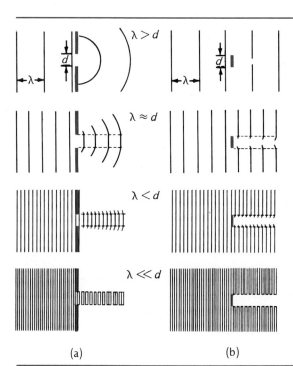

Figure 36-1. Waves of decreasing wavelength encountering (a) an aperture and (b) an obstacle of size d. Diffraction is significant for $\lambda \gtrsim d$; ray optics applies for $\lambda \ll d$.

Although we shall deal with wavelengths much shorter than the width of apertures and obstacles, we also assume that the wavelength is large compared with the microscopic objects responsible for the reflection and refraction effects. It is individual atoms and their associated electrons in a solid or a liquid that cause visible light to be reflected and refracted; but the spacing between adjacent atoms, typically of the order of a nanometer, is much less than the wavelength of visible light. By the same token, a sheet of ordinary chicken wire, porous to visible light, acts as an opaque reflector for radio waves several meters in wavelength.

36-2 The Reciprocity Principle

Suppose that we view a motion picture of a wave phenomenon, perhaps a wave traveling along the surface of a liquid. We assume no energy dissipation. Now, if the motion picture is run backward—if we view the wave motion with time reversed—the wave fronts move in the opposite directions and the directions of rays associated with these wave fronts are also reversed. Both wave motions, the first one with time running forward and the second with time reversed, are possible motions. Both are consistent with the laws governing wave motion. Thus, if we know the "path" of a wave from one point to a second point by knowing the configuration of the ray (or rays) connecting the two points, then the reverse path, from the second to the first point, is found simply by reversing the arrows on the rays. In short, if a ray goes from *A* to *B*, a ray also will go from *B* to *A* by the same route. This is the *reciprocity principle*. It asserts that any two points connected by a ray are reciprocal in the sense that the directions of wave propagation can be interchanged with no alteration in the pattern of the wave fronts. Also termed the principle of *optical reversibility*, this principle applies to all nondissipative wave motion, not merely visible light.

Consider Figure 36-2. Here we see a portion of a wave front and the corresponding rays radiating from the point source and reflecting off a parabolic reflector. (The remaining waves emitted by the point source continue to expand outward as spherical waves.) The reflected beam consists of plane waves. A parabolic reflector changes diverging rays into parallel rays; it changes a point source into a plane wave source, as it were. Reversing the ray directions (imagining time as running backward), we see that a beam of plane wave fronts incident upon the parabolic reflector is brought to a focus, and all the rays intersect at a single point. This says that a parabolic reflector can equally well be a transmitter or a receiver of a parallel beam. Even more generally, if we know the radiation pattern of a transmitting antenna—it tells how the intensity radiated varies as a function of direction—we also know the behavior of the same antenna used as a receiver of radiation incident upon it.

Another example of reciprocity is shown in Figure 36-3(a). Here light from a point source passes through a lens to form an image. On ray (or time) reversal, the image becomes the source, and the source becomes the image. Strictly, the arrows on the rays are not necessary, except to indicate for convenience which of the two possibilities is under consideration.

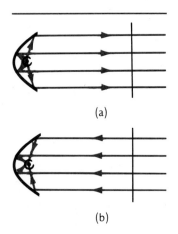

Figure 36-2. Example of "optical" reversibility: a parabolic reflector as (a) a transmitter and (b) a receiver.

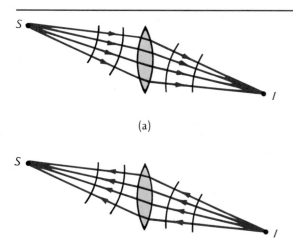

(a)

(b)

Figure 36-3. Example of the reversibility of rays. (a) Source S forms image I when rays pass through a lens. (b) Reversing the ray directions interchanges source and image.

36-3 Rules of Reflection and Refraction

Suppose that a wave encounters a boundary between two media. The incident wave is partially transmitted into the second medium and partially reflected into the first. The transmitted wave is usually bent, or *refracted,* at the interface; it is called the *refracted* ray.

First we give the rules for reflection and refraction. Later we shall see how they follow directly from fundamental principles (Sections 36-4 and 36-6). Figure 36-4 defines terms. It shows a ray incident at the interface between two media; for example, a narrow pencil of light is entering water from air. By convention, the directions of the incident ray θ_1, of the reflected ray θ_1', and of the refracted ray θ_2 are measured relative to the normal to the interface.

The fundamental facts about reflection and refraction of rays are these:

- All four of the following lines lie in a *single plane:* the incident ray, the normal, the reflected ray, and the refracted ray
- Angle of incidence equals angle of reflection:

$$\theta_1 = \theta_1' \qquad (36\text{-}1)$$

- Angles of incidence θ_1, and of refraction θ_2 are related by

$$n_1 \sin \theta_1 = n_2 \sin \theta_2 \qquad (36\text{-}2)$$

where n_1 and n_2 are constants characteristic of media 1 and 2, respectively. This relation is known as Snell's law, after its discoverer W. Snell (1591–1626). The constants, called *indices of refraction,* are related to the wave speeds and wavelengths in the respective media, as will be shown in Section 36-5. The principle of reciprocity says that we may reverse the ray directions, so that designating one angle as that of incidence and the other as that of refraction is actually arbitrary.

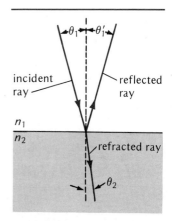

Figure 36-4. Reflection and refraction of an incident ray at an interface.

776 CHAPTER 36 Ray Optics

This is the long and short of ray optics. If you know the value of n for various materials, you can, at least in principle, trace the rays in detail as they go through a whole succession of surfaces. *Specular* reflection means reflection from a polished surface, for example, a flat or curved mirror with only small surface irregularities. *Diffuse* reflection occurs when the surface is not smooth; then the law of reflection holds exactly at any portion of the surface small enough to be considered flat and smooth. The program of ray optics is simple in principle. It involves nothing more than simple geometrical constructions following the rules given above. But the actual design of optical systems, usually with multiple lenses, is so extraordinarily tedious and difficult that nowadays it is done, trial-and-error fashion, by computers simply because one must trace an extremely large number of rays. (The simpler elements of lens design are discussed in Chapter 37.)

One aspect of reflection and refraction cannot be treated by the methods of ray optics: the relative intensities of the reflected and transmitted beams. The reflection-transmission ratio can, in fact, be computed from the n_1/n_2 ratio by applying Maxwell's equations to electromagnetic waves.

Example 36-1. A fish pond is in the shape of a right circular cylinder; its diameter is 2.00 m and its depth is also 2.00 m. The pond is first empty of water, and a person standing at a distance 170 cm from the near edge of the pond can just barely see the boundary on the opposite side between the side wall and the bottom. See Figure 36-5(a). The eye for this person is 170 cm from the bottom of his feet, so that the light ray from the boundary to his eye makes an angle of 45° with the vertical.

Now the pool is filled to the brim with water, and the observer can move farther from the edge of the pool and still see the boundary between side and bottom. See

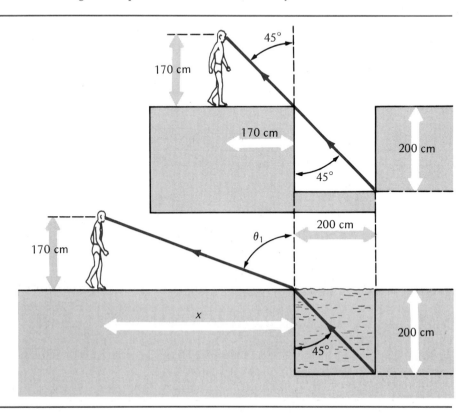

Figure 36-5. Observer just able to see the boundary between side wall and bottom of a fish pond (a) with pool empty and (b) with pool filled with water.

Figure 36-5(b). What is his maximum distance from the edge? For water, $n = 1.33$; for air, $n \simeq 1.00$.

Applying Snell's law to the refraction shown in Figure 36-5(b), we find that

$$n_1 \sin \theta_1 = n_2 \sin \theta_2$$
$$(1.00) \sin \theta_1 = (1.33) \sin 45°$$
$$\theta_1 = 70°$$

so that the maximum distance x from the pool edge is given by

$$\tan 70° = \frac{x}{170 \text{ cm}}$$
$$x = 467 \text{ cm}$$

36-4 Reflection

Let us first see that the law of reflection ($\theta_1 = \theta_1'$) can be deduced from general properties of wave propagation. Figure 36-6 shows a succession of plane wave fronts incident upon a plane surface; the angle of each wave front *relative to the surface* is also the angle of incidence θ_1. To find the reflected wave fronts, we use Huygen's principle (Section 18-4): to find a future wave front, we take the envelope of the Huygens wavelets generated along the wave front. At the interface in Figure 36-6, Huygens wavelets are generated in both the forward and the backward directions. The physical basis for Huygens's construction applied to electromagnetic waves is simply that as an electromagnetic wave travels through a medium, its electric field sets electrons in forced oscillation, and these electric oscillators generate electromagnetic waves that propagate both forward and backward. Within the refracting medium the net backward radiation is zero. At a boundary, however, the symmetry required for cancellation of the backward wave no longer obtains, so that both reflected and refracted waves are generated.

Figure 36-6 shows that the left end A of the incident wave front touches the surface when the right end B is a distance BC away from the surface. After time $t = BC/c$ has elapsed (with c as the wavespeed in the incident medium), the right end of the wave front reaches point C. At this same instant, the left end is at D, where $AD = ct$. Therefore $BC = AD$, and $\theta_1 = \theta_1'$. The reflection law has been proved.

Now consider Figure 36-7. Spherical wave fronts from a point source S are

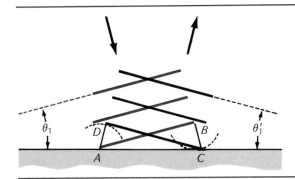

Figure 36-6. Plane wave fronts incident upon a plane reflecting surface. Incident wave front AB later becomes reflected wave front CD.

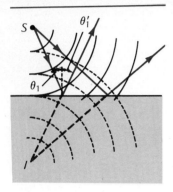

Figure 36-7. Spherical wave fronts from point source S striking a plane reflecting surface. The reflected spherical wave fronts appear to originate from virtual image I.

incident on a plane surface. The reflected rays and the reflected wave fronts are found by applying the rule $\theta_1 = \theta_1'$ to each ray. As the figure shows, the reflected rays appear to diverge from a single point I. To an observer viewing the reflected rays only, these rays and their associated spherical wave fronts seem to originate from the point *image I* rather than from their true origin, the point source S. The human eye is naïve; it interprets all rays reaching it as having always traveled in unbroken straight lines. To the eye (or a camera), the source appears (literally!) to be located at the position of the image. From the geometry of Figure 36-7, it is clear that the image is symmetrically located with respect to the source, with the reflecting boundary midway between them. The image here is said to be *virtual*. The rays appear to originate from location I; they actually come from S.

Example 36-2. An object is placed near two plane mirrors that are at right angles to one another. What images can be seen in the mirrors?

See Figure 36-8(a), where the object is represented by O and a viewer's eye by E. Applying the rule $\theta_1 = \theta_1'$ at each reflection, we find that there are three virtual images:

- Image I_1, formed by reflection from mirror 1.
- Image I_2, formed by reflection from mirror 2.
- Image I_{12}, formed by reflections from both mirrors. Image I_{12} can be described as the image in mirror 2 of the image I_1 (or equivalently, the image in mirror 1 of image I_2).

The object and its three images are located symmetrically with respect to the lines representing the mirrors. Fold the paper at the mirror lines and the three images and object fall exactly on top of one another.

Note that the ray reaching the eye from image I_{12} is parallel to the same ray leaving the object O. This means that any ray undergoing two reflections at a corner mirror always emerges antiparallel to the ray leaving the object. (The behavior is like that of a billiard ball that makes two "reflections" at a corner of a billiard table.) Three mutually perpendicular mirrors form a corner reflector. *Any* ray undergoing a reflection from each of the three mirrors emerges *parallel* to the incident ray. Any ray reflected in a corner mirror comes straight back, undeviated in direction (but displaced laterally). Corner reflectors, usually made of red glass, are used on the rear fenders of bicycles; any light shining on them is returned toward the light source. Large-scale corner reflectors are in common use as "targets" for radar signals. Corner reflectors left on the moon's

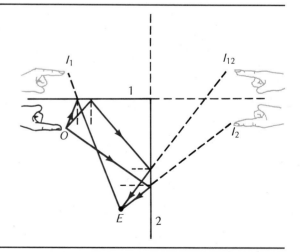

Figure 36-8. Images formed by reflections from two plane mirrors at right angles.

Using a Point Source

If you live near the San Andreas fault in California, your interest in what geologists call plate tectonics is more than casual. You'd like very much to know, if possible, when an earthquake is coming. Even small shifts in the relative positions of earth masses can provide a clue.

Actually, very small shifts in the earth's crust— even as small as 1 cm—*can* be measured with modern techniques. One way is shown in the figure. Here two radio antennas separated by the baseline distance d are pointed toward some distant point source. A microwave pulse reaches antenna A_1 first and then a time Δt later it arrives at antenna A_2, because antenna A_2 is farther from the point source than antenna A_1 by the distance $c\Delta t$. Measuring Δt to within one cycle of the microwave oscillation allows (with angle θ known) the distance d to be measured with an error of no more than about one microwave wavelength. (In more technical terms, radio-interferometry is used for long base-line geodesy, the measurement of distances between points on earth.)

What is a point source? How do you measure Δt? The point source can be a quasar, a *quasi-stellar astronomical source,* in a distant galaxy that emits pulses of radiation at microwave frequencies. Each pulse has its own characteristic shape, or signature, so that you can tell when the *same* burst of radiation reaches antennas A_1 and A_2. The signal arriving at each antenna is mixed with the signal from a continuously operating microwave oscillator of nearly constant frequency called a *maser* (where "m" for microwave replaces "l" for light in the acronym "laser"). The technique has been developed by a group from MIT,

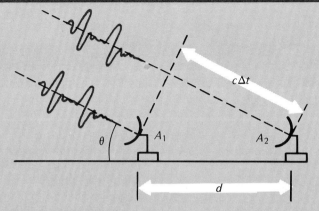

Goddard Space Flight Center, and the Haystack Observatory; the technique goes by the code name ARIES.*

Another similar technique for measuring small shifts in position between the two locations on earth involves an orbiting satellite. The satellite (code named LAGEOS), launched in 1976 and flying at an altitude of 6000 km, is covered with retroreflectors (corner mirrors); a light signal from earth is sent back in the very same direction. Light pulses (in nanoseconds) from lasers go from earth to the satellite; their return to earth, and especially the delay in their arrival at the more distant receiving site, exploits a technique that can be called "optical radar." This technique has been developed by JPL, the Jet Propulsion Laboratory at CalTech.

*See the cover story for *Physics Today* **34**, 20 (April, 1981).

surface return light pulses sent from earth and allow the distance from earth to moon to be monitored to within a few centimeters.

36-5 Index of Refraction

We first recognize what does not change when a wave travels from medium 1, in which its wave speed is v_1, to a second medium 2, in which the wave speed is v_2. The *frequency* f of the wave is the *same* in all media. With wavelengths λ_1 and λ_2 in the two media we then write

$$v_1 = f\lambda_1 \quad \text{and} \quad v_2 = f\lambda_2$$

so that

$$\frac{v_1}{v_2} = \frac{\lambda_1}{\lambda_2} \tag{36-3}$$

The wavelength is greater in the medium with the higher wave speed.

The speed of an electromagnetic wave *in vacuo* is represented by c. By definition, the index for refraction n for a medium is the ratio of c to the wave speed in that medium. Therefore,

$$v_1 \equiv \frac{c}{n_1} \quad \text{and} \quad v_2 \equiv \frac{c}{n_2} \tag{36-4}$$

where n_1 and n_2 are called the indices of refraction for media 1 and 2. By definition, the index of refraction of a vacuum, or empty space, is exactly 1. Light travels through any medium at a *lower* speed than c: the refraction index always exceeds 1. For example, the speed of light through water is 2.25×10^8 m/s, so that water's index of refraction for visible light is $n = (3.00 \times 10^8 \text{ m/s})(2.25 \times 10^8 \text{ m/s}) = 1.33$. For air near the earth's surface, n is 1.000 29, nearly the same as in a vacuum.

Visualize light going through a transparent material from a subatomic point of view. Then the electromagnetic wave actually travels *through a vacuum* at speed c, and it occasionally encounters a charged particle that scatters some of the incident beam. The combined effect of the scattering from the charged particles is to yield macroscopic propagation at a speed less than c.

Table 36-1 lists indices of refraction for some common transparent materials.

The *relative* index of refraction of medium 2 to medium 1, represented by n_{21}, is given by

$$n_{21} \equiv \frac{n_2}{n_1} = \frac{v_1}{v_2} \tag{36-5}$$

It follows that

$$n_{21} = \frac{1}{n_{12}}$$

In Section 36-6 we prove that a refraction index defined as the ratio of wave

Table 36-1

MATERIAL (at 20°C)	REFRACTIVE INDEX for yellow sodium light ($\lambda = 589.3$ nm)
Diamond	2.42
Ethyl alcohol	1.36
Glass (crown)	1.52
Glass (light flint)	1.58
Glass (heaviest flint)	1.89
Ice	1.31
Acetone	1.36
Sodium chloride	1.54
Stibnite (Sb_2S_3)	4.46
Water	1.33

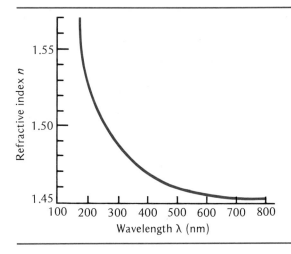

Figure 36-9. Refractive index n of fused quartz as a function of free-space wavelength λ over the ultraviolet and visible regions of the electromagnetic spectrum.

speeds in two media is the same as the refractive index appearing in Snell's law, (36-2). One can measure the value of the relative refractive index directly by observing the refraction of waves at an interface. The wavelengths in two media can be related to the respective indices of refraction by means of (36-3) and (36-4):

$$\frac{\lambda_1}{\lambda_2} = \frac{n_2}{n_1} \tag{36-6}$$

If the index is large, the wavelength is small. Suppose, for example, that blue light of wavelength 400 nm (in free space) enters glass having an index of refraction of 1.50. The wavelength in the glass of this light, still blue, is *less*, namely (400 nm)/1.50 = 267 nm. It is customary to characterize a particular color of visible light by its *wavelength in free space*, rather than by its frequency, simply because the wavelength can be measured directly, whereas the frequency is computed from a knowledge of the wave speed.

The relative index of refraction for a given transparent material usually depends on the frequency. The index of refraction usually decreases with wavelength, as shown in Figure 36-9, or increases with frequency. Thus violet light travels through glass at a lower speed (approximately 1 percent) than red light, which has a longer wavelength. Whereas all the component frequencies of white light travel through a vacuum at the same speed c, the speeds of the various wavelengths differ in a refracting medium. The phenomenon is called *dispersion*.

The surface waves on a liquid of varying depth show a similar behavior. The wave speed decreases as the depth of the water is reduced. Thus, when waves in the ripple enter a region in which the depth is reduced, the waves are compressed because of the decrease in wavelength. See Figure 36-10.

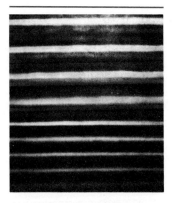

Figure 36-10. Change in wavelength arising from a change in wave speed for water waves in a ripple tank.

36-6 Refraction

Here we see that Snell's law for refraction follows directly from a change in wave speed.

Figure 36-11 shows wave fronts from medium 1, in which the wavespeed is

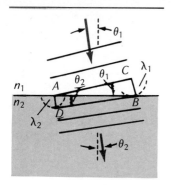

Figure 36-11. Refraction of plane wave fronts at an interface. Incident wave front AC later becomes refracted wave front DB.

v_1, the wavelength λ_1, and the refractive index n_1, going to medium 2, where the corresponding quantities are v_2, λ_2, and n_2. The angle of incidence θ_1 is also the angle between the incident wave fronts and the interface; similarly, the angle of refraction θ_2 can be measured between the interface and the wave fronts in medium 2. We concentrate on wave front AC in medium 1; it later becomes wave front DB in medium 2. Consider the time interval in which the right end of the wave front advances one wavelength λ_1 in medium 1; during the same time the left end of the wave front has advanced a smaller distance, wavelength λ_2 in medium 2. From the geometry of Figure 36-11, we have for triangles ABC and ABD, respectively,

$$\sin \theta_1 = \frac{CB}{AB} = \frac{\lambda_1}{AB} \quad \text{and} \quad \sin \theta_2 = \frac{AD}{AB} = \frac{\lambda_2}{AB}$$

Eliminating AB gives

$$\frac{\sin \theta_1}{\sin \theta_2} = \frac{\lambda_1}{\lambda_2}$$

Using (36-6), we can write this as

$$n_1 \sin \theta_1 = n_2 \sin \theta_2 \qquad \text{(36-2), (36-7)}$$

which is Snell's law.

Another form of Snell's law, written in terms of the relative refractive index [(36-5)], is

$$\frac{\sin \theta_1}{\sin \theta_2} = n_{21} = \frac{1}{n_{12}} \qquad (36\text{-}8)$$

Refraction of a ray at an interface arises from a change in the wave speed, so that we can compute the relative wave speeds in two media simply by applying Snell's law to find the relative refractive index. In 1862, J. B. L. Foucault performed an experiment highly significant in the history of the theory of light when he showed that the speed of visible light through water is less than through air.*

Consider Figure 36-12(a). It shows a ray incident on a slab of refracting material with *parallel* faces. Snell's law governs the refraction at both interfaces, and both interior angles have the same value θ_2; this means that the emerging ray is exactly parallel to the incident ray. The emerging ray is, however, displaced laterally by an amount that depends on the thickness of the slab and its refraction index, but the ray is not deviated from its initial direction. This means that objects viewed through an ordinary sheet of window glass (one with parallel surfaces) are not distorted but merely displaced laterally by the refraction of light rays.

A ray is deviated from its original direction when it passes through a slab of refracting material with *nonparallel* faces. See Figure 36-13(b). Such a device is

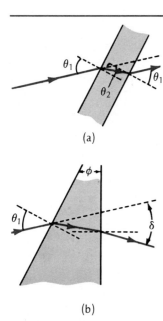

Figure 36-12. (a) Refractions of a ray through a plate with parallel faces. The emerging ray is not deviated in direction. (b) Refractions of a ray through a plate with nonparallel faces (a prism). The emerging ray is deviated through angle δ.

---

* Foucault's experiment refuted the particle theory of light, in which the refraction of a light ray is attributed to a change in the particle's velocity at the interface. Snell's law can be derived from the particle model, but this model predicts a higher speed in a refracting medium than in a vacuum. See Problem 36-40.

known as a *prism*. The angle of deviation δ between the incident and emergent rays depends, for a given incident angle, on the refractive index of the prism and its apex angle ϕ. Because the refractive index depends on wavelength, the various frequency components of white light are deviated by *different* angles, or dispersed into the visible spectrum. Index n is greater for high frequencies (violet) than for lower frequencies (red); consequently, violet light is deviated most and red least, with all intermediate colors of the visible spectrum lying between.

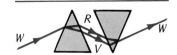

Figure 36-13. Dispersion of white light (W) by a prism into red (R) and violet (V) and its approximate recombination by a second prism.

Isaac Newton observed that in the dispersion of the visible spectrum by a prism, one single color, such as green, cannot be further resolved into component colors. He also observed that two prisms can be used first to disperse and then to reunite the various components into white light. See Figure 36-13.

A prism is commonly used in an instrument known as a prism spectrometer, for analyzing visible light into its component wavelengths. See Figure 36-14. Light from the source goes through a narrow slit and then through the prism. If the emitted light consists of certain discrete frequencies, the eye sees, or a photographic film records, a succession of "lines," where each line is an image of the slit at a particular frequency. A prism spectrometer can be used only for relative wavelength measurements; it must be calibrated against a device, such a diffraction grating (Section 38-4) that can make absolute wavelength measurements.

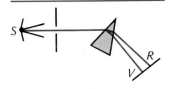

Figure 36-14. Simple elements of a prism spectrometer.

Example 36-3. Viewed from above, an object under water appears to be closer to the surface than its actual depth. Indeed, the *apparent depth* of any object immersed in a medium with relative refractive index n is given by $d' = d/n$, where d is the object's actual distance from the interface. Prove this result.

Figure 36-15 shows a point source O at depth d below the surface in a medium with refractive index n. We concentrate on a ray that emerges from O close to the vertical at angle θ_2 in the medium; this ray makes angle θ_1 relative to the normal in air. A ray directed vertically upward is undeviated. Extending these two rays backwards—as the two eyes do, in effect, in binocular vision—shows that they intersect at point I, which the eyes interpret to be the effective source, or image. The apparent depth d' is the distance of image I to the surface.

We apply Snell's law to the refracted ray, setting $n_1 = 1$ and $n_2 = n$. The angles are small enough that $\sin \theta_1 \simeq \theta_1$ and $\sin \theta_2 \simeq \theta_2$. We then have

$$n_1 \sin \theta_1 = n_2 \sin \theta_2$$

$$\theta_1 = n\theta_2$$

The geometry of Figure 36-15 shows that distances d and d' are related to angles θ_1 and θ_2 by

$$\tan \theta_1 \simeq \theta_1 = x/d' \quad \text{and} \quad \tan \theta_2 \simeq \theta_2 = x/d$$

so that eliminating x gives

$$\theta_1 = (d/d')\theta_2$$

Eliminating θ_1/θ_2 between this equation and the one above gives finally

$$d' = d/n$$

Suppose an object is immersed in water, for which $n \simeq 4/3$. Then, when viewed from above, the object will appear to be $\tfrac{3}{4} = 75$ percent of its true distance from the surface.

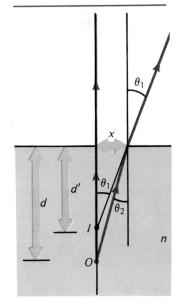

Figure 36-15.

36-7 Total Internal Reflection

When a beam is incident upon the interface between refracting media, some of its energy is transmitted and the rest is reflected. In one situation, however, the incident beam cannot be transmitted; consequently the incident wave is totally reflected.

Consider Figure 36-16. It shows a series of rays traveling from a medium of low optical density into a more dense medium (into a medium with larger n). (For simplicity, reflected rays are not shown in Figure 31-16, although there is at least some reflection.) The largest possible incident angle θ_1 is 90°. From Snell's law, the corresponding angle for θ_2 is

$$\frac{\sin \theta_1}{\sin \theta_2} = \frac{1}{\sin \theta_2} = \frac{n_2}{n_1}$$

Under these conditions, θ_2 is the *critical angle* θ_c, with

$$\sin \theta_c = \frac{n_1}{n_2} \qquad \text{where } \frac{n_1}{n_2} < 1 \qquad (36\text{-}9)$$

The angle θ_c is the largest possible angle of refraction in medium 2.

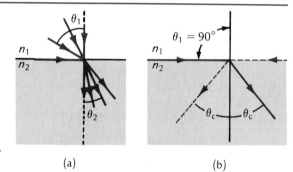

Figure 33-16. (a) The range of possible refraction angles θ_2. (b) Critical angle θ_c.

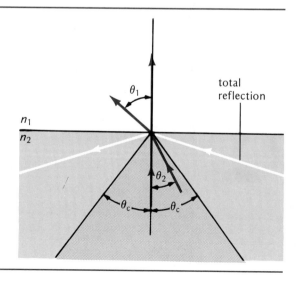

Figure 36-17. Total internal reflection occurs for $\theta_2 > \theta_c$.

Imagine now that all ray directions in Figure 36-16 are reversed, as shown in Figure 36-17. A ray incident upon the interface from medium 2 at angle θ_c travels in medium 1 along the interface. All rays incident at lesser angles go through to medium 1. But for θ_2 greater than θ_c, it is impossible for a ray to be refracted into medium 1. All rays incident at an angle greater than the critical angle will be *totally reflected* into the optically more dense medium. For example, at an air-water interface, $\theta_c = \sin^{-1}(1/1.33) = 49°$.

Suppose you are underwater in a pool with a perfectly flat water-air surface. What do you see when you look up? It would be a transparent circular hole (refraction of all rays above the surface) surrounded by a mirror (total reflection). Total internal reflection has important applications. Suppose light is incident upon a glass prism, as in Figure 36-18 at an angle greater than θ_c. The ray is reflected from the *interior* face as from a perfectly reflecting mirror. Likewise, the particular brilliance of a gem, such as a diamond, arises from its very large refractive index (2.42), its high dispersion, and especially the multiply reflected rays. An *optical fiber* is one form of a *light pipe*, a device for confining light within the refracting material by total internal reflection. See Figure 36-19.

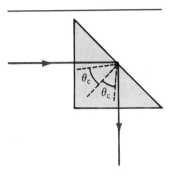

Figure 36-18. Total internal reflection in a 45°-90°-45° prism.

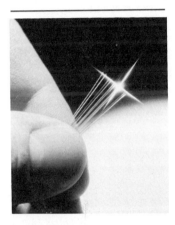

Figure 36-19. Optical fiber photograph.

Example 36-4. Will total internal reflection take place in a 45°-90°-45° prism (Figure 36-18) made of the heaviest flint glass ($n = 1.89$) and immersed in (a) air? (b) in water?

(a) The critical angle for glass (1.89) in air (1.00) is, from (36-9),

$$\theta_c = \sin^{-1}\left(\frac{n_1}{n_2}\right) = \sin^{-1}\left(\frac{1}{1.89}\right) = 31.9°$$

Since θ_c is less than 45°, total reflection does take place.

(b) The critical angle for glass (1.89) in water (1.33) is

$$\theta_c = \sin^{-1}\left(\frac{1.33}{1.89}\right) = 44.7°$$

The critical angle is just slightly less than 45°, so that with very careful alignment, total internal reflection can still take place.

Summary

Definitions

Ray optics: optical effects accounted for entirely by ray tracing, with distinctive wave effects ignored; appropriate only when wavelength is far less than dimensions of apertures or obstacles.

Index of refraction n of material in which speed of light is v:

$$n \equiv \frac{c}{v} \qquad (36\text{-}4)$$

where c is the speed of light in a vacuum.

Relative refractive index of medium 2 with respect to medium 1:

$$n_{21} \equiv \frac{n_2}{n_1} \qquad (36\text{-}5)$$

Dispersion: variation of refractive index with wavelength (typically n decreases with an increase in λ).

Total internal reflection: reflection within the slow medium of all light incident upon an interface between two media. It arises when there can be no refraction in the faster medium (condition given below).

Fundamental Principles

Reciprocity principle: If light goes from A to B by some route, light goes from B to A along the same route; more simply, arrowheads on rays may be reversed.

Laws of reflection and refraction, both of which can be deduced from Huygens's principle:

$$\text{Reflection} \quad \theta_1 = \theta_1' \qquad (36\text{-}1)$$

Angles of incidence and reflection are the same.

Refraction (Snell's law) $n_1 \sin\theta_1 = n_2 \sin\theta_2$ (36-2)

where all angles are measured relative to the normal to the interface between media and the n's are the respective refractive indices.

Important Results

When a wave goes from a medium to another, the frequency is unchanged whereas the wavelength is changed by the same factor as the wave propagation speed:

$$\frac{\lambda_1}{\lambda_2} = \frac{v_1}{v_2} \quad (36\text{-}3)$$

Critical angle θ_c (within the medium of larger refractive index) for total internal reflection:

$$\sin\theta_c = \frac{n_1}{n_2} \quad \text{where } \frac{n_1}{n_2} < 1 \quad (36\text{-}9)$$

Problems and Questions

Section 36-4 Reflection

· **36-1 P** A person views an object through a periscope as sketched in Figure 36-20. How far away from the observer is the image he sees?

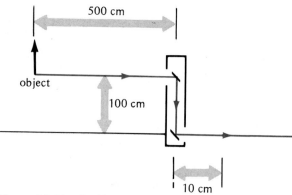

Figure 36-20. Problem 36-1.

· **36-2 P** A bird flies directly toward its image in a mirror at 10 m/s. What is the speed of the bird relative to its image?

· **36-3 P** A person 180 cm tall stands in front of a vertical wall mirror. What is the minimum height of mirror, measured from top to bottom of the mirror, that will allow him to see all of himself in the mirror?

· **36-4 Q** Suppose that when standing in front of a vertical wall mirror, you can see only half of your body. Which of the following would enable you to see more of your body in the mirror?
(A) Move closer to the mirror.
(B) Move away from the mirror.
(C) Stand on chair.
(D) None of the above.

: **36-5 Q** The image you see in an ordinary mirror is reversed left-right but not up-down. Why?

· **36-6 P** Many sensitive instruments rely on the measurement of a very small angular displacement. Frequently the angle is measured by mounting a small mirror on the rotating component and reflecting a light beam from the mirror. Show that, if the mirror rotates through angle θ, the light beam is deflected through angle 2θ.

: **36-7 P** (a) Show that the power of the signal received at the focus of a parabolic reflector is proportional to the square of the radius R of the antenna. (See Figure 36-21.) (b) The radius of the parabolic mirror in a reflecting telescope determines the intensity of the image registered on a photographic plate. If it is assumed, for simplicity, that distant stars have the same power output, how does doubling the radius of the mirror affect the range of stars that can be "seen"?

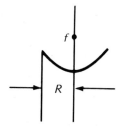

Figure 36-21. Problem 36-7.

: **36-8 Q** An ellipse can be defined as the locus of points the sum of whose distances from two fixed points is a constant. Show that any wave disturbance originating at one focus is focused at the other (Figure 36-22).

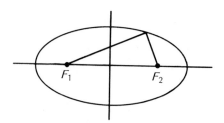

Figure 36-22. Question 36-8.

: **36-9 P** Two plane mirrors make an angle θ (Figure 36-23). Locate all the images for
(a) $\theta = 90°$, (b) $\theta = 60°$, (c) $\theta = 30°$.

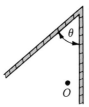

Figure 36-23. Problem 36-9.

: **36-10 P** Two adjacent walls of a room and the floor are covered with mirrors. What is the maximum number of images that can be seen with this arrangement?

: **36-11 P** A rangefinder can be constructed of two mirrors, as shown in Figure 36-24. Mirror M_1 is fixed and half-silvered; light can pass through it along path PM_1O. Light can also be reflected along path PM_2M_1O. An observer at O sees two images corresponding to light rays that have traveled the two paths. The two images can be made to coincide if mirror M_2 is rotated about a vertical axis; this will occur when the angle between the two mirrors is θ. Show that the distance d to P is then given by $d = a/(\tan \theta)$. The separation a of the mirrors and θ can be accurately measured, thereby allowing d to be determined accurately also.

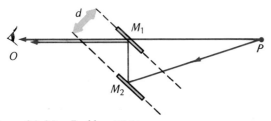

Figure 36-24. Problem 36-11.

: **36-12 P** A woman whose eyes are 150 cm above the floor stands next to a child 80 cm tall. They stand 2 m from a vertical wall mirror. What is the minimum size of mirror (from top to bottom) that will enable the woman to see all of the child?

: **36-13 P** A ray of light leaves a source at S, is reflected at a mirror surface and then travels to point O (Figure 36-25). Show that (a) the time for the light to travel this path is $t = (1/c)(y_1 \sec \theta_1 + y_2 \sec \theta_2)$. (b) Show that this time is a minimum when $\theta_1 = \theta_2$. (This illustrates a general principle called Fermat's law of least time, according to which the path followed by a light ray in going from one point to another is one for which the time is least. See also Problem 36-41.)

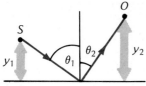

Figure 36-25. Problem 36-13.

: **36-14 P** A parallel beam of light is incident on the apex of a prism of angle ϕ (Figure 36-26). Show that the angle between the two reflected beams is 2ϕ.

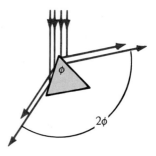

Figure 36-26. Problem 36-14.

Section 36-5 Index of Refraction

· **36-15 Q** The speed of light in water is about $\frac{3}{4}c$. When a beam of light passes from air into water, its frequency
(A) remains the same but its wavelength is reduced to three-fourths that in air.
(B) is three-fourths that in air but its wavelength remains unchanged.
(C) and wavelength are unchanged.
(D) and wavelength are reduced by a factor of $\frac{3}{4}$.
(E) and wavelength are increased by a factor of $\frac{4}{3}$.

· **36-16 P** Light has a wavelength of 460 nm in crown glass ($n = 1.52$). What are the wavelength and frequency in (a) a vacuum? (b) lithium fluoride ($n = 1.37$)? (c) magnesium oxide ($n = 1.75$)?

· **36-17 P** Light travels through carbon bisulfide at a speed of 1.84×10^8 m/s. What is the index of refraction of this material?

· **36-18 P** The index of refraction of magnesium oxide is 1.75. What is the speed of light in this material?

· **36-19 P** The index of refraction of the eye is 1.33. What is the speed of light of wavelength 540 nm when passing through the eye?

· **36-20 P** Light of wavelength 500 nm passes through glass of index of refraction 1.4. What is the frequency of the light when it is in the glass?

: **36-21 Q** When sunlight is refracted by a piece of broken glass, you can often see sparking colors in the transmitted light. Why is this?
(A) Different colors are absorbed in different amounts by the glass.
(B) Various light rays travel paths of different lengths in the glass.
(C) The speed of light in glass varies slightly with the frequency of the light.
(D) The sunlight stimulates the glass to emit fluorescent light of various wavelengths.
(E) Light of different wavelengths is scattered by different

amounts depending on the angle of scattering (that is, on the angle at which you view the light).

: **36-22 Q** At a long, straight ocean beach, you see that waves come in almost parallel to the shoreline, no matter what the wind direction on the open sea where the waves are formed. Which of the following best explains this effect?
(A) Water waves move faster in deep water than in shallow water.
(B) Water waves move faster in shallow water than in deep water.
(C) It is only the component of the wind perpendicular to the shore line that causes waves.
(D) Water wave velocity varies with wave frequency.
(E) The amplitude of a wave decreases as it moves away from its source.

: **36-23 Q** Figure 36-27 is a sketch of what appears in a satellite photograph of a harbor. The wave speed of water waves in shallow water varies with $h^{1/2}$, where h is the depth of water. What can you deduce about the configuration underwater from the wavefronts shown in this picture?

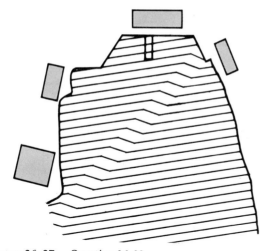

Figure 36-27. Question 36-23.

: **36-24 P** Any curve can be approximated over a small region by a circular arc whose radius R, the radius of curvature of the curve, is $R = [1 + (dy/dx)^2]^{3/2}/(d^2y/dx^2)$. The derivatives are evaluated at the point at which the curve is to be fitted. The equation of a parabola can be written $x^2 = 4Fy$, where F is the distance between the focus and the vertex of the parabola. Show that the focus of a parabola lies at $\frac{1}{2}R$, where R is the radius of curvature of the parabola at its vertex.

Section 36-6 Refraction

: **36-25 P** A penny rests at the bottom of a container that holds a 4-cm depth of water ($n = 1.33$) on top of which floats a 2-cm layer of oil ($n = 1.60$). What is the apparent depth of the penny when viewed from directly above?

· **36-26 P** A ray of light strikes a glass plate at an angle of 60° with the surface. The index of refraction of the glass is 1.45. What is the angle between the refracted ray and the surface?

· **36-27 P** A light ray strikes the surface of water in a beaker at an angle of 30° above the horizontal. At what angle with the vertical does the ray travel in the water?

: **36-28 P** Two 45°-45°-90° glass prisms are joined as shown in Figure 36-28. Their refractive indices are 1.45 and 1.60. By what angle will a light ray normally incident be deviated?

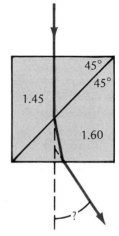

Figure 36-28. Problem 36-28.

: **36-29 P** Light is incident on a glass of index of refraction n at an angle of incidence θ. What should be the angle of incidence in terms of n for the angle of refraction to be $\frac{1}{2}\theta$?

: **36-30 P** A fish rests at the bottom of a pool of water 3 m deep and at a distance of 4 m from the edge of the pool. How far back from the edge of the pool must a fisherman 180 cm tall stand if he is not to be seen by the fish?

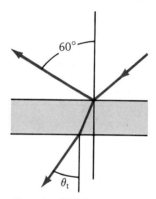

Figure 39-29. Question 36-31.

: **36-31 Q** A laser beam is incident on a plate of glass with parallel sides and index of refraction 1.4. Some light is reflected at an angle of 60° to the normal, and some transmitted through the glass. The situation is sketched in Figure 36-29, although the angles are not necessarily drawn correctly. What is the value of the angle θ_t made by the transmitted beam?
(A) 60°.
(B) 30°.
(C) 45°.
(D) 38°.
(E) θ_t cannot be determined, since the angle of incidence is unknown.

: **36-32 P** At what angle relative to the normal must a ray strike the interface between a medium with refractive index n_1 and a second medium with refractive index n_2 so that the reflected and transmitted rays will be at right angles?

: **36-33 P** A nautical buoy 1 m tall floats with 40 cm of its length projecting above a lake at a point where the water is 4 m deep. If the sun is 30° above the horizon, how long is the shadow of the buoy on the bottom of the lake? The index of refraction of water is 1.33.

: **36-34 P** Light is incident on a parallel plate of glass of thickness t and index of refraction n at an angle of incidence θ_1. Show the transmitted ray undergoes a lateral displacement $t \sin(\theta_1 - \theta_2)/\cos \theta_2$ where $\sin \theta_2 = (1/n) \sin \theta_1$.

: **36-35 P** A light ray passes through several parallel slabs of material with different refractive indices. Total internal reflection does not occur at any interface. Show that the final transmitted ray is parallel to the incident ray quite apart from the order the various slabs are arranged in; for example, 4, 3, 1, 2 gives the same effect as 1, 2, 3, 4.

: **36-36 P** A light ray is incident at 45° on a 45°-45°-90° prism as shown in Figure 36-30. The refractive index of the glass is 1.5. What is angle ϕ?

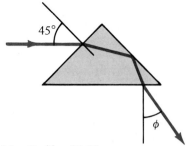

Figure 36-30.   Problem 36-36.

: **36-37 P** A ray passing through a prism is deviated by an angle that depends on the ray's angle of incidence with the first surface, on the angle of the prism, and on the index of refraction. Show that the deviation is a minimum when the ray passes symmetrically through the prism, that is, when $\theta_1 = \theta_2$ in Figure 36-31. (*Hint:* Consider reciprocity.)

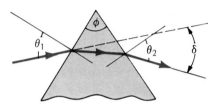

Figure 36-31.   Problems 36-37, 36-38, 36-39.

: **36-38 P** When a light ray passes symmetrically through a prism, it experiences minimum deviation (Question 36-37). Show that if δ is the angle of deviation and ϕ is the angle of the prism,

$$n = \frac{\sin \tfrac{1}{2}(\delta + \phi)}{\sin \tfrac{1}{2}\phi}$$

This provides a direct way of measuring the refractive index of the material of the prism.

: **36-39 P** Show that for a thin prism (ϕ small) with the incident light close to the normal (θ small), the deviation angle δ is independent of the angle of incidence and is given by $\delta = (n - 1)\phi$. See Figure 36-31.

: **36-40 P** Snell's law ($\sin \theta_1/\sin \theta_2 =$ a constant) may be derived from a particle model of light in which it is assumed that the particles of light travel at constant velocity within any uniform medium and that the component of a particle's velocity parallel to an interface between two refracting media is unchanged. See Figure 36-32. Note that $v_{t1} = v_{t2}$. (a) Derive Snell's law from the particle model and show that $\sin \theta_1/\sin \theta_2 = v_2/v_1$, where v_1 and v_2 are the respective particle speeds in media 1 and 2. (b) Show that, according to this particle model, the speed of light propagation through a refracting medium exceeds that through vacuum, in contradiction to the experimental findings.

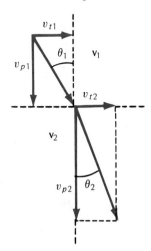

Figure 36-32.   Problem 36-40.

: **36-41 P** Ray optics—the laws of reflection and of refraction—are special examples of a remarkable general principle called the *principle of least time*, or *Fermat's principle*, and first propounded by P. Fermat (1608–1665) in 1650. In its simplest form Fermat's principle says that a ray going from point A to point B will take the path that corresponds to minimum travel time between A and B. The rectilinear propagation of light through a uniform medium is clearly consistent with this principle since a straight line between two points is the shortest distance between them and also the path of least time. To show that Snell's law is also a consequence of Fermat's principle consider Figure 36-33. Here a ray goes from point A in medium 1, to the interface at point B and finally to point C in medium 2. Think of the distance from fixed point D to point B to be the variable distance x. Write a relation for the total travel time from A to C, find the value for x corresponding to minimum travel time, and show that Snell's law is the consequence.

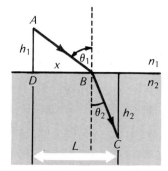

Figure 36-33. Problem 36-41.

Section 36-7 Total Internal Reflection

· **36-42 P** If you look up at the undisturbed surface of water from the bottom of a swimming pool, you will see a circular transparent hole looking out on the sky surrounded by a shiny mirror. What is the radius of the hole when your eye is 3 m below the surface?

· **36-43 P** What is the maximum angle ϕ of incidence at which light can enter the end of a glass fiber in air shown in Figure 36-34 if the light is not to escape through the sides of the fiber? The index of the fiber is 1.40, and the fiber is straight.

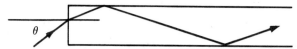

Figure 36-34. Problem 36-43.

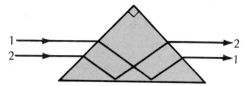

Figure 36-35. Problem 36-44.

· **36-44 Q** Figure 36-35 shows two parallel rays incident on a 45°-45°-90° prism of index of refraction 1.5. Trace the rays to show that they emerge inverted. Of what practical use is a prism of this type?

: **36-45 P** A ray of light is moving in a plane parallel to one face of a cube of glass of index n. The ray strikes another interior wall at an angle of incidence θ, as shown in Figure 36-36. To what range of values is θ restricted if the ray is to remain trapped within the cube?

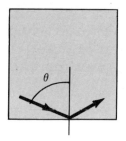

Figure 36-36. Problem 36-45.

: **36-46 Q** We can deduce that when light in material 1 is incident on the interface with material 2, total internal reflection will occur
(A) whenever $n_1 > n_2$, for all rays.
(B) whenever $n_1 < n_2$, for all rays.
(C) by examining the law of reflection, $\theta_1 = \theta_1'$.
(D) by examining the law of refraction, $n_1 \sin \theta_1 = n_2 \sin \theta_2$.
(E) independent of the values of n_1 and n_2 so long as the angle of incidence is greater than 49°, the critical angle.

: **36-47 Q** Sketched in Figure 36-37 is a light ray passing through a pair of prism binoculars. In such an instrument,
(A) the two perpendicular faces of the prism are silvered to make them reflecting.
(B) the index of refraction of the glass is not an important design consideration, so long as the prism angles are correct.
(C) the index of refraction of the prism should be greater than 1.414.

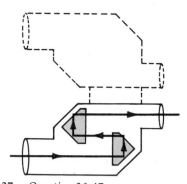

Figure 36-37. Question 36-47.

(D) the index of refraction of the prism should be greater than 1.522.
(E) the index of refraction of the prism should be less than 1.500.
(F) the prism increases the overall magnification.

: **36-48 P** Three light rays are directed into a prism with refractive index n, as in Figure 36-38. (a) What is the minimum value of n if ray No. 2 is to be totally internally reflected? Call this value n_c. (b) If ray No. 1 is to be totally reflected, is the required minimum index larger or smaller than n_c? (c) If ray No. 3 is to be totally reflected, is the required minimum n larger or smaller than n_c?

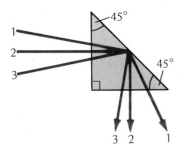

Figure 36-38.   Problem 36-48.

Supplementary Problem

36-49 P On a hot summer day you see mirages ahead of you on the surface of a highway. Typically, you see what appears to be a pool of water, but you may even see what appears to be a reflection of an approaching car. The effect is a manifestation of total internal reflection. A layer of air just above the hot, black roadway is significantly hotter than the air a meter or so above the pavement. The index of refraction of air is related to its density ρ by a relation of the form $n^2 - 1 = (\text{constant})\rho$.

Thus light angling down from above emerges from cool air (large n) and is reflected from hot air (smaller n). If the angle of incidence is large enough, the light will be totally "internally" reflected. This puts a limit on how close the near edge of the apparent "pool of water" can be to you. As you approach it, it will keep moving away. For simplicity, suppose there is a sharp temperature difference of 12 C° within a few centimeters of the pavement. Assume that the cooler air is at 300 K and has $n = 1.003$.

(a) Show that in Figure 36-39 $\phi \approx \sqrt{2(n-1)(\Delta T/T)}$. (b) If your eyes are 1.5 m above the roadway as you are driving, how far away will the near edge of the mirage appear?

Figure 36-39.   Problem 36-49.

37 Thin Lenses

37-1 Focal Length of a Converging Lens
37-2 Ray Tracing to Locate a Real Image
37-3 Ray Tracing to Locate a Virtual Image
37-4 Diverging Lens
37-5 Lens Combinations
37-6 The Lens Maker's Formula
37-7 Lens Aberrations
37-8 Spherical Mirrors (Optional)
Summary

Tracing rays through a whole succession of interfaces between transparent materials that differ in refractive index—the basic problem to be solved in designing a high-quality multiple-element thick lens—is easy in principle (just Snell's law) but tediously difficult in practice. We won't tackle it. Instead we consider only the far simpler situation with a single, very thin lens. This special case illuminates most of the fundamental ideas of image formation and allows the basic design principles of optical instruments to be understood.

37-1 Focal Length of a Converging Lens

Suppose a ray passes through a plate of glass with *parallel* surfaces. The ray is refracted at each interface but the emerging ray is *parallel* to the incident ray [Figure 37-1(a)]. The emerging ray is displaced laterally but does not deviate in direction. On the other hand, when a ray encounters a slab of glass with nonparallel faces (a prism), the emerging ray is deviated relative to the incident ray [Figure 37-1(b)].

Now consider the structure shown in Figure 37-2; it consists of two prisms

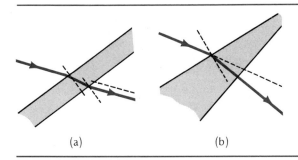

Figure 37-1. (a) Refraction through a slab with parallel faces produces a lateral displacement but no deviation. (b) A ray is deviated by a prism.

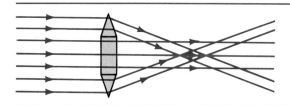

Figure 37-2. Focusing by a crude lens.

and a plate with parallel faces. A beam of horizontal rays incident from the left passes through this rather crude lens. The emerging rays intersect in a relatively small region to the right. Rays through the center are undeviated. Rays through the top prism are deviated down, and rays through the bottom prism are deviated up, and they emerge at a single deviation angle. We wish to devise a structure in which all the parallel incident rays intersect at a single point. It is clear what is required: the rays' deviation must increase gradually upward (or downward) from the center of the lens, not abruptly as in Figure 37-2. In practice, lenses are made with *spherical* surfaces; this shape is easy to grind. Here we consider only a *thin lens*. This means that its thickness is small compared with its radius, as shown in Figure 37-3.

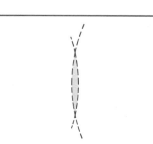

Figure 37-3. A lens with spherical surfaces.

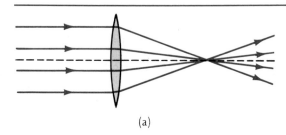

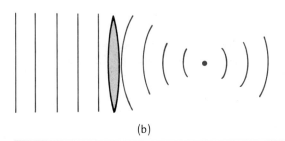

Figure 37-4. (a) Parallel incident rays and (b) plane wave, focused by a lens.

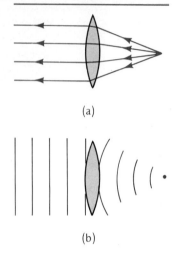

Figure 37-5. Ray and wave-propagation reversal of Figure 37-4. (a) Rays from the focal point emerge after refraction as parallel rays. (b) Diverging spherical wave fronts emerge after refraction as plane wave fronts.

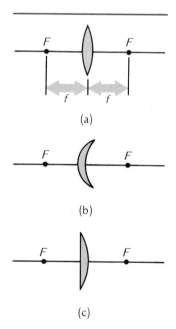

Figure 37-6. Examples of converging lenses: (a) double convex; (b) concave-convex; and (c) plano-convex.

Figure 37-4(a) shows a thin lens with two convex spherical surfaces. Incident rays initially parallel to the horizontal axis will, after passing through the lens, intersect at a single point, or at least in a very much smaller region than shown in Figure 37-2. (The proof from Snell's law is given in Section 37-6.) This intersection point is known as a *principal focal point*, or *principal focus*, of the lens. Suppose that a source is placed at an infinite (or at least a very large) distance from the lens. Then the rays incident upon the lens are effectively parallel over a small solid angle, and these parallel rays will, after traversing the lens, intersect at the principal focus. The focusing originates from the circumstance that as we go up (or down) from the center of the lens, each ray is deviated slightly more than the one below (or above) it. The behavior of wave fronts associated with the rays is shown in Figure 37-4(b). The wavefronts undergo a change in curvature at each of the two interfaces. The incident wave fronts are plane, whereas the emerging wave fronts are converging spherical wave fronts that collapse into a point at the principal focus (and thereafter expand outward as diverging spherical wave fronts). The wave fronts change shape because the relatively thicker central portion of the lens slows a wave front more than the thinner portions near its outer edge do.

Any focal point can be defined in several equivalent ways:

- The point at which *all rays intersect*.
- The point at which *wave fronts collapse to a point* (changing from converging into diverging wave fronts).
- The point at which the beam has *maximum intensity*.

Recall the principle of reciprocity, or time reversal (Section 36-2); it says that we can reverse the directions of rays (or imagine time to run backward) without changing the light paths. Applied to Figure 37-4, this means that if a point source is placed at the principal focus, rays diverge from this point and emerge from the lens as a beam of parallel rays [Figure 37-5(a)]. Equivalently, spherical wave fronts diverging from the principal focus and passing through the lens emerge as plane wave fronts [Figure 37-5(b)].

Suppose that we flip over the thin lens so that its left and right faces are interchanged. Rays then pass through the lens in opposite direction. But the incident parallel rays again converge to a focus at the same distance from the lens's center. Every lens has two principal focal points, one on each side. Both principal foci, denoted by F, are at the same distance f, the *focal length*, from the center of a thin lens. See Figure 37-6(a).

This result is general; it holds not only for the thin lens with two convex surfaces that we have concentrated on thus far, but also for a lens, such as that shown in Figure 37-6(b), with a concave and a convex surface, and for a plano-convex lens (the plane surface with an infinite radius of curvature) as in Figure 37-6(c). All these are converging lenses. The essential requirement is that the lens be thicker at its center than at its edges. (Clearly, the focal length depends on the radius of curvature of the two surfaces and on the refractive index of the lens material. The lens maker's formula, which gives f in terms of these parameters, is derived in Section 37-6.)

Imagine now that the lens is tilted a little, as in Figure 37-7(a), so that its symmetry axis (shown dashed) no longer lies along the direction of incident parallel rays. For a small-tilt angle, nothing changes; the rays focus at essentially the same point as before, a distance f from the center of the lens. Now look at the very same situation but with the lens axis, and all else, turned until it

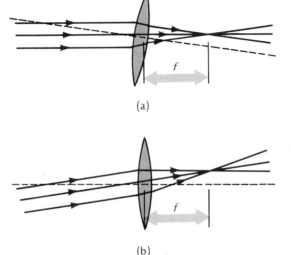

Figure 37-7. (a) The location of the focal point does not change when the lens is tilted slightly. (b) Part (a) drawn with the lens axis horizontal. It shows that oblique rays also come to a focus at distance f from the lens.

is horizontal, as in Figure 37-7(b). For what are now obliquely incident parallel rays, the focal point is displaced transversely from the lens axis. Clearly, the ray through the lens's center is undeviated; it passes, in effect, through two parallel surfaces. Thus, rays from an infinitely distant source are brought to focus in a *plane*, the *focal plane*, a distance f from the lens. This result holds, however, only if the angle between the oblique rays and the lens axis is small. Such rays, nearly parallel to the lens axis, are called *paraxial rays*. We shall assume hereafter that the lens is very thin and the rays are paraxial.

The type of lens discussed thus far is known as a *converging* lens. (Strictly, a lens thicker at its center than at its edges is a converging lens only if its material is optically more dense than its surroundings; an example is a glass lens for visible light when the lens is immersed in air.) Any converging lens increases the degree of convergence of rays and wave fronts passing through it, or decreases the degree of their divergence.

37-2 Ray Tracing to Locate a Real Image

We are now prepared to deal with a general problem. What happens to the rays from a point source located at any location along the lens axis? The object, some

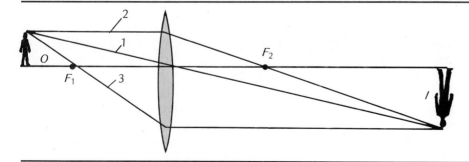

Figure 37-8. Formation of image I by rays 1, 2, and 3 from object O.

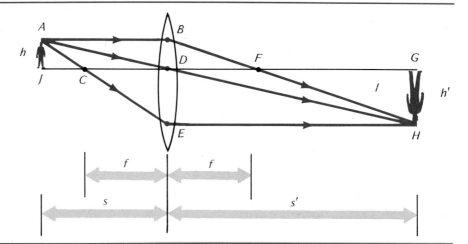

Figure 37-9. Geometrical relations among object distance s, image distance s′, and focal length f.

luminous source, is represented by O. It is located near a thin converging lens, but farther than the focal length, as shown in Figure 37-8. We want to find the place where the light from the top of the person representing object O comes to a focus after passing through the lens. Three rays are easy to trace:

• Ray 1 passes through the *center* of the thin lens *unchanged*. This ray is not deviated—the lens faces are parallel. Also, this ray is not displaced laterally by an appreciable distance—the lens is very thin.
• Ray 2 *incident parallel* to the lens axis is deviated by the lens, so that the emerging ray goes *through principal focus* F_2 on the far side.
• Ray 3, aimed to pass *through* the *near principal focus* F_1, is deviated and *emerges parallel* to the lens axis.

These three rays intersect, or focus, at a single point.* Indeed, *all* other paraxial rays from the upper tip of O intersect to form a point image of O. We have traced rays from the top of the head representing object O. When we choose some other point on an extended object lying in the same transverse plane, we can again find the corresponding image point. For paraxial rays through a thin lens, we can obtain the entire image I of object O.

Finding the image of an object is simple. All we need is a prior knowledge of the lens's focal length f; then we can locate the image by drawing the three rays, as in Figure 37-8. Ray tracing, which is strictly a geometrical procedure, is not always practicable, however. To deal with general situations, we wish to find the mathematical relation between three important quantities: the focal length f, object distance s of the object from the lens, and image distance s' of the image from the lens.

Consider Figure 37-9. It is merely Figure 37-8 redrawn, with identifying letters added. The object distance s is AB, the image distance s' is $EH = DG$, and the focal length f is $CD = DF$. The object height h is $AJ = BD$; the image height h' is $GH = DE$. Triangles AJD and HDG are similar. Therefore,

---

* The rays in Figure 37-8 are drawn as if the lens were infinitesimally thin, with a single straight ray to the plane of the lens and another single straight ray away from the plane of the lens. In actuality, every ray undergoes *two* refractions at the lens; there is a refraction at each of the two surfaces.

$$\frac{AJ}{JD} = \frac{HG}{DG}$$

$$\frac{h}{s} = \frac{h'}{s'} \tag{37-1}$$

The ratio of image-object *distances*, s'/s, is equal to the ratio of image-object *sizes*, h'/h. The ratio of the linear dimensions of image-to-object, h'/h, is known as the *lateral magnification*.

Triangles BDF and HGF are also similar. Therefore,

$$\frac{DF}{BD} = \frac{GF}{GH}$$

$$\frac{f}{h} = \frac{s'-f}{h'}$$

Using (37-1) in this relation gives, after a little algebra,

$$\frac{f}{s} = \frac{s'-f}{s'} = 1 - \frac{f}{s'}$$

and dividing by f and rearranging yield

$$\frac{1}{f} = \frac{1}{s} + \frac{1}{s'} \tag{37-2}$$

This is the formula for computing image locations formed by a thin lens. Everything we can do with it can be done equally well by ray construction [that is how we derived (37-2)]. In working any problem in which an image is formed by a lens, it is always advisable to sketch, at least roughly, the rays forming the image, as well as to do the numerical calculation.

Example 37-1. An object is 18 cm from a converging lens with a focal length of 12 cm. Where is the image? How does the object's size compare with the size of the image?

With $f = 12$ cm and $s = 18$ cm, we get from (37-2) that $s' = 36$ cm. The ray diagram is that shown in Figure 37-8 and 37-9.

Since the image distance is twice the object distance, the lateral dimensions of the image are twice the corresponding dimensions of the object. By the same token, the transverse area of the image is four times the area of the object.

Suppose that we move the object far from the lens, so that $s = \infty$ in (37-2). Then we find $s' = f$; that is, the image of a very distant object is at a principal focus, in accord with its definition. Conversely, if $s = f$, then $s' = \infty$.

The image shown in Figures 37-8 and 37-9 is said to be a *real* image; light rays actually pass through this location. If a sheet of paper were placed a distance s' from the lens, we should see an actual focused image on it. This image would be inverted in the transverse focal plane; that is, up and down would be interchanged relative to the object, and so too would be left and right. Equation (37-2) shows that as s increases, s' decreases, and the opposite, so that when the object is shifted along the lens axis, its image is displaced in the same direction. Thus, the image of a three-dimensional object, although inverted in the transverse plane, is not inverted along the lens axis.

Suppose that we apply the reciprocity principle to Figures 37-8 and 37-9. Then the rays are reversed and our original image becomes an object, and vice versa. These two locations, where object and image switch locations, are said to be *conjugate* points. Note also that when $s = 2f$, then $s' = 2f$. For this special case, the lens merely inverts the object without changing its size.

Many familiar optical devices involve a single converging lens forming a real image. Most familiar is the eye; it forms a real image on the retina. The image distance s' from the eye lens to the retina is fixed; objects at various distances from the eye are brought into focus on the central plane of the retina when the eye muscles change the focal length f of the eye lens by changing the radius of curvature of the lens surfaces. A simple camera also forms a real image; it is in the focal plane of the photographic film. On the other hand, a projector forms a much enlarged and inverted image of transparent film on the screen.

Example 37-2. A normal human eye can focus on objects from 25 cm to infinity by changing the curvature, and therefore the focal length, of the eye lens. The process is called *accommodation*. A far-sighted person can see far objects in focus, but not near ones. Suppose that the eye of a certain far-sighted person can see no objects in focus closer than 200 cm (that person's *near point*). What focal length of contact lens will bring objects as close as 25 cm into focus?

The fixed distance from the eye lens to the surface of the retina is designated s'. Then, with an object 200 cm from the eye and without a corrective lens we have, using lens equation (37-2)

$$1/f_e = 1/200 \text{ cm} + 1/s'$$

where f_e is the focal length of the unaided eye lens.

In Section 37-6 it is shown that the equivalent focal length f of two thin lenses close together is

$$1/f = 1/f_1 + 1/f_2 \qquad (37\text{-}5)$$

When a contact lens with focal length f_{cl} is used with an object 25 cm away, applying the lens equation again gives

$$1/f_{cl} + 1/f_e = 1/25 \text{ cm} + 1/s'$$

Eliminating f_e and s' from the two equations above gives

$$f_{cl} = 28.6 \text{ cm} = 0.286 \text{ m}$$

Optometrists use the reciprocal of the focal length in meters to express the *lens power* in units of *diopters*, so that the contact lens above would have a lens power of $1/0.286 \text{ m} = +3.5$ diopters.

Converging contact lenses (with positive lens power) correct the vision of a far-sighted person. Without a correcting lens the image is formed behind the retina surface; with a converging lens added, the focussed image is brought forward, closer to the eye lens, to the surface of the retina. A near-sighted person requires a diverging contact lens (with a negative lens power); here the image is moved away from the eye lens and toward the retina.

36-3 Ray Tracing to Locate Virtual Image

A new phenomenon occurs when an object is placed closer to a converging lens than the principal focal point. See Figure 37-10. We can locate the image geometrically by drawing exactly the same three rays as before:

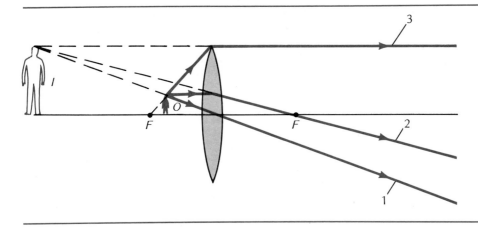

Figure 37-10. Formation of virtual image I by rays 1, 2, and 3 from object O.

- Ray 1 through the lens center, undeviated.
- Ray 2, initially parallel to the lens axis, through the far focal point.
- Ray 3, so drawn that its direction after leaving the object is the same as that of a ray starting at the near focal point and passing through the top of the object. This ray emerges from the lens parallel to the axis.

These three rays do not intersect. What sort of image is now formed? What does an eye see when looking from the right toward the lens?

The eye (or a camera) is naive in the following sense: it interprets any ray reaching it as always having traveled strictly along a straight line. Said differently, the eye recognizes only the final directions of rays entering it. Therefore, the three rays appear to have originated from that point on the left of the lens where their backward extensions intersect. They appear to come from what is called a *virtual image I*. A sheet of paper placed at the location of *I* shows no image. But a person viewing the object through the lens would see an erect, enlarged, and virtual image. Used in this fashion, with an object closer to the lens than the focal length, a converging lens is a *magnifying glass*, or *simple magnifier*.

The lens equation, (37-2), can also be used to compute the image distance s' for a virtual image. Here s' is taken to be *negative;* this can be proved in detail by analyzing the geometry of Figure 37-10 in the fashion of Figure 37-9.

Example 37-3. An object is placed 9 cm from a converging lens of 12 cm focal length. Where is the image and what is its character?

With $f = 12$ cm and $s = 9$ cm, we get by substituting in (37-2) that $s' = -36$ cm. The image is virtual, erect, and with 4 times the lateral dimensions of the object, as shown in Figure 37-10.

36-4 Diverging Lens

Thus far we have considered converging lenses; such lenses are thicker in the center than at the edges (when their material has a higher refractive index than the surrounding medium). Now consider just the reverse: a lens with spherical surfaces thinner at its center than at its edges, as shown in Figure 37-11. Such a lens is a *diverging* lens. It increases the degree of divergence of rays and wave fronts passing through it. Incident parallel rays diverge after passing through

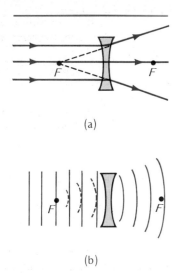

Figure 37-11. The divergence of (a) the rays and (b) the wave fronts, by a diverging lens.

the lens and appear to have originated from a point source at the principal focus on the left side of the lens (Figure 37-11(a)). Said a little differently, incident plane waves emerge from a diverging lens as diverging spherical wave fronts (Figure 37-11(b)). We can again use the basic lens equation to relate image distance, object distance, and focal length, provided that the focal length of the diverging lens is taken to be *negative*. The object distance s is, as before, taken as positive. Again there are two principal foci, one on each side of the lens, and the focusing of paraxial rays is independent of the tilt angle of the lens or of the face of the lens exposed to the object.

We locate an image produced by a diverging lens with the same ray-construction procedure as before, as shown in Figure 37-12.

- Ray 1 passes undeviated through the lens's center.
- Ray 2, parallel to the lens axis, emerges as if originating from the near principal focus F.
- Ray 3 is "aimed" to go through the far principal focus; therefore this ray is deviated, to emerge parallel to the axis.

The rays emerging from the lens appear to diverge from image I. This image is reduced, erect, and virtual. Indeed, a divergine lens always forms a virtual image of a real object. That I is virtual is again indicated by the fact that the image distance s', as computed from (37-2), is negative.

Example 37-4. An object is placed 15 cm from a diverging lens with a focal length of 12 cm. Where is the image and what is its character?

With $f = -12$ cm and $s = 15$ cm, we find in applying (37-2) that $s' = -6.7$ cm. As shown in Figure 37-12, the image is virtual, erect, and reduced in size relative to the object.

We have concentrated on lenses of the usual variety, which are made of optically dense materials, such as glass, and intended for use with visible light. Lens devices can also be made of optically light materials. For example, Figure 37-13 shows converging and diverging lenses formed by cavities within an optically dense medium. The converging lens here has *concave* surfaces, and

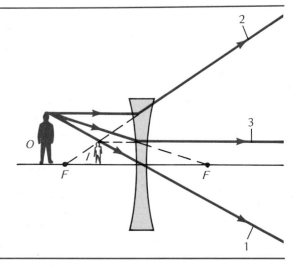

Figure 37-12. Image I formed by rays 1, 2, and 3 from object O for a diverging lens.

the diverging lens *convex* surfaces. As before, a converging lens causes plane wave fronts to collapse to a point, whereas a diverging lens causes plane wave fronts to expand outward.

The term *lens* does not need to be restricted to devices that focus visible light. There are lenses for sound waves and electromagnetic microwaves. Magnetic- and electric-field arrangements that focus a beam of charged particles are referred to respectively as magnetic and electrostatic lenses.

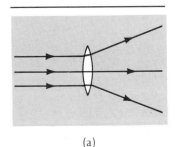

(a)

Example 37-5. The eye of a certain near-sighted person can see in focus no objects farther away than 500 cm (that person's *far point*). The image for far objects is formed in front of the retinal surface; the rays from far objects converge too much. What focal length of contact lens will bring very far objects into focus?

The fixed distance from eye lens to the surface of the retina is designated s', because s' is the image distance. Then, with an object 500 cm from the eye and without a corrective lens, we have, using the lens equation (37-2)

$$1/f_e = 1/500 \text{ cm} + 1/s'$$

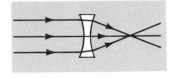

(b)

where f_e is the focal length of the unaided eye lens.

When a contact lens with focal length f_{cl} is added and the object is infinitely far away, we have, in applying the lens equation again

$$1/f_{cl} + 1/f_e = 1/\infty + 1/s'$$

where we have also used (37-5) for the equivalent focal length of two thin lenses in contact.

Figure 37-13. *Lenses formed of a material that is optically less dense than the material in which they are immersed. (a) A double convex but diverging lens. (b) A double concave but converging lens.*

Eliminating f_e and s' from the two equations above gives

$$f_{cl} = -500 \text{ cm}$$

The contact lens must be diverging with a focal length of 5.0 m; the required lens power is $-1/5.0$ m $= -0.2$ diopter.

37-5 Lens Combinations

To trace rays through two or more lenses in sequence, proceed as follows. Treat the image for the first lens as the object for the second lens, then treat the image for the second as the object for the third lens, and so forth.

Astronomical Telescope Figure 37-14 shows a combination of two converging lenses with focal lengths f_1 and f_2. Lens 1 forms a real inverted image I_1 of the object O_1. Image I_1 becomes the object O_2 for lens 2. In the particular arrangement shown here, the object O_2 falls just inside the focal point of the second lens, so that lens 2 acts effectively as a magnifying glass. Consequently, the final image I_2 formed by lens 2 is virtual; it is to the left of lens 2. The rays chosen to find the image I_1 are not continued through the second lens; instead, new rays are chosen, whose deviations through lens 2 are found in the fashion shown in Figure 37-10. Note also that for the purposes of ray construction, the lenses are taken to be of infinite transverse size.

An *astronomical telescope* has the lens arrangement in Figure 37-14. Since object O_1 is far from lens 1, its real image I_1 is examined by lens 2 acting as a magnifying glass. The final image is formed at a great distance from the lens. Lens 1 (closer to the object) is known as the *objective* lens; lens 2 (closer to the eye) is known as the *eyepiece*, or *ocular*. With object O_1 at a great distance from a

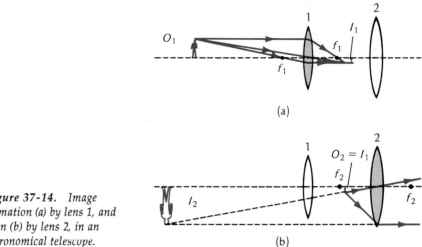

Figure 37-14. Image formation (a) by lens 1, and then (b) by lens 2, in an astronomical telescope.

telescope and a final image I_2 also at a great distance, it is clear that the total distance between the objective lens and the eyepiece is close to $f_1 + f_2$. This is the minimum length of the telescope. (One side of a binocular is basically an astronomical telescope with the rays reflected twice from prisms, as in Figure 36-47, one to reverse up-down and the other to reverse left-right.)

What matters in any optical instrument used to magnify an object is not the size of the image itself but rather the size of the image *formed on the retina of the eye*. The retinal image size is determined, in turn, by the angular spread of rays entering the eye, as shown in Figure 37-15. A proper measure of the magnification by an optical instrument is the *angular magnification*, or *magnifying power*, defined as

$$\text{Angular magnification} = \frac{\theta_I}{\theta_O} \qquad (37\text{-}3)$$

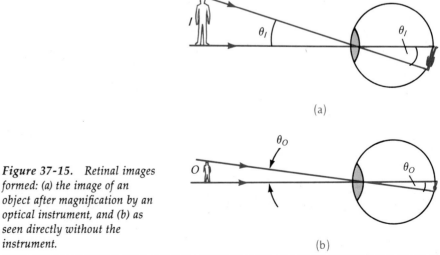

Figure 37-15. Retinal images formed: (a) the image of an object after magnification by an optical instrument, and (b) as seen directly without the instrument.

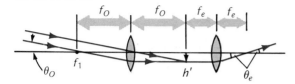

Figure 37-16. Angular magnification by an astronomical telescope.

Here angle θ_I is subtended at the eye by the rays coming from the final image, and angle θ_O is subtended at the eye by the object when viewed directly, without the aid of the instrument. Equivalently, the magnifying power is the ratio of the size of the retinal image formed with an optical instrument to the retinal image size without the instrument. Angular magnification is a more useful criterion than lateral magnification, given by (37-1), which is just the ratio of image size to object size. We can see this as follows. There is *no* angular magnification when the image size is doubled while at the same time the image is located at twice the distance of the object from the viewer.

Now we compute the angular magnification of an astronomical telescope. Figure 37-16 shows the rays incident upon a telescope from a very distant object. The angle θ_o subtended by the object at the objective lens is essentially the same as the angle subtended by the object at the unaided eye. The height of the real image of the objective lens is denoted h'. This image is formed at a distance f_o from the objective lens and a distance f_e from the eyepiece, where f_o and f_e are the focal lengths of the objective lens and eyepiece, respectively. From the geometry of Figure 37-16, we have

$$\tan \theta_o = \frac{h'}{f_o} \quad \text{and} \quad \tan \theta_e = \frac{h'}{f_e}$$

where θ_e is the angle subtended at the eye by the final image. Since the angles θ_o and θ_e are very small,

$$\frac{h'}{f_o} = \tan \theta_o \simeq \theta_o \quad \text{and} \quad \frac{h'}{f_e} = \tan \theta_e \simeq \theta_e$$

But by the definition in (37-3), we have

$$\text{Angular magnification} = \frac{\theta_e}{\theta_o} = \frac{f_o}{f_e} \tag{37-4}$$

A telescope's magnifying power is simply the ratio of the focal lengths of the objective and the eyepiece. High magnification means using a long-focal-length objective lens with a short-focal-length eyepiece.

Microscope The lens combination shown in Figure 37-17 is that for a compound microscope. Here the object O_1 is placed close to but outside of the

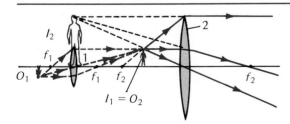

Figure 37-17. A compound microscope.

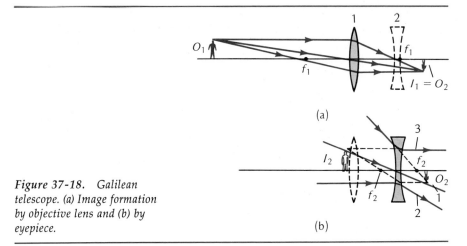

Figure 37-18. Galilean telescope. (a) Image formation by objective lens and (b) by eyepiece.

principal focus f_1 of the objective lens; its image I_1 is enlarged and real. The eyepiece is used as magnifying glass to form a still further enlarged but virtual image I_2. How do we examine an object to see it most clearly? We must make the size of the image *on the retina* as large as possible, so that we bring the object as close to the eye as the focusing properties of the eye will permit, typically 25 cm from the eye. Therefore, a microscope is also most effective if the final image I_2 is located at 25 cm from the eye. The angular magnification of a microscope is θ_I/θ_O, where θ_O is the angle subtended by the object held 25 cm from the unaided eye.

Galilean Telescope A special situation arises when the image formed by the first lens does not lie between two lenses in combination. Consider Figure 37-18(a). If lens 2 were not present, the image I_1 would be formed at the location shown. To locate the final image after rays traverse the second lens, we must regard this image I_1 as a *virtual object* for lens 2. We find the final image I_2 of the virtual object $O_2 = I_1$ in Figure 37-20(b) as follows:

• Ray 1 is so "aimed" that it passes through the center of the diverging lens without deviation.
• Ray 2, initially parallel to the lens axis and aimed at I_1, is deviated, so as to be directed from the near focal point f_2.
• Ray 3 is aimed to pass through the far focal point and therefore emerges from lens 2 parallel to the axis.

The final image I_2 is greatly enlarged, erect, and virtual. The optical device shown in Figure 37-18 comprising a converging objective lens and a diverging eyepiece, is a *Galilean telescope,* or opera glass.

37-6 The Lens Maker's Formula

Here we derive the lens maker's formula. It gives the focal length of a thin lens in terms of the radius of curvature of its two spherical surfaces and the relative index of refraction of its material.

The underlying assumptions are these:

- The lens is very thin.
- All rays are *paraxial;* that is, the angle θ of any ray relative to the lens axis is so small that we can assume $\sin\theta \simeq \theta$ and $\tan\theta \simeq \theta$.

To simplify the derivation, we imagine a doubly convex lens to be split into two plano-convex lenses, as shown in Figure 37-19. We concentrate first on the right half with a curved surface of radius R_1. See Figure 37-20, where the incident ray strikes the plane surface along the normal at a distance h from the lens axis. This ray is therefore undeviated as it continues into the lens material. Relative to the normal of the curved surface, this ray makes an angle of incidence θ_1. The emerging ray makes angle θ_2 relative to the normal to the exterior curved surface, and it intercepts the lens axis at a distance f_1 from the lens. (Since the lens is assumed to be very thin, we need not specify more precisely exactly from what point f_1 is measured.)

Assume that the lens has refractive index n and is immersed in a vacuum. Then θ_1 and θ_2 are related by Snell's law:

$$n \sin\theta_1 = \sin\theta_2$$

All angles are assumed small, so that $\sin\theta_1 \simeq \theta_1$, and $\sin\theta_2 \simeq \theta_2$. Snell's law reduces to

$$n\theta_1 = \theta_2$$

We see in Figure 37-20 that the emerging ray makes angle $\theta_2 = n\theta_1$ with the

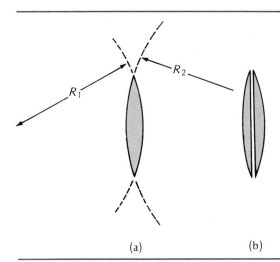

Figure 37-19. (a) A doubly convex lens with radii R_1 and R_2. (b) This lens imagined as a composite of two plano-convex lenses.

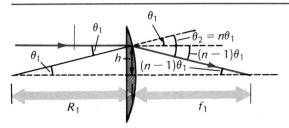

Figure 37-20. Refraction of a ray by a plano-convex lens.

normal. The angle of this ray relative to the horizontal lens axis is $\theta_2 - \theta_1 = n\theta_1 - \theta_1 = (n-1)\theta_1$.

From the geometry of Figure 37-20, we have for small angles

$$\theta_1 = \frac{h}{R_1} \quad \text{and} \quad (n-1)\theta_1 = \frac{h}{f_1}$$

Eliminating θ_1 and h from these equations then yields

$$\frac{1}{f_1} = \frac{n-1}{R_1}$$

The other plano-convex lens of Figure 37-19 has a focal length f_2, given by

$$\frac{1}{f_2} = \frac{n-1}{R_2}$$

Now imagine that the two plano-convex lenses are reassembled to form the original doubly convex lens. We apply the general lens equation to each of the two parts as follows:

$$\frac{1}{s_1} + \frac{1}{s_1'} = \frac{1}{f_1} = \frac{n-1}{R_1}$$

and

$$\frac{1}{s_2} + \frac{1}{s_2'} = \frac{1}{f_2} = \frac{n-1}{R_2}$$

But the image of the first lens is the object of the second one, so that $s_1' = -s_2$ (the minus sign appears because the real image of the first lens becomes a *virtual* object for the second lens). Adding the two equations above and dropping the subscripts give

$$\frac{1}{s} + \frac{1}{s'} = (n-1)\left(\frac{1}{R_1} + \frac{1}{R_2}\right)$$

$$\frac{1}{s} + \frac{1}{s'} = \frac{1}{f_1} + \frac{1}{f_2}$$

The general thin-lens equation is

$$\frac{1}{s} + \frac{1}{s'} = \frac{1}{f} \tag{37-2}$$

so that the equivalent focal length f for two thin lenses with focal lengths f_1 and f_2 in contact is

$$\frac{1}{f} = \frac{1}{f_1} + \frac{1}{f_2} \tag{37-5}$$

and the focal length of the doubly convex thin lens is given by

$$\frac{1}{f} = (n-1)\left(\frac{1}{R_1} + \frac{1}{R_2}\right) \tag{37-6}$$

This relation applies for a lens of refractive index n immersed in a vacuum.

More generally, if a lens with refractive index n_2 is immersed in a medium with index n_1, we replace n in (37-6) by the relative refractive index $n_{21} = n_2/n_1$.

Radii of curvature R_1 and R_2 are taken to be positive quantities for convex surfaces. The radius of curvature for a concave surface is then negative in (37-6). Clearly, then, a doubly concave lens (both R_1 and R_2 negative) has a negative f and is diverging. More generally, the sign of the focal length is controlled by the relative magnitudes and signs for R_1 and R_2 and by whether $n > 1$ or $n < 1$. A plane surface corresponds to an infinite radius of curvature.

37-7 Lens Aberrations

An ideal lens forms an image that is, apart from being inverted or magnified, an exact replica of the object. All parts of the image are in exact focus, with every point in the object rendered as a point in the image. All colors are faithfully rendered. The shape is not distorted. No such ideal lens exists. A variety of aberrations attributable to the lens cause the image to differ from the object not only in size, but in clarity, in color, and in shape.

The most fundamental limitation comes from the circumstance that a lens focuses *waves*. As a consequence of *diffraction*, a point in the object is rendered not as a point, but as a smeared image, the size of the smear depending on the wavelength. As we shall see in Section 39-6, diffraction effects are determined by the ratio of the wavelength to the outside radius of the lens aperture. The shorter the wavelength, the smaller the diffraction. Using short-wavelength violet light, rather than long-wavelength red light, reduces diffraction. The limitation arising from diffraction is obvious when you recognize that you cannot "see" any smaller dimension in the object than one wavelength of the waves used to "look" at the object.

Diffraction is an inherent lens aberration. The following lens aberrations can, however, be corrected to some degree by using two or more lenses in combination.

- *Chromatic aberration* has its origin in dispersion. For transparent materials in the visible region of the spectrum, the refractive index depends on the freespace wavelength, so that the refraction angle at an interface depends on the color of light. As Figure 36-9 shows, the refractive index is higher for violet light than for red. Consequently, if polychromatic light from an object refracts through a converging lens, the violet component will be imaged closer to the lens than the red component, as shown in Figure 37-21.
- *Spherical aberration* arises because a spherical lens surface, although easy to grind, is not the ideal shape for focusing all rays originating at the lens axis. As the angle between the rays and the lens axis increases, so that rays are not all paraxial, the outer edges of the lens focus at a different point from that formed by rays through the central portion of the lens, as shown in Figure 37-21. Because of spherical aberration, a point in the object is rendered as a diffuse circular disc in the image plane.
- Still other aberrations occur when the object is not on the lens axis: *Coma*, an extension of spherical aberration, makes the images of objects off the lens axis appear comet-shaped. *Astigmatism* arises because object points in a single transverse plane are imaged on a spherical surface. *Distortion* comes from the

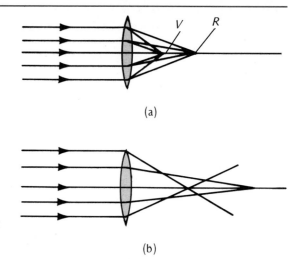

Figure 37-21. (a) Chromatic aberration. (b) Spherical aberration.

circumstance that the magnification depends to some degree on the object's distance from the lens axis.

37-8 Spherical Mirrors (Optional)

A mirror consisting of a small portion of a spherical surface has focusing properties similar to those of a thin lens. First consider the beam of parallel rays incident on the concave spherical mirror in Figure 37-22a. The rays intersect at principal focus F at a distance $\frac{1}{2}R$ from the mirror surface, where R is the radius of curvature of the mirror. This result is obvious from the geometry of Figure 37-22b, where a ray is incident parallel to the mirror's symmetry axis, strikes at point P, and makes an angle of incidence i with the normal to the surface, which is also a radius R of the sphere, whose center is at C. The reflection angle

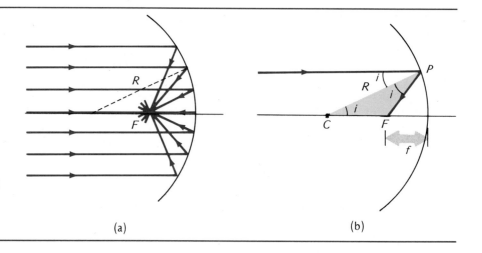

Figure 37-22. (a) Spherical concave mirror with radius of curvature R brings paraxial rays to focus at F; (b) ray incident on point P is reflected through principal focus F at distance f from the mirror surface.

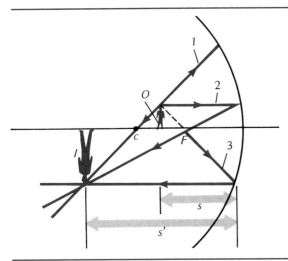

Figure 37-23. Object O produces real image I.

is also i and the ray crosses the symmetry axis at focal point F. The two small angles in triangle CPF are equal to i, so that sides CF and FP are equal. For very small i (paraxial rays), each small triangle side is approximately $\frac{1}{2}R$. The focal length f from principal focal point F to the mirror surface is $f = \frac{1}{2}R$.*

A concave spherical mirror is very much like a thin converging lens in focusing an incident beam of parallel rays. The significant difference is this: rays go through a lens, but rays are bent back to the same side as the incident rays after reflection in the mirror.

Just as for image formation by a thin lens, it is easy to draw three principal rays to locate the image I of object O, as illustrated in Figure 37-23.

- Ray 1 passes through (or is directed away from) the *center of curvature C*. This ray necessarily strikes the mirror surface along the normal, and it is reflected back along the *same straight line*.
- Ray 2 is *incident parallel* to the mirror axis so that it is reflected to pass *through the principal focus F*.
- Ray 3 is aimed to pass *through* (or to originate from) the *principal focus*, so that this ray is reflected to go *parallel* to the lens axis.

Not only is the ray-construction procedure the same as for a thin converging lens; the mathematical relation for focal length f, object distance s, and image distance s' is also the same:

$$1/f = 1/s + 1/s' \qquad (37\text{-}2)$$

The proof is implicit in the geometry of Figure 37-23. A concave spherical mirror is assigned a *positive* focal distance; image distance s' is also positive for a real image.

A virtual image is formed when the object is closer to the surface of a concave mirror than the principal focus. See Figure 37-24, where the same principal rays are used to locate the image. A virtual image—also erect and

---

* Only paraxial rays focus at point F; rays far from the mirror axis undergo spherical aberration. A parabola (or, strictly, a parabola turned around its symmetry axis to form a paraboloid) focuses all rays incident along the symmetry axis at the parabola's focal point.

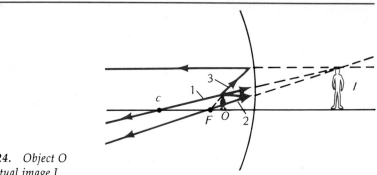

Figure 37-24. Object O produces virtual image I.

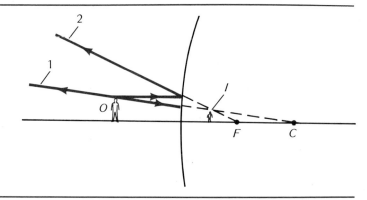

Figure 37-25. Convex spherical mirror with center of curvature C and principal focus F produces a virtual image I of object O.

enlarged—corresponds to a negative value for image distance s' in the relation above.

Just as the concave spherical mirror is analogous to the converging thin lens, a convex spherical mirror is the analog of the diverging thin lens. Rays incident on a convex spherical mirror along the lens axis diverge after reflection and appear to originate from a focal point on the far side of the mirror, a distance $\frac{1}{2}R$ from the surface. A negative focal length is assigned to the convex mirror. As Figure 37-25 shows, a convex mirror always forms a virtual image of an object at any distance s from the mirror surface.

Summary

Definitions

Principal focal point (for converging lens): location at which rays from infinitely distant point source intersect, or wave fronts collapse to a point, or intensity has maximum value.

Focal length: distance of principal focal point from thin lens.

Important Results

Thin lens equation:

$$\frac{1}{s} + \frac{1}{s'} = \frac{1}{f} \qquad (37\text{-}2)$$

where object and image distances, s and s', and focal length f, are measured from the lens.

Sign conventions:

$f > 0$,	Converging lens	$f < 0$,	Diverging lens
$s' > 0$,	Real image	$s' < 0$,	Virtual image
$s > 0$,	Real object	$s < 0$,	Virtual object

Lens-maker's formula for the focal length f of a lens of relative refractive index n:

$$\frac{1}{f} = (n-1)\left(\frac{1}{R_1} + \frac{1}{R_2}\right) \qquad (37\text{-}6)$$

where R_1 and R_2 are the radii of curvature (positive for a convex surface when $n > 1$).

The equivalent focal length f of two thin lenses in contact with focal lengths f_1 and f_2 is given by

$$\frac{1}{f} = \frac{1}{f_1} + \frac{1}{f_2} \qquad (37\text{-}5)$$

Problems and Questions

Section 37-2 Ray Tracing to Locate a Real Image

· **37-1 Q** Which of the following statements best describes the underlying reason that a lens is able to form an image of an object?
(A) The velocity of light is independent of the motion of the source and of the observer.
(B) The angle of incidence equals the angle of reflection.
(C) Two light rays interfere constructively or destructively, depending on the phase difference between them.
(D) Light has both a particle nature and a wave nature.
(E) Light travels at different speeds in different media.

· **37-2 P** A simple box camera has a focal length of 50 mm. When the camera is focused to take a picture of an object 100 m away, what is the distance from the film to the center of the lens?

· **37-3 P** A camera has a focal length of 50 mm. What is the size of the image on the film of a building that is 4 m tall and 40 m from the camera?

· **37-4 P** A projector is used to form an image of a slide on a screen. The slide has dimensions 3×3 cm and is 20 cm from a lens. How far from the lens must the screen be if the image is to be 300×300 cm?

: **37-5 P** A point isotropic radiator of light is located at $x = 0$ along the axis of a converging lens of focal length 20 cm and located at $x = 40$ cm. Sketch along the x axis from $x = 1$ cm to $x = 80$ cm, the intensity of light as a function of x.

: **37-6 Q** A real object is placed on the axis of a converging lens at a point far away from the lens. As the object is brought closer to the lens, the image
(A) is initially at the focal point but recedes from the lens as the object approaches.
(B) is initially at the focal point but approaches the lens as the object approaches.
(C) remains at the focal point but changes in size.
(D) moves away from the lens in such a way that the distance from the object to the image stays constant.

: **37-7 P** If an object is a distance $2f$ in front of a converging lens of focal length f, the image is thus
(A) twice the size of the object and inverted.
(B) twice the size of the object and erect.
(C) of the same size as the object and inverted.
(D) of the same size as the object and erect.
(E) four times the size of the object.

: **37-8 Q** The radius of curvature R is the same for each of the two lenses in Figure 37-26. The lenses are made of the same kind of glass. One is a thin lens and the other a thick lens. Can you reason qualitatively something about their focal lengths?
(A) They should have equal focal lengths.
(B) The thin lens should have greater focal length.
(C) The thick lens should have greater focal length.
(D) Whether the thick lens has a greater or a smaller focal length than the other lens depends on whether the ratio t/R is greater than 1 or less than 1.

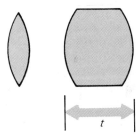

Figure 37-26. Problem 37-8.

: **37-9 P** Two thin lenses of focal lengths f_1 and f_2 are placed very close to each other. Show that they are equivalent to a single thin lens whose focal length is approximately $f = f_1 f_2/(f_1 + f_2)$.

: **37-10 P** A screen is placed 10 m from a converging lens that is to be used to project an image on a screen with magnification $\times 4$. (a) Where should the object be placed relative to the lens, and (b) what focal length must the lens have?

: **37-11 P** An object is located 4.0 m from a thin converging lens of focal length 1.0 m. (a) Where is the image formed? Draw the principal rays. (b) Where should the object be placed to form a real image 150 cm from the lens? (c) Where should the object be placed to form a virtual image 25 cm from the lens?

: **37-12 P** A camera lens of focal length 55 mm is positioned so that a distant object is in focus. (a) How far must the lens be moved to focus on an object that is 3 m distant? (b) Must the lens be moved toward the film or away from the film to bring the close object into focus?

: **37-13 P** Show that the lens formula

$$\frac{1}{f} = \frac{1}{s} + \frac{1}{s'}$$

can be written $f^2 = xx'$, where $x =$ distance from the first focal point to the object, and $x' =$ distance from the second focal point to the image.

: **37-14 P** A simple thin lens has a focal length of 30 cm. Find the image and determine whether it is real or virtual, and erect or inverted, for each of the following object distances: (a) 15 cm; (b) 30 cm; (c) 45 cm; (d) 60 cm.

: **37-15 P** The "f-stop" number of a lens is the ratio of the lens focal length to the diameter of the aperture used. On a typical camera, the f stops are numbered 1.4, 2.0, 2.8, 4.0, 5.6, 8.0, 11, 16, 22. (a) By what factor does the amount of light striking the film change when the f stop is changed from f5.6 to f8? (b) Why are the particular f number settings listed above used on cameras?

: **37-16 Q** Shown in Figure 37-27 are three light rays emanating from a source S and being imaged on a screen. If all three left the source at the same time, which would arrive at the screen first?
(A) A.
(B) B.
(C) C.
(D) All would arrive at the same time.
(E) This question cannot be answered without more information.

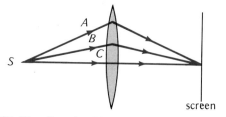

Figure 37-27. Question 37-16.

Section 37-3 Ray Tracing to Locate a Virtual Image

· **37-17 Q** Is it possible to tell just by looking at an image with your eye if the image is real or virtual?
(A) Yes, because real images are always turned upside down.
(B) Yes, because one can see right through a virtual image (since it is not actually there), whereas a real image looks like a solid object.
(C) Yes, because a real image can be seen from the side, whereas a virtual image can only be seen head-on.
(D) No. They both look the same to the eye.

· **37-18 Q** The bright filament of a lamp is placed 30 cm from a converging lens of focal length 20 cm. A piece of white cardboard is held at the proper position on the side of the lens opposite the lamp, so that an image of the filament is seen on the cardboard. What happens to this image if the cardboard is then removed?
(A) Nothing; that is, the image is still at the same place as when the cardboard was present.
(B) The image no longer exists.
(C) The image is present, but now it is at the surface of the lens.
(D) The image is still present, but now it is a virtual image instead of a real image.
(E) The image is still present, but now it is a real image instead of a virtual image.

: **37-19 P** An object 1 cm tall is placed 30 cm from a converging lens, as shown in Figure 37-28. The lens of diameter 2 cm has a focal length of 40 cm. Sketch rays emanating from either end of the object and thereby determine the region of space to which an eye is limited if it is to see the entire virtual image formed by the lens.

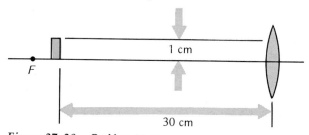

Figure 37-28. Problem 37-19.

Section 37-4 Diverging Lens

· **37-20 Q** A virtual image of a real object formed by a thin lens
(A) always occurs between the lens and a focal point.
(B) can be formed by a diverging lens but not by a converging lens.
(C) cannot be used to expose film and thereby make a photograph.
(D) may be erect or inverted, depending on the position of the object.
(E) cannot be seen with the eye since it does not actually exist.

· **37-21 Q** A diverging lens will always give an image that is
(A) virtual and reduced.
(B) virtual and enlarged.
(C) real and reduced.
(D) real and enlarged.
(E) None of the above is true consistently.

: **37-22 P** For a normal eye, the far point is at infinity and the near point is at 25 cm. What focal-length lens should be used to correct the vision of a person who can read a book clearly when it is 80 cm away but not when it is closer?

· **37-23 P** A girl can see most clearly objects placed 25 cm from her eyes when she wears glasses of power +1.2 diopters. Where does she see objects most clearly when not wearing her glasses?

: **37-24 P** A point source of light is placed at $x = 0$ on the axis of a diverging lens of focal length 10 cm and placed at $x = 20$ cm. Sketch along the x axis from $x = 1$ cm to $x = 40$ cm, the intensity of light as a function of x.

: **37-25 Q** A certain person can see clearly an object 25 cm from his eyes only when he wears eyeglasses with converging lenses. If this person takes off his glasses, the image formed in his eye is
(A) 25 cm farther behind the eye.
(B) farther from the lens of his eye than from the retina.
(C) closer to the lens of his eye than to the retina.
(D) virtual.
(E) erect, rather than inverted.

: **37-26 Q** Which of the following is more likely to be able to see clearly underwater?
(A) A person with normal vision.
(B) A farsighted person.
(C) A nearsighted person.
(D) A person with stigmatism.
(E) All of the above could see equally well.

: **37-27 Q** You are sitting in a theatre behind a person wearing eyeglasses. You notice that when you view the scene on stage through one of the eyeglasses of the person ahead of you, the image is smaller than when you view the scene directly. Is the person ahead of you near-sighted or far-sighted?

Section 37-5 Lens Combinations

· **37-28 Q** Suppose that the top half of the objective lens of a telescope is covered. What effect, if any, will this have on the image you see when looking through the telescope?
(A) The top half of the image will be blacked out.
(B) The bottom half of the image will be blacked out.
(C) The image will look the same as before except that it will be dimmer.
(D) The magnification will be half as large as before.
(E) There will be no noticeable effect.

: **37-29 Q** See Figure 37-29, where a convex-convex lens of focal length 8 cm is placed 2 cm to the left of a concave-concave lens of focal length 6 cm. A 1-mm beam of parallel light rays is incident on the convex lens from the left. As a result, the transmitted beam will
(A) converge.
(B) diverge.
(C) form a parallel beam of diameter 1 mm.
(D) form a parallel beam of diameter greater than 1 mm.
(E) form a parallel beam of diameter less than 1 mm.

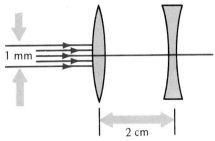

Figure 37-29. Question 37-29.

: **37-30 P** An object 2 mm tall is placed at the origin. A diverging lens of focal length 20 cm is placed on the x axis at $x = 30$ cm. A converging lens with focal length 24 cm is placed on the x axis at $x = 50$ cm. Determine the position and size of the final image formed by these lenses.

: **37-31 P** A telephoto lens consists of a converging lens of focal length $f = +6.0$ cm placed 4 cm in front of a diverging lens of focal length $f = -2.5$ cm. Determine the position of the image (relative to the diverging lens) of a very distant object.

: **37-32 P** Two thin converging lenses, each of 30-cm focal length, are separated by 15 cm. An object is placed 60 cm from the first lens. Where is the final image located?

: **37-33 P** An astronomical telescope is to be constructed with magnification ×120. Its eyepiece lens has a focal length of 2.0 cm. (*a*) What is the required focal length of the objective lens? (*b*) What will be the overall length of the telescope?

: **37-34 P** Assume that a magnifying glass of focal length f (in centimeters) is placed immediately adjacent to the eye when it is viewing an object located close to the principal focal point. The image formed by the magnifying glass is 25 cm distant from the eye, the so-called near point for the most distinct, comfortable vision for most people. Show that the angular magnification of this simple magnifier is given approximately by $25/(f - 1)$.

: **37-35 P** A microbiologist observes a specimen through a microscope whose tube length is 16 cm. The objective lens has a focal length of 0.4 cm. If the overall magnifying power is ×900, what focal length of the eyepiece is required?

: **37-36 P** An object 2 mm tall is placed at the origin. A diverging lens of focal length 20 cm is placed on the x axis at $x = 30$ cm. A converging lens with $f = 24$ cm is placed on the x axis at $x = 50$ cm. Determine the size and location of the image formed by the lenses.

: **37-37 Q** Is it ever possible to get a real image from a diverging lens?
(A) No, because the rays always diverge from a point behind the lens.
(B) No, because a diverging lens forms no image at all; it

merely spreads out a light beam. Only converging lenses form images.
(C) Yes. A real image is formed whenever the object is placed inside the focal point.
(D) Yes. A real image can be formed if the object distance is negative, that is, if the image from a converging lens is used as the object for the diverging lens.

: **37-38 P** Suppose that you have two lenses with focal lengths of 2 cm, two with focal lengths of 20 cm, and two with focal lengths of 150 cm. If you can use any combination of lenses, (a) which two would you use for a telescope? (b) Which two for a microscope?

: **37-39 P** An astronomical telescope's objective lens has a focal length of 50 cm, and the eyepiece has a focal length of 2 cm. (a) How far apart should the lenses be placed to form an image at infinity? (b) For this case, what is the magnification? (c) What should be the separation of the lenses so that an image will be formed 25 cm from the eyepiece?

: **37-40 P** Does the order in which two lenses are arranged make any difference in the final image they form? Carry out the following exercise to see. An object is placed at the origin. A converging lens of focal length 20 cm is placed at $x = 30$ cm, and a diverging lens of focal length -20 cm is placed at $x = 100$ cm. (a) Determine the position of the image formed by the two lenses. (b) Determine the position when the two lenses are interchanged.

Section 37-6 The Lens Maker's Formula

· **37-41 Q** The focal length of a thin converging lens is
(A) equal to the radius of curvature of the lens.
(B) the distance from the lens to the point at which parallel rays converge.
(C) twice the radius of curvature of the lens.
(D) the distance between the two focal points of the lens.
(E) half the radius of curvature of the lens.

· **37-42 Q** A certain thin lens has one convex surface and one concave surface. Both have the same radius of curvature. What is the focal length of the lens?

: **37-43 Q** Which of the lenses shown in Figure 37-30 has the smallest positive focal length? (All are made of the same kind of glass.)

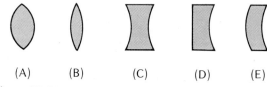

Figure 37-30. Question 37-43.

: **37-44 P** You have flat plates of glass with a refractive index of 1.50, and a device for grinding a convex spherical surface with a radius of curvature of 5.0 cm. What are the focal lengths of the lenses you can produce?

: **37-45 P** A certain doubly-convex thin lens made of glass with refractive index 1.60 has a focal length 50 cm. What is (a) the type of lens and (b) the focal length of this lens when it is immersed in water?

: **37-46 P** You wish to design a lens that is to be used underwater with light. The lens is to be converging, its focal length is to be 50 cm, and the two spherical surfaces are to have the same radius of curvature. The lens will consist of very thin transparent material enclosing air. (a) Should the surfaces be convex or concave? (b) What is their required radius of curvature?

: **37-47 P** Measuring the radius of curvature of the convex surface of a lens is not easy without special instruments, but it is relatively easy to measure the thickness of a lens at its center and its outer diameter. Suppose that a doubly convex thin lens has equal radii of curvature, a thickness t and an outer diameter d, and its glass has refractive index n. Find the relation for the focal length of the lens in terms of t, d, and n.

Section 37-7 Lens Aberrations

: **37-48 Q** If you look closely at an object's image, formed in sunlight by a lens, you will see a fuzzy rainbow around the edge of the image. This effect (called chromatic aberration) originates because
(A) rays that are not close to the lens axis, compared with rays near the axis, are focused at a different position.
(B) different light frequencies travel at different speeds in glass.
(C) some wavelengths interfere constructively and some destructively when they pass through the lens.
(D) the lens creates a diffraction pattern similar to that due to a circular aperture.
(E) light is a transverse wave, as opposed to a longitudinal wave, like sound.

: **37-49 Q** Why is it that using a bright light helps you to read fine print?
(A) A bright light increases the number of photons hitting your retina each second.
(B) A bright light produces a brighter image on your retina.
(C) A bright light enables you to use your rods rather than your cones.
(D) A bright light enables you to relax your eye and focus on infinity.
(E) A bright light enables you to use the central part only of the lens of your eye.

Supplementary Problems

37-50 Q A converging lens can be used to focus the light from the sun onto a very small spot. Such a "burning glass" was once widely used to start fires. The temperature one can achieve presumably depends on the temperature of the surface of the sun and on the diameter and focal length of the lens. Would it be possible, if a sufficiently large and powerful lens were used, to achieve, at least in principle, a spot hotter than the sun?

37-51 Q An automobile has just been polished to a very high gloss, and then it rains briefly. Large water drops sit on the polished surfaces as the sun comes out and shines brightly. You later notice that a small spot appears on the waxed surface at the center of each large water drop. Why?

37-52 Q One obvious difference in operation between the eye and a 35-mm single-lens reflex camera is the method for detecting the light. What are other differences?

38 Interference

38-1 Superposition and Interference of Waves
38-2 Interference from Two Point Sources
38-3 More on Interference from Two Point Sources
38-4 Young's Interference Experiment
38-5 The Diffraction Grating
38-6 Coherent and Incoherent Sources
38-7 Reflection and Change in Phase
38-8 Interference with Thin Films
38-9 The Michelson Interferometer
Summary

38-1 Superposition and Interference of Waves

The term *interference,* with which we're stuck because of its long entrenched usage, is an unhappy one. Interference phenomena have to do with the combined effect of waves from two or more sources, and the basic observation is that one wave does not influence (interfere with) the other. The basic proposition governing what happens when two waves arrive at the same location at the same time, the *superposition principle* (Section 17-3), says that the resultant wave disturbance at any location is simply the sum of the instantaneous individual wave disturbances. Thus, the resultant electric field from two electromagnetic waves is given by $E_r = E_1 + E_2$, where the electric fields, E_1 and E_2, of the individual waves are superposed as vectors.*

---

* We could equally well choose to find the resultant magnetic field **B**. It is conventional, however, to choose the electric rather than the magnetic field for an electromagnetic field, especially since the force on charged particles from **E** in a material typically greatly exceeds the force from **B**.

A consequence of the superposition principle (and the best evidence for its validity) is this: if two waves "collide," with both passing through the same region of space at the same time, each wave emerges from the "collision" unchanged in shape. It is just as if the other wave had not been present at all. The progress of one wave is completely independent of the presence of the other. Neither wave affects the other. See Figure 38-1. The superposition principle for electromagnetic waves is illustrated in quite ordinary circumstances. Suppose that you look at some bright object. What you see when emitted or reflected light reaches your eye is totally unaffected—in shape, color, brightness—by any other electromagnetic radiation, light or otherwise, that happens to pass through the line connecting your eye with the sighted object.

As we have seen (Chapter 18), the superposition principle applies also to mechanical waves propagated through a deformable medium to the degree that the relationship between deformation magnitude and deformation force magnitude is strictly linear.

Merging waves are said to interfere. If the resultant wave disturbance is *greater* than the disturbance from either one alone, for example, one wave crest on a second wave crest, we have *constructive interference*. If the resultant wave disturbance is *less* than that from either one separately, for example, one wave's crest on another wave's trough, the waves exhibit *destructive interference*. A point in space at which the component waves always interfere destructively, so that there is always a zero resultant wave disturbance, is called a *node*.

Hereafter in this chapter we shall be concerned primarily with the interference of sinusoidal electromagnetic waves from two sources of the same frequency. We shall concentrate on the electric fields and intensities of radio waves and visible light. The analysis holds, however, for all types of waves. For example, applying the analysis to sound waves requires merely that we replace the electric field (vector) by the pressure difference (scalar).

This is the primary problem to be solved. Suppose that you know how sinusoidal wave sources of the same frequency are placed or arranged. Then, how does the time-average intensity of the combined wave, the average power per transverse area—for light, a measure of brightness—vary with position?

First, why time average? If each source and the waves it generates oscillate sinusoidally with time, so also does the intensity of the resultant wave at any location. We are interested typically not in the instantaneous intensity, but in

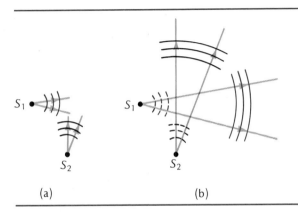

Figure 38-1. When the superposition principle applies, the waves generated by two independent sources do not interact. (a) Before "collision" (b) After.

its average over some period. After all, the human eye does time averaging— we don't see the discrete still frames of a motion picture or a television picture because of the persistence of vision; in visible light, with a frequency of more than 10^{14} Hz, the eye samples a signal over many oscillation cycles; a photographic negative gives an even longer time average of visible light.

As we have seen, the time average of the intensity $\bar{I}$ of a sinusoidal electromagnetic wave at any location is directly proportional to the *square* of the *resultant* electric-field amplitude E_r at that location:

$$\bar{I} \propto E_r^2 \tag{35-11}$$

For mechanical waves, the intensity is also proportional to the square of the amplitude of the resultant wave disturbance.

What determines E_r? The superposition principle says that

$$\mathbf{E}_r = \mathbf{E}_1 + \mathbf{E}_2$$

where $\mathbf{E}_1$ and $\mathbf{E}_2$ are the electric fields from two sources. The relative magnitudes of $\mathbf{E}_1$ and $\mathbf{E}_2$ depend in turn on the phase difference ϕ between the two waves at the observation point. And this phase difference depends in turn on two factors:

- The *path difference* from the observation point to each of the wave sources.
- Any *source phase difference* between the oscillations at the sites of the two sources.

A simple example in Section 38-2 illustrates these features.

38-2 Interference from Two Point Sources

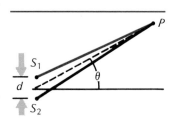

Figure 38-2. The resultant radiation from two point sources, S_1 and S_2, oscillating in phase, is observed at distant point P.

The situation, shown in Figure 38-2: two identical sources, S_1 and S_2, oscillating in phase with respect to one another; each source radiates waves of wavelength λ uniformly in all directions (in the plane of Figure 38-2) and the sources are separated vertically by a distance $d = \lambda/2$. (We can think of the sources as electric-dipole radio antennas.) The question: How does the time-average intensity $\bar{I}$ vary with angular position θ (at a large fixed distance from the sources)? Observation point P in Figure 38-2 is taken to be so far from S_1 and S_2 that the lines S_1P and S_2P can be taken as essentially parallel; each line then makes the same angle θ with the perpendicular bisector of the line joining S_1 and S_2.

Consider first point P_1, in Figure 38-3(a). It is far to the right of S_1 and S_2, and here $\theta = 0$. The path lengths from S_1 and S_2 to P are the same. The resultant field E_r at this location is the sum of E_1 and E_2, which are the electric-field amplitudes at P_1 from the individual sources. Fields E_1 and E_2 arrive in phase at P_1, so that we have constructive interference there. It is helpful to represent the sinusoidally oscillating electric fields by phasors (see Section 33-3). The magnitude of a phasor represents the amplitude of the associated simple harmonic oscillation, and the direction of the vector represents the phase of the oscillation. Therefore, phasors $\mathbf{E}_1$ and $\mathbf{E}_2$ are parallel at P_1. These two fields have essentially the same magnitude, so that the resultant field E_r has a magnitude

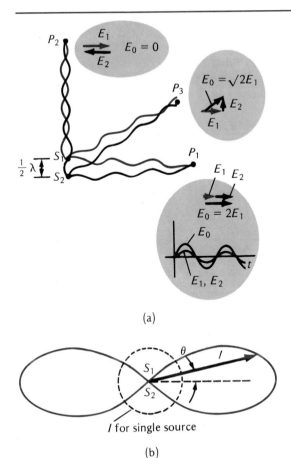

Figure 38-3. (a) Radiation from two identical point sources oscillating in phase and separated by $\frac{1}{2}\lambda$. For P_1, $\theta = 0$; for P_2, $\theta = 90°$; for P_3, $\theta = 30°$. (b) The radiation pattern, in which the length of vector **I** gives the magnitude of the radiated intensity.

twice that of a field from a single source. It follows that at P_1 the intensity, proportional to E_r^2, is four times that for a single source.

Now consider location P_2 ($\theta = 90°$) in Figure 38-3(a). The path lengths to P_2 differ by $\frac{1}{2}\lambda$. The wave from S_2 travels $\frac{1}{2}\lambda$ farther than the wave from S_1 to reach P_2. At P_2, the electric fields from S_1 and S_2 are out of phase by 180°. Again, the separate fields have essentially the same magnitude, so that we have nearly complete destructive interference at this location. At P_2, the intensity is zero.

Now consider point P_3 in Figure 38-3(a), where now $\theta = 30°$. The situation here is a little more complicated because the interference is neither completely constructive nor completely destructive. From the geometry of the figure, we see that path difference is $d \sin \theta = (\frac{1}{2}\lambda) \sin 30° = \frac{1}{4}\lambda$. The electric fields at P_3 are 90° out of phase; this corresponds to the two phasors at right angles. As Figure 38-3(a) shows, the amplitude of the resultant electric field is $\sqrt{2}E_1$, so that the intensity here is proportional to $(\sqrt{2}E_1)^2 = 2E_1^2$. At $\theta = 30°$, we have an intensity twice that from a single oscillator.

We could readily compute the intensity at still other angles to find $\bar{I}$ as a function of θ. The results, displayed in the polar diagram of Figure 38-3(b), show $\bar{I}$ as a polar vector plotted as a function of θ. We see that electromagnetic energy from the two sources is radiated primarily in the directions left and right ($\theta = 0$ and $\theta = 180°$) and none is radiated up or down ($\theta = 90°$ and $\theta =$

$-90°$). The radiation pattern of the two identical antennas separated by a half wavelength is said to consist of two *lobes*. Each antenna by itself radiates a circular pattern. Interference between the two identical sources does not change the total energy radiated (it is twice the energy from one source alone), but interference is responsible for redistributing how this total energy is radiated in various directions.

Can the radiation pattern be changed without actually moving the sources? Yes. A change in the relative phases of oscillation at the sources will do this. Suppose now that the two identical antennas are again separated vertically by $\frac{1}{2}\lambda$ but now with the sources oscillating 180° out of phase. The phase difference at observation point P_1 ($\theta = 0$) is also 180° because there is no path difference to P_1. At P_2 ($\theta = 90°$), on the other hand, the $\frac{1}{2}\lambda$ path difference from S_1 and S_2 to P_2 is just compensated by the 180° phase difference at the sources, so that the two waves arrive at P_2 *in phase*. Under these circumstances, the intensity is zero at $\theta = 0$ and four times the intensity from one source alone at $\theta = 90°$. (Query: What is the intensity at $\theta = 30°$? Is it still twice that from one antenna? Is the radiation pattern of Figure 38-3(b) merely rotated 90°?)

Notice what can be accomplished by changing the relative phase of oscillation of the two antennas: the direction in which the energy is beamed outward in space can be changed without changing the physical location of the antennas. This also means, through the reciprocity principle, that a receiving antenna can be swept through space simply by changing the relative phases of the receivers at the antennas.

Now that we've seen the basic features of the interference effects from two continuous sinusoidal point sources of waves, let us summarize how one generally finds the time-average intensity as a function of position. Because of the interference between the continuous (coherent) waves from the two sources, the observed intensity pattern is not merely the arithmetic sum of the intensity patterns radiated by the two sources separately. For electromagnetic waves, the time-average intensity $\bar{I}$ at any location P is proportional to the square of the resultant electric-field amplitude E_r at P:

$$\bar{I} \propto E_r^2 \tag{38-1}$$

where the resultant field is merely the vector sum of the electric fields at P from the two sources:

$$\mathbf{E}_r = \mathbf{E}_1 + \mathbf{E}_2 \tag{38-2}$$

If the two sources are approximately the same distance from P, then $\mathbf{E}_1$ and $\mathbf{E}_2$ have nearly the same magnitude. Then the net field at P is determined solely by the relative phase of the two arriving sinusoidal signals. The relative phase difference ϕ at the observation point P is in turn determined by:

- The *path difference* Δr, the difference between the distances from P to each of the two sources and P.
- The *source phase difference* Φ between sinusoidal oscillations of the two sources.

As we know, a path difference of $\Delta r = \frac{1}{2}\lambda$ produces destructive interference and corresponds to a phase difference at the observation point of π radians. The general relation between path difference and the corresponding phase difference is

$$\frac{\text{phase difference}}{2\pi} = \frac{\text{path difference}}{\lambda} \qquad (38\text{-}3)$$

Furthermore, if the two sources are the same distance from P, so that there is no path difference, we can still have destructive interference if the sources oscillate 180° out-of-phase so that $\Phi = \pi$ radians.

From (38-3), the general relation for the relative phase ϕ at the observation point is then

$$\phi = \frac{2\pi}{\lambda}\Delta r + \Phi \qquad (38\text{-}4)$$

If Δr represents the additional path length from source 2 to point P compared with the path length from source 1, then Φ is the phase lag of source 2 behind source 1.

Example 38-1. Three identical omnidirectional radio antennas oscillating in phase and radiating waves of wavelength λ are arranged as shown in Figure 38-4. Source 1 is a distance of $\frac{1}{2}\lambda$ north of source 2, and source 3 is $\frac{1}{2}\lambda$ east of source 2. A receiving antenna, located far to the east of the transmitting antennas, registers absorbed power of 3.0 mW when all three antennas are turned on. (a) Suppose that the source driving antenna 3 is turned off. What power is then registered at the receiving antenna? (b) How must the relative oscillation phases of the three antennas be adjusted to have maximum power delivered to the receiving antenna?

(a) With all three transmitting antennas turned on, waves to the east from sources 2 and 3 destructively interfere because of their $\frac{1}{2}\lambda$ separation; in effect, only the radiation from *one* antenna reaches the receiving antenna. But suppose that antenna 3 is turned off. Then we have constructive interference to the east of sources 1 and 2. The amplitude of the resultant wave arriving at the receiving antenna is doubled. The received power, which is proportional to the intensity at the receiving antenna and therefore proportional to the square of the resultant amplitude of the electric field, is increased by a factor 4. The power received is 4(3.0 mW) = 12.0 mW.

(b) Changing the phase of the sinusoidal signal driving antenna 3 by 180° (switching the two lead wires, for example) will mean that the path difference of $\frac{1}{2}\lambda$ between antennas 2 and 3 is exactly compensated by a source phase difference of $\Phi = \pi$. All three antennas then will constructively interfere for waves to the east. The resultant wave amplitude will be three times the amplitude of one alone, and the power at the receiving antenna is then up by a factor 9 to $3^2(3.0 \text{ mW}) = 27.0$ mW.

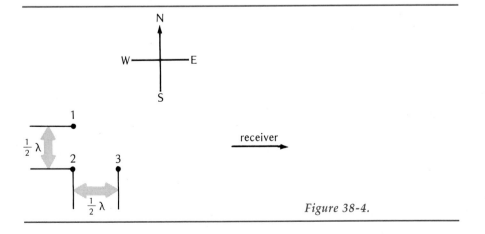

Figure 38-4.

38-3 More on Interference from Two Point Sources

In Section 38-2, we considered the interference pattern for a special case: two identical point oscillators separated by $\frac{1}{2}\lambda$. Here we treat the more general case of two identical point oscillators or sources S_1 and S_2 separated by any distance d (where d typically is large compared with λ). See Figure 38-5. Again, observation point P is so far from S_1 and S_2 (compared with d) that the two rays to P make the same angle θ with the perpendicular bisector of the line joining S_1 and S_2. The geometry of Figure 38-5 then shows that the path difference is

$$\Delta r = d \sin \theta \tag{38-5}$$

Using (38-4), we can express the phase difference ϕ at P in terms of the path difference

$$\phi = \frac{2\pi}{\lambda} \Delta r = \frac{2\pi d}{\lambda} \sin \theta \tag{38-6}$$

Constructive interference corresponds to

$$\text{Path difference:} \quad \Delta r = m\lambda \tag{38-7}$$

or equivalently,

$$\text{Phase difference:} \quad \phi = m(2\pi) \quad \text{where} \quad m = 0, 1, 2, 3, \ldots$$

so that from (38-5), the directions θ for maximum radiated time-average intensity are given by*

$$\text{Max } \bar{I}: \quad \Delta r = m\lambda = d \sin \theta \tag{38-8}$$

where $m = 0, 1, 2, 3, \ldots$. Destructive interference occurs when the path difference is an odd multiple of $\lambda/2$ or the phase difference is an odd multiple of π. Point P in Figure 38-5 corresponds to a path difference of 2λ; the locus of points, all with this same path difference, defines a curve of maximum intensity. The curve is a hyperbola; this follows from the definition of a hyperbola as the locus of points whose distances from two fixed points (here S_1 and S_2) differ by a constant (here 2λ). The asymptotes of the hyperbola make the angle θ relative to the bisector of the line connecting S_1 and S_2. A series of such hyperbolas gives the locations of maximum intensity. A second set of hyperbolas represents the lines of zero intensity, or *nodal lines*. The path difference for these curves is an odd multiple of $\lambda/2$.

The interference pattern from two sinusoidally oscillating point sources can easily be demonstrated with water-surface waves generated by transverse oscillators on the water surface. A photograph of such a ripple-tank interference pattern is shown in Figure 38-6. To observe the interference of visible light from two sources requires special arrangements (to be examined in Section 38-4).

---

* This is the first of what will be a series of equations, in this chapter and the next, that give the conditions for either zero intensity or maximum intensity for various interference or diffraction effects. These equations all look pretty much alike, and they can easily be confused. It's never a good idea to try to learn physics by memorizing specialized formulas, and it is especially risky for interference and diffraction relations. Far better, then, to know the fundamental principles and derive a specialized relation from scratch when it is needed.

SECTION 38-3 More on Interference from Two Point Sources   823

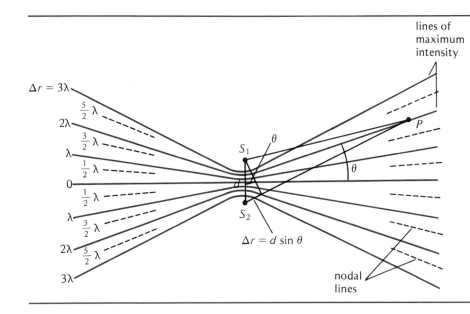

Figure 38-5. Lines of maximum intensity (solid) and nodal lines (dashed) for two point sources oscillating in phase.

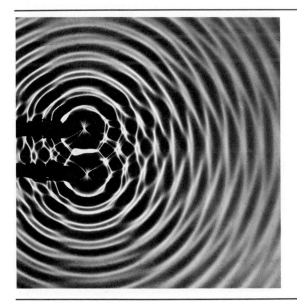

Figure 38-6. Photograph of a water-ripple-tank surface, showing the interference from two point sources.

Example 38-2. Derive the general expression for the time-average intensity $\bar{I}$ as a function of θ.

Phasors for the electric fields $\mathbf{E}_1$ and $\mathbf{E}_2$ differ in phase by ϕ, as shown in Figure 38-7. Vector $\mathbf{E}_2$ leads $\mathbf{E}_1$ by ϕ. The magnitudes are the same, $E_1 = E_2$. The sum of the horizontal components of the vectors in Figure 38-7 is given by

$$E_1 + E_2 \cos \phi = E_1(1 + \cos \phi) = 2E_1 \cos^2 \frac{\phi}{2}$$

From Figure 38-7, we can see that the horizontal component of resultant field $\mathbf{E}_r$ is $E_r \cos(\phi/2)$. Therefore, equating horizontal components gives

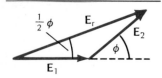

Figure 38-7. Resultant electric field $\mathbf{E}_r$ from the phasors representing electric fields $\mathbf{E}_1$ and $\mathbf{E}_2$ out of phase by ϕ.

$$E_r \cos \frac{\phi}{2} = 2E_1 \cos^2 \frac{\phi}{2}$$

$$E_r = 2E_1 \cos \frac{\phi}{2}$$

But

$$\bar{I} \propto E_r^2$$

so that

$$\bar{I} \propto \cos^2 \frac{\phi}{2}$$

From (38-6), the phase difference ϕ is

$$\phi = \frac{2\pi d}{\lambda} \sin \theta$$

Substituting this result in the relation for I gives

$$\bar{I} = I_0 \cos^2 \left(\frac{\pi d}{\lambda} \sin \theta \right) \tag{38-9}$$

where the maximum value for $\bar{I}$ is written as I_0. For small angles, $\sin \theta \approx \theta$, so that the intensity varies directly with the square of the cosine, as shown in Figure 38-8.

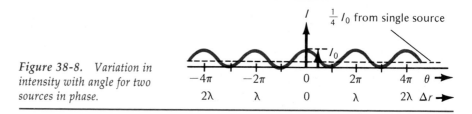

Figure 38-8. Variation in intensity with angle for two sources in phase.

38-4 Young's Interference Experiment

The British physician and scientist Thomas Young showed in an epochal experiment in 1800 that because visible light exhibits interference effects, it consists of waves.

The essential parts of a double-slit interference apparatus are shown in Figure 38-9. Light from a distant ordinary light source S passes first through a narrow single slit.* Wave fronts spreading from the single slit fall on the double slits, S_1 and S_2. The two parallel slits separated by distance d act as wave sources that oscillate in phase and produce circular wave fronts. A pattern of equally spaced interference fringes appears on the screen a large distance D from the slits.† Let y represent the distance on the observation screen from the central maximum. Then, if the interference fringes are closely spaced, and angle θ is small, we can write $\theta \approx y/D$. The relation (38-8), for the positions of maxima in the interference pattern, then becomes, with $\sin \theta \approx \theta \approx y/D$,

---

* The single slit is required for the incoherent light from an ordinary light source. The distinction between coherent and incoherent light is given in Section 38-6; how a single slit diffracts light is given in Section 39-2.
† Young, in his original interference experiment, used pinholes instead of slits.

SECTION 38-4 Young's Interference Experiment   825

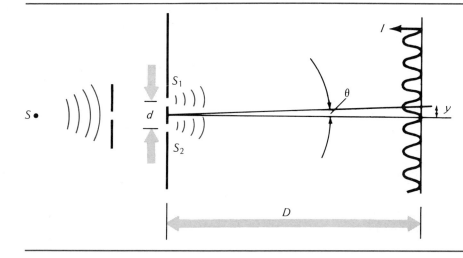

Figure 38-9. Young's double-slit experiment. Ordinary light first passes through a narrow single slit, then through the parallel slits S_1 and S_2, and finally to the distant observation screen.

$$\text{Max } \bar{I}: \quad m\lambda = \frac{yd}{D} \qquad (38\text{-}10)$$

where $m = 0, 1, 2, 3, \ldots$.

The crucial observation is that the pattern on the screen does not look at all like two parallel narrow openings; the regular alternations in the brightness of light make sense only when light is assumed to consist of waves. In particular, at the location on the screen exactly midway between the two slits, we find a *bright* line, not darkness.

Figure 38-10 is a photograph of the interference fringes from parallel slits. The intensity falls off in both directions from the center because of diffraction effects associated with the finite width of the two slits. (The combined effects of interference, and diffraction by a double slit, are treated in Section 39-6.)

Example 38-3. The red light from a helium-neon laser falls on two parallel narrow slits separated by 0.50 mm. The interference pattern on a screen 1.00 m away shows that the distance between the dark lines bordering ten interference fringes is 1.27 cm. What is the wavelength of the light?

From (38-10), we have

$$\lambda = \frac{yd}{mD} = \frac{(1.27 \times 10^{-3} \text{ m})(0.50 \times 10^{-3} \text{ m})}{(1)(1.00 \text{ m})} = 6.35 \times 10^{-7} \text{ m} = 635 \text{ nm}$$

It turns out that the single slit (preceding the double slit in Figure 38-9) is not required when the double slit is illuminated by light from a laser. The reasons are given in Section 38-6.

Figure 38-10. Photograph of the central interference fringes from a double slit.

Hey, Phenomenal!

They called him "Phenomenal"—literally. That was his nickname in college.

He deciphered the Rosetta stone. This made it possible to read Egyptian hieroglyphics in pyramids and on other ancient Egyptian structures. The discoverer: Thomas Young, M.D. (English, 1773–1829). Young's other big discovery—the experiment and the explanation showing that light consists of waves because of interference effects.

An infant prodigy, Young grew up to be an adult prodigy. He could read at age 2, he had read through the Bible twice by the time he was 6, and by age 20 he had mastered 13 languages on his own.

Although Young was trained as a physician, he had such a wide range of scholarly interests and such a poor bedside manner that he practiced medicine only part-time and with indifference. In his first scientific work, done while he was still in medical school, he showed that the eye can accommodate (focus on objects at various distances) because the eye muscles can change the radius of curvature of the lens of the eye.

His observations on the interference effects of light, made at age 28, involved first two pinholes and later what has since been known as the Young double-slit experiment. He interpreted the fringes on the basis of

The Rosetta stone, now on view in the British Museum, London, has inscriptions dating from about 200 B.C. in two languages and three alphabets. The top is in Egyptian hieroglyphics; the same message is written in the middle in demotic characters, a cursive form of Egyptian hieroglyphics, and at the bottom in Greek. Found in 1799 near the town of Rosetta in Egypt, the Rosetta stone was deciphered by Thomas Young in 1814.

38-5 The Diffraction Grating

The diffraction grating, invented by J. Fraunhofer (1786–1826), is a device for measuring wavelengths of light with high precision. A *transmission* grating consists of a large number N of parallel slits separated from one another by a distance d (greater than wavelength λ). For example, a few thousand closely spaced lines may be scratched over a distance of about 1 cm onto a plate of glass. Another example of a transmission grating, one suitable for use with millimeter-wavelength waves of sound, is a Venetian blind. A *reflection* grating may be produced by a precision ruling machine that scratches parallel grooves with a diamond point on a smooth surface of metal. The device is called a *diffraction* grating because diffraction effects (to be discussed in Section 39-7) enter its more complete description.

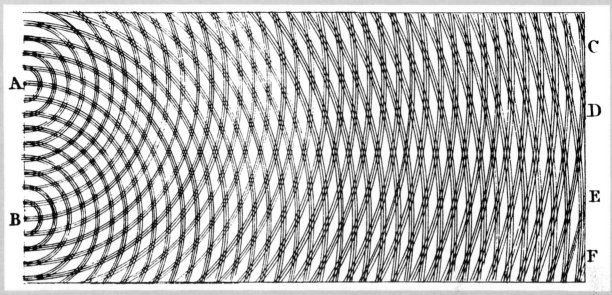
Thomas Young's original diagram showing how waves from two sources produce an interference pattern on a screen.

wave theory, and measured the wavelength of light. But he could not persuade his British contemporaries that the wave theory of light was superior to the particle theory; they stood in awe of Isaac Newton, and could not imagine him to have been wrong in advocating a particle theory of light. What is now known as the superposition principle for waves was first formulated by Young. He was also the first to use the term *energy* in its present meaning.

Young produced a three-color theory for the response by the eye's retina to all colors of light; three colors now serve as the basis for color photography and color television. *Young's modulus* is the name given to the elastic modulus that quantitatively measures a material's susceptibility to being stretched or compressed.

The range of Young's talents and expertise, developed primarily in studying on his own, is staggering. Among other things, he was a consultant on nautical matters to the British Admiralty; he established the formula for mortality tables used by insurance companies; he wrote 60 articles for the fourth edition of the *Encyclopedia Britannica* on topics ranging from bathing to hieroglyphics.

It is easy to see that a diffraction grating produces:

• Intensity maxima at the *same angular locations as a double slit* (with the same separation distance d), namely,

$$\text{Max } I: \quad m\lambda = d \sin \theta \qquad (38\text{-}8)$$

where $m = 0, 1, 2, 3, \ldots$.

• *Very narrow and intense* interference peaks.

Consider Figure 38-11, which shows N point sources all oscillating in phase at the same frequency arranged along a line with *grating spacing* d ($\gg \lambda$) between adjacent sources (or slits). As before, observation angle θ is measured from the normal to the line of sources, and the observation point P is far from the line of sources. The angle θ in Figure 38-11 is that for the first-order ($m = 1$)

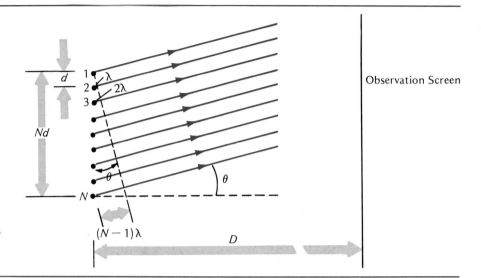

Figure 38-11. Arrangement of sources in a diffraction grating ($d > \lambda$).

interference peak, because the path difference from sources 1 and 2 is λ. The path difference is also λ between any other pair of adjacent sources, so that all point sources interfere constructively at distant point P. The condition here for constructive interference is the same as that for two sources.

A simple argument, based on energy conservation, shows that the interference peaks produced by a diffraction grating are much stronger and narrower than peaks from just two slits. When the waves from all N slits interfere constructively at some distant point, the resultant electric field there is N times the electric field from just one slit. Therefore, the intensity is up by a factor N^2. But the total energy radiated per unit time is up by a factor of only N, not N^2. Proportionately more energy (up by factor N) goes to the peak, so that proportionately less energy (down by a factor N) must go in the region between peaks. The inescapable conclusion: the width of interference peaks from a diffraction grating of N slits is smaller by a factor N than the width of peaks from just two slits. The intensity variation for two slits is shown in Figure 38-9; smooth variation (with the square of the cosine) from peak to the adjoining zeros with I dropping to half its peak value midway between the peak and the nearby zero. As more detailed arguments (Section 39-4) show, for a grating the zero is brought closer to the peak by a factor N.

As (38-8) shows, the angle θ at which intensity peaks are radiated depends on the wavelength λ. This means that if the radiation incident upon a diffraction grating consists of many wavelengths, each component wavelength will be deviated differently; the larger wavelengths at the red end of the visible spectrum will be deviated through a larger angle than the shorter wavelengths at the violet end. For $m = 0$, we have $\theta = 0$ for all wavelengths; the undeviated zero-order "spectrum" consists of all wavelengths, which is white light. But the first-order ($m = 1$), second-order ($m = 2$), and still higher-order spectra are increasingly dispersed and may overlap appreciably. The wavelength can readily be computed from the number of lines per unit length on the grating ($= 1/d$) and the measured deviation angle θ. The special advantages of diffraction grating over the prism spectrometer (Section 36-6) for observing a spectrum are:

- The wavelength is directly computed from readily measured quantities.
- The lines are sharp.
- The spectrum can have a large angular spread.

A disadvantage of the diffraction grating is that the incident light is dispersed into several spectra to the left and right of the zero-order central bright white line and thus are not as bright as the single spectrum in the prism spectrometer.

Example 38-4. A certain diffraction grating has 5000 rulings over 1.0 cm. What is the angular separation in the first order of the extremes in the visible spectrum (from 400 to 700 nm)?

We have $d = \frac{1}{5000}$ cm^{-1}. From (38-8), we have for $m = 1$ and $\lambda = 400$ nm $= 400 \times 10^{-9}$ m,

$$\theta = \sin^{-1}\left(\frac{m\lambda}{d}\right) = \sin^{-1}(4.0 \times 10^{-7} \text{ m})(5.0 \times 10^5 \text{ m}^{-1}) = 11.5°$$

Similarly, $\theta = 20.5°$ for $\lambda = 700$ nm, so that the entire visible spectrum lies within the 9 degrees between the violet and red limits.

38-6 Coherent and Incoherent Sources

Suppose that two waves of the same amplitude and wavelength arrive at observation point P in phase. The waves interfere constructively, and location P corresponds to an intensity maximum. Of course, the intensity at P will remain maximum only so long as the phase relationship between the two waves is fixed. If the phase of the source of one wave were to change by π, the interference at P would become destructive and the intensity there would fall to zero. In short, a steady interference pattern is maintained only if the two interfering waves have a fixed phase relationship. Two waves or two sources between which the phase relationship remains *constant* are said to be *coherent* with respect to one another. Examples of coherent sources of waves of the same frequency: two radio antennas driven by the same continuously operating oscillating electric circuit; two loudspeakers driven by the same audio signal generator.

Visible light from an ordinary source is *not* coherent. Ordinary sources consist of many individual atoms emitting short bursts of light at random. An example of an incoherent light source is the heated filament of an incandescent bulb. Here the atoms are thermally excited to upper atomic energy states, from which the atoms make downward transitions discontinuously and randomly. A typical atom emitting light is "on," that is, radiating light, for only about 10^{-8} s. Although this time interval is very short, it is long compared with the oscillation period of about 10^{-15} s for typical visible light. Of course, the radiation from any two emitting atoms *is* coherent over a period of about 10^{-8} s [or over a distance of $(3 \times 10^8$ m/s$)(10^{-8}$ s$) = 3$ m], but since we register the intensity of visible light over a time much longer than 10^{-8} s, the light appears essentially incoherent.

A coherent source of visible light is a *laser,* whose name is an acronym for Light Amplification by the Stimulated Emission of Radiation. A thoroughgoing explanation of laser operation is possible only with the quantum theory;

what follow here are merely some general features. In a laser, the atoms typically are brought to excited states in preparation for radiation, not by thermal excitation, but by *optical pumping,* a process in which light of higher frequency than the light emitted by the atoms is absorbed by the active material, or by an electric discharge. Atoms remain in the excited energy states for much longer than 10^{-8} s. When light of the frequency to be amplified by the laser enters the material, it stimulates the atoms in upper energy states to make transitions to lower energy states. The emitted light is the same frequency as the "stimulating" radiation.

Most important, the waves of light emitted by stimulated atoms are exactly in phase with the waves stimulating the emission. Therefore, the emitted light is coherent with the light stimulating the emission. There is light amplification. In practice, the active material that has been optically pumped to the excited states is held between two parallel reflecting boundaries and the emitted light is made to traverse the region between the boundaries repeatedly through multiple reflections. The intensity of the wave grows because the light first present causes additional atoms to emit light in coherence. The useful light—highly monochromatic, unidirectional, intense, and coherent—leaves the laser by passing through a partially reflecting end mirror. Many technological applications of lasers derive from the fact that they produce, in the visible region, electromagnetic radiation that has coherence properties heretofore available only in radio waves. The laser is treated in more detail in Section 42-6.

38-7 Reflection and Change in Phase

The procedures of classical electromagnetic theory provide a way of working out what happens to the phase of a sinusoidal electromagnetic wave that encounters a boundary between two media. Since these procedures are beyond our immediate scope, what we do here instead is to consider the analogous properties of reflected mechanical waves. This is not a proof, but it does give a useful mnemonic for remembering the results.

First, compare the two traveling wave pulses shown in Figure 38-12. Their shapes are the same, but one is positive and the other negative; the negative pulse is in fact just the positive pulse flipped over (about a horizontal axis). The labels "positive" and "negative" are not useful for a sinusoidal wave. As Figure 38-13 shows, a sine wave flipped over becomes a negative sine wave. Another way of describing the reversed, or flipped, sinusoidal wave is to say that its *phase* has been *shifted* by 180°, or by *one half-wavelength.* In short, the inversion in any wave disturbance can also be described as a 180° phase shift.

In the mechanical-optical analogy, low speed for a mechanical wave corre-

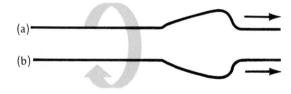

Figure 38-12. (a) A wave pulse on a taut string traveling to the right. (b) The same wave pulse inverted (flipped up-down).

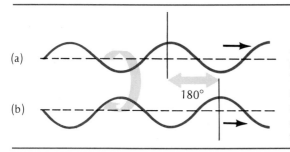

Figure 38-13. (a) A sinusoidal wave traveling to the right. (b) The same wave inverted (flipped up-down). Note that the inverted wave is 180° out of phase with the uninverted wave.

sponds to low speed (or higher refractive index) for an electromagnetic wave. A mechanically hard boundary corresponds to an optical mirror or a conducting surface.

The easily remembered results for a transverse wave pulse encountering a boundary on a stretched string under tension are shown in Figure 38-14 (like Figure 17-8), where the wave speed is lower in the more massive string:

- No phase change in the transmitted wave.
- For slow to faster medium (optically dense to less dense), no phase change in the reflected wave; for fast to slower medium, a 180° phase change in the reflected wave.
- Reflection from a conductor (with no transmitted wave), a 180° phase change in the reflected wave.

The phase change for a reflected wave is illustrated in the interference arrangement known as *Lloyd's mirror*, named for its originator, H. Lloyd, and shown in Figure 38-15. Here we have *two* waves arriving at the observation point P but only *one* source S_1. Waves reach point P both through a direct path from S_1 and through a second path involving reflection in a mirror. The re-

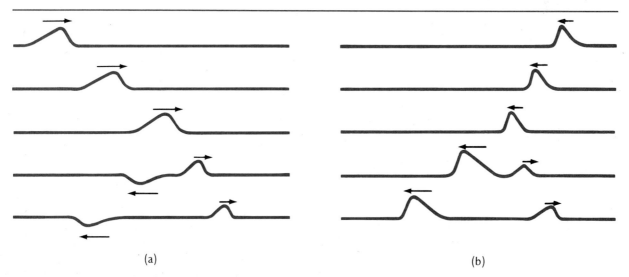

Figure 38-14. (a) Wave pulse approaches and is reflected from the boundary layer of a medium of lower wave speed. (b) Wave pulse is incident in low-speed medium and encounters medium of higher wave speed.

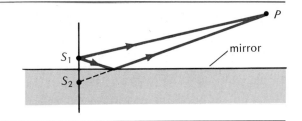

Figure 38-15. *Rays from point source S_1 and its mirror image S_2 reach distant point P.*

flected ray may be thought to originate from virtual source S_2, located below the mirror surface at the position of S_1's mirror image. The arrangement is just like that for two point sources (Figure 38-5) except that the wave from S_2 undergoes a 180° phase shift; therefore, the intensity pattern is like that for two point sources but with maximum and zero intensity locations interchanged.

Example 38-5. See Figure 38-16. A radar unit (S_1) is located 2.0 m ($\frac{1}{2}d$) above the surface of a smooth lake. It sends out microwave pulses of 10-cm wavelength toward an airplane flying at altitude y above the lake surface and at a horizontal distance of $D = 2.0$ km from the radar transmitter. The radar registers no signal reflected from the airplane. What is the minimum elevation of the airplane above the lake surface?

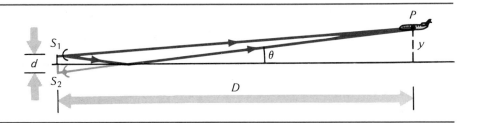

Figure 38-16.

First, if no signal reaches P from S_1, there can be no echo signal back at S_1. Waves from S_1 and S_2 arrive out of phase at P. Because of the 180° phase shift in the reflected waves, the relation (38-8), which gives the intensity maxima for two sources separated by d and oscillating in phase, will now give the locations of the intensity zeros.

$$\text{Zero I:} \quad m\lambda = d \sin\theta = d\frac{y}{D}$$

Minimum elevation corresponds to $m = 1$, so that we have

$$y = \frac{\lambda D}{d} = \frac{(10 \text{ cm})(2 \text{ km})}{(4 \text{ m})} = 50 \text{ m}$$

The lowest altitude is 50 m.

38-8 Interference with Thin Films

Everyone is familiar with the variegated colors one observes from a film of oil on water. The effect comes from interference between the light waves reflected from the top and bottom surfaces of the oil film.

Consider the situation shown in Figure 38-17. Here light waves are incident, for simplicity, nearly along the normal to two parallel interfaces separating media with refractive indices n_1, n_2, and n_3. We are concerned with interference at some distant observation point between the reflected rays from the two interfaces separated by distance d. Whether the interference is constructive or destructive depends on the following circumstances:

- The *film thickness d*. For normal incidence, the path difference for the two rays is $2d$.
- The wavelength λ_m *within the medium* of refractive index n. From (36-6),

$$\lambda_m = \frac{\lambda}{n}$$

where λ is the free-space wavelength.

- The *change in phase*, if any, *of the reflected ray*. Recall (Section 38-7) that the reflected wave undergoes a 180° phase change when light goes from a medium of low refractive index to one of higher (or from a high-speed medium to a low-speed medium), but the reverse situation does not obtain.

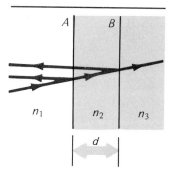

Figure 38-17. Incident rays are reflected from the boundary between media 1 and 2 and from the boundary between media 2 and 3.

The *optical path length* is a useful quantity for light in a refractive medium. It is defined as the geometrical path length multiplied by the refractive index of the medium. As so defined, the optical path length takes into account the shrinking of a wavelength in an optically dense medium and makes the optical path length equivalent to the actual path length in a vacuum. For example, red light with a wavelength of 600 nm in a vacuum becomes, in glass with refractive index 1.5, red light with wavelength of 600 nm/1.5 = 400 nm. A path length in a vacuum of 0.6 mm contains 1000 wavelengths. In glass, the same 1000 wavelengths extend over only 0.4 mm. The optical path length in glass is (0.4 mm)(1.5) = 0.6 mm; this geometrical distance in a vacuum contains the same number of wavelengths.

Consider Figure 38-18(a), in which two flat plates of glass are inclined slightly to produce a wedge-shaped film of air between them. The observed interference pattern, consisting of equally spaced, parallel interference fringes, is shown in Figure 38-18(b). The path difference changes by one wavelength (and the air thickness by $\lambda/2$) as one goes from one fringe to an adjoining one. This arrangement can be used to test for optical flatness (within a fraction of a wavelength, or better than a micron); departure from a perfectly flat plane surface is manifest in the interference pattern as a departure from equally spaced, parallel fringes.

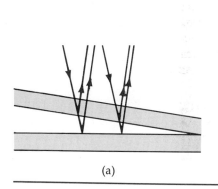
(a)

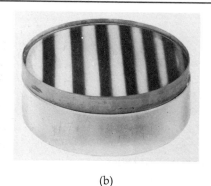
(b)

Figure 38-18. (a) Interference from an air wedge. (b) The observed fringes.

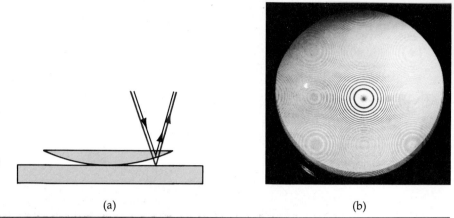

Figure 38-19. (a) Interference producing Newton's rings. (b) The observed interference pattern.

A modification of this arrangement is shown in Figure 38-19(a). Here a plano-convex lens is in contact with a flat plate of glass. The air wedge has cylindrical symmetry and the interference fringes now appear as concentric circles. The phenomenon, known as *Newton's rings,* was discovered by Robert Hooke (of Hooke's law) and studied, but not explained fully, by Isaac Newton. Figure 38-19(b) is a photograph of light reflected from the flat and spherical surfaces in contact. Note one especially interesting feature; the central region is dark, not bright. At the point of contact, there is no path difference but the two reflected rays are out of phase. At one interface light goes from glass to air; at the other light goes from air to glass.

Example 38-6. Lenses are often coated with a thin film of transparent magnesium fluoride ($n = 1.38$) to reduce reflection. How thick should the coating be to minimize reflection at the center (550 nm) of the visible spectrum?

When there is minimum reflection from the lens, there must be maximum transmission into the lens. Minimum reflection implies destructive interference between the rays reflected from the two faces of the film. The overall path difference must then be one half-wavelength within the coating material; this means that the coating has a thickness of $\frac{1}{4}\lambda_m$.

What about phase changes in the reflected rays? There is a 180° phase change at the outer surface (air to coating, $n = 1$ to 1.38) and also a phase change at the inner surface (coating to glass, $n = 1.38$ to 1.5). The two phase changes produce no net effect, and the required film thickness then is

$$t = \frac{1}{4}\lambda_m = \frac{\lambda}{4n} = \frac{550 \text{ nm}}{4(1.38)} = 99.6 \text{ nm} \approx 0.1 \text{ } \mu\text{m}$$

Coated lenses on cameras and binoculars have a purplish look. The thickness of the coating is chosen to transmit yellow-green light (0.55 μm) in the middle of the visible spectrum. This thickness is not $\frac{1}{4}\lambda_m$ for the red and the violet light at the ends of the spectrum and therefore some of the red and violet light is reflected.

38-9 The Michelson Interferometer

An *interferometer* is an instrument for measuring displacements in terms of wavelength by using the interference between two beams. The Michelson

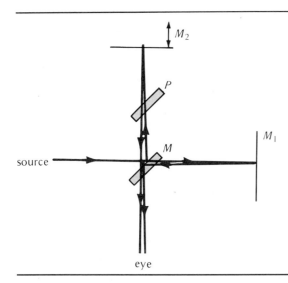

Figure 38-20. Elements of a Michelson interferometer: M, a partially silvered mirror; P, a compensator plate; M_1, a fixed mirror; and M_2, a movable mirror.

interferometer, invented by the first American Nobel laureate in physics, A. A. Michelson (1852–1931), is commonly used in measuring distances to within a fraction of a wavelength of visible light. Basically, a beam is split into two beams that travel separate paths and then are recombined; a change in the path length of one beam produces a change in the interference pattern of the recombined beams. Specifically, as shown in Figure 38-20, incident light from the source is split by a partially silvered mirror M (half reflected, half transmitted) into two beams that travel separate perpendicular paths to mirrors M_1 and M_2; the reflected beams are then recombined to interfere as they enter the observer's eye. (Compensator plate P, parallel to M and equal to it in thickness, is placed as shown in the figure to ensure that the two optical path lengths, up-down and left-right, are at least approximately equal and the two beams are coherent.) This produces an interference pattern much like that of Figure 38-19(b). Suppose that the central spot in the viewing screen is dark; the two waves arrive there out of phase. Now suppose that M_2 is shifted upward just $\frac{1}{4}\lambda$. The wave reflected from M_2 travels an additional overall path length of $\frac{1}{2}\lambda$, so that the central spot in the viewing screen is now bright. It follows that a shift from one bright fringe to the adjoining bright fringe corresponds to a shift in one mirror of only $\frac{1}{2}\lambda$, so that the Michelson interferometer allows displacements to be measured to within a fraction of a wavelength.

Summary

Definitions

Newton's rings: concentric circular interference fringes produced by interference between waves reflected from flat and spherical surfaces in contact.

Interferometer: device for measuring displacement with high precision by interference effects.

Important Results

The basic principle underlying the interference between two waves: The time-average intensity $\bar{I}$ at observation point P is proportional to the square of the resultant electric field at P:

$$\bar{I} \propto E_r^2 \qquad (38\text{-}1)$$

where

$$\mathbf{E}_r = \mathbf{E}_1 + \mathbf{E}_2 \qquad (38\text{-}2)$$

and $\mathbf{E}_1$ and $\mathbf{E}_2$ are the separate electric fields from the two sources.

How E_1 and E_2 combine depends on the phase difference ϕ at P, where

$$\phi = \frac{2\pi}{\lambda}\Delta r + \Delta\Phi \qquad (38\text{-}4)$$

Here Δr is the path difference from the two sources to P and Φ is the phase difference between the two sources.

Constructive interference: $\phi =$ integral multiple of 2π.
Destructive interference: $\phi =$ half-integral multiple of 2π.

Two point sources separated by distance d, in phase, and producing waves of wavelength λ (or a double slit with separation d):

$$\text{Max } I: \quad \Delta r = m\lambda = d\sin\theta \qquad (38\text{-}8)$$

where $m = 0, 1, 2, 3, \ldots$.

Angle θ is measured from the perpendicular bisector of the line joining the two point sources.

Diffraction grating, with many slits (or sources) separated by d (where $d > \lambda$): same relation (38-8) for intensity maxima as for the double slit.

Change in phase for reflected electromagnetic beam: *180° phase change* for beam reflected from *interface* leading to a *medium of higher refractive index* (or reflection from a conductor); otherwise, no phase change.

Problems and Questions

Section 38-2 Interference from Two Point Sources

· **38-1 Q** A simple test to determine whether the two speakers in a stereo system are properly wired is this. Compare the bass response with the speakers wired in one sense with that when the wires to one speaker are reversed. The correct wiring corresponds to the louder bass response. Why?

· **38-2 P** The two speakers of a stereo system are in phase and connected to the same 165-Hz source, which produces sound of 2.0-m wavelength. The speakers are separated by 2.0 m. How far back from one speaker must a listener move (along a line perpendicular to the line joining the speakers) until he hears a minimum in sound intensity?

: **38-3 P** Four identical oscillators oscillate in phase, emitting radiation of wavelength λ. They are arranged as shown in Figure 38-21, and two receivers R_1 and R_2 are placed at large (but equal) distances from the sources. (a) Which receiver detects the strongest signal? (b) Which receiver detects the strongest signal if source B is turned off? (c) if source D is turned off? (d) Which receiver can detect which source, B or D, has been turned off?

: **38-4 P** A plane electromagnetic wave of wavelength 30 cm traveling in a horizontal plane is detected by two vertical electric dipole antennas separated by 10 cm along an east-west line. Measurements show that the signal received by the eastern antenna lags in phase by 60° with respect to the signal picked up in the other antenna. What are the possible directions of propagation of the wave?

: **38-5 P** The radiation pattern of two identical omnidirectional radio oscillators consists of exactly six lobes. What is the separation distance between the oscillators if they radiate at a frequency of 20 MHz?

Section 38-3 More on Interference from Two Point Sources

: **38-6 P** Two omnidirectional microwave transmitting antennas are separated by 7.5 cm and they radiate at a frequency of 10 GHz. What is the angular separation between the lobes of their radiation pattern?

: **38-7 P** Two coherent sources interfere to produce an interference pattern on a distant screen (Figure 38-2). The intensity detected at the screen due to S_1 alone is I_0 and the intensity detected due to S_2 alone is $2I_0$. Sketch the resulting intensity of the interference pattern.

: **38-8 P** Two omnidirectional sources radiate at wavelength λ and in phase. They are placed at the points $(\tfrac{1}{2}d, 0)$ and $(-\tfrac{1}{2}d, 0)$ in the xy plane. (a) Show that the condition for maximum intensity at the point (x, y) is

$$\sqrt{(x+\tfrac{1}{2}d)^2 + y^2} - \sqrt{(x-\tfrac{1}{2}d)^2 + y^2} = m\lambda,$$

where $m = 0, 1, 2, \ldots$

(b) Prove that this curve is a hyperbola, and sketch it for the cases $m = 1$, $m = 2$, and $m = 3$.

Section 38-4 Young's Interference Experiment

· **38-9 Q** In Young's double-slit experiment, monochromatic light is incident on two slits and light and dark fringes

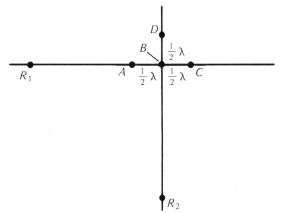

Figure 38-21. Problem 38-3.

are observed on a distant screen. If the distance between the two slits is halved, the distance between neighboring dark fringes is
(A) doubled.
(B) halved.
(C) unchanged.
(D) increased by a factor of 4.
(E) decreased by a factor of 4.

· 38-10 P Two narrow slits separated by 0.30 mm are illuminated with light of wavelength 496 nm. (a) How far are the first three bright fringes from the center of the pattern if observed on a screen 120 cm distant? (b) How far are the first three dark fringes from the center of the pattern?

· 38-11 P Two slits separated by 1.2 mm are placed 1.40 m from a screen. Light of wavelength 480 nm and 640 nm is incident on the slits. What is the separation distance on the screen of intensity maxima of the second order?

: 38-12 Q White light passes through a double slit and an interference pattern is formed on a screen. What color will appear on the screen at a node for red light?
(A) Black (that is, no light).
(B) Yellow.
(C) Blue.
(D) Red.
(E) White (that is, a mixture of all colors).

: 38-13 P In a double-slit experiment performed with light of wavelength 620 nm, an interference pattern is observed on a screen 1.2 m distant from the slits. Adjacent bright fringes near the center of the pattern are separated by 0.5 mm. (a) What is the separation of the two slits? (b) What is the separation between adjacent fringes if light of wavelength 540 nm is used?

: 38-14 P Light of wavelength 580 nm is incident on two slits separated by 1.0 mm. An interference pattern is observed on a screen 2.0 m distant. What percentage of error is made in locating the eighth-order maximum if it is not assumed that $D \gg d$ (or equivalently, that $\sin\theta \simeq \tan\theta$)?

: 38-15 P The *Fresnel biprism* produces interference effects analogous to those observed with a double slit. Figure 38-22 shows a point monochromatic source placed a distance b from a small-angle prism. When the source is viewed from the opposite side of the prism, two virtual coherent sources are seen. The light from these two equivalent sources can interfere just as the light from two slits does. Show that the separation of the two sources is $d = 2b\phi(n-1)$.

: 38-16 P In a double-slit experiment with monochromatic light, a thin sheet of transparent material of thickness t and index of refraction n is placed over one of the slits. What now is the condition for the maxima observed on a distant screen?

Section 38-5 The Diffraction Grating

· 38-17 Q Which wavelengths appear closer to the undeviated beam direction when white light is incident on a diffraction grating?
(A) Short wavelengths.
(B) Long wavelengths.
(C) Either long or short wavelengths, depending on the particular grating spacing used.
(D) Either long or short wavelengths, depending on the angle at which the incident light strikes the grating.

· 38-18 P A Venetian blind has adjacent slats separated by 4.0 cm. What wavelength of sound will produce a strong diffraction peak at 30° off the normal to the plane of the blind?

· 38-19 P Violet light of wavelength 400 nm is incident on a grating with 8000 lines per cm. At what angle will the (a) first-order spectrum be located? (b) second-order angle spectrum be located? (c) third-order spectrum be located?

· 38-20 P How many lines per centimeter are required in a diffraction grating that will spread the visible spectrum, from 440 nm to 680 nm, through an angle of 30° in first order?

· 38-21 P A diffraction grating has 5400 lines/cm. It is illuminated by monochromatic light of wavelength 570 nm. At what angle from the normal to the grating will one observe the (a) first-order spectrum? (b) second-order spectrum? (c) third-order spectrum?

: 38-22 P When monochromatic light is being used with a certain diffraction grating, it is found that the first-order maximum occurs at 8°. What is the highest order of maximum that can be observed?

: 38-23 P Monochromatic light of wavelength 510 nm is incident on a diffraction grating with 5000 lines/cm. What is the highest-order spectrum that can be observed?

: 38-24 Q White light is incident on the "black box" apparatus sketched in Figure 38-23. The light striking the screen consists of numerous brightly colored bands. Within the black box there is probably
(A) a prism.
(B) a chromatic disperser.

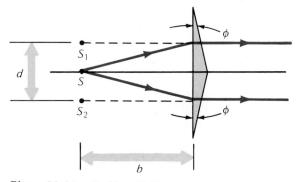

Figure 38-22. Problem 38-15.

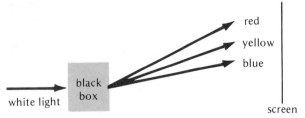

Figure 38-23. Question 38-24.

(C) a diffraction grating.
(D) an interference filter.
(E) a lens.
(F) a polarizer.

: **38-25 Q** If one wants to disperse light to get a monochromatic light source, why is it better to use a diffraction grating instead of a simple double-slit arrangement?
(A) The slits are closer together in a grating.
(B) The slits can be made narrower in a grating.
(C) The intensity maxima are sharper for a grating.
(D) More orders of interference can be obtained with a grating.
(E) Greater angles of dispersion are obtainable with a grating.

: **38-26 Q** A light beam consists of two components of light whose wavelengths differ very slightly. An experimenter tries to resolve them with a grating he has available, but can't. How could he improve the resolution, that is, increase the separation between adjacent maxima?
(A) Use a grating with more lines per centimeter.
(B) Use a grating with more total lines and the same number of lines per centimeter.
(C) Use either of the above approaches.
(D) Use a brighter beam.
(E) Move the screen farther from the grating.

: **38-27 P** Light of wavelength 540 nm falls on a diffraction grating, and a series of lines is observed. When light of unknown wavelength is shone on the same grating, it is found that the sixth-order maximum of the 540-nm light falls at the same place on the screen as the fifth-order maximum for the light of unknown wavelength. What is the wavelength of the unknown?

: **38-28 P** An experimenter wants to design a grating that will spread the visible spectrum (from 430 nm to 680 nm) through an angular range of 20° in the first order. How many lines per centimeter should the grating have?

: **38-29 Q** Suppose that a source, a diffraction grating, and a screen were all immersed in water. What effect would this have on the deviation, compared with the normal situation, when the apparatus is used in air? The deviation is the angle through which a given maximum is deviated from the normal to the grating.
(A) The deviation would be increased.
(B) The deviation would be decreased.
(C) The deviation would be unchanged.
(D) The deviation would be increased for some wavelengths and decreased for others.

: **38-30 Q** Suppose that you are using one diffraction grating 2 cm wide and another 4 cm wide, each with the same number of lines per centimeter. Monochromatic light produces an interference pattern on a distant screen. At a given maximum, the amplitude for the wider grating will be twice as great, since there are twice as many slits. Thus the intensity will be four times as great as for the narrower grating. But only twice as much light enters the wide grating, compared with the narrower one. How do you explain this?
(A) There are only half as many maxima for the wider grating.
(B) Each maximum for the wide grating is sharper than the maxima for the narrow grating.
(C) Each maximum for the wide grating is correspondingly wider than the maxima for the narrow grating.
(D) If a grating is made twice as large, four times as much light will enter it.

: **38-31 P** Light consisting of two components, of wavelengths λ and $\lambda + \Delta\lambda$, is to be resolved in mth order by a diffraction grating with slit separation d. Show that the angular separation is $\Delta\theta = \Delta\lambda / \sqrt{(d/m)^2 - \lambda^2}$.

: **38-32 P** Plot intensity as a function of distance from the central maximum for the interference pattern observed on a screen 1 m away, for narrow slits of spacing 0.1 mm and wavelength 500 nm, and numbering (a) two; (b) eight.

Section 38-6   Coherent and Incoherent Sources

: **38-33 P** Explain why two coherent beams of light must travel in parallel (or nearly parallel) paths for complete destructive interference to occur.

: **38-34 P** Two point sources, S_1 and S_2, are separated by a distance of 1.2×10^{-6} m, with S_2 east of S_1 along an east-west line. Each is radiating light of wavelength 500 nm. An observer at point O, far to the east of the two point sources, detects a light intensity of 3 μW/m² when either of the sources is radiating alone. Calculate (a) the intensity from both sources if they are incoherent; (b) the intensity from both sources if they are coherent. (c) The minimum distance S_2 should be moved toward the observer to get the maximum possible intensity at the observer.

Section 38-7   Reflection and Change in Phase

· **38-35 P** A source positioned a distance y above a plane mirror (Lloyd's mirror arrangement) emits light of wavelength 590 nm. See Figure 38-15. An interference pattern is observed on a screen 1.2 m from the source. Fringes 0.8 mm apart are observed on the screen. Determine y.

: **38-36 P** A microwave transmitter is placed 3.0 cm above a large horizontal conducting sheet. It emits 1.2-cm microwaves, which are detected by a receiver at a distance of 36 cm (measured horizontally). What is the minimum height above the sheet (different from zero) at which no signal will be picked up by the receiver?

Section 38-8 Interference with Thin Films

- **38-37 P** A researcher is studying artificial fibers. She measures the diameter of a fiber by placing it between two flat glass plates 30 cm long, as sketched in Figure 38-24. She illuminates the plates from above with green light of wavelength 546 nm and finds that 5 bright fringes are separated by a lateral distance of 8.2 mm. What is the diameter of the fiber?

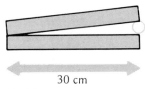

Figure 38-24. Problem 38-37.

: **38-38 P** When you observe the reflected interference colors in a soap film held in a vertical plane, you see a series of brightly colored horizontal bands. These bands keep moving as the liquid in the film gradually sinks downward. Finally the film becomes too thin to support its own weight and it breaks. Just before the film breaks, the colored bands disappear and the film looks black. Which of the following underlies this phenomenon?
(A) Incoherent waves will, on the average, interfere destructively.
(B) Light undergoes a 180° phase shift when reflected from a more dense medium.
(C) As the film thickness goes to zero, all interference effects must vanish.
(D) A film whose thickness is of the order of one wavelength will experience resonance vibrations and absorb all light incident on it.
(E) All the incident light will be transmitted through a very thin film, and none will be reflected.

: **38-39 P** A soap film formed in a loop of wire is, because of its weight, not exactly uniform in thickness, but wedge-shaped in cross section. When the film is illuminated with 546 nm light, one sees four interference fringes spread over a distance of 2.0 cm of the film. By how much does the film differ in thickness over this distance?

: **38-40 P** Light of wavelength 480 nm travels through a liquid of index of refraction 1.42. Two rays initially in phase travel different paths from point A to point B; one path is 2.4 mm longer than the other. What is the phase difference between the two rays when they are brought together at B?

: **38-41 P** Newton's rings are observed when a plano-convex lens is placed on an optical flat plate of glass (Figure 38-19). Show that the area between adjacent bright rings of high order ($m \gg 1$) is approximately constant and equal to $\pi R \lambda$, where R is the radius of curvature of the lens and λ is the wavelength of the light.

: **38-42 P** When a plano-convex lens is placed on a glass optical flat one sees Newton's rings with a dark central spot when viewing from above (Figure 38-19). Show that the radius of the dark interference fringes is given approximately by $r = \sqrt{mR\lambda}$, where $m = 0, 1, 2, \ldots$ and R is the radius of curvature of the lens.

: **38-43 P** An experimental setup using Newton's rings is carried out with two plano-convex lenses, as sketched in Figure 38-25. Show that if the two lenses have radius of curvature R_1 and R_2, then the radius of the maxima of the interference rings is given by

$$r = \sqrt{\lambda\left(m + \frac{1}{2}\right)\frac{R_1 R_2}{R_1 + R_2}}, \quad \text{where } m = 0, 1, 2, \ldots$$

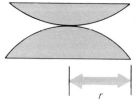

Figure 38-25. Problem 38-43.

: **38-44 P** An antireflection coating is designed to give no reflection at a wavelength of 550 nm, in the middle of the visible spectrum. By what fraction is the intensity diminished at wavelengths of (a) 450 nm? (b) 700 nm?

Section 38-9 The Michelson Interferometer

- **38-45 P** When one mirror of a Michelson interferometer was moved through a distance d, it was observed that 624 fringes passed through the field of view. Light of 579-nm wavelength was used. How far was the mirror moved?

- **38-46 P** If one mirror in a Michelson interferometer is shifted 0.204 mm, 686 fringes are seen to pass the field of view. What is the wavelength of the light used?

: **38-47 P** An evacuated transparent chamber 9.0 cm long is placed in one arm of a Michelson interferometer. As gas is slowly admitted to the chamber, fringes are observed slowly passing the field of view. When a pressure of 1 atm is reached, 88 fringes of light of wavelength 589 nm have passed. What is the index of refraction of the gas?

: **38-48 P** When a thin film of material of refractive index 1.40 is placed in one beam of a Michelson interferometer, it causes a shift of 38 fringes for light of 540-nm wavelength. What is the thickness of the film?

: **38-49 P** A transparent air-filled chamber 8.0 cm long is placed in one arm of a Michelson interferometer. As the air is evacuated from the chamber, one observes 80 fringes move past the field of view. The light wavelength is 600 nm. What is the index of refraction of air under these conditions?

Supplementary Problems

38-50 P A laser beam is coherent and it can be modulated much as is done with radio waves. One could envision using a light wave as a carrier and modulating it with the range of audio frequencies needed for speech communication (say a bandwidth of 2400 kHz). How many nonoverlapping channels of communication with this bandwidth could be accommodated in the visible spectrum, from 480 nm to 680 nm? Would there be enough such communication bands to assign a specific one to everyone on earth?

38-51 P A microwave oven operates at a frequency of 2.45 GHz. (*a*) What is the purpose of the fan with metal blades that rotates when the microwave oven is turned on? (*b*) Litton Industries, a principal manufacturer of microwave ovens, petitioned the Federal Communications Commission in 1976 to let it operate ovens at a frequency of 10.6 GHz; Litton said that the higher frequency would be much better for browning meat and heating small items. Why should this be? (The FCC denied the request.)

38-52 P A *phased-array radar* antenna is fixed in position, yet it can electronically steer a narrow outgoing beam (or be sensitive to an incoming signal from a minute distant source) because the phase of oscillations to the many individual antenna elements of which it is comprised can be shifted. The PAVE PAWS is the code name for a phased array radar operated by the U.S. Air Force, one in Cape Cod and another in California; it can give early warning for submarine-launched ballistic missiles. The antenna has nearly 3600 individual radiating elements separated from one another by about 0.3 m on each of two 30-m wide faces. The beam direction can be shifted rapidly, within a few microseconds, from one target to another. The PAVE PAWS has a range of about 5000 km; it can detect a target with a frontal area of only 10 m²; and it can keep track of hundreds of targets. See "Phased-Array Radars," by Eli Brookner, in *Scientific American* **252**, 94 (Jan., 1985).

When all radiating elements of a phased-array antenna are evenly spaced over a plane and oscillate in phase, a strong signal is radiated along the outward normal to the plane. Suppose that the microwave wavelength is 10 cm, and that the beam is to be directed 30° off the normal. What is the required phase shift in the radiating elements?

38-53 Q Ideally, a "stealth" aircraft cannot be detected by reflected radar signals. Assume for simplicity that a stealth aircraft is designed that will foil radar signals of a *single* microwave frequency. How might this be done? Why is designing a broad-band stealth aircraft, one that is immune to radar pulses of microwave wavelengths over a broad band, very difficult?

Diffraction

39

39-1 Radiation from a Row of Point Sources
39-2 Single-Slit Diffraction
39-3 The Double Slit Revisited
39-4 Diffraction and Resolution
39-5 X-Ray Diffraction
39-6 $I(\theta)$ for Single Slit (Optional)
39-7 The Diffraction Grating Revisited (Optional)
Summary

The term *diffraction* is used to describe the distinctive wave phenomena that result from the interference of many (even an infinite number of) waves from point sources oscillating coherently. Diffraction is nothing more than the interference of waves from many sources. The physics is the same as for just two sources; the geometry may be more complicated.

As we shall see, the distinctive diffraction effects of alternating bright and dark bands arise whenever a wave front is impeded by an obstacle or an aperture.

Diffraction effects, like interference effects, may be applied to measure wavelengths with high precision.

39-1 Radiation from a Row of Point Sources

Consider Figure 39-1, where a large number N (here, 12) of identical point oscillators are arranged in a row of total width w. Each source is separated from neighboring sources by the distance d. One might, for example, have 12 equally spaced electric-dipole radio antennas. All point sources oscillate at the

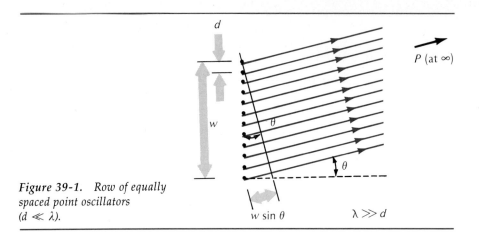

Figure 39-1. Row of equally spaced point oscillators ($d \ll \lambda$).

same frequency and are in phase. The sources generate waves of length λ; this wavelength is large compared with the distance d.

We wish to find the radiation pattern of this array of equally spaced oscillators. More specifically, we wish to find the relative intensity, observed at some very distant point P, a fixed distance from the oscillators, as a function of the angle θ between the normal to the line of oscillators and the line joining any of the oscillators to the point P. Any phase difference ϕ at a distant observation point arises solely from the difference Δr in path length between sources. When $\theta = 0$, all rays are drawn horizontally to an infinitely distant point. These effectively horizontal rays all have the same length, so that the sources interfere constructively. Therefore, the resultant electric field at the angle $\theta = 0$ is N (here, 12) times the electric field of any one single oscillator.

At what angle is the intensity first zero? As we shall see, it is much easier to find the angular positions for zero intensity than for maxima in the intensity. The strategy for locating zeros is this: group the oscillators into pairs so that the resultant field at P for every such pair is zero. This means that we must so choose the oscillators and angle that the difference in path length between the pair is $\frac{1}{2}\lambda$ (or an odd multiple of $\frac{1}{2}\lambda$); then each such pair of oscillators will interfere destructively. Suppose, then, that the angle θ is such that the difference in path length between oscillator 1 and oscillator 7 is $\frac{1}{2}\lambda$. At a distant point, the resultant electric field from this pair is zero. But by the geometry of Figure 39-1, we see that oscillator 2 and oscillator 8 then also differ in path length by $\frac{1}{2}\lambda$. Indeed, we can match all the oscillators in pairs — 1 and 7, 2 and 8, 3 and 9, and so on — so that the resultant electric field at point P from every oscillator pair is zero. The path difference between the oscillators at the top and the bottom of the array is, from Figure 39-1, equal to $w \sin \theta$. The path difference between the first and seventh oscillators is $(w/2) \sin \theta$, so that destructive interference between waves from these two implies that $\frac{1}{2} w \sin \theta = \frac{1}{2} \lambda$.

We then have

First-intensity zero: $w \sin \theta = \lambda$

The angle for second-intensity zero is found in similar fashion. Now we divide the array into four *zones*, or groups of oscillators: oscillators 1 through 3, 4 through 6, 7 through 9, and 10 through 12. Angle θ must now be larger, so large, in fact, that the path difference between oscillator 1 and oscillator 4 is $\frac{1}{2}\lambda$.

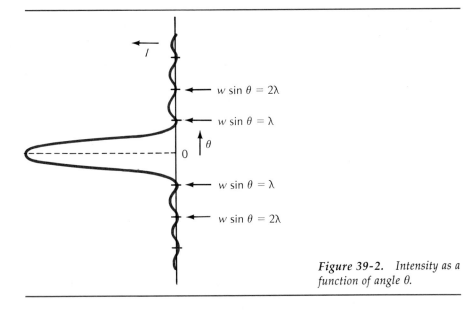

Figure 39-2. Intensity as a function of angle θ.

This pair of sources then interfere destructively at an infinite distance. Similarly, we match oscillators 2 and 5, 3 and 6, and so on, so that again the resultant field of the array is zero. The path difference $w \sin \theta$ between the sources at the extremes of the array is now 2λ, and we have

Second-intensity zero: $w \sin \theta = 2\lambda$

It is apparent that the angles for zero intensity at an infinitely distant observation point are given, in general, by

Intensity zeroes: $w \sin \theta = m\lambda$ (39-1)

where $m = 1, 2, 3, \ldots$ (but not zero). The number of sources N must be large, and they must be spaced closely, less than one wavelength apart.

Figure 39-2 shows the radiated intensity I as a function of θ. (The derivation of the relation for $I(\theta)$ is given in Section 39-6.) This *diffraction pattern* consists of an intense maximum at $\theta = 0$, twice the width of the relatively weaker

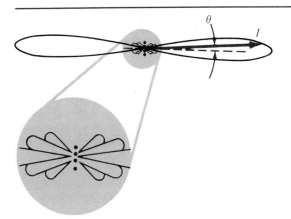

Figure 39-3. Radiation pattern from a row of equally spaced point oscillators. The magnitude of the radius vector from the origin is proportional to the intensity of the radiation in that direction.

secondary maxima that appear symmetrically to its sides. The angular locations of the intensity zeros are given by (39-1). The same information is portrayed in a different graphical form in Figure 39-3, which is the radiation pattern of the array of oscillators; the pattern consists of two strong narrow central lobes, into which most of the radiated energy is directed, together with small side lobes. We see that a linear array of equally spaced antennas (with $d \ll \lambda$) will radiate strongly only in the directions perpendicular to the line of antennas.

39-2 Single-Slit Diffraction

Suppose that monochromatic waves illuminate a single narrow slit with parallel straight sides. What is the intensity pattern of the light falling on a far distant screen? We have already solved this problem! We merely recognize that when a plane wave front of light impinges upon an opening in an otherwise opaque plane, we can imagine each point on the wave front across the slit as a new point source of radiation. All these coherent point sources oscillate in phase. In effect, we have an array of equally spaced point sources spread over the width of the slit. The intensity pattern on the distant screen, therefore, is just what we have already derived for a row of closely spaced oscillators (Section 39-1, Figure 39-2).

An arrangement for observing single-slit diffraction is shown in Figure 39-4. The angular positions of the zeros in the intensity pattern are given as before by

$$\text{Intensity zeros:} \quad w \sin \theta = m\lambda \tag{39-1}$$

where λ is the wavelength, w is now the slit width, and $m = 1, 2, 3, \ldots$ (but not zero). If y is the displacement from the central maximum on the screen and D the distance from slit to screen, we can write for small angular displacements that $\theta \approx y/D$. Equation (39-1) can then be written as

$$w \frac{y}{D} = m\lambda \tag{39-2}$$

where y gives the locations of intensity zeros.

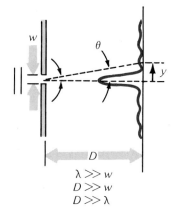

Figure 39-4. Diffraction from a single slit.

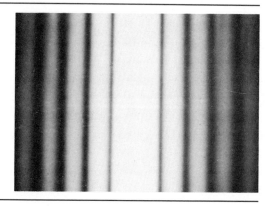

Figure 39-5. Single-slit Fraunhofer diffraction pattern.

Figure 39-5 is a photograph of a single-slit diffraction pattern; it corresponds to the intensity variation shown in Figure 39-2.

Note especially what happens if the slit width is reduced: as (39-1) indicates, with the wavelength fixed, the diffraction pattern expands. Light goes far outside the geometrical shadow of the slit edges. Indeed, the diffraction pattern on the distant screen bears no resemblance to the pattern predicted by ray optics: intensity constant over the slit but falling abruptly to zero at the slit's sharp edges.

The wavelengths of visible light are very small compared with the dimensions of ordinary objects. That is why diffraction of light is a fairly subtle effect and you don't ordinarily see diffraction fringes.* Diffraction of much longer wavelengths is commonplace, however. That is why you can often hear a sound source around a corner even though you can't see the source.

The diffraction effects we have described thus far are officially known as examples of *Fraunhofer-type diffraction*. This special type of diffraction applies when both the source and the observation screen are infinitely distant from, or at least a long way from, the diffracting object (here a slit); equivalently, Fraunhofer diffraction implies that wave fronts encountering the diffraction object are plane wave fronts. These special conditions can be met when the source and the screen are actually physically close to a slit. You simply place the light source at the principal focus of a converging lens so that the spherical or cylindrical wave fronts diverging from the source become plane wave fronts as they emerge from the lens and strike the slit. See Figure 39-6. Similarly, suppose that the observation screen is located at the far principal focus of a second converging lens. Then plane wave fronts leaving the slit are brought to a focus on the screen.

The term *Fresnel diffraction* is used to denote the general case in which there are no restrictions on distances or wave fronts. In this chapter we concentrate on the simpler but special case of Fraunhofer diffraction. One important case of Fresnel diffraction is, however, shown in Figure 39-7. Here we see the diffraction pattern produced by a straight edge (one-half of a slit) with the observation screen located very close to the straight edge. We note, first, that the shadow of a sharp straight edge is not perfectly sharp, as expected from ray optics.† Instead, there are variations in intensity—the fringes always characteristic of diffraction—near the edge. Note further that the intensity falls gradually to zero on the shadow side beyond the geometrical edge of the shadow region.

Figure 39-6. Conditions for Fraunhofer diffraction achieved with source and screen at finite distances from the slit by use of converging lens.

Example 39-1. Yellow light from atomic sodium with a wavelength of 589 nm illuminates a single slit. The central dark fringes in the diffraction pattern are found to be separated by 2.2 mm on a screen 1.0 m from the slit. What is the slit width?

From (39-2), we have

$$w \frac{y}{D} = m\lambda$$

---

* To see alternating bright and dark diffraction fringes easily, simply look at a bright source of light through the narrow slit formed when two fingers are pressed together and held close to your eye.
† The first recorded observation of the diffraction of visible light was by F. M. Grimaldi, who noted in 1655 that the shadow of the straight edge was not sharp.

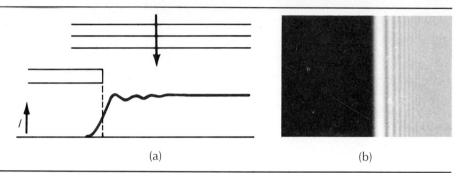

Figure 39-7. Fresnel diffraction pattern for plane wave diffracted at a straight edge: (a) intensity plot; (b) photo.

Here $2y = 2.2$ mm, and $m = 1$. The slit width is then computed as

$$w = \frac{m\lambda D}{y} = \frac{(1)(589 \times 10^{-9} \text{ m})(1.0 \text{ m})}{1.1 \times 10^{-3} \text{ m}} = 5.4 \times 10^{-4} \text{ m} = 0.54 \text{ mm}$$

39-3 The Double Slit Revisited

The simple interference pattern from two parallel slits, as shown in Figure 38-8 — peaks equally spaced and with *equal*-intensity maxima — applies only if the slits are infinitesimally narrow. Here we consider a pair of real slits each of *finite* width w, with their centers separated by distance d, illuminated by

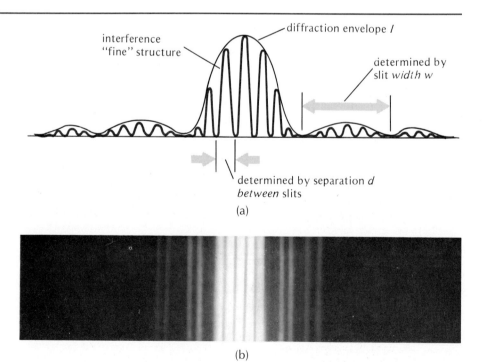

Figure 39-8. (a) Intensity variation for interference and diffraction by a double slit. (b) Photograph of a double-slit diffraction pattern.

plane waves of wavelength λ. The pattern on a distant screen shows diffraction as well as interference effects.

Imagine first that we cover one slit and leave the other exposed to the incident waves. The intensity pattern on a distant screen is certainly that for single-slit diffraction: a broad, intense central peak flanked by weak, equally spaced diffraction fringes (Figure 39-2). And if we cover the second slit and expose the first, we see the *same* diffraction pattern on the screen (shifted by distance d). The pattern when both slits are exposed is as shown in Figure 39-8:

- *Slow* variations in the *envelope* of the intensity pattern arising from *diffraction* through the slits and controlled by the slit width w. The intensity variation for diffraction is given by (39-7).
- *Rapid variations* in the intensity arising from *double-slit interference* and controlled by the slit separation distance d. The interference intensity variation is given by (38-9).

Another way of describing Figure 39-8 is to say that the interference pattern is "modulated" by the diffraction pattern. The envelope is controlled by diffraction, the fine structure by interference.

39-4 Diffraction and Resolution

Suppose that a point source of light is far distant from an opaque surface with a circular hole of diameter d. Plane wave fronts arrive at the hole, and a diffraction pattern appears on a screen, also far from the aperture. See Figure 39-9(a). The diffraction pattern is as shown in Figure 39-9(b): a bright central, circular spot surrounded by concentric diffraction fringes. The pattern for the circular hole is rather like the diffraction pattern for a single slit (Figure 39-5) but turned, so to speak, in a circle. The first zero in intensity for a single slit of width w is off center by the angle θ, where $\theta \simeq \sin\theta = \lambda/w$. The angular location of the first zero off center for a circular hole is similar. A detailed analysis shows that*

$$\sin\theta \simeq \theta = 1.22\frac{\lambda}{d} \qquad (39\text{-}3)$$

Here again we see that, as the opening is reduced (d is reduced) for a constant wavelength, the diffraction pattern expands (θ increases). Further, for a hole of fixed size, the diffraction pattern shrinks as the wavelength is reduced.

The essential fact is that a point source of light far from one side of a circular opening is rendered on a distant screen on the other side not as a bright point, but as a smeared bright circle ringed by diffraction fringes. If you know about this diffraction effect, there is no surprise when you see the image of a single point source as a smeared bright circle of light wreathed with fainter rings. Furthermore, two point sources well separated from one another produce a

---

* The number 1.22 is the smallest root of the first-order Bessel function divided by π. The Bessel function enters into the mathematical solutions for arrangements with cylindrical symmetry just as sinusoidal functions appear in arrangements with rectangular symmetry.

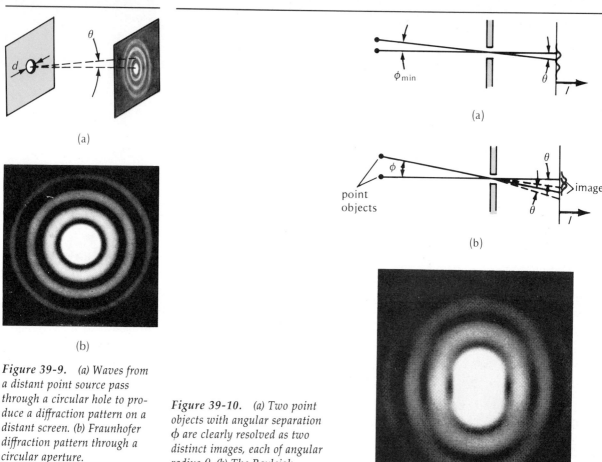

Figure 39-9. (a) Waves from a distant point source pass through a circular hole to produce a diffraction pattern on a distant screen. (b) Fraunhofer diffraction pattern through a circular aperture.

Figure 39-10. (a) Two point objects with angular separation ϕ are clearly resolved as two distinct images, each of angular radius θ. (b) The Rayleigh criterion for resolution, $\phi_{min} = \theta$. (c) Intensity pattern corresponding to part (b).

nonoverlapping pair of such circular diffraction patterns. But what if two point sources are so close together (in angular position) that their diffraction patterns overlap?

Take a specific example; consider two distant stars with angular separation ϕ. The light from the two point sources goes through an aperture, such as the objective lens of an astronomical telescope.* The images are two smeared diffraction disks of angular "radius" θ, given by (39-3), with their centers separated by ϕ. As shown in Figure 39-10(a), angle ϕ is large enough that the images are clearly resolved. What is the minimum angular separation ϕ_{min} that will permit us to say that there are two stars, not just one? Figures 39-10(b) and (c) show the images for two point sources just barely resolved. The angular separation ϕ here is just equal to the angular radius θ of either image separately. Said somewhat differently, the bright center of one image alone falls

---

* One always views through an aperture, if only that of the eye's pupil.

Figure 39-11. Photograph of the Very Large Array radio telescope in Socorro, New Mexico. There are 27 giant antennas (diameter, 25 m; mass, 2×10^5 kg), which can be shifted along a Y-shaped track laid out on the desert. Signals from operating antennas are fed to a processing center. The array functions with a resolution like that of a single antenna with a 27 km diameter.

exactly at the first dark ring of the other. You can just barely tell that there are, in fact, two images. This criterion for resolution, which Lord Rayleigh first proposed, is known as *Rayleigh's criterion for resolution*; it can then be written as

$$\phi_{\min} = \theta = \frac{1.22\lambda}{d} \qquad (39\text{-}4)$$

This resolution criterion applies generally, to any group of sources, not merely to two point sources. After all, any "picture" can be regarded as consisting of point sources.

Resolution is controlled generally by the ratio λ/d, where d is the characteristic dimension of an aperture (the width of a slit, the diameter of a circular hole). To improve resolution, we must use shorter wavelengths or bigger apertures or both. For example, the resolution of a microscope is improved by using short-wavelength, blue light. Telescopes are made big, not to produce more magnification, but to capture more radiation and thereby give a bright picture, and also to reduce diffraction effects and thereby give a sharp picture. Figure 39-11 shows a radio telescope with widely separated receiving "dishes." If the signals from two or more receiving antennas are combined, in a procedure known as long base-line interferometry, the sources can be located with a resolution that is controlled by the greatest distance separating the individual antennas. Although this result is stated without detailed proof, its reasonableness is seen at once when we recall that two separated transmitting antennas (Section 38-2) can concentrate the net radiation into a small angular region.

Example 39-2. What is the size of the smallest detail you can see directly in a hand-held photograph? (Take the diameter of the pupil of a normal eye to be about 4 mm.)

Consider what you do when you examine any object closely. You hold it about 25 cm from your (normal) eyes. At this distance, the image *on the eye's retina* not only is focused but has maximum dimensions. For an object held farther from the eye than 25 cm, the retinal image is smaller; when an object is brought closer than 25 cm, a normal eye cannot bring the image to focus on the eye's retina.

What, then, is the separation distance x between two point sources of light 25 cm away from a circular aperture (the eye's pupil) that will just be resolved as two distinct sources? The angular separation of the two sources measured from the plane of the aperture is $\phi = x/25$ cm. But the minimum value of ϕ for resolution is, from (39-4), $\phi_{min} = 1.22\lambda/d$. For visible light, diffraction effects will first be evident in the long-wavelength (red) end of the spectrum, $\lambda \approx 700$ nm. We then have

$$\phi_{min} = \frac{x}{25 \text{ cm}} = 1.22 \frac{\lambda}{d}$$

$$x = \frac{(25 \text{ cm})(1.22)(700 \times 10^{-9} \text{ m})}{4 \text{ mm}} \approx 0.05 \text{ mm} = \frac{1}{20} \text{ mm}$$

Therefore, in a hand-held photograph the eye can just barely discern details only as small as $\frac{1}{20}$ mm. There is really no point in having even finer details in the print. If a photographic negative is to be used to produce a larger print, or a picture is to be projected on a screen, then the resolution requirements may be far more stringent.

Example 39-3. Show that the smallest separation distance you can "see" with waves of wavelength λ is of the order of λ.

Suppose that we are examining two point objects separated by distance x with a lens of diameter d_1 held a distance d_2 from the objects. See Figure 39-12. (If the lens is used as a simple magnifier, distance d_2 is close to its focal length.) Rays passing through the center of the lens are not deviated, so that rays from the point objects to the center of the lens make angle ϕ, both before entering and after leaving the lens. From the geometry,

$$x = \phi \, d_2$$

and at the limit of resolution

$$\phi = 1.22 \frac{\lambda}{d_1} = \frac{x}{d_2}$$

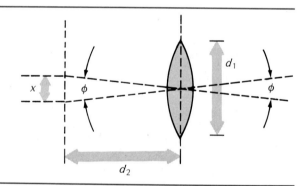

Figure 39-12.

If the lens is to form a bright image and therefore capture a significant fraction of the light from the point source, distances d_1 and d_2 must be comparable. The relation above then yields $x \sim \lambda$.

39-5 X-Ray Diffraction

The wavelengths of x-rays ($\lambda \approx 0.1$ nm) are of the same order of magnitude as the distance between adjacent atoms in a solid. A crystalline solid is a material in which atoms are arranged in a regular geometric array. It may be used, in

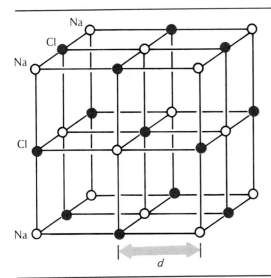

Figure 39-13. Cubic crystal structure of NaCl.

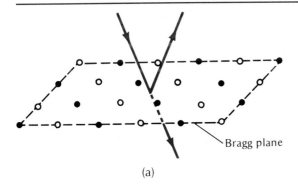

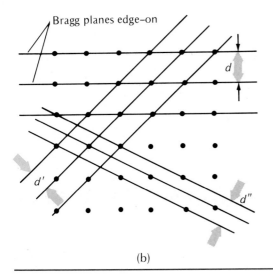

Figure 39-14. (a) Atoms in a Bragg plane. (b) Three sets of Bragg planes with different grating spacings.

effect, as a three-dimensional diffraction grating for measuring x-ray wavelengths. Similarly, x-rays of known wavelengths may be used to deduce the atomic arrangements in crystals.

Sodium chloride has a particularly simple cubic crystalline structure. See Figure 39-13. Sodium ions and chloride ions are located at alternate corners of cubes. Each atom acts as a scattering center for x-rays; strictly, the electrons around the nucleus of each atom are responsible for scattering and diffracting an incoming wave. Strong diffraction involves the cooperative scattering from many atoms, especially those atoms that lie in parallel planes, called Bragg planes, as shown in Figure 39-14(a). The distance between adjacent parallel planes of atoms is the *lattice spacing d*. As Figure 39-14(b) shows, there may be a variety of Bragg planes with differing values of d for a single type of crystalline structure.

Consider now what happens when a wave is incident at angle θ measured with respect to the Bragg planes (not measured with respect to the normal to these planes). Scattering by atoms all lying in a plane is equivalent to the partial reflection of the incident beam from that Bragg plane. Reflected beams from two adjoining parallel planes will interfere constructively when the path difference is an integral multiple of the wavelength. From the geometry of Figure 39-15, we see that the overall path difference (shown with brackets) is $2d \sin \theta$ for two adjoining Bragg planes, so that the condition for strong diffraction is

$$2d \sin \theta = m\lambda \quad \text{where } m = 1, 2, 3, \ldots \quad (39\text{-}5)$$

This, the so-called *Bragg relation*, is named after W. L. Bragg, who first derived it.

When polychromatic x-rays illuminate a crystal, only those wavelengths satisfying (39-5) will be strongly diffracted. Or when a single wavelength of

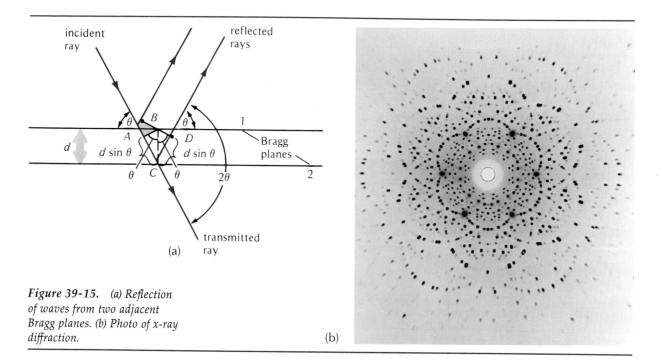

Figure 39-15. (a) Reflection of waves from two adjacent Bragg planes. (b) Photo of x-ray diffraction.

x-rays illuminates a crystal whose orientation in space can be varied, only those particular lattice spacings satisfying the Bragg relation produce strong diffraction.

Such ordinary particles as electrons have wave properties according to the quantum theory; and a crystal exposed to a beam of electrons can also exhibit electron diffraction, an effect that is altogether analogous to x-ray diffraction.

Example 39-4. The lattice spacing for the principal Bragg planes of a sodium chloride crystal is 0.282 nm. For what wavelength x-rays will the first-order diffracted beam be deviated from the incident x-ray beam by 60°?

As Figure 39-15 shows, the angle between the incident and strongly diffracted beams is $2\theta = 60°$, so that the Bragg relation, (39-5), gives

$$\lambda = \frac{2d \sin \theta}{m} = \frac{2(0.282 \text{ nm}) \sin 30°}{1} = 0.282 \text{ nm}$$

39-6 $I(\theta)$ for Single Slit (Optional)

Here we derive the relation for intensity I as a function of observation angle θ for Fraunhofer diffraction through a single slit. We must first find the resultant field $\mathbf{E}_r$ at a distant point; then $I \propto E_r^2$. The resultant field $\mathbf{E}_r$ is computed most simply by representing the continuum of simple harmonic sources spread evenly across the slit of width w by phasor vectors (Section 33-3).

Figure 39-16 shows a number of phasors of equal magnitude E_1 arranged with a constant phase difference ϕ_1 between adjoining vectors. These electric-field vectors represent the contributions from separate oscillators, as in Figure 39-1. The constant phase difference ϕ_1 arises from the constant path difference $\Delta r = d \sin \theta$ between any two adjoining oscillators. The total phase difference ϕ from all N oscillators (with $N - 1$ intervals) is given by

$$\phi = (N-1)\phi_1$$
$$= (N-1)\frac{2\pi \Delta r}{\lambda} = \frac{2\pi w}{\lambda} \sin \theta \qquad (39\text{-}6)$$

As (39-6) shows, the angle ϕ for the observed total phase difference between oscillators at the two extremes of the array increases as the space angle θ increases.

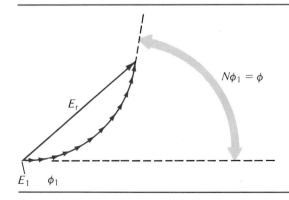

Figure 39-16. Electric-field vectors (phasors) arranged according to their relative phase differences.

We wish to find the magnitude of $\mathbf{E}_r$, the resultant electric field at a distant point. Before computing $\mathbf{E}_r$ in detail, let us see qualitatively how the separate vectors must be arranged at the maxima and minima of the intensity pattern of Figure 39-2 and thereby deduce general features of the I versus θ curve.

At the central peak, $\theta = 0$; therefore, $\Delta r = 0$ and $\phi = 0$. The little vectors are all aligned parallel, and one has the maximum possible E_r, namely $E_r = NE_1$. See Figure 39-17. Now, as θ increases, so does ϕ, and the electric fields become progressively out of phase. The vertices of the vectors lie on a circular arc, and the magnitude of $\mathbf{E}_r$ is now less than NE_1. Suppose that the total phase difference ϕ is 2π. The vectors now complete one circle; $\mathbf{E}_r$ is zero, as the intensity is also. Equation (39-6) shows that when $\phi = 2\pi$, then $w \sin \theta = \lambda$, in accord with our earlier finding. Note also that when $\phi = 2\pi$, the vectors for oscillators 1 and 7, and 2 and 8, and so on are antiparallel, corresponding to a 180° phase difference, or a $\tfrac{1}{2}\lambda$ path difference, for each such pair.

The maximum in the first secondary peak occurs very nearly at that space angle θ for which the little vectors make $1\tfrac{1}{2}$ turns. The magnitude of the resultant $\mathbf{E}_r$ is now much less than NE_1, so that the intensity, $I \propto E_r^2$, at the secondary peak is much less than the intensity of the central maximum. The second zero of intensity corresponds to that space angle θ and phase angle ϕ for which the little electric-field vectors complete two circles.

We can compute the magnitude of $\mathbf{E}_r$ in general by referring to Figure 39-18. The radius of the circle is R, the angle subtended by the chord of length E_r is ϕ, and the length of the corresponding circular arc is NE_1. In either triangle

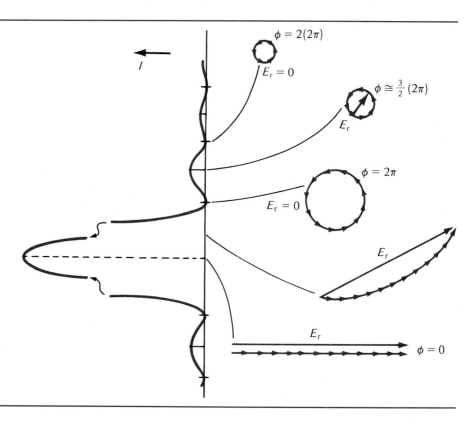

Figure 39-17. Arrangements of electric-field vectors (phasors) for various locations in the intensity pattern.

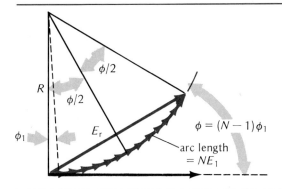

Figure 39-18.

with angle $\phi/2$, we have

$$\sin\frac{\phi}{2} = \frac{\frac{1}{2}E_r}{R}$$

From the definition of the angle ϕ in radians, we have

$$NE_1 = R\phi$$

This relation holds exactly only if N is *infinite*; the little vectors truly form a circular arc, rather than a polygon, only if their number is infinite.

Eliminating R from the two relations above yields

$$E_r = NE_1 \frac{\sin(\phi/2)}{\phi/2}$$

Therefore, we have for the intensity, $I \propto E_r^2$,

$$I = I_0 \left[\frac{\sin(\phi/2)}{\phi/2}\right]^2 \tag{39-7}$$

where I_0 is the intensity with $\phi = 0$ (at $\theta = 0$). Phase angle ϕ is related to the space angle θ through

$$\phi = \frac{2\pi w}{\lambda}\sin\theta$$

Figure 39-2 is a plot of (39-7). We readily verify from this equation that $I = 0$ for $w \sin\theta = m\lambda$, where $m = 1, 2, 3, \ldots$, in agreement with (39-1). Using (39-7), one can also show that the intensities of the secondary peaks relative to the central peak (as 1.00) are 0.045, 0.016, 0.008, ... The central peak is much more intense than the secondary peaks; in fact, more than half of the radiated energy falls within the middle half of the central peak.

39-7 The Diffraction Grating Revisited (Optional)

A grating with many parallel slits is called a *diffraction* grating because the strong and narrow intensity peaks it produces are due to diffraction.

Using an argument like that for a single slit, we show here that the principal

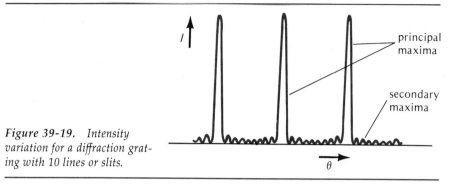

Figure 39-19. Intensity variation for a diffraction grating with 10 lines or slits.

peaks in a diffraction-grating intensity pattern are very narrow. We again consider Figure 38-11, where N sources are separated from one another by distance d (with $d > \lambda$). Angle θ in Figure 38-11 is such that the path difference between adjoining sources is λ. This means that the path difference between the first slit and the Nth slit is $(N-1)\lambda$. (With N slits, there must be $N-1$ spaces between adjoining slits.)

Suppose now that angle θ is made just slightly larger, so that the path difference between the first and Nth slit is $N\lambda$, rather than $(N-1)\lambda$. (The change in angle is very small indeed for a typical value for N, say a few thousand.)

We have already dealt with a situation of this sort in finding the zeros in intensity for an array of oscillators (Section 39-1 and Figure 39-1). Here again we imagine the slits divided into two groups, a top half and a bottom half. At the new, slightly larger angle, the ray from the uppermost slit in the top half will have a path difference of $\frac{1}{2}\lambda$ with respect to the uppermost slit in the lower half. These two rays will destructively interfere. So will every pair of corresponding top-half and bottom-half slits. The principal interference peak falls to zero for a very small change in angle.

Figure 39-19 shows the intensity pattern for a diffraction grating with ten slits. The detailed analysis yielding the curve is like that given in Section 39-6. The principal maxima are more intense at their peaks than the secondary maxima by a factor N. The principal maxima have twice the width of the secondary maxima, and there are $N-2$ secondary peaks between adjoining principal peaks.

Summary

Definitions

Diffraction: interference from a large number of wave sources.

Fraunhofer-type diffraction: sources and observation screens at infinite distances from diffracting object; equivalently, plane waves incident upon the diffracting object.

Fresnel-type diffraction: sources and observation screens at finite distances from diffracting object.

Rayleigh criterion for the resolution of two adjacent point sources: first zero in the diffraction pattern of one point source alone coincides with the central maximum in the diffraction pattern of the other point source.

Bragg plane: in a crystalline solid, a plane containing many atoms.

Important Results

Single-slit diffraction (Fraunhofer condition) for plane waves of wavelength λ incident upon a slit of width w. The angular locations of the zeros in the diffraction pattern are given by

$$\text{Zero } I: \quad \sin\theta = \frac{m\lambda}{w} \qquad (39\text{-}1)$$

where $m = 1, 2, 3, \ldots$. This relation also applies to an array of many point sources, equally spaced and oscillating in phase, spread over a distance w.

Circular opening of diameter d. The first zero in the circular diffraction pattern has an angular location given by

$$\text{First zero } I: \quad \theta = 1.22\frac{\lambda}{d} \qquad (39\text{-}3)$$

The *resolution of adjacent point objects* as two distinct objects is limited ultimately by diffraction effects. Resolution is improved by reducing the ratio λ/d, where d is a characteristic dimension of an aperture or obstacle.

X-ray diffraction, the Bragg law: strong diffraction from parallel Bragg planes separated by lattice spacing d for waves of wavelength λ incident at angle θ (measured relative to the Bragg planes) corresponds to

$$2d\sin\theta = m\lambda \qquad (39\text{-}5)$$

where $m = 1, 2, 3, \ldots$.

Problems and Questions

Section 39-1 Radiation from a Row of Point Sources

· **39-1 Q** What is the difference between interference and diffraction?
(A) Interference requires coherent light; diffraction does not.
(B) Diffraction describes the interaction of light waves that have passed through some kind of aperture, whereas interference does not.
(C) Interference always occurs between waves of the same amplitude, whereas this is not a requirement for diffraction.
(D) In interference, the waves interacting are either in phase or 180° out of phase, whereas in diffraction the phase difference may have any value.
(E) They are essentially the same thing, with the term *diffraction* used when many sources are involved.

· **39-2 Q** Why are diffraction effects more noticeable for sound waves than for light waves in everyday life?
(A) Humans can detect sound more easily than they can detect light.
(B) Because sound intensities are generally much greater than common light intensities.
(C) Because sound wavelengths are so much greater than light wavelengths.
(D) Because sound wavelengths are so much smaller than light wavelengths.
(E) They are not. Diffraction effects for light are much more noticeable.

: **39-3 P** Consider a straight line of 24 antennas with a separation of 50 cm between adjacent antennas. What is the angular width of the central band in the reception pattern for incoming waves of wavelength (a) 5.0 m? (b) 2.5 m?

: **39-4 P** Six identical microwave oscillators with frequencies of 10 GHz are aligned along a north-south line with constant spacing of 1.0 cm between adjacent oscillators. Show by a sketch what the radiation pattern is at a great distance from the oscillators (in a horizontal plane).

: **39-5 P** Four identical radio oscillators generate radiation with a 10-m wavelength. The oscillators are aligned along a north-south line. The spacing between adjacent oscillators is 1 m. Sketch the intensity of the radiation at a great distance in a horizontal plane as a function of θ, measured from the east-west line.

Section 39-2 Single-Slit Diffraction

· **39-6 Q** We can understand the origin of the diffraction pattern due to a single slit by considering
(A) the interference of the light passing through one part of the slit with light passing through another part of the same slit.
(B) the slit to be simply a double slit with zero spacing between the slits.
(C) the variation in phase across a given wave front.
(D) the distortion produced when light does not travel along the optic axis of the slit.
(E) the 180° phase shift that occurs when light is diffracted into a more dense medium.

· **39-7 Q** The first maximum away from the central maximum in the diffraction pattern due to a single slit
(A) has approximately the same intensity as the central maximum.
(B) has an intensity about half that of the central maximum.
(C) occurs halfway between the central maximum and the second maximum from the center.
(D) occurs approximately halfway between the first two minima.

· **39-8 P** Monochromatic light of wavelength 589 nm falls on a single slit. The first dark fringes are observed at 20° from the central maximum. What is the slit width?

· **39-9 P** What is the angular full width of the central diffraction maximum for a slit whose width is (a) λ? (b) 2λ? (c) 5λ? (d) 10λ?

· **39-10 P** The first minimum of a diffraction pattern of a slit falls at 90° when light of wavelength 580 nm illuminates the slit. What is the slit width?

39-11 Q Indicate for each of the following phenomena whether or not coherence of the light is important. (a) Reflection. (b) Refraction. (c) Interference. (d) Diffraction.

39-12 P A light beam consists of two components, one of unknown wavelength and the other of 524-nm wavelength. When this light is incident on a single slit, a diffraction pattern shows that the fifth secondary maximum of the unknown coincides with the sixth secondary minimum of the 524-nm light. What is the unknown wavelength?

39-13 P A slit of width 0.010 mm is illuminated with light of wavelength 550 nm, and Fraunhofer diffraction is observed by using a thin converging lens of 60-cm focal length to project the pattern on a screen. (a) What is the angular separation from the central maximum to the first minimum? (b) How wide is the central maximum (in centimeters) observed on the screen?

39-14 Q Every particle with momentum $p = mv$ has a wavelength $\lambda = h/p$, where h is the quantum constant, so a beam of particles can exhibit diffraction effects. Suppose that a beam of monoenergetic electrons passes through a narrow slit, and a diffraction pattern is observed on a distant screen. What can be done to narrow the width of the central maximum (that is, to reduce the separation between the two central minima of intensity)?
(A) Increase the electron wavelength.
(B) Reduce the electron speed.
(C) Reduce the potential difference through which the electrons were initially accelerated.
(D) Increase the kinetic energy of the electrons.
(E) Change the width of the slit.

39-15 P A slit of width 0.02 mm is illuminated with light of wavelength 616 nm, and a diffraction pattern is observed on a screen 2 m distant. (a) What is the variation in phase for light coming through various segments of the slit at a point on the screen 6 cm from the central maximum? (b) What is the ratio of the electric-field amplitude at this point to its value at the central maximum? (c) What is the ratio of the light intensity at this point to its value at the central maximum?

Section 39-3 The Double Slit Revisited

39-16 P Light from a helium-neon laser (632.8 nm) is incident on two slits, each 1.6×10^{-6} m wide and separated by 0.038 mm. How many bright fringes are contained within the central diffraction maximum?

39-17 Q In our first simple analysis of the two-slit interference pattern formed on a screen, we deduced that all the maxima had the same amplitude. In fact, their intensity decreases rapidly in a direction away from the central maximum. This is because
(A) the slits have finite width, whereas we initially assumed they had negligible width.
(B) the slits have finite spacing, whereas initially we assumed they had negligible separation.
(C) we had not made the approximation $\sin \theta \simeq \theta$.
(D) in real experiments, plane waves are not used.
(E) the light used is not actually monochromatic.

39-18 P The interference pattern of two slits is observed and diffraction is significant. More specifically, you observe 13 maxima (arising from interference between the two slits) within the central diffraction maximum. How many interference maxima would you observe within the diffraction maximum adjoining the central maximum?

39-19 P Two slits, each of width 0.1 mm, have their centers separated by 0.8 mm. They are illuminated by light of wavelength 500 nm. On a distant screen, one observes the two-slit interference pattern modulated by the single-slit diffraction pattern, with the result that certain maxima of the interference pattern are "missing" (that is, they have zero intensity). What is the first such missing maximum in this case?

39-20 P A source of light of wavelength 542 nm is placed 1.2 m directly behind a slit, as shown in Figure 39-20. A second slit is located a distance d from the first. The light passing through these slits is detected at a point P on the screen that is equidistant from the two slits. In an inspection of the equipment, the intensity at the screen is measured with slit S_1 covered, with slit S_2 covered, and finally with both slits open. In the last case, the intensity is found to equal the sum of the intensities for each single slit. What is the separation of the two slits?

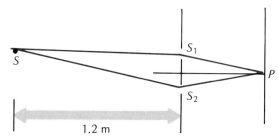

Figure 39-20. Problem 39-20.

Section 39-4 Diffraction and Resolution

39-21 Q Light from a small light source passes through a circular hole and falls on a very distant screen. The intensity pattern observed on the screen is shown in Figure 39-21. It is not certain from the appearance of the spot whether the source is a single oblong source or two closely spaced point sources. One way to decide between these alternatives is to
(A) use light of longer wavelength.
(B) make the hole larger.
(C) make the hole smaller.

Figure 39-21. Question 39-21.

(D) bring the screen closer to the hole.
(E) move the screen farther from the hole.

· **39-22 Q** To resolve the smallest cell structure under a microscope, which of the following combinations would be most effective?
(A) Large-diameter objective lens, red light.
(B) Small-diameter objective lens, blue light.
(C) Large-diameter objective lens, blue light.
(D) Large-diameter objective lens, with as much magnification as possible.
(E) Small-diameter objective lens, with as much magnification as possible.

: **39-23 Q** Diffraction limits the resolving power of a lens. Does this limitation apply to the image formed by a mirror, for example, a telescope with a mirror as objective rather than a lens?

: **39-24 Q** The useful magnification of optical microscopes is limited to about ×400 because
(A) lenses suitable for greater magnification cannot be made.
(B) the length of a microscope with greater magnification would be unwieldy.
(C) the electron microscope is a more suitable instrument for higher magnifications.
(D) of the limit in resolving power due to diffraction.
(E) there is nothing worth looking at with greater magnification.

: **39-25 P** Birds of prey are reputed to have very keen eyesight. Using resolution criteria, estimate the greatest altitude at which an eagle could fly and still see distinctly a rodent 8 cm long. Assume a maximum iris diameter of 10 mm for the eagle's eye and a wavelength of 540 nm.

: **39-26 P** Suppose the surface of the moon is studied with a terrestrial telescope that has an objective diameter of 50 cm. What is the minimum separation of two objects on the moon if they are just barely to be resolved? Assume a wavelength of 500 nm.

: **39-27 P** A surveyor looks at two small objects 200 m away using a transit with a 3.4-cm diameter objective. What is the minimum separation of the objects if they are to be resolved using light of wavelength 550 nm?

: **39-28 P** The world's largest radiotelescope, at Arecibo, Puerto Rico, is a 1000-ft diameter "dish" that fits into a natural mountain basin. (a) What is the angular resolution of this telescope when it reflects 10-cm wavelength microwaves? (b) What is the angular resolution of the Mt. Palomar reflecting telescope, which has a parabolic mirror with an outer diameter of 200 in., for visible light of 550-nm wavelength?

: **39-29 P** We have examined diffraction effects that result when light passes through a circular aperture in an opaque screen. Discuss qualitatively the nature of the diffraction effects to be expected for the inverse situation, that is, an opaque circular disk placed in a light beam with plane wave fronts in front of a screen.

Section 39-5 X-Ray Diffraction

· **39-30 P** A polychromatic beam of x-rays is incident on a KCl crystal whose lattice spacing is 0.314 nm. What wavelengths will be predominantly diffracted at a scattering angle of 30°?

: **39-31 P** Monochromatic waves have a wavelength equal to the lattice spacing on a crystal. At what angle relative to the direction of the incident beam do diffraction beams occur?

: **39-32 P** A neutron, like any other particle with mass m and velocity v, has a wavelength $\lambda = h/mv$, according to the quantum theory, where h is the basic quantum constant (Planck's constant). A beam of neutrons directed at a crystal can show diffraction effects. The first two diffraction peaks for the Bragg diffraction of thermal neutrons with a wavelength of 0.144 nm are observed at scattering angles (2θ) of 18° and 36° with a crystal of MnO. (a) What is the lattice spacing? (b) How many additional diffraction peaks are there for this set of planes?

Supplementary Problems

39-33 P What is the right distance to sit away from a 25-in. television picture tube? In the United States the standard television picture consists of 525 horizontal lines. If you sit too close to the picture tube, you see horizontal lines; if you sit too far away from the picture tube, you throw away details your eye cannot see because of the limit of resolution. Take the eye aperture to have a 6-mm diameter; assume light with 550-nm wavelength; a 25-in. picture tube (measured diagonally) has a picture height of about 15 in.

39-34 P A reconnaissance satellite tries to resolve the image of two vehicles separated by 5 m. If diffraction limits the resolution available, what minimum diameter lens must be used when the satellite is at an elevation of 150 km? Assume that the light wavelength is 550 nm.

39-35 Q *Poisson's Spot.* The French physicist, Simeon Denis Poisson (1781–1840) argued that the wave theory of light developed by Augustin Jean Fresnel (1788–1827) could not possibly be right because the following effect was predicted by the wave theory: if a round object is placed in a beam of light, then behind the object at the center of the shadow region you expect to see, not darkness, but a bright spot. The bright spot—often referred to as Poisson's spot—*is* observed. Why?

39-36 Q Diffraction effects show up whenever a regular array of apertures or opaque objects interrupt a beam of light. For example, an ordinary window screen and a distant street lamp produce a readily seen diffraction pattern. What other easily available objects can produce easily observed diffraction effects?

40 Special Relativity

40-1 The Constancy of the Speed of Light
40-2 Relativistic Velocity Transformations
40-3 Space and Time in Special Relativity
40-4 Relativistic Momentum
40-5 Relativistic Energy
40-6 Mass-Energy Equivalence and Bound Systems
40-7 The Lorentz Transformations (Optional)
Summary

The theory of relativity, primarily the creation in 1905 of Albert Einstein (1879–1955), ranks as one of the two great advances in twentieth-century physics and as one of the greatest triumphs of the human intellect of all time.

Often thought to be esoteric and recondite, the principal features of the theory of special relativity can be set forth using mathematics no more sophisticated than algebra. Relativity theory is no longer conjectural; even its most bizarre predictions have been amply confirmed by experimental test. Many of these predictions conflict with our common sense; indeed, relativity theory shows classical physics to be downright wrong when applied to high-speed phenomena.

40-1 The Constancy of the Speed of Light

All observers measure the speed of light through a vacuum to be the same constant c. This was the starting point for Einstein when he first formulated the theory. (He may not have known of the very experimental evidence that supported this curious postulate.)

The postulate is curious because it claims that all observers will measure the speed of electromagnetic waves through empty space as the same constant value, quite apart from the state of motion of observer or of the source of electromagnetic radiation. In this respect, light differs drastically from other types of waves. Consider sound waves. Their speed through air at room temperature is 340 m/s relative to the medium — air — in which the sound waves propagate. An observer at rest in air measures the pressure disturbance of a sound pulse as advancing a distance of 340 m in 1 s. But if the observer is in motion relative to the air at, say, 40 m/s, in the same direction as that in which the pulse of sound moves, he finds that the speed of the sound pulse is, relative to him, $340 - 40 = 300$ m/s. Only when the observer is at rest in the medium propagating sound does he measure its speed to be 340 m/s.

Physicists used to think that propagating light required a medium. This conjectured medium — the ether — would then constitute the only reference frame in which the speed of light would be c. An observer in motion relative to the ether — for example, an observer attached to the earth as it circled the sun — would then necessarily find the speed of light greater than or less than c by the magnitude of his speed relative to the ether. The speed would be exactly c only if he happened to be at rest in the ether. In short, if electromagnetic waves were like other waves, then their speed would differ from c whenever the observer was in motion relative to the unique reference frame in which the ether was at rest.

Measuring the speed of light is difficult; measuring a small change in c is extraordinarily difficult and requires very subtle experimental procedures. The first significant test was made by A. A. Michelson (1852–1931) and E. W. Morley (1838–1923) in the famed Michelson-Morley ether-drift experiment of 1887. Its basis is this: If an ether exists and is not rigidly fixed to the earth, then surely at some time during a year the earth will, because of its orbital motion about the sun at a speed of 3×10^4 m/s, be drifting through the ether at this speed, and the speed of light along or against the direction of drift will differ from c by 3×10^4 m/s, or 1 part in 10^4. At the same time, the speed of light when directed at right angles to the drift direction will be unaffected. As a consequence, the round-trip travel time for a pulse of light going "upstream" and then "downstream" (or the reverse) will differ from the round-trip travel time for a pulse of light over the same distance but at right angles to the drift direction. The very minute difference in travel time for the two routes was measured indirectly by Michelson and Morley. Their method involved examining the interference effects between two beams of light that were sent outward at right angles in a Michelson interferometer and then recombined; more specifically, they looked for a shift in the interference pattern arising from a rotation of the instrument and the concomitant interchange of the upstream-and-downstream and right-angle routes.

The Michelson-Morley experiment showed a null effect — no effect attributable to the ether. There is no necessity for assuming that an ether exists. The speed of light is the same for all observers. Other more recent experiments also confirm this result, with higher precision.

40-2 Relativistic Velocity Transformations

A light source at rest in our reference frame directs a beam of light to the right. Some observer moves left at speed $0.2c$ toward the light source. What is the speed of light relative to this observer?

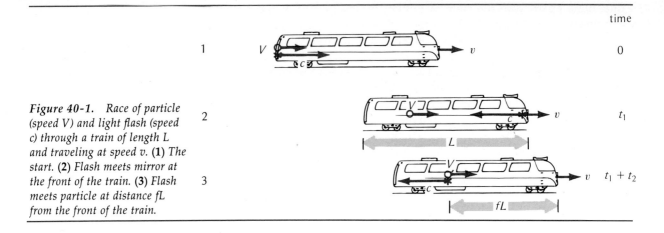

Figure 40-1. Race of particle (speed V) and light flash (speed c) through a train of length L and traveling at speed v. (1) The start. (2) Flash meets mirror at the front of the train. (3) Flash meets particle at distance fL from the front of the train.

• *Prerelativity physics* says (using the classical velocity transformation rules, Section 4-6) that the speed should be $c + 0.2c = 1.2c$. (If the observer were to move away from the source, the answer would then be $c - 0.2c = 0.8c$.)

• *Relativity physics* says that there is only one possible value, c.

The classical rules for combining relative velocities are wrong, or at least applicable only for low speeds. Here, we wish to find the relativistic velocity transformations, applicable to all possible speeds.*

Consider this situation, pictured in Figure 40-1. All the following quantities are measured relative to reference frame S, which we can for definiteness imagine to be at rest on earth. A train of length L travels right at speed v. A particle moves to the right at speed V, starting from the back end of the train. A flash of light also travels right at speed c, also starting from the back of the train. This light flash is reflected from a mirror at the front end of the train and then meets the more slowly moving particle. The distance, measured from the front of the train, at which flash meets particle is fL; this means that f is the fractional length of the train at which the two meet.

The time it takes for the light flash to travel from the train's back end to the mirror in front is t_1. The additional time it takes for the flash to travel from the mirror back to the particle is t_2.

The following three statements relate distances, times, and velocities, using their ordinary meanings.

• The total distance traveled by the particle $V(t_1 + t_2)$ from **1** to **3** is just the distance ct_1, traveled by the flash going to the right, less the distance ct_2 the flash travels going left:

$$V(t_1 + t_2) = c(t_1 - t_2) \qquad \text{(A)}$$

• The distance ct_1 covered by the flash as it goes from the back **1** to the front

---

* The simple derivation of the relativistic-velocity-transformation rule is due to N. David Mermin; it appears in *American Journal of Physics* **51**, 1130 (December 1983). How the relativistic-velocity-transformation relations can be derived from the Lorentz coordinate transformation relations (underived) is given in Optional Section 40-7.

2 of the train is just the train's length L plus the distance vt_1 the train advances during time t_1:

$$ct_1 = L + vt_1 \qquad \text{(B)}$$

- We consider the flash during the time it moves left from the train's front end at **2** to the place where it meets the particle at **3**. The distance ct_2 covered by the light flash equals the distance fL from the train's front to the meeting point, reduced by the distance vt_2 by which the train has advanced during time t_2:

$$ct_2 = fL - vt_2 \qquad \text{(C)}$$

We wish to express fraction f in terms of velocities V, v, and c:*

$$f = \frac{(c+v)(c-V)}{(c-v)(c+V)} \qquad \text{(F)}$$

This result applies for the train moving at speed v relative to frame S. But the train need not be moving. Suppose that we now become observers riding with the train in a second reference frame we call S'; in S', the train is at rest. What changes are required?

- The train is observed at rest in S', so that we put $v = 0$.
- The particle's velocity relative to the new reference frame S' is V'.

But the following two items do not change:

- All observers agree on the spot in the train where flash meets particle, so that f, the fractional length measured from the train's front end, is the same as before. (In more formal terms, f is invariant.)
- The speed of light in S' or any other reference frame is still exactly c. Here is where we invoke the basic assumption of special relativity. (The speed of light is an invariant.)

In (F) we put $v = 0$ and replace V by V'. The result is

$$f = \frac{c - V'}{c + V'} \qquad \text{(G)}$$

Finally we equate (F) and (G) and get, after some fairly messy but basically easy algebra, the final result:

$$V = \frac{V' + v}{1 + V'v/c^2} \qquad \text{(40-1)}$$

---

* This is the first of a number of places in this chapter in which intermediate steps in an algebraic development are, in the interest of brevity, relegated to a footnote. To be clear that the derivation works in detail, be sure to check this by filling in the relatively straightforward missing steps.

We wish to eliminate t_1, t_2, and L from (A), (B), (C). First, solving (A) for t_2/t_1 gives

$$\frac{t_2}{t_1} = \frac{c - V}{c + V} \qquad \text{(D)}$$

We get another relation for t_2/t_1 by eliminating L between (B) and (C):

$$\frac{t_2}{t_1} = \frac{f(c - v)}{c + v} \qquad \text{(E)}$$

Equating (D) and (E) yields the result shown in the main text above as (F).

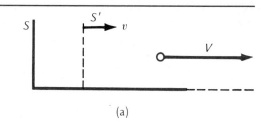

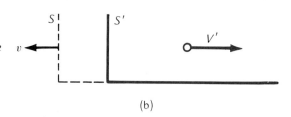

Figure 40-2. (a) Relative to S, the particle speed is V and the speed of reference frame S' is v. (b) Relative to S', the particle speed is V' and the speed of reference frame S is −v.

Let us be clear on the meaning of the terms in this equation, which applies not merely for the train in Figure 40-1, but in general. First, we are relating velocities for two reference frames S and S'; their x axes coincide, and all motion is along x.

- v is the velocity of S' relative to S. (By the same token, $-v$ is the velocity of S relative to S'. To switch from S to S', we merely change the sign of their relative velocity v.)
- V is the velocity of some point or particle measured by an observer at rest in S.
- V' is the velocity of the very same point or particle measured by an observer at rest in S'.

See Figure 40-2.

Equation (40-1) gives V in terms of V' and v. To get the inverse relation, we merely interchange V and V' and replace v by $-v$, to get

$$V' = \frac{V - v}{1 - (Vv/c^2)} \qquad (40\text{-}2)$$

The distinctive relativistic effect comes from the denominator. If any of the speeds is small compared with c, or equivalently, if we can imagine the speed of light to become effectively infinite, then the second term in the denominators of (40-1) and (40-2) is negligible compared with 1, and the relativistic relations reduce, as they must, to the simpler classical relations.*

Another consequence of the velocity rule is that velocities no longer can be combined by the simple rule of vector addition. Instead, the more complicated

---

* The numerators alone in (40-1) and (40-2) are what we get from the classical velocity transformation relations (Section 4-6). For example, with double subscripts for denoting object in motion and reference frame relative to which the velocity is measured, (40-1) becomes, for a low-speed particle p,

$$v_{pS} = V'_{pS'} + v_{S'S}$$

(40-1) and (40-2) relations must be invoked. These relations apply only when the direction of relative motion and the particle in motion are parallel.

Here, as with later relativistic results, we see that classical physics is, in effect, the low-speed limit of relativistic physics.

Example 40-1. A light source is at rest in S; the speed measured in this reference frame is c. What is the speed of light relative to reference frame S', in motion along the direction of the light beam at velocity v?

We use (40-2) with $V = c$:

$$V' = \frac{V-v}{1-Vv/c^2} = \frac{c-v}{1-vc/c^2} = \frac{c-v}{c-v}c = c$$

Example 40-2. A particle moves east at the speed 24×10^7 m/s ($=0.8c$) relative to the earth. What is this particle's speed as measured by an observer in a spaceship traveling west relative to the earth at 15×10^7 m/s ($=0.5c$).

Let S be a reference frame attached to the earth, and S' a reference frame attached to the spaceship. We are then given that

$$V = 0.8c$$
$$v = -0.5c$$

Therefore from (40-1), we have

$$V' = \frac{V-v}{1-vV/c^2} = \frac{(0.8+0.5)c}{1+(0.5)(0.8)} = \frac{1.3c}{1.4} = 0.93c$$

or

$$V' = 28 \times 10^7 \text{ m/s}$$

The classical velocity combination rule (inapplicable) would have yielded the observed particle speed as $0.8c + 0.5c = 1.3c$, a speed in excess of the speed of light. The relativistic relation ensures that the particle speed not exceed c (here, 93 percent of c).

40-3 Space and Time in Special Relativity

Special relativity reverses the absolute and the relative. In classical physics, the speed of light is relative to a hypothetical medium (the ether); in relativity physics, the speed of light is absolute. In prerelativity physics, time intervals and space intervals are taken as obviously absolute, in agreement with our common sense. Here we see that because the speed of light is absolute, time and space intervals are relative and depend on the state of motion of the observer.*

Time Dilation The speed of a particle means the spatial interval it traverses divided by the corresponding time interval, both measurements made by the same observer. But if light has the same speed for all observers, then space intervals and time intervals may not have unique values for all observers. Said differently, if c is absolute, space intervals and time intervals cannot be absolute; they may depend on the observer's state of motion.

---

* In Optional Section 40-7, it is shown how the space-contraction and time-dilation effects can be derived from the Lorentz coordinate transformation relations, given there without proof.

866 CHAPTER 40 Special Relativity

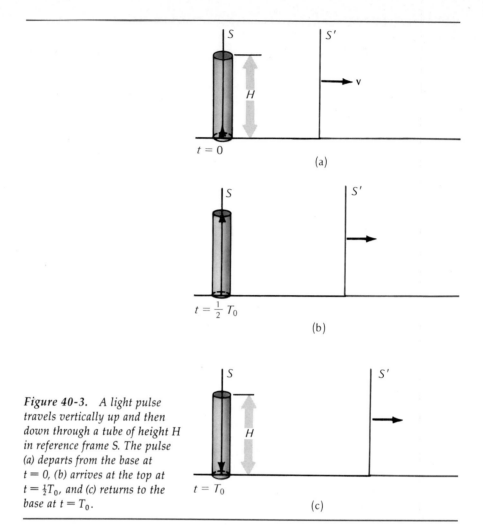

Figure 40-3. A light pulse travels vertically up and then down through a tube of height H in reference frame S. The pulse (a) departs from the base at $t = 0$, (b) arrives at the top at $t = \frac{1}{2}T_0$, and (c) returns to the base at $t = T_0$.

In prerelativity physics, the absolute character of space and time is taken as axiomatic and self-evident. If one observer measures the distance between two separated points as, say, 1 m, then other observers will agree that their separation distance is exactly 1 m. Or if one observer clocks a time interval between two events as, say, 1 s, other observers will likewise measure the time interval as precisely 1 s. These seemingly obvious claims, certainly in accord with experience and common sense for all ordinary speeds, are actually fundamentally incorrect.

We wish to find the relationship for time intervals between the same two events as measured in two reference frames, S and S'. Consider the following hypothetical experiment. Observer S has a tube of length H at rest and aligned along the y axis; he sends a pulse of light from the base of the tube along the tube axis until it reaches a mirror at the top end and is then reflected to the base. See Figure 40-3. The departure of the light pulse and its later return to the base take place at the same location in reference frame S. To emphasize that the two events (departure and return of the pulse) are measured by an observer who is at rest with respect to their location, we indicate the time interval S measures as T_0. The pulse of light traverses the total round-trip distance of 2H in the time

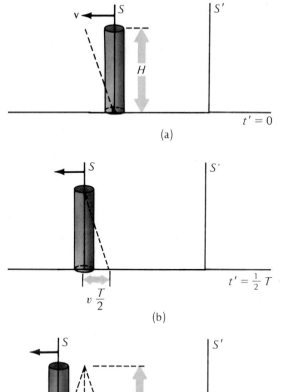

Figure 40-4. The same events as in Figure 40-3, but now as observed in reference frame S'. The pulse departs from the base at $t' = 0$, (b) arrives at the top at $t' = \frac{1}{2}T$, and (c) returns to the base at $t = T$, all as registered on the clock of observer S'. Over the round trip, the tube advances to the left relative to S' through a distance vT. The speed of the light pulse over the two oblique path segments is also c relative to S'.

interval T_0 at the speed c. So we have

$$c = \frac{2H}{T_0} \quad (40\text{-}3)$$

Now consider the same events as seen by observer S', who travels to the right at speed v relative to S. The path of the light pulse is now two oblique lines. See Figure 40-4. As S' sees things, the pulse does not return to its starting location. Overall, the path length S' observes exceeds that ($2H$) which S observes. But the speed of the light pulse must be the same for both observers. It follows at once that the time interval T recorded by S' between the departure and the return of the pulse to the base of the tube must exceed T_0. As Figure 40-4 shows, the tube moves left a total distance vT in the time interval T, or a distance $vT/2$ in time $T/2$. During the half-time $T/2$, the light pulse covers the oblique distance $cT/2$. From the geometry of the triangle in Figure 40-4, the distances are related by

$$H^2 + \left(\frac{vT}{2}\right)^2 = \left(\frac{cT}{2}\right)^2$$

Using (40-3) to eliminate H from the above equation,* we have

$$T = \frac{T_0}{\sqrt{1-(v/c)^2}} \qquad (40\text{-}4)$$

This is the fundamental *time-dilation* equation. Keeping straight the meaning of the terms in it is crucial. The time interval T_0 is between two events that occur at the *same location* and are measured on the clock of an observer at rest at this location; T_0 is termed the *proper time* (or rest time). On the other hand, T is the time interval between the *very same two events* but registered on the clock of an observer traveling at speed v relative to the location at which the two events take place (and who therefore sees the two events take place at different locations in his reference frame). Of course, the clocks of the two observers when compared at rest with respect to one another give identical readings. Equation (40-4) shows that in general $T > T_0$. Relative to a moving observer, time intervals are increased, or dilated. The effect is a result of the constancy of the speed of light; it is not attributable to a physical cause, but reflects the relativistic properties of time (and space). The phenomenon is called *time dilation*.

Time-dilation effects are significant only at speeds approaching that of light. Suppose $v = 0.1c$ (a speed about 4000 times greater than that of a satellite orbiting the earth), then

$$T = \frac{T_0}{[1-(0.1)^2]^{1/2}} = \frac{T_0}{\sqrt{0.99}} = 1.005 T_0$$

and T exceeds T_0 by only half of 1 percent. At speeds close to c, the effects are dramatic, however. Suppose now that $v = 0.98c$. Then

$$T = \frac{T_0}{[1-(0.98)^2]^{1/2}} = \frac{T_0}{\sqrt{0.04}} = 5T_0$$

and T exceeds T_0 by a factor of 5. To see what this means in a particular situation, suppose that the timer for a bomb is set to make the bomb explode in one hour (when the bomb is at rest). But the bomb is set in motion at $0.98c$. So an observer seeing it move by at this speed finds that as read on his clock (and on the clocks of his associates stationed throughout his reference frame), it takes 5 h for the bomb to explode.

No one has observed time-dilation effects with bombs moving at high speeds. But an exactly analogous effect has been observed, and the time-dilation effect confirmed, in experiments with high-speed unstable subatomic particles, such as muons. A muon is created (born) when another unstable particle (a pion) decays; a muon decays (dies) in turn to an electron (together with two uncharged, massless particles called neutrinos). You can't tell when any one muon will be born or die. But the average lifetime of a large number of muons can be given precisely. The half-life of a muon is found to be 1.52×10^{-6} s. This means that if you have a large number of muons at some initial time and these muons remain at rest, one-half will have survived after $1.52 \times$

---

* The vertical distance H, at right angles to the direction of relative motion of the two reference frames, is assumed to be the same for both observers. Although spatial intervals along the direction of relative motion are not the same for all observers, the transverse distances are unchanged.

10^{-6} s has elapsed, while the other half will have decayed. But what if the muons are moving at high speed? Say that 10,000 muons are in motion at $v = 0.98c$. The number of surviving muons is 5000 only after the elapsed time is $T = 5T_0 = 5(1.52 \times 10^{-6} \text{ s}) = 7.60 \times 10^{-6}$ s. This means that muons in flight at high speed live longer than muons at rest. The increased lifetime arising from time dilation can be observed directly by noting that if a high-speed unstable particle lives longer than the same particle at rest, it will travel a correspondingly greater distance before decaying.*

The increased lifetime of an unstable particle in motion at high speed reflects the properties of time itself (or more properly, space-time), not any physical mechanism. Time dilation applies to any processes taking place with time, including those in biochemical systems. Consider the famous twin paradox, first introduced by Einstein. There are two identical twins. One stays home while the other goes on a round trip in a spaceship to and back from some distant point at such a high speed that time-dilation effects are significant, and is finally reunited with her stay-at-home sister. Time is dilated for the traveling twin, and when the sisters compare their ages on being reunited, they agree that the stay-at-home twin is older than the traveling twin, for whom the time-dilation effect has introduced a shorter period of elapsed time. (The effect is not reciprocal. Whereas the stay-at-home twin has an uneventful history, the traveling twin experiences three profound shocks—the first as the spaceship takes off suddenly; the second as it comes suddenly to rest on arriving at its far destination and immediately reaccelerates to start the homeward portion of the trip; and the third as the spaceship arrives home and is suddenly brought to rest.)

Space Contraction A time interval between two events depends on the observer's state of motion. Similarly, a spatial interval, or length, may, because of the constancy of c, depend on the state of motion of the observer.

We use the same arrangement as before (Figures 40-3 and 40-4). A tube through which a light pulse travels up and down is at rest in reference frame S. As observed by S', the tube and observer S travel left at speed v, as shown in Figure 40-5. To make the events more picturesque, suppose that observer S' places his meter stick along his x' axis, and that the light pulse makes one burn mark on this meter stick when it leaves the base of the tube and a second burn mark when it returns to the base. The distance between the two burn marks on the meter stick of S' is L_0. The subscript zero emphasizes that this length is measured by observer S', who is at rest relative to his own meter stick.

As indicated earlier, the time interval between the markings of the two spots, as measured by S', is the dilated time T. Reference frame S moves left at speed v over a distance L_0 in time interval T. (All these quantities are measured by S'.) So we have

$$v = \frac{L_0}{T}$$

---

* Significant numbers of high-speed muons, created in the upper portions of the earth's atmosphere when energetic particles from outer space (cosmic radiation) strike the earth, can be observed at sea level because of the time-dilation phenomenon. Without this effect, most muons would not have survived decay long enough, and therefore traveled far enough, to reach the bottom of the atmosphere at sea level.

"The Italian Navigator Has Just Landed."

The moment was truly awesome. Forever after, history would be divided into what happened before or after. At 5:30 A.M. on July 20, 1945, at a remote spot on a desert in New Mexico, the sudden flash was so bright, so blinding, that observers nearly 10 miles from the site of the first nuclear explosion would later struggle for words adequate to describe it.

One of the observing physicists counted seconds from the time of the flash to know when the shock wave, produced by the sudden compression of air at the explosion site, and traveling outward at the speed of sound, would reach him. When it did, he released pieces of paper and watched them being buffeted. He was able to tell at once—because of a computation he had made in advance—that the total "yield" of the plutonium bomb was close to 10^{13} J (the equivalent of about 20 kilotons of TNT).*

Such was the special genius of Enrico Fermi (1901–1954); his penetrating insight could make tough problems look simple.

Fermi mastered analytical geometry at age 10. It was not until he was 14, however, that he first studied physics—he learned it by himself from a sixty-year-old textbook written in Latin. He got his Ph.D. in the very shadow of the place where an earlier Italian physicist had done important physics, at the University of Pisa. World renowned by age 25, Fermi was installed as the occupant of the first chair in physics at the University of Rome. There he formed what amounted to a school of physics with other Italian colleagues. He did fundamental theoretical work: on quantum statistics, on basic processes in the beta decay of unstable nuclei.

He also began systematic experimental studies on the effect of slow neutrons on various elements. He started with hydrogen and worked his way up the periodic table. When he got to uranium he noticed that peculiar things happened when this element was bombarded with slow neutrons—he thought that transuranic elements were being produced. The correct explanation, that uranium was undergoing nuclear fission, eluded him.

He was given permission to leave Mussolini's fascist Italy to pick up the 1938 Nobel prize in physics at Stockholm. Fermi never returned. Instead, he came to the U.S.A., where his slow-neutron experiments continued. Indeed, the famous letter from Albert Einstein to Franklin D. Roosevelt of August 2, 1939, in which Einstein urged the U.S. President to give prompt governmental attention to the implications of the large energy release in nuclear fission, began with the words "Some recent work by E. Fermi and L. Szilard. . . ."

Fermi was made director of the Manhattan Project at the University of Chicago to produce a controlled self-sustaining nuclear reaction. If energy could be released in a controlled way, it could most likely also be released suddenly. The site of the experiment was the squash courts in the unused football stadium (the University of Chicago had given up intercollegiate football). The big day was December 2, 1942; then

*From *Lawrence and Oppenheimer*, Nuel Pharr Davis (Simon and Schuster, New York, 1968), page 241.

Now consider the distance between the same two burn marks measured by S. Observer S sees S' and his meter stick in motion to the right at speed v, as shown in Figure 40-5. We call L the distance between the two burn marks on the moving meter stick, as measured by observer S. (Observer S, in measuring the length of an object in motion with his own meter stick, must be sure to mark the locations of the two burn marks *simultaneously*.) Relative to S, reference frame S' advances to the right at speed v over a distance L; moreover, this occurs in the *undilated* time interval T_0. Therefore, observer S may write

$$v = \frac{L}{T_0}$$

Eliminating v from the two equations above, we have

the reactor first "went critical," with the uranium generating more energy than it consumed.

The news was telephoned in code by Arthur H. Compton, also of the University of Chicago, to James B. Conant, a leading chemist, president of Harvard University, and another principal figure in the Manhattan Project.

Compton: "Jim, you'll be interested to know that the Italian navigator has just landed in the New World."
Conant: "Were the natives friendly?"
Compton: "Everyone landed safe and happy."

Contemporary physics is replete with references, direct and indirect, to Fermi:

- Fermium is the transuranic element of atomic number 100.
- The fermi is a unit of distance (1 fm = 10^{-15} m that just happens to agree with the official SI unit, femtometer).
- Fermi-Dirac statistics governs the quantum behavior of particles with half-integral spin (called fermions).
- The *Fermi energy*, and the *Fermi surface* show up repeatedly in solid-state physics.
- The Fermi prize (first recipient, Enrico Fermi) is awarded annually by the U.S. Department of Energy for outstanding achievements in nuclear energy.
- Fermilab, near Batavia, Illinois, is a large high-energy research establishment devoted to the study of elementary particles. Protons with a kinetic energy up to 1 TeV (10^{12} eV), the most energetic particles produced by man, are hurled head-on at antiprotons with the same kinetic energy in the huge accelerating machine known as the Tevatron.

Fermi was a person of prodigious and diverse talents. He did not need to keep an extensive library of books because he found that, without effort, he would pretty much memorize everything he had read. He could recite long stretches of Dante's *Divine Comedy* verbatim. Once, when his automobile broke down in a remote place and it was taken to a small repair shop, *he* fixed the car promptly (and was offered a job as repairman on the spot by the owner). He liked sports requiring physical stamina—especially swimming, mountain climbing, skiing.

Fermi delighted above all in posing interesting problems and solving them. A problem in physics that may at first seem very hard but can be solved—with insight—in just a few deft steps has become known as a "Fermi-type" problem. (Ph.D. candidates in physics at the University of Chicago were once given a test written by Fermi with just one question: "How deep a hole can you dig?")

Fermi also used to pose and solve questions of a more general type. For example, "Within an order of magnitude, how many piano tuners are there in Philadelphia?" How can you possibly do a problem like that? Be bold, but also be reasonable. Use whatever knowledge you have from past experience; if you're not sure, you make a sensible estimate, or still better, a couple of estimates from different, independent approaches. Try it, it works. One of the nice things about this type of question (Philadelphia piano tuners) is that, after you have your final number, you can look up the answer in the back of the book—the Yellow Pages of the Philadelphia Telephone Directory.

$$L = L_0 \left(\frac{T_0}{T}\right)$$

and using the time-dilation relation (40-4), we have finally

$$L = L_0 \sqrt{1 - (v/c)^2} \qquad (40\text{-}5)$$

This is the basic *space-contraction* relation. The terms mean this: L_0 is the spatial separation, or *proper length*, between two points (lying along the line of relative motion of S and S') and measured by an observer at rest with respect to these points. The contracted length L is the distance between the same two points as measured by an observer traveling at speed v relative to them. We emphasize again that to measure properly the length of an object in motion, the

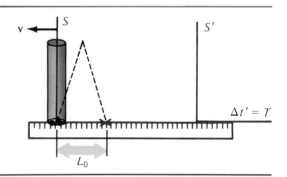

Figure 40-5. The light pulse of Figure 40-3 makes burn marks on a meter stick attached to reference frame S' as the light flash departs from and returns to the base of the tube. The separation distance is L_0 and the elapsed time interval is T, both relative to S'.

two ends must be marked *simultaneously*. Lengths are contracted along the direction of relative motion, but not at right angles to this direction.

As (40-5) shows, in general $L < L_0$. Length contraction is significant only for high speeds; for example, with $v = 0.1c$, $L = 0.995L_0$, but for $v = 0.98c$, we have $L = \frac{1}{5}L_0$. At low speeds ($v/c \to 0$), the relative spatial intervals and time intervals of relativity physics become effectively the absolute space and time intervals of classical physics: $L = L_0$ and $T = T_0$.

The term *space* contraction is used to emphasize that the effect is not due to a physical cause, for example, shrinking because of external pressure or a drop in temperature. The effect reflects instead that space and time intervals are fundamentally changed by the requirement that all observers measure the same speed for light.

Example 40-3. A long, fast spaceship passes an observer stationed at a post fixed to the earth. The spaceship travels at $0.8c$ relative to the earth observer at the post, who notes that the back end of the spaceship passes the post 42 μs after the front end. (a) What is the time interval, relative to observers at rest within the spaceship, elapsing between the instant when the front end aligns with the post and the later instant when the back end is aligned with it? (b) How long is the spaceship, as measured by the crew at rest within the spaceship? (c) What is the length of the spaceship, as measured by observers on earth?

(a) There are basically two events in this problem:

- Front end aligns with post.
- Back end aligns with post.

Relative to the earth observer, the two events take place at the same location, separated by 42 μs. Therefore, this is the *rest* time T_0. The dilated time interval for the same two events relative to spaceship observers is

$$T = \frac{T_0}{\sqrt{1 - (v/c)^2}} = \frac{42 \text{ μs}}{\sqrt{1 - (0.8)^2}} = 70 \text{ μs}$$

(b) Relative to spaceship observers, the post travels by at $0.8c$ for 70 μs. The length of the spaceship, clearly at rest relative to them, is

$$L_0 = (0.8c)(70 \text{ μs}) = (0.8 \times 3.0 \times 10^8 \text{ m/s})(70 \times 10^{-6} \text{ s}) = 16.8 \text{ km}$$

(c) The earth observer at the post sees the spaceship in motion, and therefore contracted, with a length of

$$L = L_0 \sqrt{1 - (v/c)^2} = (16.8 \text{ km}) \sqrt{1 - (0.8)^2} = 10.1 \text{ km}$$

What this means in more detail is this. The observer at the post and a colleague also at rest on earth and separated from him by 10.1 km would find that the two ends of the spaceship were at their respective locations at the same time; that is, when the two earth observers got together later to compare notes, they would find that each of their two watches read the same time when an end of the spaceship was at their location.

40-4 Relativistic Momentum

The constancy of the speed of light fundamentally changes time intervals, space intervals, and the ways in which velocities combine. What about mechanics and such a basic quantity as momentum, which clearly depends on speed? How if at all is it altered for high speeds? What is its appropriate relativistic form?

The starting point is a second postulate of special relativity theory. This postulate, which is also fundamental to classical mechanics, is often unstated or ignored, not because it is so subtle, but because it is so transparently obvious. It is this. *The laws of physics are the same in all inertial frames of reference.* An inertial frame is, of course, a frame of reference in which the law of inertia, or Newton's first law of motion, holds: In an inertial frame an undisturbed object has a constant velocity.* Now the invariance of the laws of physics for all inertial frames means simply this. Such fundamental propositions as the momentum conservation principle, and the mass conservation principle, if they are to be truly laws of physics—propositions that are universally valid—cannot depend on the particular inertial frame in which they are applied. One inertial frame must be just as good as any other.

We consider a very simple collision in two different inertial reference frames. We shall insist that in each of the two reference frames, the total momentum before the collision equals the momentum after and also that mass before equals mass after.

The collision is shown in Figure 40-6, in (a) the center-of-mass frame S' and (b) a laboratory frame S. In more detail we have:

- (a) CM frame. Two identical particles are fired head-on, each at speed V'. They stick together and form a composite particle with mass M_0. This composite object must be at rest if total momentum after the collision is to equal total momentum before. The result also follows simply from the requirements of symmetry; if what happens on the left is mirrored on the right, the composite can show no preference for left or right.
- (b) Lab frame. Here we have the very same collision, but now as seen by an observer in a laboratory reference frame S in which the particle on the right is initially at rest. This means that the composite particle must have velocity V' as measured in the lab.† The mass of the composite is labeled M, *different* from its rest mass M_0; here we anticipate that if we are still to write relativistic momentum as the product of mass and velocity, the mass thus defined may not be independent of speed. In similar fashion, the mass of the single particle at rest is written m_0, and the mass of the single particle in motion is m.

Applying in turn mass and momentum conservation to the collision in Figure 40-6(b) yields:

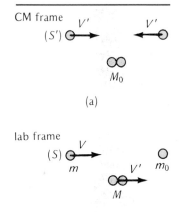

Figure 40-6. (a) In the center-of-mass reference frame (S'), two identical particles moving initially at speed V' collide head-on and produce the composite particle of mass M_0 at rest. (b) The same collision as viewed from the laboratory reference frame (S), in which the particle on the right is initially at rest with mass m_0. The composite particle with mass M has speed V'. The other particle in motion with mass m has speed V.

---

* The special theory of relativity is restricted to inertial frames; the general theory of relativity includes accelerated reference frames as well.
† This result follows simply from the fact that if A has velocity V' relative to B, then B has velocity $-V'$ relative to A.

Mass conservation: $\quad m + m_0 = M \quad$ (40-6)

Momentum conservation: $\quad mV + m_0(0) = MV' \quad$ (40-7)

If we were applying classical mechanics, we should say simply that $M = M_0 = 2m_0$ and that $V = 2V'$. But these results do not hold for relativistic speeds. Eliminating M from the two equations above yields

$$\frac{V}{V'} = 1 + \frac{m_0}{m} \qquad (40\text{-}8)$$

Velocities V and V' are also related to one another by the general velocity transformation relation

$$V' = \frac{V - v}{1 - Vv/c^2} \qquad (40\text{-}2)$$

In Figure 40-6, V and V' are the velocities in S and S' of the particle on the left. The relative velocity between the two reference frames here is $v = V'$, so that the relation above becomes

$$V' = \frac{V - V'}{1 - VV'/c^2} \qquad (40\text{-}9)$$

We find after some manipulation that this equation can be written as*

$$\frac{V}{V'} = 1 + \sqrt{1 - \left(\frac{V}{c}\right)^2} \qquad (40\text{-}10)$$

Comparing (40-10) with (40-8), we see that

$$\frac{m_0}{m} = \sqrt{1 - \left(\frac{V}{c}\right)^2}$$

or solving for m,

$$m = \frac{m_0}{\sqrt{1 - (V/c)^2}} \qquad (40\text{-}11)$$

Keep in mind what *relativistic* mass m means. It is the quantity by which

---

* Equation (40-9) can be rearranged to become

$$V' - \frac{VV'^2}{c^2} = V - V'$$

Now multiply both sides by c^2/V^3 and collect terms:

$$\left(\frac{V'}{V}\right)^2 + \left(\frac{-2c^2}{V^2}\right)\left(\frac{V'}{V}\right) + \frac{c^2}{V^2} = 0$$

This is a quadratic equation in V'/V, whose solution is

$$\frac{V'}{V} = \frac{1 \pm \sqrt{1 - (v/c)^2}}{(V/c)^2}$$

Taking the reciprocal of this equation and multiplying numerator and denominator on the right by $1 \pm \sqrt{1 - (V/c)^2}$ give finally (40-10). The plus sign was chosen; this ensures that for $V/c \ll 1$, we get the simple classical result $V/V' = 2$.

velocity must be multiplied to yield the quantity, relativistic momentum **p**, that is conserved in every collision.

Therefore,

$$\mathbf{p} = m\mathbf{V} = \frac{m_0 \mathbf{V}}{\sqrt{1 - (V/c)^2}} \quad (40\text{-}12)$$

Equation (40-12), which relates the momentum to rest mass m_0, is the fundamental relation; (40-11), which relates m to m_0, is not. How m varies with speed is shown in Figure 40-7. At low speed, $m \simeq m_0$; at speeds close to c, mass m becomes infinite. This means that relativistic momentum increases with velocity at a higher rate than the classical relation $\mathbf{p} = m_0 \mathbf{V}$, to which it reduces for low speeds.

An electrically charged particle moves at high speed at right angles to the field lines of a uniform magnetic field. We take the force on the particle to be the time rate of change of momentum, so that

$$\mathbf{F} = \frac{d(m\mathbf{V})}{dt} = \frac{dm}{dt}\mathbf{V} + m\frac{d\mathbf{V}}{dt}$$

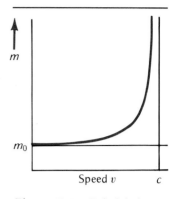

Figure 40-7. Relativistic mass m of a particle as a function of its speed v.

The particle's speed is constant and so is its relativistic mass m, so that $dm/dt = 0$. Then the relation above becomes

$$\mathbf{F} = m\frac{d\mathbf{V}}{dt} = m\mathbf{a}$$

For charge q at speed V in uniform magnetic field **B**, we have

$$F = ma$$

$$qVB = m\frac{V^2}{r}$$

$$mV = \frac{m_0 V}{\sqrt{1 - (V/c)^2}} = qrB \quad (40\text{-}13)$$

The relativistic momentum mV is directly proportional in the magnitude of **B** and radius of curvature r. Equation (40-13) provides a simple means of measuring relativistic momentum; simply measure the particle's curvature in a known magnetic field.

40-5 Relativistic Energy

To find the relation for relativistic kinetic energy E_k, we merely do what is done in classical physics: find the work done by a force in bringing a particle from rest to the final speed v.* We write F as $d(mv)/dt$ and write $ds = v\, dt$:

$$E_k = \int_0^s F\, ds = \int_0^s \frac{d}{dt}(mv)\, ds = \int_0^t \frac{d}{dt}(mv)v\, dt$$

$$= \int v\, d(mv) = \int (v^2\, dm + mv\, dv)$$

---

* Henceforth particle speed is given by a lowercase v.

To integrate the right side, we must recognize that both m and v are variables. The dependence of m on v is given by (40-11). It is simpler to express v in terms of m and then integrate with respect to m as variable. The required relation is*

$$mv \, dv = (c^2 - v^2) \, dm \tag{40-14}$$

Substituting for $mv \, dv$ in the equation for the kinetic energy E_k, we then have

$$E_k = \int_{m_0}^{m} [v^2 \, dm + (c^2 - v^2) \, dm] = c^2 \int_{m_0}^{m} dm = mc^2 - m_0 c^2$$

$$E_k = (m - m_0)c^2 \tag{40-15}$$

The relativistic kinetic energy can be regarded as the increase in mass, arising from the particle's motion, multiplied by c^2. At high speeds, the relativistic energy is markedly different from the classical kinetic energy, $\frac{1}{2}m_0 v^2$. The relativistic energy must reduce to the familiar $\frac{1}{2}m_0 v^2$ for $v/c \ll 1$. This is shown in the footnote to Section 9-1. Be careful; relativistic kinetic energy is *not* given by $\frac{1}{2}mv^2$, where m is the relativistic mass. Let us write (40-15) as

$$E_k = E - E_0 = mc^2 - m_0 c^2 \tag{40-16}$$

Here E represents the particle's *total energy*,

$$E \equiv mc^2 = \frac{m_0 c^2}{\sqrt{1 - (v/c)^2}} \tag{40-17}$$

and E_0 is the particle's *rest energy*,

$$E_0 \equiv m_0 c^2 \tag{40-18}$$

(For a system of particles, rest energy E_0 and rest mass m_0 are the system's total energy and mass when its center of mass is at rest.)

Equation (40-17) is the famous Einstein relation. It implies an equivalence of energy and mass, in which mass and energy are interpreted as different manifestations of the same physical entity. A particle at rest has rest mass m_0 and rest energy $m_0 c^2$; in motion, its mass and energy are m and mc^2. Mass and energy need not be regarded as having separate conservation laws of energy and of mass; instead, we consider the two combined into a single, simple law, the conservation law of mass-energy.

Equation (40-17) gives E in terms of v. It is often convenient to express energy E in terms of p. We find the relation by squaring (40-11) and multiplying both sides by $c^4[1 - (v/c)^2]$. We get

$$m^2 c^4 - m^2 v^2 c^2 = m_0^2 c^4$$

---

* Equation (40-11) may be written as

$$1 - \left(\frac{v}{c}\right)^2 = \left(\frac{m_0}{m}\right)^2$$

Taking the differential of this relation gives

$$\frac{-2v \, dv}{c^2} = \frac{-2m_0^2 \, dm}{m^3}$$

Combining this relation with the one immediately above it gives (40-14).

This equation can immediately be written more simply as

$$E^2 = (pc)^2 + E_0^2 \qquad (40\text{-}19)$$

Here are the relations in dynamics for the two extreme limits of particle speed:

- Very low speeds, $v/c \ll 1$ (classical limit):

$$m \simeq m_0 \qquad E_k = \tfrac{1}{2}m_0 v^2$$
$$p = m_0 v \qquad E_0 \gg E_k$$

- Very high speeds, $v/c \simeq 1$ (extreme relativistic limit):

$$m \gg m_0 \qquad E_k \simeq E \simeq pc$$
$$p \simeq E/c \qquad E_0 \ll E_k$$

Here are the most appropriate forms and units for expressing relativistic quantities for particles:

- Speed, relative to the speed of light: v/c.
- Energy in electron volts (or such related units as kilo-, mega-, or giga-electron-volts).
- Mass in unified atomic mass units, where

$$1 \text{ u} = \tfrac{1}{12} \text{ mass of carbon-12 atom}$$
$$= 1.6606 \times 10^{-27} \text{ kg} = 931.5 \text{ MeV}/c^2$$

- Momentum as energy in eV divided by the speed of light, or eV/c.

Example 40-4. (a) A proton ($E_0 = 0.938$ GeV) is accelerated from rest across an electric potential difference of 500 V. What is its momentum?

(b) A proton is accelerated in the high-energy accelerator at Fermi National Laboratory, Batavia, Illinois, so that its final kinetic energy is 500 GeV. What is the momentum of such a proton?

(a) The proton's final kinetic energy is

$$E_k = qV = e(500 \text{ V}) = 500 \text{ eV}$$

Since the proton's kinetic energy is much less than its rest energy (500 eV versus $0.938 \text{ GeV} = 0.938 \times 10^9$ eV), we can use the classical kinetic-energy relation, $E_k = p^2/2m_0$:

$$p = \sqrt{2m_0 E_k} = \frac{\sqrt{2(m_0 c^2) E_k}}{c}$$

$$= \frac{\sqrt{2(0.938 \times 10^9 \text{ eV})(500 \text{ eV})}}{c} = 0.97 \text{ MeV}/c$$

(b) The kinetic energy of a 500-GeV proton is much greater than its rest energy, so that the relation $p = E/c$, which applies for very high energies, can be applied here.

$$p = \frac{E}{c} = \frac{E_k}{c} = \frac{500 \text{ GeV}}{c} = 5.0 \times 10^5 \text{ MeV}/c$$

Example 40-5. As shown in Example 40-4, a 500-eV proton has a momentum of $0.97 \text{ MeV}/c$. What is the radius of curvature of the proton's path when it enters a uniform magnetic field of 0.40 T at right angles to the field lines?

From $p = mv = qrB$, (40-13), we get

$$r = \frac{p}{qB} = \frac{(0.97 \times 10^6 \text{ eV}/c)(c/3.0 \times 10^8 \text{ m/s})}{(1.6 \times 10^{-19} \text{ C})(0.40 \text{ T})(1 \text{ eV}/1.6 \times 10^{-19} \text{ J})} = 8.1 \text{ mm}$$

40-6 Mass-Energy Equivalence and Bound Systems

To see the significance of the conservation law of mass-energy, we consider two situations: unbound systems and bound systems.

Unbound Systems Two particles, each with a rest mass m_0, are projected toward one another, each with speed v relative to an observer in the center-of-mass reference frame. The collision is perfectly inelastic, and the two particles stick together to form a single, composite particle, with rest mass M_0. How is rest mass M_0 of the composite particle related to the rest mass m_0 of each of the separate incident particles? Classical physics would say that M_0 equals $2m_0$ exactly. But in relativity this is not true.

The total energy of the two particles before collision, $2mc^2$, must equal the total energy M_0c^2 of the composite particle after the collision. The total energy of the composite after collision is entirely rest energy, since this object is at rest. The total energy of the particles before collision is, however, their rest energy plus their kinetic energy. Mass-energy conservation then gives

$$M_0c^2 = 2mc^2 = \frac{2m_0c^2}{\sqrt{1-(v/c)^2}}$$

or

$$M_0 = \frac{2m_0}{\sqrt{1-(v/c)^2}}$$

The rest mass M_0 of the composite object exceeds the total rest mass $2m_0$ of the incident particles. What has effectively happened is that the kinetic energy of the two particles has become a part of the rest energy of the combined particles after the collision.

Example 40-6. Two satellites, each with a rest mass of 4000 kg, travel in orbits in opposite directions at a speed of 8.0 km/s with respect to an earth observer. They happen to collide head-on and stick together. What is the change in the total rest mass of the system?

The satellites have equal but opposite momenta; their total momentum is zero. After the collision, the composite object is at rest. The kinetic energy of the incident satellites is converted to rest mass, where

$$\text{Increase in rest mass} = \Delta m = \frac{2E_k}{c^2}$$

where E_k is the initial kinetic energy of each satellite. The speed of each satellite, 8.0 km/s, is much less than the speed of light, and we can properly use the classical expression $E_k = \frac{1}{2}m_0v^2$ for the kinetic energy. Therefore,

$$\Delta m = \frac{2(\frac{1}{2}m_0v^2)}{c^2} = m_0\left(\frac{v}{c}\right)^2 = (4000 \text{ kg})\left(\frac{8.0 \times 10^3 \text{ m/s}}{3.0 \times 10^8 \text{ m/s}}\right)^2 = 2.8 \text{ mg}$$

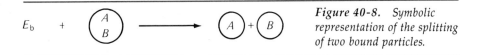

Figure 40-8. *Symbolic representation of the splitting of two bound particles.*

If the two satellites could collide and form a single composite object whose mass could, at least in principle, be measured on a balance, then one would find it to be nearly 3 mg greater than the mass of the two satellites taken separately.

Bound Systems Now consider two particles A and B bound together to form a bound system. To break the composite object into its component parts requires work. We must add energy to the system. The rest mass of the composite system is M_0; the rest masses of the individual particles are m_{0A} and m_{0B}.

The breaking up of the bound system into separated parts is shown symbolically in Figure 40-8. Here E_b is the energy that must be added to the system in order to separate the particles completely. Energy E_b is called the *binding energy*.

Applying mass-energy conservation, we have

$$M_0 + \frac{E_b}{c^2} = m_{0A} + m_{0B} \qquad (40\text{-}20)$$

Since the system is bound and $E_b > 0$, it follows from the equation above that $M_0 < m_{0A} + m_{0B}$. The rest mass of the bound system is *less* than the sum of the rest masses of the individual particles when separated.* The binding energy E_b can be computed simply by knowing the system's rest mass and the rest masses of its constituents. Only for particles bound by the very strong nuclear forces within an atomic nucleus is the binding energy sufficiently great that the mass difference can actually be measured.

Example 40-7. (a) It takes 13.6 eV to ionize a hydrogen atom. By what fraction is the mass changed during the ionization?

(b) For nuclei, the process analogous to ionizing hydrogen is separating a deuteron nucleus into a proton and a neutron. It takes 2.2 MeV. By what fraction is the nuclear mass changed in this process?

(a) If energy is added to a hydrogen atom to separate it into a proton and an electron, then the sum of the masses of proton and electron exceeds the hydrogen mass by $\Delta m = \Delta E/c^2$; here $\Delta E = 13.6$ eV. The rest energy of a proton (and of a hydrogen atom) is about 0.94 GeV. The fractional increase in mass is then

$$\frac{\Delta m}{m_0} = \frac{\Delta E/c^2}{m_0} = \frac{\Delta E}{m_0 c^2} = \frac{13.6 \text{ eV}}{0.94 \text{ GeV}} = 1.4 \times 10^{-8}$$

The difference is only 1.4 parts in 10^8, a typical value for a chemical process and far too small to be measurable in the masses of the particles.

(b) The rest energy of a proton or of a neutron is approximately 0.94 GeV, and the mass of a deuteron is approximately twice this amount. Strictly, the total mass of the separated proton and neutron exceeds the deuteron mass because energy must be added to the bound system to separate the particles. Here the fractional mass increase is

---

* Compare with the unbound systems above, in which the rest mass of the composite exceeded the rest masses of the separated particles.

$$\frac{\Delta m}{m_0} = \frac{\Delta E}{m_0 c^2} = \frac{2.2 \text{ MeV}}{2(0.94 \text{ GeV})} = 1.2 \times 10^{-3}$$

The mass difference for this nuclear binding of particles, 1.2 parts in a thousand, is so large that it is easily discernible in the masses of the particles.

40-7 The Lorentz Transformations (Optional)

Here, for the record, are the relativistic transformation relations between the three space and one time coordinates (x, y, z, t) in reference frame S and the corresponding coordinates (x', y', z', t') for a second reference frame S' in motion with velocity v relative to S. They are called the Lorentz transformations, after H. E. Lorentz, who first introduced them in 1903 (before Einstein's relativity theory).

Here are the assumptions:

- Corresponding axes are parallel, x with x', y with y', z with z'.
- The two origins coincide at the times when $t = t' = 0$.
- Relative velocity v is of S' along the x axis relative to S. (By the same token, S has velocity $-v$ relative to S'.)
- The transformation relations are so constructed that all observers measure the same speed c for light.

The Lorentz coordinate transformation relations are:

$$x' = \frac{x - vt}{\sqrt{1 - (v/c)^2}} \quad (40\text{-}21\text{a})$$

$$y' = y \quad (40\text{-}21\text{b})$$

$$z' = z \quad (40\text{-}21\text{c})$$

$$t' = \frac{t - (v/c^2)x}{\sqrt{1 - (v/c)^2}} \quad (40\text{-}21\text{d})$$

Simply by looking at (40-21) we can see the following.

- Space and time coordinates are not independent: x' depends on t, as well as on x and v. Much more surprising, however, is that t' depends on x, as well as on t, as shown by (40-21d). What S' reads on his clock actually depends, not only on the reading (t) on the clock of S, but also on where (x) S is located.
- For $v \ll c$, the relativistic coordinate transformation relations reduce to the intuitively obvious low-speed relations for space and time: $x' = x - vt$; $y = y'$; $z = z'$; and $t' = t$.
- Equation (40-21) gives x', y', z', t' in terms of x, y, z, t. To get the inverse transformation relations, replace every primed quantity by the corresponding unprimed coordinate (and conversely) while also replacing v by $-v$.

Some important consequences follow immediately.

- The *relativity of simultaneity*. Two spatially separated events that are simultaneous in one reference frame occur in sequence in a second reference frame that is in motion relative to the first. Let two simultaneous events take place in S, one at time $t = 0$ at the origin and the second at the coordinate x. Then from (40-21d), the times of these same two events measured in S' differ by the time interval $(v/c^2)x/\sqrt{1-(v/c)^2}$. Einstein considered this the most profound result in relativity.

- *Time dilation.* A clock is at rest at the origin in S, so that its coordinate is $x = 0$; and we can take $t = T_0$, the rest time. The time interval for the same events as observed by S' is $t' = T$. Equation (40-21d) then gives

$$T = \frac{T_0}{\sqrt{1-(v/c)^2}} \tag{40-4}$$

- *Space Contraction.* Let a rod be at rest, with one end at the origin of S' and the other at coordinate x'. The rod's rest length is $x' = L_0$. The rod is in motion in S with its two ends observed simultaneously (same t), so that the rod's contracted length is $x = L$. Equation (40-21a) then gives

$$L = L_0\sqrt{1-(v/c)^2} \tag{40-5}$$

- *Relativistic velocity transformations.* Let the velocity components of a particle observed in S be V_x, V_y, and V_z. The velocity components of the same particle observed in S' carry primes. By definition $V_x = dx/dt$ and $V' = dx'/dt'$, and so on for the other components. Taking the differential of both sides of (40-21a) and (40-21d) gives

$$dx' = \frac{dx - v\,dt}{\sqrt{1-(v/c)^2}} \quad \text{and} \quad dt' = \frac{dt - (v/c^2)\,dx}{\sqrt{1-(v/c)^2}}$$

Dividing the first relation by the second and applying the definitions yields (40-22a):

$$V'_x = \frac{V_x - v}{1 - (v/c^2)V_x} \tag{40-22a}$$

$$V'_y = \frac{V_y\sqrt{1-(v/c)^2}}{1 - (v/c^2)V_x} \tag{40-22b}$$

$$V'_z = \frac{V_z\sqrt{1-(v/c)^2}}{1 - (v/c^2)V_x} \tag{40-22c}$$

Equation (40-22b) results from taking the differentials of (40-21b) and (40-21d); and similarly for (40-22c).

Summary

Definitions
Rest energy E_0 of a particle with rest mass m_0:

$$E_0 \equiv m_0 c^2 \tag{40-18}$$

Total relativistic energy E:

$$E \equiv mc^2 \tag{40-17}$$

where relativistic mass m:

$$m \equiv \frac{m_0}{\sqrt{1-(v/c)^2}} \tag{40-11}$$

Fundamental Principles
Postulates of the special theory of relativity:

- The speed of light has the same value for all observers, whatever the state of motion of the light source or observer.

- The laws of physics have the same form for all observers in inertial reference frames.

Important Results

Relativistic velocity transformation relation:

$$V' = \frac{V - v}{1 - (v/c^2)V} \quad (40\text{-}2)$$

where V' is particle velocity relative to S',
V is particle velocity relative to S, and
v is velocity of S' relative to S.
All three velocities are along the same line.

Time dilation: T_0 = time interval between two events taking place at the same location in reference frame S;

T = time interval between same two events as observed in a reference frame with velocity v relative to S:

$$T = \frac{T_0}{\sqrt{1 - (v/c)^2}} \quad (40\text{-}4)$$

Length contraction: L_0 = length (along direction of relative motion) in which two end points are at rest in S;

L = length between same two points as measured simultaneously in reference frame with a velocity v relative to S:

$$L = L_0 \sqrt{1 - (v/c)^2} \quad (40\text{-}5)$$

Relativistic dynamics: The relativistic forms of dynamical quantities assure that the momentum and mass-energy conservation laws are satisfied in all inertial reference frames.

Relativistic momentum **p**:

$$\mathbf{p} = m\mathbf{v} = \frac{m_0 \mathbf{v}}{\sqrt{1 - (v/c)^2}} \quad (40\text{-}12)$$

Relativistic kinetic energy E_k:

$$E_k = mc^2 - m_0 c^2 = E - E_0 \quad (40\text{-}15)$$

Relation between E, E_0, and p:

$$E^2 = E_0^2 + (pc)^2 \quad (40\text{-}19)$$

Charged particle in magnetic field (with $\mathbf{v} \perp \mathbf{B}$):

$$p = mv = qrB \quad (40\text{-}13)$$

The relativistic momentum $p = mv$ is proportional to the particle charge q, the radius r of curvature of the path, and the magnitude B of the magnetic field.

To separate a composite particle with rest mass M_0 into its constituent parts with rest masses m_{0A} and m_{0B}, binding energy E_b must be added where

$$M_0 + \frac{E_b}{c^2} = m_{0A} + m_{0B} \quad (40\text{-}20)$$

Problems and Questions

Section 40-1 The Constancy of the Speed of Light

· **40-1 Q** A fundamental postulate on which the theory of relativity is based is that
(A) light in vacuum always moves with speed c, independent of the velocity of the source or of the observer.
(B) everything is relative.
(C) mass and energy must be equivalent.
(D) the form of the equations of physics depends on the velocity of the reference frame used.
(E) what appears to be vacuum is not really empty space but instead, a substance called ether.

: **40-2 Q** Consider the following thought (*gedanken*) experiment. A flashlamp is placed at the exact center of a boxcar that is traveling at high constant velocity. A man stationed in the car, by using photocells, is able to measure the time at which a light pulse emitted from the lamp strikes each end of the car. A woman at rest on earth can also measure the time at which the light pulses strike the ends of the moving boxcar. By considering such an experiment, Einstein was able to conclude that
(A) the velocity of light is independent of the motion of the source and of the observer.
(B) moving clocks run slow.
(C) a moving boxcar is shortened in its direction of motion.
(D) events that are simultaneous in one inertial frame may not be simultaneous in another.
(E) the laws of physics are the same in all inertial frames.

Section 40-2 Relativistic Velocity Transformations

· **40-3 P** Relative to some observer, A moves east at $0.8c$ and B moves west at $0.8c$. What is the velocity of B relative to A?

: **40-4 P** A particle has a velocity of $0.8c$ to the east relative to a train. The train has a velocity $0.7c$ to the east relative to earth. What is the velocity of the particle relative to the earth?

: **40-5 P** Particle A travels north at $0.1c$. Particle B travels south at $0.1c$. What is the speed of B relative to A computed from (a) the relativistic velocity transformation relations and (b) the classical velocity transformation relations?

: **40-6 P** Any attempt to "piggy-back" speeds to reach a value over c is frustrated by the relativistic velocity transformation relations. To visualize a specific situation, consider the following arrangement. A ball is thrown at speed $0.9c$ relative to a cart. The cart has speed $0.8c$ relative to a train. The train has speed $0.7c$ relative to the earth. All velocity vectors are in the same direction. What is the velocity of the ball (a) relative to the train and (b) relative to the earth?

Section 40-3 Space and Time in Special Relativity

· **40-7 Q** You want to travel to a star 100 light-years distant. Assume your maximum life span to be only 90 years. Is it possible for you to reach the star?
(A) No, because nothing can travel faster than light.
(B) No, because your rocket ship will grow shorter and shorter as it approaches the speed of light, so that eventually it will be essentially standing still.
(C) No, because time will pass much faster for you than for a person at rest back on earth.
(D) Yes. Time will pass more slowly for you than for a person back on earth, so as your speed approaches c, the trip could last less than 90 years.
(E) Yes, because speed is only relative. Though you appear to be going at less than the speed of light to an observer on earth, you actually may be moving at speeds much greater than the speed of light.

· **40-8 Q** The proper time is
(A) time measured in any inertial frame.
(B) the shortest time interval between two events.
(C) any time interval measured in seconds, as opposed to time measured in days, months, years, and so on.
(D) greater or less than a time interval measured in another inertial frame moving with respect to the proper frame.

: **40-9 Q** If you were traveling with respect to the distant stars at a speed close to the speed of light, you could detect this by
(A) the increase in your own mass that you would experience.
(B) a change in your pulse rate.
(C) a change in your physical dimensions.
(D) all of the above means.
(E) none of the above means.

· **40-10 P** An astronaut in a satellite circles the earth at a speed of 8200 m/s for the duration of a 14-day mission (as measured by Mission Control in Houston). By how much will his age differ from that of earthbound persons when he returns?

· **40-11 P** A rocket 20 m long when at rest is 16 m long when moving past stationary observers. How fast is the rocket moving?

: **40-12 Q** An important result of special relativity is that we must give up the concept of an "absolute time" independent of the state of motion of the person measuring the time. The principal reason is that
(A) it has not been possible to determine an "origin" of time in the universe.

(B) time is simply a fourth dimension, not unlike spatial dimensions.
(C) energy and mass are equivalent.
(D) the velocity of light is the same for all observers in inertial reference frames, independent of their motion or the motion of the light source.
(E) moving objects are shortened along their direction of motion.

: **40-13 P** How fast would an object have to move so that its length appears to have decreased by a factor of 2?

: **40-14 P** A spacecraft receding from earth with a speed of $0.98c$ emits pulses of radio waves at the rate of 10,000 per second. At what rate are the pulses received on earth?

: **40-15 Q** The timer on a bomb is set to explode the bomb after a time τ (with the bomb at rest). The bomb is set in motion with velocity v. The bomb will then explode, according to a stationary observer, after a time
(A) $\dfrac{\tau}{\sqrt{1-(v/c)^2}}$
(B) $\dfrac{1}{\tau}\sqrt{1-\left(\dfrac{v}{c}\right)^2}$
(C) $\tau\sqrt{1-(v/c)^2}$
(D) τ
(E) $c\tau$

: **40-16 P** How long would a jet plane flying 1000 km/h have to fly before its clocks had lost 1 s with respect to clocks on earth?

: **40-17 P** Pions have a half-life of 2.2×10^{-8} s as measured by an observer at rest with respect to them. An observer in the laboratory sees a beam of pions traveling at $0.995c$. As the beam passes him, he counts 1000 pions. How many pions will be left after the beam travels 10 m farther, as measured in the lab?

: **40-18 P** Electrons in the Stanford Linear Accelerator (SLAC) achieve a final speed that is less than c by only 3 parts in 10^{10}. The accelerator is 2 miles long. (a) How long is the accelerator as measured in the rest frame of an electron traveling at the final speed? (b) How long does it take, as measured by an observer traveling with the electron, for an electron at the final speed to travel the length of the accelerator? (c) How long does it take for an electron with the final speed to travel the length of the accelerator, as measured by an experimenter in the lab?

: **40-19 P** Two satellite space stations are separated by a fixed distance of 1000 km as measured by an observer on one of them. What is the separation of the space stations as measured by an observer in a rocket flying between them with a velocity of $0.8c$?

: **40-20 P** An object in the xy plane moves along the x axis at speed $0.8c$. A stationary observer notes that the moving shape is a square 2.0 cm on a side and that its sides make an angle of 45° with its line of motion. What is the area of the object as measured in its rest frame?

: **40-21 P** A rod of length L_0 is attached to the roof of a car at rest and it makes an angle θ_0 with the horizontal. (a) What is the length of the rod, as determined by a stationary observer on the roadway, when the car is moving with very high speed v? (b) What is the angle between the rod and the horizontal as observed by the stationary observer on the roadway?

Section 40-4 Relativistic Momentum

· **40-22 P** By what factor does the momentum of a particle change when its speed (a) goes from $0.04c$ to $0.08c$ and (b) goes from $0.4c$ to $0.8c$?

· **40-23 P** Calculate the momentum (in MeV/c) of a proton moving with speed (a) $0.01c$; (b) $0.1c$; (c) $0.5c$; (d) $0.9c$.

: **40-24 P** An electron is moving at right angles to a magnetic field of 1.2 T in a path with a radius of 3.0 cm. What is the magnitude of the electron's momentum (expressed in units of MeV/c)?

: **40-25 P** A particle of rest mass m_1 and velocity $v_1 = 0.8c$ collides head-on with a particle of rest mass m_2 moving toward it with speed $v_2 = 0.6c$. The two stick together and are at rest in the laboratory reference frame. Find the ratio of the rest masses, m_1/m_2.

Section 40-5 Relativistic Energy

· **40-26 P** What is the momentum (in units of MeV/c) of an electron of kinetic energy (a) 10 eV and (b) 10 GeV?

· **40-27 P** (a) Through what electric potential difference must an electron be accelerated from rest for its relativistic mass to exceed its rest mass by 10 percent? (b) What would be the electron's final speed?

· **40-28 Q** Confirm that the relation between relativistic momentum p and relativistic energies E_0 and E (40-19) can be represented by a right triangle with sides pc and E_0 and hypotenuse E. Also check that this triangle, which can serve as a useful mnemonic, yields the correct results in the classical and extreme relativistic limits.

: **40-29 P** How much energy is needed to double the momentum of an electron whose velocity is 1.8×10^8 m/s?

: **40-30 P** A particle with speed $0.49c$ has its speed doubled. (a) By what factor does its momentum change? (b) By what factor does its kinetic energy change? Repeat the calculation for a particle with initial speed $10^{-5} c = 3000$ m/s. (c) By what factor does its momentum change when the speed is doubled? (d) By what factor does the kinetic energy change?

: **40-31 P** (a) What is the momentum of a 100-MeV electron? (b) What result do you get for the momentum if you use the incorrect classical relations $E_k = \frac{1}{2}m_0v^2$ and $p = m_0v$?

: **40-32 P** An electron has speed 2.4×10^8 m/s. Compute the following quantities using both the correct, relativistic expression and the classical approximation. (a) Total energy. (b) Kinetic energy. (c) Momentum.

: **40-33 P** Determine the energy (in MeV) needed to accelerate an electron from (a) $0.10c$ to $0.90c$; and (b) $0.90c$ to $0.99c$.

: **40-34 Q** An electron is accelerated from rest by an electric potential difference. Then the electron enters a uniform magnetic field at right angles to the magnetic-field lines and travels in a circular arc. Sketch a plot of the radius of the arc as a function of the accelerating potential over a range that includes the classical and extreme relativistic regions.

: **40-35 P** An electron is accelerated from rest through a potential difference of 25 kV in a TV tube. (a) What is its final speed? (b) What is the electron's final kinetic energy measured in units of the particle's rest energy?

: **40-36 P** Verify that

$$\frac{1}{\sqrt{1 - v^2/c^2}} = 1 + \frac{E_k}{m_0 c^2}$$

: **40-37 P** The most energetic cosmic rays detected are protons with energies of the order of 10^{13} MeV. Our galaxy is about 100,000 light-years across. How long would it take such a particle to traverse the Milky Way as measured in the reference frame of (a) the galaxy? (b) the particle?

: **40-38 Q** If a particle's kinetic energy is equal to its rest energy, E_0, the particle's momentum is
(A) E_0/c
(B) $2E_0/c$
(C) $E_0/2c$
(D) $\sqrt{3}E_0/c$
(E) none of the above.

: **40-39 Q** Indicate whether or not each of the following equations is a valid relativistic relation: (a) $E_k = \frac{1}{2}mv^2$; (b) $p = mv$; (c) $F = ma$; (d) $F = dp/dt$; (e) $E = m_0c^2 + \frac{1}{2}m_0v^2$.

: **40-40 P** An electron is accelerated until its total energy changes from $2m_0c^2$ to $4m_0c^2$. By what factor do each of the following change? (a) Kinetic energy. (b) Speed. (c) Momentum. (d) Mass.

: **40-41 P** In calculating, in terms of m_0c^2, what work must be done to accelerate an electron from rest to the speed $0.9c$, what result do you get (a) using Newtonian mechanics? (b) Using relativistically correct relations?

: **40-42 P** A particle of rest mass M_0 moves with speed V. It decays to two identical particles, each with rest mass m_0 and velocity v. These particles move off symmetrically to the original line of motion, making angles $+\theta$ and $-\theta$ with this direction. (a) Show that

$$m_0 = \frac{M_0(1 - V^2/c^2 \cos^2\theta)^{1/2}}{2(1 - V^2/c^2)^{1/2}}$$

(b) Determine the maximum allowed value of θ when $V = 0.6c$. (c) Evaluate m_0 when $V = 0.6c$ and $\theta = 45°$.

: 40-43 P An electron moving with speed $0.8c$ in the x direction enters a region where there is a uniform electric field in the y direction. Show that the x component of velocity of the electron will *decrease*.

Section 40-6 Mass-Energy Equivalence and Bound Systems

· 40-44 Q Tell whether the mass of the indicated system increases or decreases for each of the following processes: (a) a battery loses its charge; (b) a capacitor is charged; (c) a block of ice melts; (d) an inductor has a current established in it; (e) a spring is stretched; (f) a soldering iron is turned on.

· 40-45 P How much does the mass of one liter of water increase when it is heated 100 C°?

· 40-46 Q Suppose a nuclear bomb were exploded in a container strong enough that it could contain the explosion. All the reaction products, including radiation, are trapped within the box. Under these circumstances, the mass of the box and its contents, relative to its value just before the explosion, would
(A) increase.
(B) decrease.
(C) stay the same.
(D) any of the above, depending on the state of motion of the observer relative to the box.

· 40-47 P Two lumps of clay, each with a rest mass of 0.100 kg and a speed of $0.6c$, collide head-on and stick together. What is the rest mass of the resulting composite lump?

· 40-48 P What mass of uranium is consumed by a nuclear power plant that produces power at an average rate of 1000 MW for one year?

: 40-49 P The yield of a nuclear bomb is typically measured in megatons of TNT (that is, in terms of the energy released when 10^6 tons of TNT is detonated). One kilogram of TNT releases about 1 kcal of energy. To what rest mass is one megaton of TNT equivalent?

: 40-50 P A carbon-12 atom has an atomic mass of exactly 12 u. The $^{12}_{6}C$ nucleus consists of six protons (mass, 1.007 825 u) and six neutrons (mass, 1.008 665 u). (a) How much energy is required to dismember the $^{12}_{6}C$ entirely, into individual protons and neutrons? (b) How much energy is required to "cut" the $^{12}_{6}C$ nucleus into three helium atoms (mass of $^{4}_{2}He$, 4.002 603 u)? (All masses given above include the requisite number of electrons to make the object electrically neutral.)

: 40-51 P Each fission of a $^{235}_{92}U$ nucleus produces 200 MeV. How many uranium nuclei must undergo nuclear fission per second to yield a power of 1 kW?

: 40-52 P When a $^{235}_{92}U$ nucleus undergoes nuclear fission in a nuclear reactor, about one part in 10^3 of its rest mass is converted into kinetic energy. How much fissionable uranium-235 is needed during a one-year period for a nuclear power plant that has an average electrical output of 100 MW? Only about 30 percent of the kinetic energy of the fission particles is converted to electric energy in a nuclear power plant (a fossil-fuel plant has an efficiency of about 40 percent because it can operate at somewhat higher temperatures).

Supplementary Problems

40-53 P In principle, time-dilation effects could be used to make possible interstellar travel that would otherwise be impossible in human life spans. To accomplish this, one is faced with accelerating a spaceship to speeds sufficiently near the speed of light that slowing down the astronauts' biological clocks would allow them to make in 10 years, say, a trip that appears to an earthbound observer to last 10^6 years. Some formidable practical problems will probably prevent realization of this idea for a long time, however. First, humans do not like to accelerate at rates much greater than g, so it would take a very long time to get up to speed. Second, huge amounts of energy would be needed. To confirm this, calculate the energy needed for a ship of mass 10^5 kg for the above trip. In order to gauge adequately the enormous magnitude of the required energy, express your answer in terms of the energy used by the entire world in one year, about 10^{20} J.

40-54 P Serious proposals have been advanced for spaceships propelled by starlight falling on huge mirrored sails. The intensity of light from our star (the sun) near the earth's orbit is 1.4 kW/m². (a) What maximum force could this radiation exert on a sail 1 km × 1 km? (b) What speed could a spaceship of 10,000-kg mass achieve if it were accelerated from rest by this force for one year? (c) A spaceship with a solar sail would be attracted to the sun but repelled by the radiation force on the sail. Show that if the two forces balance at one distance from the sun, the forces will balance at any other distance from the sun if the sail size and orientation are unchanged.

41 Quantum Theory

41-1   Quantization
41-2   Photoelectric Effect
41-3   X-Ray Production and Bremsstrahlung
41-4   Compton Effect
41-5   Pair Production and Annihilation
41-6   Matter Waves
41-7   Probability Interpretation of the Wave Function
41-8   Complementarity Principle
41-9   Uncertainty Principle
41-10 The Quantum Description of a Confined Particle
       Summary

Relativity physics is one of the two great ideas in twentieth-century physics. Quantum physics is the other. Relativity physics is the physics of the very fast; it changes our conceptions, not only of space and time, but also of mass and energy. Quantum physics is sometimes called the physics of the very small. Certainly atomic and subatomic structure can be comprehended only with quantum theory. But quantum effects also show up in macroscopic phenomena; superconductivity, for example, is basically a quantum effect. Quantum physics revises classical conceptions just as radically as relativity does. As we shall see, the perfect predictability, the clockwork universe, that characterizes Newtonian mechanics is gone in quantum theory. We must deal with probabilities and uncertainties.

41-1   Quantization

One aspect of quantum theory is not entirely new—we find it also in classical physics. It is the idea of *quantization,* whereby quantities come only in certain discrete amounts. Here are examples:

- People. They come only in integers.
- Sides of a coin. We have only heads or tails.
- Electric charge. For an observable particle, always an integral multiple of the basic charge e.
- Frequencies of waves trapped between boundaries. Consider the simplest case, a string attached at both ends. Then the allowed standing-wave patterns are those for which the frequency is exactly an integral multiple of the fundamental frequency (Section 17-6).

One important aspect of quantum theory is that physical quantities that were thought in classical physics to have a continuous range of possible values, are, in fact, quantized.

Quantum theory began in 1900 with Max Planck, who attempted to give a theoretical interpretation of the electromagnetic radiation from a blackbody, a perfect absorber and radiator, and especially of how the intensity of the emitted radiation depended on wavelength. Planck found that only by postulating energy quantization could he produce satisfactory agreement between experiment and theory. A detailed analysis of blackbody radiation is complicated and involves sophisticated arguments. Therefore, we shall introduce quantum concepts through the simpler and in many ways, more compelling arguments that arise in the photoelectric effect.

41-2 Photoelectric Effect

How the photoelectric effect was discovered is an irony of history. Heinrich Hertz in 1887, during the experiments that confirmed Maxwell's theoretical prediction (1864) of the existence of continuous, classical electromagnetic waves, found the following: a charged object loses its charge more readily when it is illuminated by violet light.

The photoelectric effect is this: electromagnetic radiation shines on a clean metal surface and electrons are released from the surface. Conduction electrons in a metal are relatively free to move about the interior but they are bound to the metal as a whole. These electrons may become photoelectrons. The radiation supplies an electron with energy that equals or exceeds the energy that binds the electron to the surface and thereby allows the electron to escape. What matters in releasing an electron from the metal is, in the view of classical electromagnetic theory, simply whether enough energy has reached the initially bound electron. But experiment shows different behavior. The *frequency* of the radiation determines, for any particular kind of emitting surface, whether electrons are released, quite apart from how intense the radiation may be. Unless a certain radiation frequency, characteristic of the material, is exceeded, no electrons are released.

To see how the quantum theory accounts for this circumstance, we put on record the basic properties of the *photon*.

Electromagnetic waves are quantized. They consist of discrete *quanta*, called photons. Each photon has an energy E that depends only on the frequency ν (or on the wavelength λ) and is given by

$$E = h\nu = h\frac{c}{\lambda} \qquad (41\text{-}1)$$

where h is a constant. Indeed h is the fundamental constant of the quantum theory, and it is called *Planck's constant*. Its value was first determined and its significance first appreciated by Planck in 1900. The present value of Planck's constant is

$$h = 6.626\ 176 \times 10^{-34}\ \text{J}\cdot\text{s}$$

According to the quantum theory, a beam of monochromatic light of frequency v consists of particlelike photons. Each has energy hv. A photon travels at the speed of light. It must, on the basis of relativity theory, then have a zero rest mass, and its energy must be entirely kinetic. As long as it exists, a photon moves at speed c. Indeed, the only thing a photon can do is to travel through space at speed c. When it interacts with any object, it ceases to exist. Thus, when a photon strikes an electron bound in a metal, it relinquishes its *entire* energy hv to the single electron it strikes. See Figure 41-1. If the energy the bound electron gains from the photon exceeds the energy binding it to the metal surface, the electron is freed, with the excess energy appearing as kinetic energy of the photoelectron. If the photon has less energy than that binding the electron, it simply cannot be dislodged.

Let ϕ be the energy with which an electron is bound to metal; ϕ is often called the *work function* of the material. The kinetic energy of the released photoelectron is E_k. Then energy conservation yields

$$hv = \phi + E_k \tag{41-2}$$

The left side of this equation is the energy initially carried by the incoming photon. The right side tells what happened to it; part of it goes to unbind the electron and the rest appears as the particle's kinetic energy. (Strictly, E_k is the maximum kinetic energy; some energy may be gained by the solid material.)

Clearly, unless $hv > \phi$, no photoelectron can be produced. The *threshold frequency* v_0 for photoemission is then, with $E_k = 0$, from (41-2),

$$hv_0 = \phi \tag{41-3}$$

Both v_0 and ϕ are characteristic of the particular photoemitter. Equation (41-2) can also be written as

$$hv = hv_0 + E_k \tag{41-4}$$

The basic equation of the photoelectric effect lends itself well to graphical interpretation, as shown in Figure 41-2. A plot of photoelectron kinetic energy, $E_k = hv - \phi$, against photon frequency is a straight line with slope h. The energy intercept gives the work function ϕ; the frequency intercept is the threshold frequency v_0. For a typical photoemitter, ϕ is a few electron volts and v_0 corresponds to violet or ultraviolet light (for example, for potassium, $\phi = 2.30$ eV and $\lambda_0 = c/v_0 = hc/\phi = 539$ nm).

Determining the maximum photoelectron kinetic energy is straightforward; it is measured directly by the stopping electric potential V that brings the most energetic photoelectrons to rest (and therefore the photocurrent to zero). More specifically, E_k (max) $= eV$. The number of photons incident upon a photoemitter is proportional to the intensity of incident radiation; similarly, the number of electrons released per unit time is proportional to the photocurrent (electric current from photoelectrons). Since one photon is extinguished for each photoelectron released, the photocurrent is directly proportional to the incident intensity. This effect is used in practical applications of the photoelectric effect.

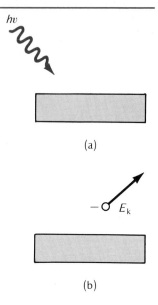

Figure 41-1. The photoelectric effect: (a) A photon with energy hv strikes a photoemitting material; and (b) an electron is released with kinetic energy E_k.

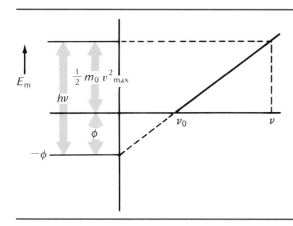

Figure 41-2. Photon energy plotted as a function of frequency.

Although the term *photoelectric effect* means, strictly, the release of electrons from a metallic surface, this same effect is seen in more general circumstances, whenever a bound particle is released by the absorption of a photon (see Examples 41-3 and 41-4).

The photoelectric effect provides a fundamentally new insight into the nature of electromagnetic radiation; it is quantized and consists of photons. With the frequency v of the radiation specified, a photon can have but one energy, hv, and the total energy of a monochromatic beam is always precisely an integral multiple of the energy of a single photon. See Figure 41-3.

The granularity of electromagnetic radiation is not conspicuous in ordinary observations; this is simply because the energy of any one photon is very small, and because the number of photons in a light beam of moderate intensity is enormous. The situation is like that found in the molecular theory. The molecules are so small and their numbers so great that the molecular structure of all matter is disclosed only in very subtle observations.

The ideas of wave and particle are apparently mutually incompatible, even contradictory. An ideal particle has vanishing dimensions and is completely localizable. On the other hand, an ideal wave, one with a perfectly defined wavelength and frequency, has infinite extension in space. In the photoelectric effect, light behaves as if it consisted of particles or photons, but this does not mean that we dismiss the incontrovertible experimental evidence of the wave properties of light. Both descriptions must be accepted. How this dilemma is resolved is discussed in Section 41-8, after we have explored more fully other quantum attributes of light.

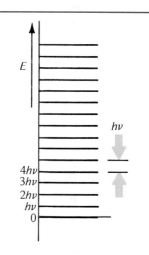

Figure 41-3. Quantization of monochromatic electromagnetic radiation—the allowed energies are an integral multiple of the energy of a single photon.

Example 41-1. A radio transmitter broadcasts a continuous power output of 6.0 W at a frequency of 600 kHz. How many radio-frequency photons leave the transmitting antenna each second?

The power output P can be written as the number N of photons emitted per second multiplied by the energy hv per photon. We then have

$$P = Nhv$$

so that

$$N = \frac{P}{hv} = \frac{6.0 \text{ W}}{(6.63 \times 10^{-34} \text{ J} \cdot \text{s})(600 \times 10^3 \text{ s}^{-1})}$$

$$= 1.5 \times 10^{28} \text{ photons/s}$$

Example 41-2. A small and fairly bright sodium lamp with an output of 1.0 W of yellow light at a wavelength of 590 nm is 1.0 km away from an observer. The observer looks directly at the light source; the diameter of the pupil of his eye is 2.0 mm. How many yellow photons strike the eye's retina per second?

The total number N of photons emitted from the source per second is, as given in Example 41-1, $N = P/h\nu = P\lambda/hc$. The fraction of photons entering the pupil of radius r at a distance R from an effective point source is $\pi r^2/4\pi R^2$, since the surface area of a sphere of radius R is $4\pi R^2$. Therefore, the number n entering the eye per unit time is

$$n = N \frac{r^2}{4R^2}$$

$$= \frac{P\lambda}{(hc)} \frac{r^2}{(4R^2)}$$

$$= \frac{(1.0 \text{ W})(590 \times 10^{-9} \text{ m})(1.0 \times 10^{-3} \text{ m})^2}{(6.6 \times 10^{-34} \text{ J} \cdot \text{s})(3.00 \times 10^8 \text{ m/s})(4)(1.0 \times 10^3 \text{ m})^2}$$

$$= 7.4 \times 10^5 \text{ photons/s}$$

Example 41-3. It takes 13.61 eV to ionize a hydrogen atom. What is (a) the energy and (b) the wavelength of a photon that will ionize hydrogen?

(a) The ionization energy of 13.61 eV for a hydrogen atom is, in effect, its work function,

$$h\nu_0 = \phi = 13.61 \text{ eV}$$

(b) The wavelength is from

$$E = h\nu_0 = h\frac{c}{\lambda_0}$$

given by

$$\lambda_0 = \frac{hc}{E} = \frac{(6.6 \times 10^{-34} \text{ J} \cdot \text{s})(3.0 \times 10^8 \text{ m/s})}{(13.61 \text{ eV})(1.60 \times 10^{-19} \text{ J/eV})} = 91.1 \text{ nm}$$

or an ultraviolet photon.

A more general form for the relation between photon energy E in electron volts and wavelength λ in nanometers is

$$\lambda = \frac{hc}{E} = \frac{1239.852 \text{ eV} \cdot \text{nm}}{E}$$

Example 41-4. The masses of a proton, a neutron, and a deuteron (the nucleus of a heavy hydrogen atom ^2_1H) are as follows:

Proton (p):	1.007 825 u
Neutron (n):	1.008 665 u
p + n:	2.016 490 u
Deuteron:	2.014 104 u

(Strictly, the mass of the "proton" includes also the mass of an electron, so that 1.007 825 u is the mass of a neutral hydrogen atom. Similarly, the "deuteron" mass includes an electron, so that 2.014 104 u is actually the mass of the deuterium atom.)

(a) What is the deuteron binding energy? (b) What is the minimum photon energy and maximum wavelength that will separate a deuteron into a proton plus a neutron, a process known as *photodisintegration*?

(a) Finding the deuteron binding energy is strictly an exercise in applying mass-energy conservation. We see above that the total mass of the separated particles exceeds that of the bound system; mass-energy must be added to $_1^2$H to yield $_1^1$H + n. We have

$$\Delta m = (2.016\,490 - 2.014\,104)\,\text{u} = 0.002\,386\,\text{u}$$

so that the binding energy is*

$$E_b = \Delta mc^2 = (0.002\,386\,\text{u})c^2\,(931.5\,\text{MeV}/\text{u}c^2) = 2.224\,\text{MeV}$$

and the corresponding wavelength is

$$\lambda = \frac{hc}{E} = \frac{1239.852\,\text{eV}\cdot\text{nm}}{2.224\,\text{MeV}} = 5.57 \times 10^{-4}\,\text{nm} = 557\,\text{fm}$$

---

* Actually, the threshold photon energy for dissociating a deuteron initially at rest into a proton and a neutron exceeds the binding energy slightly. The reason? A photon with an energy of 2.22 MeV has a momentum of 2.22 MeV/c. Therefore, the proton and the neutron cannot be at rest; they must be in motion and also have a total momentum of 2.22 MeV/c. The threshold photon energy is, it can be shown, given by $h\nu = E_b/(1 - E_b/M_d c^2)$, where M_d is the deuteron mass.

41-3 X-Ray Production and Bremsstrahlung

In the photoelectric effect, a photon transfers energy to an electron. The inverse effect is this: an electron loses kinetic energy and creates a photon. The process is most clearly illustrated in the production of x-rays.

When a fast-moving electron comes close to the positively charged nucleus of an atom and is deflected thereby, the electron is accelerated. The accelerated electric charge radiates electromagnetic energy. But in the quantum theory, this radiated electromagnetic energy consists of photons. In short, a deflected electron radiates one or more photons, and the electron leaves the collision site with reduced kinetic energy.

The radiation produced in such a collision is often referred to as *Bremsstrahlung* ("braking radiation" in German). A Bremsstrahlung collision is shown schematically in Figure 41-4. An electron approaches the deflecting atom with a kinetic energy K_1; it recedes with a kinetic energy K_2, having produced a single photon of energy $h\nu$. From energy conservation, we have

$$K_1 - K_2 = h\nu$$

(We can ignore the very small energy of the recoiling atom.)

X-rays were discovered and first investigated in 1895 by Wilhelm Roentgen, who assigned this name because the true nature of the radiation was at first unknown. X-rays, now known to consist of electromagnetic waves, or photons, having wavelengths of about 0.1 nm, pass readily through many materials that are opaque to visible light.

Suppose an electron is accelerated through an electric potential difference V, of several thousand volts, and then strikes a target. The electron acquires a kinetic energy of $K = eV$. We have ignored the electron's kinetic energy as it left the cathode, typically much less than Ve. When the electron strikes the target, it acquires an additional energy, the energy that binds it to the target surface; but this binding energy is also only a few electron volts, and it too can be ignored. When the electron strikes the target, it is brought essentially to rest in a single collision. The most energetic photon that can be produced in a single Bremsstrahlung collision is one whose energy $h\nu_{\text{max}}$ is

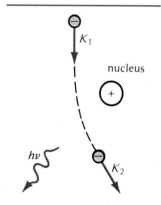

Figure 41-4. Bremsstrahlung collision of an electron with a nucleus, causing the creation of a photon.

$$hv_{max} = eV = K$$

where v_{max} is the maximum frequency of the x-ray photons produced. More typically, an electron loses its energy at the target by heating it or by producing two or more photons, the sum of whose frequencies will then be less than v_{max}. There will be a distribution in photon energies with, however, a well-defined maximum frequence v_{max} or minimum wavelength $\lambda_{min} = c/v_{max}$, given by

$$K = hv_{max} = \frac{hc}{\lambda_{min}} = eV \tag{41-5}$$

The intensity of x-rays emitted has in fact an abrupt cutoff at the limit v_{max}; this limit is determined solely by the accelerating potential V applied to the x-ray tube, not by the chemical identity of the target material.

41-4 Compton Effect

What happens when a monochromatic electromagnetic wave impinges on a charged particle whose size is much less than the wavelength of the radiation? The charged particle is accelerated principally by the wave's sinusoidally varying electric field. In fact, the particle oscillates in simple harmonic motion at the same frequency as that of the incident radiation. And since the charged particle is accelerated continuously, it radiates electromagnetic radiation of the same frequency. This is what classical theory predicts: scattered radiation with the same frequency as the incident radiation. The charged particle acts as transfer agent; it absorbs energy from the incident beam and reradiates it at the same frequency but scattering it in all directions. Classical scattering theory agrees with experiment for visible light and other, longer-wavelength radiation. A simple example is this: Light reflected from a mirror (a collection of scatterers) undergoes no apparent change in frequency.

To consider scattering from the point of view of quantum theory, we first need the relation for the momentum of a photon. We can get this from relativistic dynamics by regarding the photon as a particle that, because it always travels at speed c, has a zero rest mass and zero rest energy. The general relativistic relation between momentum p and total energy E is

$$E^2 = E_0^2 + (pc)^2 \tag{40-19}$$

and it yields with $E_0 = 0$,

$$p = E/c$$

This result follows also from classical electromagnetic theory (Section 35-6), where it was shown that the linear momentum of a beam of electromagnetic radiation is the energy of the beam divided by c (35-19).

It follows that the momentum of a photon with frequency v and wavelength λ is

$$p = \frac{E}{c} = \frac{hv}{c} = \frac{h}{\lambda} \tag{41-6}$$

The direction of wave propagation is the direction of a photon's momentum.

A photon's momentum increases with frequency. The momentum of a high-frequency (or high-energy) photon, such as a gamma (γ) ray, will exceed by far the momentum of a low-frequency (and low-energy) photon, such as a radio photon. The distinctive feature introduced by the quantum theory is this: electromagnetic momentum occurs not in arbitrary amounts, but only in integral multiples of the momentum h/λ carried by a single photon.

Now consider a photon incident upon a free charged particle at rest. Basically, we have a photon colliding with a particle, and the laws of energy and momentum conservation apply. Figure 41-5 shows the photon and free particle before and after collision. The special advantage in applying the conservation laws is that we need not be concerned with the details of what happens when photon meets electron, but merely with the total energy and momentum going into and coming out of the collision.

As Figure 41-5 shows, if a photon carries momentum and energy into the collision, the struck particle must gain some energy and momentum.

We take the particle to have rest mass m_0 and rest energy $E_0 = m_0c^2$; it is free and initially at rest. Energy conservation applied to the collision of Figure 41-5 gives

$$h\nu + E_0 = h\nu' + E \qquad (41\text{-}7)$$

Here E is the relativistic energy of the recoiling particle after collision. The energies of the incident and scattered photons are $h\nu$ and $h\nu'$, respectively. The particle's final energy E (rest energy plus kinetic energy) exceeds initial energy E_0. We immediately see from (41-7) that $h\nu' < h\nu$. The scattered photon has *less* energy, a *lower* frequency, and a *longer* wavelength than the incident photon, a result quite different from classical physics. The incident and scattered photons have different frequencies, so that the scattered photon is not to be thought of as merely the incident photon moving in a different direction with less energy. Rather, in the collision the incident photon is annihilated, and the scattered photon is created.

Momentum conservation is implied by the vector triangle of Figure 41-5(c). Here $\mathbf{p} = m\mathbf{v}$ is the relativistic momentum of the recoiling particle. The magnitudes of the momenta of the incident and scattered photons are, respectively, $p_\lambda = h\nu/c = h/\lambda$ and $p_{\lambda'} = h\nu'/c = h/\lambda'$. Scattering angle θ is the angle between the directions of $\mathbf{p}_\lambda$ and $\mathbf{p}_{\lambda'}$, the directions of the incident and scattered photons.

The law of cosines applied to the triangle in Figure 41-5(c) yields

$$p_\lambda^2 + p_{\lambda'}^2 - 2p_\lambda p_{\lambda'} \cos\theta = p^2 \qquad (41\text{-}8)$$

We wish to solve (41-7) and (41-8) for the change in wavelength $\lambda' - \lambda = \Delta\lambda$. After some manipulation,* we get

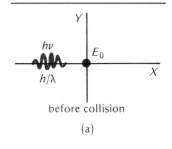

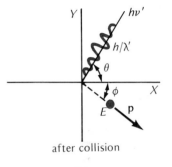

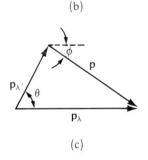

Figure 41-5. *A Compton collision: (a) Before and (b) after the collision. (c) Momentum vectors for the incident and scattered photons and for the electron.*

---

* Multiply both sides of (41-8) by c^2 and use $pc = h\nu$, and we have

$$h^2\nu^2 + h^2\nu'^2 - 2h^2\nu\nu' \cos\theta = p^2c^2$$

We get a similar relation from (41-7) as follows. Put $h\nu$ and $h\nu'$ on one side of the equation and E and E_0 on the other; then square. We get

$$h^2\nu^2 + h^2\nu'^2 - 2h^2\nu\nu' = E^2 + E_0^2 - 2EE_0$$
$$= 2E_0^2 + p^2c^2 - 2EE_0$$

Now subtract the two equations above, and get

$$h^2\nu\nu'(1 - \cos\theta) = E_0(E - E_0) = m_0c^2(h\nu - h\nu')$$

Using $\nu = c/\lambda$ and $\nu' = c/\lambda'$, we finally get (41-9).

$$\Delta\lambda = \lambda' - \lambda = \frac{h}{m_0 c}(1 - \cos\theta) \qquad (41\text{-}9)$$

This is the basic equation for the Compton effect. It gives the increase $\Delta\lambda$ in the wavelength of the scattered photon over that of the incident photon. Note that $\Delta\lambda$ depends only on the rest mass m_0 of the recoiling particle, Planck's constant h, the speed c of light, and the angle θ of scattering; $\Delta\lambda$ is independent of the incident photon's wavelength λ. The quantity $h/m_0 c$, appearing on the right-hand side of (41-9) and having the dimensions of length, is known as the *Compton wavelength*. Given the scattering angle θ, we can compute the wavelength increase unambiguously, but we cannot predict in advance the angle at which any one photon will emerge.

Suppose the recoiling particle is a free electron, or one that is only very loosely bound to a parent atom. Then $m_0 = 9.11 \times 10^{-31}$ kg, and we compute $h/m_0 c = 2.43$ pm. As (41-9) shows, when $\theta = 90°$, the wavelength change is $\Delta\lambda = h/m_0 c = 2.43$ pm. When θ is $180°$ and the scattered photon travels in the backward direction, and the recoil electron straight forward, so that the collision is effectively "head-on," the wavelength change is a maximum. Then the electron's kinetic energy is also a maximum.

As (41-9) shows, the increase in wavelength of the scattered photon relative to the incident photon does not depend on the wavelength of the incident photon. *All* photons scattered at $\theta = 90°$ have a wavelength shift of 2.43 pm. For visible light the shift is so small as to be virtually unobservable. An observable shift, one of at least a few percent, can occur for x-rays. For example, an incident x-ray photon with a wavelength of $\lambda = 0.1000$ nm will, when scattered through $90°$, produce a scattered photon with a wavelength of 0.1024 nm, or a wavelength increase of 2.4 percent.

The shift in wavelength in the scattering of x-rays was observed first by A. H. Compton in 1922. Figure 41-6 shows schematically the experimental arrangement. At any fixed angle θ, the x-ray detector measures the scattered intensity as a function of wavelength. Figure 41-6 shows two wavelengths for scattered photons. The photon with the expected wavelength *shift* comes from the collision of any incident photon with an essentially *free* electron in the target material. The scattered photons with the *same* wavelength as incident photons come from a photon that has collided with a tightly bound electron. Since a tightly bound electron cannot move without also moving the entire atom, the bound electron's mass is effectively that of the whole atom.

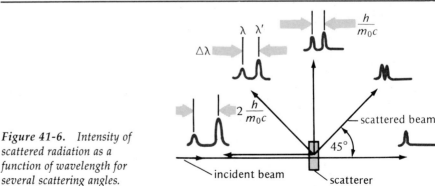

Figure 41-6. Intensity of scattered radiation as a function of wavelength for several scattering angles.

41-5 Pair Production and Annihilation

Can a photon's energy be converted into rest mass? The answer is yes, and it is illustrated most directly in the phenomenon of pair production.

Pair Production What is the minimum energy required to create a single particle? The electron has the smallest nonzero rest mass of all known particles, it requires the least energy for its creation. But a photon has zero electric charge. So the law of electric-charge conservation precludes the creation of a single electron from a photon. But an electron pair, consisting of two particles with opposite electric charges, would be possible. A positively charged particle, called the *positron* and the *antiparticle* of the electron, is in fact observed. The electron and the positron are similar in all ways except in the signs of their charges, $-e$ and $+e$ (and the effects of this difference). Clearly, the minimum energy $h\nu_{min}$ to create an electron-positron pair is

$$h\nu_{min} = 2m_0 c^2 \qquad (41\text{-}10)$$

Since the rest energy $m_0 c^2$ of an electron or a positron is 0.51 MeV, the threshold energy $2m_0 c^2$ for pair production is 1.02 MeV. The photon wavelength corresponding to this threshold is 1.2 pm. Electron pairs can be produced only by γ-ray photons. The general phenomenon in which a particle and its antiparticle are created from electromagnetic radiation is called *pair production*. It is a very emphatic demonstration of the interconvertibility of mass and energy.

If a photon's energy exceeds the threshold energy $2m_0 c^2$, the excess appears as kinetic energy of the created pair.

Pair production cannot occur in empty space. We prove this by showing that energy and momentum cannot simultaneously be conserved in particle-antiparticle production unless the photon is near some massive particle, such as an atomic nucleus. Suppose, for the sake of argument, that a pair has been created in empty space and that we, the observers, are at rest with respect to the center of mass of this two-particle system. Then the total momentum of the pair is zero. But the photon creating the pair would have had some nonzero momentum in this reference frame, since a photon always moves at speed c, whatever the reference frame. Under these imagined circumstances, we should have the momentum of the photon before collision but zero momentum after, clearly a violation of momentum conservation. A photon cannot decay spontaneously to an electron-positron pair in free space; the process can take place only if the photon encounters a massive particle that acquires the requisite (but negligible) momentum.

Figure 41-7 is a schematic drawing of pair production, and Figure 41-8 is a cloud-chamber photograph showing electron-positron pairs. The paths of the charged particles are visible because the charged particles produce ionization effects at which bubbles are formed as the particles travel through the liquid. The oppositely charged particles are deflected into oppositely directed circular arcs by a uniform magnetic field.

Positrons were predicted on theoretical grounds by P. A. M. Dirac in 1928. Four years later, C. D. Anderson observed directly and identified a positron. Electron-positron pairs are now commonly observed whenever high-energy photons interact with matter. Proton-antiproton, neutron-antineutron, and

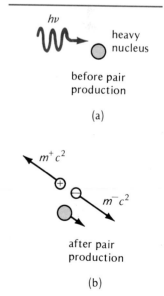

Figure 41-7. Schematic diagram for pair production.

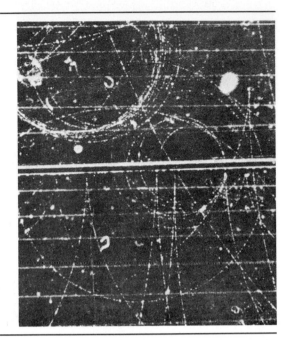

Figure 41-8. Photograph of pair production (in a cloud chamber).

still other types of pairs can be created. They have higher threshold energies than electron-positron pairs because of their larger rest mass.

Pair Annihilation A particle-antiparticle pair can annihilate each other and create photons. The process is the inverse of pair production. Suppose an electron and a positron are close together and essentially at rest. Their total linear momentum is initially zero; therefore, a single photon cannot be created. That would violate momentum conservation. Momentum can, however, be conserved if *two* photons, moving in opposite directions with equal momenta, are created. Such a pair of photons would also have equal frequencies and energies. See Figure 41-9. Actually, three or more photons can be created, but with a much smaller probability than for two photons.

Energy conservation implies

$$m_0^+ c^2 + m_0^- c^2 = 2m_0 c^2 = 2h\nu_{min}$$

with an electron and a positron at rest initially. The minimum energy of one photon created by electron-positron annihilation is $h\nu_{min} = m_0 c^2 = 0.51$ MeV.

Annihilation is the ultimate fate of positrons. When a high-energy positron appears, as in pair production, it loses its kinetic energy in collisions as it passes through matter, and finally moves at low speed. Then it combines with an electron and forms a bound system, called a *positronium*; this "atom" decays quickly (10^{-10} s) to two photons of equal energy. Thus, the death of a positron is signaled by the appearance of two annihilation quanta, or photons, of about $\frac{1}{2}$ MeV each. The transitoriness of positrons is due not to an intrinsic instability, but to the high risk of their collision and subsequent annihilation with electrons.

In our part of the universe there is a preponderance of electrons, protons, and neutrons; their antiparticles, when created, quickly combine with them in

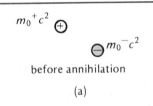

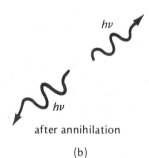

Figure 41-9. Schematic diagram showing pair annihilation with the creation of two photons.

annihilation processes. It is conceivable, although at present purely conjectural, that there exists a part of the universe in which positrons, antiprotons, and antineutrons predominate.

Example 41-5. A highly energetic photon creates an electron-positron pair in a uniform magnetic field of 1.5 T. The two tracks registered in a bubble chamber lie in a plane perpendicular to the magnetic field, and the two radii of curvature are 10 cm and 14 cm. What is the photon energy?

If the electron's kinetic energy far exceeds its rest energy (~ 0.5 MeV), we can write its energy as $E = pc$, where p is the momentum. Likewise for the positron. The momentum of a charged particle in a magnetic field is, from (40-13), given by $p = qrB$, so that

$$h\nu = E^+ + E^- = (p^+ + p^-)c = qBc(r^+ + r^-)$$

$$= \frac{(1.6 \times 10^{-19} \text{C})(1.5 \text{ T})(3.0 \times 10^8 \text{ m/s})(10 \text{ cm} + 14 \text{ cm})}{(1.6 \times 10^{-19} \text{ J/eV})} = 108 \text{ MeV}$$

The assumption initially that the electron rest mass is negligible is justified.

41-6 Matter Waves

As we have seen, a photon of wavelength λ has a momentum given by $p = h/\lambda$. Louis de Broglie in 1924 posed the question, based on the conjectured symmetry of nature, Does every particle with momentum p have associated with it a wavelength λ given by the same relation? The answer of observation is an emphatic yes. Every particle has a wave character, and its wavelength is given by the *de Broglie relation*,

$$\lambda = \frac{h}{p} = \frac{h}{mv} \qquad (41\text{-}11)$$

where p is the particle's relativistic momentum, m is its relativistic mass, and v is its speed.

What is the wavelength of an ordinary object, say, a pitched baseball ($m = 0.2$ kg, $v = 40$ m/s)? Equation (41-11) yields $\lambda \simeq 10^{-34}$ m. This wavelength is so extraordinarily small (smaller than the size of a proton by a factor of about 10^{19}) that one cannot observe diffraction or interference effects for a baseball. For example, a baseball pitched through an open window — really a single slit — is not diffracted, or deviated off-center. It is, after all, by observing such distinctive wave phenomena as diffraction and interference that we can tell whether energy traveling through space is, in fact, a wave. That is certainly how we confirm that visible light consists of waves. That is also how such short-wavelength electromagnetic radiation as x-rays is confirmed to consist of waves. Recall that x-rays show diffraction effects (Section 39-4) when sent through ordinary crystalline solids; the x-ray wavelength (~ 0.1 nm) is comparable to the distance between adjacent atoms situated in geometrical arrays in a crystal, and diffraction effects are relatively easy to observe.

The key to observing the wave character of ordinary particles is to have the particle wavelength be comparable to the size of a possible diffracting object. What is the kinetic energy of an electron, for example, that yields $\lambda = 0.10$ nm? For an electron of electric charge e accelerated from rest by electric potential difference V, we have, using (41-11),

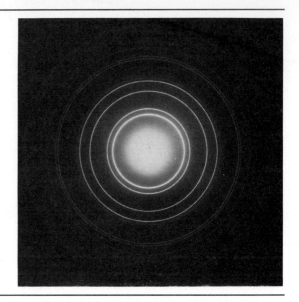

Figure 41-10. Photograph of electron diffraction.

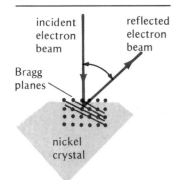

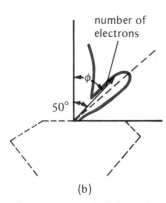

Figure 41-11. (a) Schematic for the Davisson-Germer experiment. (b) Number of electrons as a function of angle ϕ.

$$eV = \frac{1}{2}mv^2 = \frac{p^2}{2m} = \frac{(h/\lambda)^2}{2m}$$

so that with $\lambda = 1.00 \times 10^{-10}$ m, we have

$$V = \frac{h^2}{2\,me\,\lambda^2} = \frac{(6.63 \times 10^{-34}\text{ J}\cdot\text{s})^2}{2(9.11 \times 10^{-31}\text{ kg})(1.60 \times 10^{-19}\text{ C})(1.00 \times 10^{-10}\text{ m})^2}$$

$$= 150\text{ V}$$

A 150-eV electron has a wavelength 0.1 nm. Observing electron diffraction should be and is very much like observing x-ray diffraction. For electron diffraction, the "bright" spots are those locations where many electrons are observed and the "dark" spots are those where few if any are observed. See Figure 41-10. Indeed, essentially all the wave effects that can be demonstrated for visible light—diffraction through a single slit, for example—can also be observed for electrons. The wave properties were first observed in 1927 by C. Davisson and L. H. Germer at AT&T Bell Laboratories. A beam of monoenergetic (and therefore monochromatic) electrons was directed at a single crystal of nickel. The beam was diffracted by reflection from Bragg planes (Section 39-4) within the crystal, and a very pronounced peak was observed, which is explained by electron diffraction. See Figure 41-11.

Not merely electrons show wave effects. The wave character of neutrons, atoms, even molecules is also confirmed in detail. Indeed, all the interference and diffraction effects observed for electromagnetic radiation have been duplicated for particles.

Example 41-6. A nuclear reactor provides a copious supply of neutrons. When a neutron is in thermal equilibrium at temperature T, it has, like a molecule in a gas, an average kinetic energy of $\frac{3}{2}kT$ (Section 20-3). A thermal neutron is one with $T = 293$ K, room temperature. Thermal neutrons are especially useful in the phenomenon of neutron diffraction, which is used to study the structure of crystalline materials. Unlike electrons, neutrons are electrically uncharged, so that their behavior is unaffected by electric forces. Show that the wavelength of a thermal neutron is about 0.1 nm.

We have

$$\frac{3}{2} kT = \frac{p^2}{2m} = \frac{(h/\lambda)^2}{2m}$$

$$\lambda = \frac{h}{\sqrt{3mkT}} = \frac{6.6 \times 10^{-34} \text{ J} \cdot \text{s}}{\sqrt{3(1.7 \times 10^{-27} \text{ kg})(1.4 \times 10^{-23} \text{ J/K})(293 \text{ K})}} \approx 0.1 \text{ nm}$$

so that a thermal neutron, a 150-eV electron, and a typical x-ray all have wavelengths of the same order.

41-7 Probability Interpretation of the Wave Function

What is the wave associated with a photon? It is, of course, the oscillating electric field E (or magnetic field B) of an electromagnetic wave.

But what is waving when we associate a wavelength λ with an ordinary material particle? The very bland name *wave function*, typically represented by ψ, is given to the mathematical quantity that represents a particle's wave character. Although ψ cannot be observed directly, the square of the wave function ψ^2 can; it is, as we shall see, proportional directly to the *probability of observing a particle*.*

To see why ψ^2 is proportional to the probability of observing a particle, consider the analogous situation for a photon. A photon's wave function is the electric field E of the associated electromagnetic wave, so that the probability of observing the photon at any location in space would be proportional to E^2. To see that E^2 must indeed represent the probability of observing the photon, recall that the relation between the intensity I of an electromagnetic wave and E^2 is

$$I = \epsilon_0 c E^2 \tag{35-11}$$

so that

$$I \propto E^2$$

Now it is easy to see that in the quantum view of electromagnetic radiation, the intensity, the energy per unit time per unit transverse area, is itself proportional to the probability of observing a photon. Certainly the intensity may be written as

$$I = Nh\nu \tag{41-12}$$

where N is the number of photons, each of energy $h\nu$, passing through a unit transverse area per unit time. But the *number* of photons per unit time per unit area in a beam of radiation must mean really the *average* number. This would certainly be true for a beam of very low intensity, one in which the intensity was so low that on the average, N had the value of, say, 0.5 photon/cm²·s. There is no such thing as half a photon, so that this numerical value of N here

---

* It turns out that the wave function ψ is always a complex mathematical quantity with both a real and an imaginary part. Strictly, ψ^2 represents $\psi \cdot \psi$, where $\psi*$ is the complex conjugate of ψ, and $\psi* \psi$ is then always a real quantity.

Particles, Fields

How can you keep cars from speeding on campus? The problem finally landed on the desk of the Chancellor of Washington University in St. Louis, Missouri. His solution—neat in its simplicity—was the speed bump.* The Chancellor, Arthur H. Compton (1892–1962), was good at analyzing things that bump. He had, after all, been awarded a Nobel prize in physics in 1927 for studying what happens when a photon collides with a particle.

The Compton effect is just one of the basic electron-photon interactions. It, together with the other basic interactions between a photon and a charged particle such as an electron, are shown in the figure. The pictures are those of that branch of contemporary physics known as *quantum electrodynamics* (QED for short), or more generally and simply, *field theory*.

Read each figure with time going forward from left to right. A straight line represents an electron, a wiggly line a photon. A positron corresponds to a straight line with its arrow reversed (a positron regarded as an electron running backward in time).

The interactions are these:

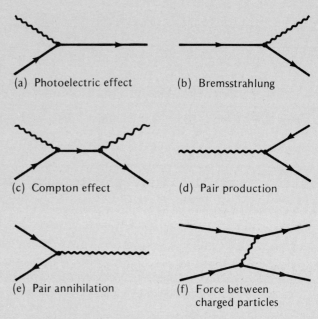

(a) Photoelectric effect
(b) Bremsstrahlung
(c) Compton effect
(d) Pair production
(e) Pair annihilation
(f) Force between charged particles

- (a) The primitive photoelectric effect: an electron absorbs a photon and the electron's momentum and energy are changed.

- (b) Bremsstrahlung: an electron emits a photon and the electron's momentum and energy change, just the reverse of process (a).

- (c) Compton effect: this process takes place as two closely spaced events—first the absorption of the incident photon and then the emission of the scattered photon.

*The story of Compton's design of the speed bump is given in the St. Louis *Globe Democrat* of 7 April 1953.

---

means that over a 2-s interval we observe about one photon on the average crossing a 1-cm² transverse area. In short, N and therefore I and therefore also E^2 are proportional to the probability of locating a photon. In exactly the same way, ψ^2 is proportional to the probability of observing a material particle. See Table 41-1.

We state this important result in more detail. If ψ represents the wave function at the location x, the probability of observing the particle's being between x and $x + dx$ is given by $\psi^2(x)\,dx$:

Table 41-1

	WAVE FUNCTION	PROBABILITY OF OBSERVING ENTITY
Photon	E	E^2
Particle	ψ	ψ^2

- (d) Pair production: a photon becomes an electron and a positron.
- (e) Pair annihilation: an electron-positron pair becomes a photon.
- (f) The electromagnetic force between two charged particles: here one electron emits a photon that is absorbed by the second electron, and thereby each electron has its momentum and energy changed. The force between charged particles is attributed to their trading a photon, the field particle of the electromagnetic interaction.

Look again at the various interactions in the figure, and you see that there is just *one* fundamental process—a vertex, where an incoming and outgoing electron line joins a photon line. It all boils down to this: a photon bumping an electron.

Should we worry that the fundamental laws of momentum and energy conservation seem not to be satisfied in these processes? For example, in (b) we see a free electron blithely coasting at constant velocity and then suddenly emitting a photon. The uncertainty principle says that it's all right. Momentum conservation *may* be violated, but only over a limited region of space. Energy conservation *may*, according to the uncertainty principle, also be violated, but again only over a limited interval of time controlled by Planck's constant.

Take another look at the force between charged particles in (f). The conservation laws of momentum and energy *are* violated when the first electron emits a photon but the momentum and energy debt is repaid when this photon—an unobservable, or *virtual photon*—is absorbed by the second electron. Since the Coulomb force drops off with distance but can extend, even if feebly, to charges separated by an infinite distance, the virtual photon for such a very-long-range interaction must have almost zero energy. Indeed, the infinite range of the electromagnetic force corresponds exactly to the circumstance that the photon as field particle can have zero energy.

This is all very picturesque. But does it work? Of course, the detailed theoretical analysis is far more complicated than the simple diagrams that summarize it. It turns out that these ideas in quantum field theory for electromagnetism produce the very best theory there is in physics. Nothing else shows such nearly perfect agreement between theory and experiment—better than a part in 10^7, by a variety of tests.

Because QED works so well, physicists have been emboldened to use it as the model for *all* interactions between elementary particles. Indeed, the recent discovery of the W and Z particles, described in the panel in chapter 35, is just one example of the validity of this approach.

*To get the latest word on elementary particles and quantum field theory look at recent issues of *Scientific American*. This magazine for the non-specialist interested in science has at least one or two articles each year on this topic. Another source, probably available in your physics library, on latest developments at high-energy accelerating machines, is the *CERN Courier*.

$$\text{Probability of observing a particle in the interval } dx \quad \propto \quad \psi^2 dx$$

Just as the electric field of a photon will generally be a function of both position and time, so too generally will the wave function ψ.

The probability interpretation of waves associated with particles was first given in 1926 by Max Born. That branch of quantum physics that deals with finding the values of ψ is known as *wave mechanics*, or *quantum mechanics*. The two principal originators of the wave mechanics of particles were Erwin Schrödinger (in 1926) and Werner Heisenberg (in 1925), who independently formulated quantum mechanics in different but equivalent mathematical forms.

Maxwell's electromagnetic theory is summarized in the Maxwell's equations; they are the basis for computing values of E. The wave mechanics of matter is governed by the Schrödinger equation; it is the basis for computing values of ψ in any problem in quantum physics. Here the parallel stops,

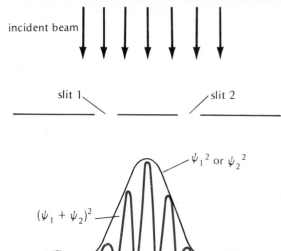

Figure 41-12. Double-slit diffraction for particles (vertical distances drastically compressed). Wave functions ψ_1 and ψ_2 give the diffraction pattern with either slit 1 or 2 open. The superposed wave function $(\psi_1 + \psi_2)$ gives the pattern when both slits are open.

however. The electric field has its origin in electric charges, and E gives not only the probability of observing a photon but also the electric force on a unit positive electric charge. But the wave function ψ of the Schrödinger equation is not directly measurable or observable. It does, however, give the most information one can extract concerning any system of objects; and all measurable quantities, such as the energy and momentum, as well as the probability of location, can be derived from it.

Consider the interference of waves that go through two parallel slits. When either of the two slits is closed, the pattern on a distant screen is the typical single-slit diffraction pattern: a broad, central maximum flanked by weaker, secondary maxima (Figure 39-2). When both slits are open, the pattern is as shown in Figure 41-12: interference fine structure within a diffraction envelope. The pattern is not merely two single-slit diffraction patterns superposed; the interference between waves traveling through both of the slits is responsible for the rapid variations in intensity. Here waves (or particles) can take two or more routes from a source to an observation point. We first superpose the wave function from the two separate routes to find the resultant wave function; then we square the resultant wave function to find the probability (or intensity). That is to say, if ψ_1 and ψ_2 represent the wave functions for passage through slits 1 and 2 separately, then $(\psi_1 + \psi_2)^2$, not $\psi_1^2 + \psi_2^2$, gives the probability of observing a particle on the screen. In this view, when a single electron or photon is directed toward a pair of slits, we cannot say through which of the two slits it will pass. When we speak in the language of waves we say, in effect, that the particle passes through *both* slits.

41-8 Complementarity Principle

How can we speak simultaneously of an electron as a particle — a point object — and an electron as a wave — an object spread far over space? Or how can electromagnetic radiation be viewed as a wave phenomenon and also as a collection of particlelike photons? The answer: We can never do the impossible and describe a particular entity *simultaneously* as particle and wave. The two extreme descriptions are mutually incompatible and contradictory.

According to the *principle of complementarity*, enunciated by Niels Bohr in 1928, the *wave and particle aspects* of electromagnetic radiation and of material particles *are complementary*. In any one experiment, we choose either the particle or the wave description. The two aspects are complementary in that our knowledge of the properties of electromagnetic radiation or of particles is partial unless both wave and particle aspects are known. The choice of one description, imposed by the nature of the experiment, precludes the simultaneous choice of the other. The quantities in quantum theory are more complicated than can be comprehended in the simple and extreme notions of wave and particle, notions borrowed from our direct, ordinary experience with large-scale phenomena.

The complementarity principle applied to electrons says this: Electrons in a cathode-ray tube follow well-defined paths and indicate their collisions with a fluorescent screen by very small, bright flashes. A particle model is used to describe electrons in cathode ray-experiments because all the electron energy, momentum, and electric charge is assigned at any one time to a small region of space. The particle nature of electrons is revealed in the cathode-ray experiments; and therefore, by the principle of complementarity, the wave nature of electrons must be suppressed.

The wave nature of electrons shows up in electron diffraction. Here electrons are propagated as waves with an indefinite extension in space, and it is necessarily impossible to specify the location of any one electron. In short, the electron-diffraction experiments exhibit the wave nature of electrons, and by complementarity, the particle nature is necessarily suppressed.

41-9 Uncertainty Principle

Consider this hypothetical experiment. You want to find the location of an electron initially at rest by firing a single photon at it. You can tell whether the photon hits the electron by observing a scattered photon. What kind of photon would be best? Since you can never "see" details smaller than wavelength λ, short-wavelength photons, or gamma rays, will give the highest resolution. But there is a serious complication. Directing a short-wavelength, high-frequency—and therefore high-energy and high-momentum—photon at an electron will deflect the electron, so that you end up learning where the electron *was*, not where it now is. On the other hand, if the electron is to remain nearly undisturbed, you must use long-wavelength light and therefore settle for only a fuzzy idea of its location. This simple illustration of a *gamma-ray microscope*, first discussed in 1927 by Werner Heisenberg, shows that it is fundamentally impossible to specify simultaneously with complete precision certain pairs of physical quantities. More specifically, one important formulation of the Heisenberg *uncertainty principle*, or *principle of indeterminacy*, is this:

$$\Delta p_x \, \Delta x > \frac{\hbar}{2} \qquad (41\text{-}13)$$

Here Δx is the uncertainty in a particle's location along the x axis and Δp_x is the

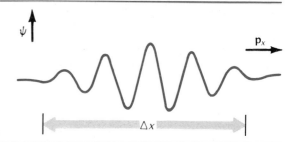

Figure 41-13. A wave packet.

uncertainty in the momentum component in that direction. The product of the uncertainties can be no smaller then $\hbar/2$, which is comparable to h, the fundamental constant of quantum theory.* If you know very precisely where a particle is located, the particle's momentum component must be highly uncertain. Or if you know very precisely a particle's momentum (and therefore also its wavelength $\lambda = h/p$), you pay for this knowledge by high uncertainty about where the particle is located. A second formulation of the uncertainty relation is

$$\Delta E \, \Delta t > \hbar/2 \qquad (41\text{-}14)$$

Here ΔE and Δt are the respective uncertainties in energy and time interval. We can see qualitatively the basis for (41-14) as follows. Suppose an oscillation is observed only over a finite time interval Δt. Then the angular frequency of oscillation is uncertain by $\Delta \omega \simeq 1/\Delta t$. But an uncertainty in the frequency of oscillation of a photon also implies an uncertainty in its energy. Again, when either energy or time interval is small, the other quantity must be correspondingly large by an amount controlled by $\hbar$.

Suppose that we want to represent an electron by a wave and yet localize it in space to some degree. It cannot be a single, monochromatic sinusoidal wave; such a wave would necessarily extend to infinity and it certainly would not be localized. We can, however, superpose a number of sinusoidal waves differing in frequency over a range of frequencies $\Delta \nu$ and so have a *wave packet*. The component waves constructively interfere over a limited region of space Δx, identified as the somewhat uncertain location of the "particle," and so yield a resultant wave function ψ of the sort shown in Figure 41-13. Because there is a range of frequency and a range in wavelength, $\Delta \nu$ and $\Delta \lambda$, the associated momentum and energy are necessarily uncertain, and it is impossible to predict precisely where or when the wave packet will reach another point and what the momentum and energy will then be.

Since the uncertainty relation implies an uncertainty in energy ΔE over a time interval Δt, it also implies that the law of energy conservation may actually be violated—by that amount, $\Delta E = \frac{1}{2}\hbar/\Delta t$, but only for the time interval Δt. The greater the amount of energy "borrowed" or "discarded," the shorter the interval over which nonconservation of energy may occur.

To see the uncertainty principle illustrated in a specific situation, consider

---

* The constant on the right side of (41-13), here $\hbar/2$, depends on the precise definitions of Δp_x and Δx. Uncertainties Δp_x and Δx represent the root-mean-square values of several independent measurements. Symbol $\hbar$ (read as "aitch bar") stands for $h/2\pi$. For example, a photon of angular frequency ω has energy $\hbar \omega$.

waves diffracted by a single, parallel-edged slit of width w, as shown in Figure 41-14. The diffraction pattern is formed on a distant screen. The location of the points of zero intensity is given by $\sin \theta = m\lambda/w$, as shown in Section 39-2.

We have not yet specified what sort of wave passes through the slit. If it is electromagnetic radiation, the intensity of the diffraction pattern is proportional to E^2, the square of the electric field at the screen. If on the other hand, the wave consists of a beam of monoenergetic electrons, the intensity is proportional to ψ^2, which is the square of the electron wave function at the screen and gives the probability of finding an electron at any point along the screen. Whatever the wave type, diffraction effects are pronounced only when the wavelength is comparable to the slit width; at the limit of vanishing wavelength, the intensity pattern on the screen corresponds to a geometrical shadow cast by the edges of the slit.

Suppose now that we reduce drastically the amount of incident radiation or the number of electrons, as the case may be. Then, on the screen we no longer

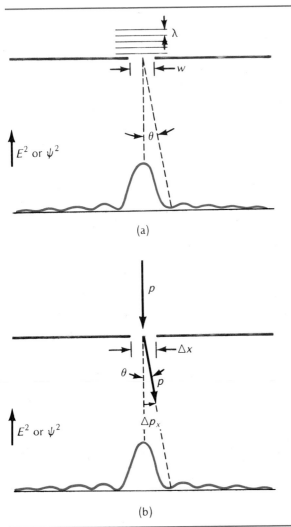

Figure 41-14. (a) Monochromatic waves diffracted by a single slit of width w. (b) Uncertainties Δx in position and Δp_x in momentum.

see smooth variations but instead, photons or electrons arriving one by one. The quantity plotted in Figure 41-14 represents the *probability* that a particle will strike a certain spot on the screen. At very low illumination, bright flashes appear over a large area of the screen. As time passes, more and more particles accumulate on the screen, and the distinct bright flashes merge and form the smoothly varying intensity pattern predicted by wave theory.

There is no way of telling in advance where any one electron or photon will fall on the screen. All that wave mechanics permits us to know is the probability of a particle's striking any one point. Before the particles pass through the slit, their momentum is known with complete precision both in magnitude (monochromatic waves) and in direction (vertically down in this case). When they pass through the slit, their location along x, completely uncertain before they reached the slit, is now known with an uncertainty $\Delta x = w$, the slit width. What we don't know, however, is precisely where any one particle will strike the screen. Any particle has approximately a 75 percent chance of falling within the central region. There is an uncertainty in the x component of momentum p_x.

Suppose that Δx is very large. With a wide slit, the uncertainty in position is increased. We are less certain about where an electron is located along x. The uncertainty in the momentum is reduced correspondingly; the diffraction pattern shrinks, and in the limit, essentially all electrons fall within the geometrical shadow. On the contrary, if the slit width is reduced and Δx becomes very small, the diffraction pattern is expanded along the screen. For the increase in our certainty of the electron's position, we must pay by a correspondingly greater uncertainty Δp_x in its momentum.

For a slit width much greater than the wavelength, particles pass through the slit undeviated to fall within the geometrical shadow. This agrees with classical mechanics, where the wave aspect of material particles is ignored. Thus, there is a close parallel in the relationship of wave optics to ray optics, and of wave mechanics to classical mechanics. Ray optics is a good approximation of wave optics whenever the wavelength is much less than the dimensions of obstacles or apertures that the light encounters; similarly, classical mechanics is a good approximation of wave mechanics whenever a particle's wavelength is much less than the dimensions of obstacles or apertures encountered by material particles. Symbolically, we can write

$$\lim_{\lambda/w \to 0} \text{(wave optics)} = \text{ray optics}$$

$$\lim_{\lambda/w \to 0} \text{(wave mechanics)} = \text{classical mechanics}$$

No ingenious subtlety in the design of the diffraction experiment will remove the basic uncertainty. We do not have here, as in the large-scale phenomena encountered in classical physics, a situation in which the disturbances on the measured object can be made indefinitely small by ingenuity and care. The limitation here is rooted in the fundamental quantum nature of electrons and photons; it is intrinsic in their complementary wave and particle aspects.

Example 41-7. What is the uncertainty in momentum of a 100-eV electron whose position is uncertain by no more than 1.0×10^{-10} m, about the size of an atom. From $\Delta p_x = \hbar/2\,\Delta x$, we compute $\Delta p_x = 5.3 \times 10^{-25}$ kg·m/s. Now compare this uncertainty with the particle's momentum. The fractional uncertainty is $\Delta p_x/p_x =$ about 10 percent. From these numbers, we see that it is impossible to specify the momentum of an electron confined to atomic dimensions with even moderate precision.

Consider now the uncertainty involved when a 10.0-gm body moves at a speed of 10.0 cm/s; that is, an ordinary-sized object is moving at an ordinary speed. Suppose further that the position of the object is uncertain by no more than 1.0×10^{-3} mm. We are interested again in the fractional uncertainty in momentum. We find $\Delta p_x = 5.3 \times 10^{-29}$ kg·m/s and $p = 1.0 \times 10^{-3}$ kg·m/s; therefore, $\Delta p_x/p_x = 5.3 \times 10^{-26}$! The fractional uncertainty in the momentum of a macroscopic body is so extraordinarily small as to be negligible compared with all possible experimental limitations. The uncertainty principle imposes an important limitation on the certainty of measurements only in the microscopic domain. In the macroscopic domain, the uncertainties are essentially trivial.

Figure 41-15(a) shows the momentum p_x of the electron in our example above,

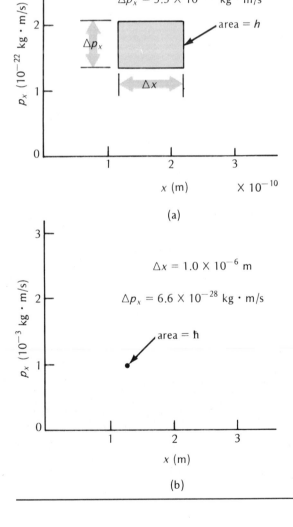

Figure 41-15. (a) Uncertainties in position and momentum for an electron confined to atomic dimensions. (b) Plotted to the same scale, the uncertainties in position and momentum for a 10-gm particle. (On the scale of the drawing, the small dot representing $\hbar$ is too big by a factor of about 10^{26}.)

plotted against its position x. The uncertainty principle requires that the shaded area in this figure, the product of the uncertainties in the momentum and the position, be equal in magnitude to $\hbar/2$. If the position is known with high precision, the momentum is rendered highly uncertain; if the momentum is specified with high certainty, the position must necessarily be highly indefinite. It is therefore impossible to predict and follow in detail the future path of an electron confined to essentially atomic dimensions. Newton's laws of motion, which are completely satisfactory for giving the paths of large-scale particles, cannot be applied here. To predict the future course of any particle, one must know not only the forces that act on the particle but also its initial position and momentum. Because *both* position and momentum cannot be known simultaneously without uncertainty, it is not possible to predict the future path of the particle in detail. Instead, wave mechanics must be used to find the probability of locating the particle at any future time.

Now consider again the 10.0-gm body moving at 10.0 cm/s. Figure 41-15(b) shows its momentum and position. Area $\hbar/2$, representing the product of the uncertainties in momentum and position, is so extraordinarily tiny in these macroscopic circumstances that it appears as an infinitesimal point of the figure. Here the classical laws of mechanics may be applied without entailing appreciable uncertainty.

The finite size of Planck's constant is responsible for quantum effects. Quantum effects are subtle because Planck's constant is very small—but not zero. Recall that the relativity effects are subtle because the speed of light is very large—but not infinite. If somehow Planck's constant were zero, the quantum effects would disappear. Thus classical physics may be thought of as the limit of quantum physics as h is imagined to approach zero. Symbolically,

$$\lim_{h \to 0} (\text{quantum physics}) = \text{classical physics}$$

41-10 The Quantum Description of a Confined Particle

A completely free particle moves in a straight line; it has constant momentum. In wave mechanics, a particle with constant, well-defined momentum must be represented by a monochromatic sinusoidal wave.

Now suppose that such a particle is confined between two infinitely high, hard walls. The particle moves freely back and forth along the x axis, but encounters an infinitely hard wall at $x = 0$ and another at $x = L$; it is, then, confined between these boundaries. The infinitely hard walls correspond to an infinite potential energy V for all values of x less than zero and greater than L. The particle is free between zero and L; therefore its potential energy V in this region is constant. For convenience, we choose the constant potential energy to be zero. The situation we have described is that of a *particle in a one-dimensional box*, or a particle in an infinitely deep potential well. Because the walls are infinitely hard, the particle imparts none of its kinetic energy to them, its total energy remains constant, and it continues to bounce back and forth between the walls unabated.

From the point of view of wave mechanics, we can say that if the particle is confined within the limits stated, then the probability of finding it outside these limits is zero. Therefore, the wave function ψ, whose square represents this probability, must be zero for $x < 0$ and $x > L$. Only those wave functions that satisfy the boundary conditions are allowed. Since the particle has constant momentum magnitude, it is represented by a sinusoidal wave. To satisfy the boundary conditions, only those wavelengths are allowed that permit an integral number of half-wavelengths to be fitted between $x = 0$ and $x = L$. The condition for the existence of stationary, or standing, waves is then

$$L = n \frac{\lambda}{2} \tag{41-15}$$

where λ is the wavelength and n is the *quantum number* having the possible values 1, 2, 3, ...

Figure 41-16 shows the potential well, the wave function ψ, and the probability ψ^2 plotted against x for the first three possible stationary states of the particle in the box. Note that whereas ψ can be negative as well as positive, ψ^2 is always positive. Note also that ψ^2 is always zero at the boundaries. For the first or ground state, $n = 1$, the most probable location of the particle is the point midway between the two walls, at $x = L/2$; for the second state, $n = 2$, however, the least probable location is this point, where, in fact, $\psi = 0$, which is to say that it is impossible for the particle to be located there!

Basically, the boundary conditions on ψ, the fitting of the waves between the walls, restricted the wavelength of the particle to the values given by (41-15). Now if only certain wavelengths are permitted, the magnitude of the momentum also is restricted to certain values, since $p = h/\lambda$. Therefore, the permitted momenta are given by

$$p = \frac{h}{\lambda} = \frac{nh}{2L}$$

Finally, the kinetic energy E_k (and therefore the total energy E of the particle, since the potential energy is zero) is given by

$$E_k = E = \frac{1}{2} mv^2 = \frac{p^2}{2m} = \frac{(nh/2L)^2}{2m}$$

$$E_n = n^2 \frac{h^2}{8mL^2} \tag{41-16}$$

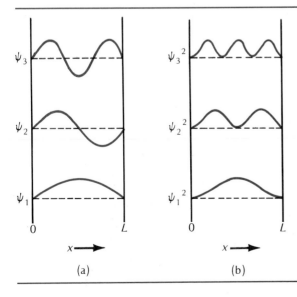

Figure 41-16. The first three stationary states for a particle in a one-dimensional box with infinitely high sidewalls: (a) Wave functions. (b) Probability distributions.

where m is the particle's mass (for nonrelativistic speeds). The subscript n signifies that the possible values of the energy depend only on the quantum number n for fixed values of m and L. The *energy* of the particle in the one-dimensional box is *quantized*. The particle cannot assume just any energy but only those particular energies that satisfy the boundary conditions placed on the wave function.

What are the possible values of the energy if an electron with $m = 9.1 \times 10^{-31}$ kg is constrained to move back and forth within $L = 4 \times 10^{-10}$ m? Setting these values in (41-16) gives for the energy of the first state, $n = 1$, the value $E_1 = 2.3$ eV. Because $E_n = n^2 E_1$, the next possible energies of the particle are $4E_1, 9E_1, 16E_1, \ldots$. The permitted energies of the electron in the atom-sized box are shown in Figure 41-17(a), called an *energy-level diagram*. An electron that is confined to atomic dimensions has possible energies in the range of a few electron volts.

Now consider the allowed energies of a relatively large object confined in a relatively large box. Let us take $m = 9.1$ mg $= 9.1 \times 10^{-6}$ kg, and $L = 4$ cm $= 4 \times 10^{-2}$ m. Equation (41-16) shows that for these values $E_1 = 2.3 \times 10^{-41}$ eV, a fantastically small amount of energy! Figure 41-17(b) is the energy-level diagram for these circumstances [with the energy plotted to the same scale as in Figure 41-17(a)]. The spacing between adjacent energies is so very small that energy is effectively continuous. That is why we never see any obvious manifestation of the quantization of the energy of a macroscopic particle; the quantization is there, but it is too fine to be discerned. This result agrees with the classical requirement that the actually discrete energies of a bound system appear continuous in large-scale phenomena.

The lowest possible energy of a particle in an infinitely deep box is not zero, but E_1. This is in accord with the uncertainty principle. If the particle's energy were zero, with the particle at rest somewhere within the box ($\Delta x = L$), both the momentum p_x and the uncertainty in the momentum Δp_x would be zero. This would violate the uncertainty relation, since the product $\Delta p_x \Delta x$ would then be zero.

For an electron confined within atomic dimensions, the energy of the *ground state* is a few electron volts. The electron is never at rest but bounces

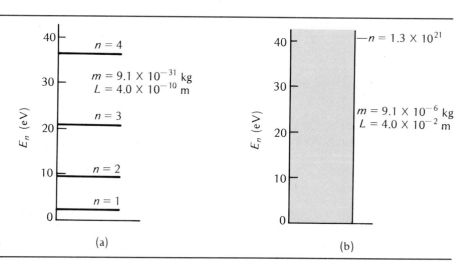

Figure 41-17. Allowed energies of a particle in a one-dimensional box: (a) Electron in a box of atomic dimensions. (b) A 9.1-mg particle in a 4-cm box.

back and forth between the confining walls with its lowest possible energy, the so-called *zero-point energy*. This is, of course, true of any confined particle, for example, the one shown in Figure 41-17(b). The minimum speed of the 9.1-mg particle restricted to 4 cm is easily computed to be 9.0×10^{-28} m/s, or a mere 10^{-8} nm per millennium. The particle is effectively at rest.

The problem of the particle in the box is somewhat artificial, but it is important because it reveals the quantization of the energy. Energy quantization occurs basically because only certain discrete values of the wavelength can be fitted between the boundaries.

Summary

Definitions

Photon: A particlelike quantum of electromagnetic radiation.

Work function: The energy required to release an electron from a metallic surface.

Bremsstrahlung: A process whereby an accelerated charged particle radiates photons.

Compton wavelength: For a particle of rest mass m_0, equal to $h/m_0 c$.

Positron: Electron antiparticle with electric charge $+e$.

deBroglie wavelength: Wavelength $\lambda = h/p$ associated with a particle with momentum p.

Wave function ψ: Contains all information concerning a particle or system; ψ^2 is proportional to the probability of observing the particle.

Energy-level diagram: Horizontal lines representing the allowed energies of a quantized system.

Fundamental Principles

Particle properties of electromagnetic waves: Electromagnetic radiation of frequency v and wavelength λ consists of particlelike photons, where a photon has the properties:

Energy: $E = hv = hc/\lambda$ (41-1)

Speed: $v = c$

Rest Energy: $E_0 = 0$

Momentum: $p = E/c = hv/c = h/\lambda$ (41-6)

where Planck's quantum constant is $h \approx 6.63 \times 10^{-34}$ J·s.

Wave properties of particles: According to the quantum theory every particle with momentum p has an associated wavelength λ, where

$$\lambda = h/p = h/mv \quad (41\text{-}11)$$

Uncertainty principle: The uncertainty Δx in the location of a particle along x and the uncertainty Δp_x in the component of the particle's momentum along x are related in quantum theory by

$$\Delta p_x \, \Delta x \geq \hbar/2 \quad (41\text{-}13)$$

where $\hbar \equiv h/2\pi$. A similar relation governs uncertainties in energy ΔE and time interval Δt:

$$\Delta E \, \Delta t \geq \hbar/2 \quad (41\text{-}14)$$

Important Results

Photoelectric effect: A photon of energy hv dislodges a particle bound with energy ϕ (the work function, $\phi = hv_0$, where v_0 is the threshold frequency) and the particle leaves with maximum kinetic energy E_k.

$$hv = hv_0 + E_k \quad (41\text{-}4)$$

X-ray production, Bremsstrahlung (breaking radiation): A particle emits one or more photons as it is accelerated. If a particle loses all of its kinetic energy E_k in a single collision, the maximum frequency v_{max} and minimum wavelength λ_{min} of the photon are given by

$$E_k = hv_{max} = hc/\lambda_{min} \quad (41\text{-}5)$$

Compton effect: The quantum theory of the scattering of electromagnetic radiation in which a photon collides with a free particle at rest; after the collision the particle acquires kinetic energy and the scattered photon travels at angle θ relative to the direction of the incident photon. The increase in wavelength $\Delta \lambda$ of the scattered photon over the incident photon is related to the scattering angle θ and the rest mass of the particle by

$$\Delta \lambda = (h/m_0 c)(1 - \cos \theta) \quad (41\text{-}9)$$

Pair production: A photon is annihilated and a particle-antiparticle pair (for example, an electron-positron pair) is created. The photon energy for creating particles each with a rest energy $m_0 c^2$ is

$$hv_{min} = 2m_0 c^2 \quad (41\text{-}10)$$

Pair annihilation: A particle-antiparticle pair is annihilated and (typically) two photons of equal energy traveling outward in opposite directions are created.

Probability interpretation of the wave function: If the wave function ψ of a particle has the value $\psi(x)$ at location x, the probability of observing the particle between x and $x + dx$ is proportional to $\psi^2 dx$.

Principle of complementarity: The wave and particle descriptions in quantum situations, while they are mutually contradictory, are complementary. If one description is chosen by the experimental arrangements, the other description is suppressed.

Particle in a (one-dimensional) box: The quantization of energies for a system arises basically from the fitting of wave functions between the boundaries. For infinitely high walls at the boundaries of a constant potential, the allowed wave functions are sinusoidal with zeros at the boundaries. The allowed energies of the system with a one-dimensional box of length L are

$$E_n = n^2 h^2 / 8mL^2 \qquad (41\text{-}16)$$

Problems and Questions

Section 41-2 Photoelectric Effect

· **41-1 Q** The photoelectric effect was crucial in the development of modern physics because it demonstrated the
(A) wave nature of the electron.
(B) localization of most of the mass of an atom in the nucleus.
(C) wave nature of light.
(D) particle nature of light.
(E) validity of the special theory of relativity.

· **41-2 P** The visible spectrum covers a range of wavelengths from approximately 450 nm to 680 nm. What are the energies, in electron volts, of photons at the extremes of this range?

· **41-3 P** A 50-kW radio transmitter operates at a frequency of 1.0 MHz. How many photons does it emit each second?

· **41-4 P** What is the number of photons emitted per second in the beam of a 1.0-mW laser with a wavelength of 562 nm.

· **41-5 P** Sodium has a work function of 2.28 eV. What is the longest wavelength of electromagnetic radiation that can release photoelectrons from sodium?

· **41-6 P** Light falls on a metal surface and 2.0×10^{13} photoelectrons are released each second. Suppose that all these electrons are collected at another metal electrode. What is the photocurrent?

: **41-7 P** A radio station transmits 50,000 W with a carrier frequency of 980 kHz. (a) How many photons per second are emitted? (b) A sensitive receiver detects a signal of 6 μW from this station. How many photons per second is this?

: **41-8 P** In a photoelectric experiment the following values for the stopping potential for photoelectrons were found as a function of the wavelength of the incident light:

Wavelength (nm):   250   284   357   382   429   547
Cut-off voltage (V):   3.22   2.45   1.72   1.30   0.95   0.45

(a) Plot the data and determine the corresponding value of h, Planck's constant. (b) Determine the work function of the metal.

: **41-9 P** The work function for cesium, 1.81 eV, is much lower than the work function for other metals. (a) What is the photoelectric cut-off wavelength? (b) What is the maximum electron kinetic energy when cesium is illuminated with light of 512-nm wavelength? (c) What is the speed of such an electron?

: **41-10 Q** Photons of frequency ν and wavelength λ release photoelectrons with kinetic energy E_k from a metal. Which statement is not correct?
(A) There is a minimum value of λ below which the electrons will not be emitted.
(B) The number of photoelectrons emitted is proportional to the number of incident photons.
(C) The maximum velocity of the photoelectrons is different for different metals.
(D) For different frequencies, E and ν are linearly dependent.
(E) The kinetic energy of a photoelectron will be less than the energy of an incident photon.

: **41-11 P** A mercury arc lamp emits 0.12 W of UV radiation with wavelength 253.7 nm. The arc acts like a point source, and some of the light is incident on a potassium photocathode 1 m from the lamp. The cathode has an effective area of 3.0 cm². The work function for potassium is 2.22 eV.

(a) Experimental measurements show that a photocurrent is emitted within less than 10^{-12} s of the time the potassium is illuminated. Suppose, however, that light did not consist of photons. Assume that a potassium atom is a small sphere of radius 0.2 nm that absorbs light. Determine how long it would take for an atom to collect enough energy to emit an electron. Is the result close to the experimental value? (b) What is the energy, in electron volts, of the photons from the lamp? (c) How many photons hit the photocathode each second? (d) What is the saturation current that would be emitted if the photoconversion efficiency (the probability that a given photon will eject an electron) is 4 percent? (e) What is the stopping potential required to prevent any flow of current?

Section 41-3 X-Ray Production and Bremsstrahlung

· **41-12 P** A helium nucleus (an alpha particle) with charge $+2e$ is accelerated from rest through a potential difference of 10 kV. What would be the maximum energy of a photon created when the helium collides with a massive target?

: **41-13 P** The conservation laws of energy and of momentum must hold for every quantum effect. Use this requirement to show that it is impossible for a moving, unbound, single charged particle to slow down and emit a photon. (Hint: View the system in a reference frame in which the particle is initially at rest.)

Section 41-4 Compton Effect

· **41-14 P** An Air Force laser weapon produces a 3-MW pulse of light with a pulse duration of 1 μs. What is the total linear momentum of the pulse of light?

: **41-15 P** A beryllium-8 nucleus in an excited state ($^8_4$Be*) is initially at rest. It then decays to its ground state ($^8_4$Be) with the emission of a 17.6-MeV gamma ray. (a) What is the momentum of the gamma ray? (b) With what momentum does the nucleus (mass $\simeq$ 8 u) recoil?

· **41-16 P** A 12-keV photon collides with a stationary free electron. The scattered photon has a momentum of 10 keV/c. What is the final kinetic energy of the electron?

· **41-17 Q** A photon collides with a stationary electron. After the collision, a scattered photon travels at θ with the direction of motion of the incident photon, and the electron is set in motion. The speed of the scattered photon, relative to an observer at rest with respect to the moving electron, is
(A) c
(B) $(1 - \cos \theta)h/m_0 c$
(C) $(1 - \cos \theta)c$
(D) $(1 + \cos \theta)c$
(E) $0.5c$

· **41-18 Q** Which of the following statements concerning the Compton effect is incorrect?
(A) The wavelength of the scattered photon is equal to or larger than the wavelength of the incident photon.
(B) The electron can be given an energy equal to the energy of the incident photon.
(C) The energy of the incident photon equals the kinetic energy of the electron plus the energy of the scattered photon.
(D) The energy the electron acquires is largest when the incident and scattered photons move in opposite directions.
(E) Both energy and momentum are conserved in the process.

: **41-19 P** Assuming a dust particle to have a specific density of 3, determine the minimum size of dust particle that will not be pushed out of the solar system by radiation pressure from the sun. Just outside the earth's atmosphere, the intensity of electromagnetic radiation from the sun is 1.4 kW/m².

: **41-20 P** Cesium-137 emits a gamma ray of energy 662 keV. (a) This gamma ray is Compton-scattered through 90°. What is the energy of the scattered photon? (b) What is the minimum energy of the scattered photon, all possible scattering angles considered?

: **41-21 P** In a Compton-scattering experiment, photons with wavelength λ are scattered at 90°. The scattered photons are found to have a wavelength 1.5λ. What is the wavelength of the incident photons?

: **41-22 P** A 4.2-MeV photon is backscattered ($\theta = 180°$) by an electron. (a) What is the wavelength of the scattered photon? (b) What is the kinetic energy of the electron?

: **41-23 P** In the scattering of x-rays from a crystal of NaCl, it is assumed that the scattered x-rays undergo no change in wavelength. Show that this is a reasonable assumption by calculating the order of magnitude of the Compton wavelength for a sodium atom and for a chlorine atom and comparing it with a typical x-ray wavelength of 0.10 nm.

: **41-24 P** What is the wavelength of a photon that can impart a kinetic energy of up to 50 keV to an electron in Compton scattering?

: **41-25 P** In a Compton collision of a photon with an electron, the scattered photon can create an electron-positron pair if it is sufficiently energetic. Show that, no matter how energetic the incident photon, no photon scattered by more than 60° can create an electron-positron pair.

: **41-26 P** The Compton effect occurs for protons as well as for electrons. (a) What is the value of the Compton wavelength for a proton? (b) A 1.0-GeV photon collides with a single proton at rest and the proton recoils in the forward direction. What is the energy of the scattered photon?

Section 41-5 Pair Production and Annihilation

· **41-27 Q** A beam of 0.8-MeV photons is incident on a thin slab of material. Fewer photons emerge from the slab, in the same direction as the incident photons, than enter the slab. A process that can account for this is
(A) Bremsstrahlung.
(B) the Compton effect.
(C) x-ray production.
(D) pair production.
(E) pair annihilation.

· **41-28 Q** Electromagnetic radiation must be assumed to consist of particlelike photons in all of the processes except
(A) x-ray diffraction.
(B) Compton scattering.
(C) the photoelectric effect.
(D) pair production.
(E) pair annihilation.

: **41-29 P** A positron and an electron, each with a kinetic energy of 1.0 MeV, collide head-on and annihilate each other. What is the wavelength of the resulting photons?

: **41-30 P** A 2.0-MeV photon creates an electron-positron pair. If the resulting electron has a kinetic energy of 0.25 MeV, what is the kinetic energy of the positron? (Note that a third heavy particle was present; it acquired some momentum, but not much energy, in this process.)

: **41-31 P** An electron collides head-on with a positron, and two photons are created. The electron and positron each has a speed of $0.8c$. What is the energy of one of the photons?

: **41-32 Q** A high-energy photon creates an electron-positron pair when it comes close to a very massive nucleus. The electron and the positron have equal kinetic energies, each equal to the rest energy of an electron. Then the electron emits a single photon when it comes to rest in a collision. If the initial photon creating the electron-positron pair had a wavelength λ_1, the wavelength of the photon produced in the electron collision is
(A) λ_1
(B) $2\lambda_1$
(C) $4\lambda_1$
(D) $\tfrac{1}{4}\lambda_1$
(E) $\tfrac{1}{2}\lambda_1$

: **41-33 P** By applying the conservation principles of energy and of linear momentum, show that a single free particle cannot absorb a single photon.

: **41-34 P** An electron and positron at rest mutually annihilate and produce *three* photons. What are (a) the minimum and (b) the maximum photon energies?

: **41-35 P** A positron with kinetic energy of 2.0 MeV collides with an electron at rest. The electron and positron are annihilated and two photons with equal energies are created. What is the angle between the photon momenta?

Section 41-6 Matter Waves

· **41-36 Q** The so-called duality of light refers to the fact that light
(A) is characterized by either a frequency or a wavelength.
(B) may be considered as either electric or magnetic, depending on how it is detected.
(C) behaves in some ways like a wave motion and in some ways like particles.
(D) can create an electron-positron pair, or an electron and a positron can combine to create a photon.
(E) has both energy and linear momentum.

· **41-37 Q** Which of the following is not a feature that photons have in common with electrons?
(A) Ability to transport energy.
(B) Momentum.
(C) Electric charge.
(D) An associated wavelength.
(E) Diffraction.

· **41-38 Q** A photon, an electron, and a helium atom all have the same momentum. The particle(s) with the largest de Broglie wavelength is (are)
(A) the photon.
(B) the electron.
(C) the helium atom.
(D) the electron and the helium atom.
(E) All three have the same wavelength.

· **41-39 P** A golf ball with a mass of 49 gm can be given a velocity of 80 m/s with a good drive. What is the de Broglie wavelength of such a ball?

: **41-40 P** Show that if the kinetic energy of a particle is much greater than its rest energy, it has nearly the same de Broglie wavelength as a photon of the same total energy.

: **41-41 P** A photon and a particle have the same wavelengths. How do the following properties of the two compare? (a) their momenta? (b) the particle's total energy and the photon's energy? (c) the particle's kinetic energy and the photon's energy?

: **41-42 P** What is the energy and wavelength of a photon that has the same momentum as a (a) 1.0-MeV electron? (b) 1.0-MeV proton?

: **41-43 P** What is the wavelength of an electron that is accelerated from rest through a potential difference of 25 kV in a color TV set?

: **41-44 P** What is the de Broglie wavelength of a 5-MeV electron? Note that for such an energetic particle Newtonian mechanics does not apply.

: **41-45 P** Show that the de Broglie wavelength of a particle of rest mass m_0 and kinetic energy K can be written

$$\lambda = \frac{hc}{\sqrt{K(K-2m_0c^2)}}$$

: **41-46 P** In an electron microscope a beam of electrons replaces a beam of light, and electric and magnetic focusing fields replace refracting lenses. The resolving power of a microscope—the smallest distance that can be seen—is approximately equal to the wavelength used in the microscope. A typical electron microscope might use 80-keV electrons. (a) What is the minimum distance that can be resolved with a 80-keV electron microscope? (b) What is the energy of a photon that has the same wavelength as a 80-keV electron? (c) What is the momentum of such a photon? (d) What is the momentum of a 80-keV electron? (e) The photon and electron above have the same wavelengths and thereby would produce the same resolving power. Why is the electron microscope used rather than a photon microscope of the same wavelength?

Section 41-9 Uncertainty Principle

· **41-47 Q** Suppose that we lived in a universe where Planck's constant had the very large value of 1.0 J·s (as opposed to the value of 6×10^{-34} J·s in this universe). A pitching machine fires a stream of baseballs, each of rest mass 0.10 kg, with speed 10 m/s perpendicularly through a

window 1 m wide. The balls then strike a distant wall. Under these circumstances,
(A) the kinetic energy of any one ball is appreciably more than 5 J.
(B) all the balls strike the wall within a 1-m-wide strip (that is, within the geometrical shadow of the window).
(C) a significant fraction of the balls will strike the wall outside the 1-m-wide shadow of the window.
(D) the motion of a ball is still well described by Newtonian mechanics.
(E) the momentum of a ball is appreciably larger than 1 kg·m/s.

· 41-48 P An electron in an atom of hydrogen is confined to a region of space on the order of 0.1 nm. What is the order of magnitude of the minimum momentum of the electron?

: 41-49 P A nucleus has a size of the order of 5×10^{-15} m. What is the minimum kinetic energy, consistent with the uncertainty principle, of a proton confined to this region?

: 41-50 P An electron has a velocity of 300 m/s. If this value is accurate to ± 0.01 percent, what is the lower limit to the accuracy with which one can locate the position of this electron? Under these circumstances, is it valid to regard the electron as a point object?

: 41-51 P Sodium emits a spectral line with wavelength 589 nm. Because of the uncertainty in the lifetime of the excited state from which the decay takes place, the spectral line has a half-width (that is, width at half-maximum) of 1.16×10^{-5} nm. This is the effective uncertainty in the wavelength. (a) What is the uncertainty in the energy of the photon? (b) What is the mean lifetime of the excited state from which the emitting electron decayed? (c) What is the "size" of the emitted photon (that is, the length of the emitted wave train)?

: 41-52 P The light emitted from an ordinary light source (such as a mercury arc lamp) consists of numerous relatively short wave trains (each perhaps a meter in length). Interference in such a device as the Michelson interferometer can be observed only if the path differences involved do not exceed the length of this wave train (called the *coherence length*). We can use the uncertainty principle to estimate the coherence length. For the 546-nm line of mercury, the uncertainty in the wavelength is about 0.0005 nm. Estimate the corresponding coherence length.

: 41-53 P Starting with $\Delta x \, \Delta p \approx \hbar/2$, deduce the relation $\Delta E \, \Delta t \approx \hbar/2$ for a free particle with an energy $E = p^2/2m$. The uncertainty in time Δt is related to the uncertainty in position Δx by $\Delta t = \Delta x/v$, where $p = mv$.

: 41-54 P The gamma ray emitted when a Cs-137 nucleus decays has an energy of 662 keV with a line width of 53.0 keV. (a) What is the uncertainty in the wavelength of this photon? (b) What is the uncertainty in the time of emission of this photon? (c) What is the "size" of this photon, that is, the length of the wave train associated with it?

: 41-55 P Show that if the wavelength of a photon is uncertain by $\Delta\lambda$, the corresponding uncertainties in energy and momentum are

(a) $\quad \Delta E = -\dfrac{hc \, \Delta\lambda}{\lambda^2}$

(b) $\quad \Delta p = -\dfrac{h \, \Delta\lambda}{\lambda^2}$

: 41-56 P The energy of a simple harmonic oscillator can be written

$$E = \frac{p^2}{2m} + \frac{1}{2} kx^2$$

where the natural angular frequency is $\omega = \sqrt{k/m}$. Use the uncertainty principle to estimate the minimum energy of such an oscillator.

Section 41-10 The Quantum Description of a Confined Particle

· 41-57 Q A particle of mass m is trapped in a one-dimensional box whose walls are infinitely high and separated by a distance L. The particle's wave function is shown in Figure 41-18. Thus the particle's momentum is
(A) $L/3h$ (D) $3h/L$
(B) $2L/h$ (E) $3h/2L$
(C) $h/2L$

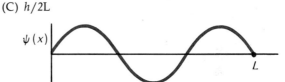

Figure 41-18. Question 41-57.

· 41-58 Q A particle of mass m is contained in an infinite one-dimensional square well of width L. Sketched in Figure 41-19 is a plot of the probability of finding the particle at a given value of x when it is in a particular energy state of the system. What is the particle's kinetic energy in this state?
(A) $8mL^2/h^2$ (D) h^2/mL^2
(B) $3h^2/8mL^2$ (E) $9h^2/2mL^2$
(C) $9h^2/8mL^2$

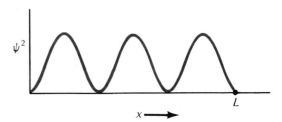

Figure 41-19. Question 41-58.

: **41-59 Q** If the ground-state energy for a particular particle trapped in a one dimensional box with infinitely high side walls is 2.0 eV, what is the next-highest energy the particle can have?
(A) 3 eV
(B) 4 eV
(C) 6 eV
(D) 8 eV
(E) This energy cannot be determined without knowing the width of the box and the mass of the particle.

: **41-60 P** Consider an electron in a box of width 3×10^{-10} m. What wavelength photon will induce a transition from the ground state to the $n = 3$ state?

: **41-61 P** What are the first three energy levels (in electron volts) for a neutron contained in a box of width 10^{-14} m?

: **41-62 P** Plot the first four energy levels for a particle in a box as a function of the width of the box as the width is varied from a value L to $4L$.

: **41-63 P** Consider a particle in a box in one dimension. The width of the box is L. Determine approximately (without integrating) the probability that a particle in the ground state is within a region of width $0.01L$ centered at (a) $x = 0$; (b) $x = 0.25L$; (c) $x = 0.50L$; (d) $x = 0.75L$; (e) $x = L$.

: **41-64 P** A small pellet with a mass of 10^{-4} gm is bouncing back and forth in a box of width 1 cm with a speed of 10 m/s. Considering this as a quantum-mechanical problem of a particle in a box, estimate the value of n for this state of motion.

Atomic Structure

42

42-1 Nuclear Scattering
42-2 The Hydrogen Spectrum
42-3 Bohr Theory of Hydrogen
42-4 The Four Quantum Numbers for Atomic Structure
42-5 Pauli Exclusion Principle and the Periodic Table
42-6 The Laser
    Summary

42-1 Nuclear Scattering

An atom as a whole is ordinarily electrically neutral; if we were to remove all the electrons, what would remain would have all the positive electric charge and essentially all the mass. How are the mass and the positive charge distributed? From a variety of experiments, we know that an atom has a "size" (diameter) of the order of 0.1 nm. The positive charge and the mass are confined to at least this small a region, but it is impossible by any direct measurement to see and observe any details of the atomic structure. Indirect measurement must be resorted to. One of the most powerful methods of studying the distribution of matter or of electric charge — in fact, one of the few ways of studying matter of subatomic dimensions — is scattering. It was by the α-particle-scattering experiments of Rutherford that the existence of small, massive atomic nuclei was established.

Here is a simple example of a scattering experiment. We have a large black box; we can't look inside, but we are to determine how the mass is distributed within the box. At the two extremes, the box might be filled completely with some material of relatively low density, such as wood, or be only partly filled with some material of high density. To find out which is the actual distribution, we can use a very simple expedient: shoot bullets into the box and see what happens to them. If all bullets emerge in the forward direction with reduced speeds, then we might infer that the box is filled throughout with material that deflects the bullets only slightly as they pass through. Suppose that on the other hand, we find a few bullets deflected through a large angle from their original paths. Then we might conclude that they collided with small, hard, and massive objects dispersed throughout the box. Notice that we don't aim the bullets; the shots may be fired randomly over the front of the box. This is the essence of the particle-scattering experiments in atomic and nuclear physics.

Ernest Rutherford suggested in 1913 that the positive charge and the mass of an atom are a point charge and a point mass, which compose a *nucleus*. He suggested that his hypothesis be tested by shooting high-speed, positively charged particles (the bullets) through a thin, metallic foil (the black box), and then examining the distribution of the scattered particles. At the time of Rutherford, the only available suitable charged particles were α particles, with energies of several MeV from radioactive materials. An α particle is a doubly ionized helium atom, with a mass several thousand times larger than that of an electron, yet much smaller than the mass of such a heavy atom as gold.

Figure 42-1 shows the essentials of a scattering experiment. A collimated beam of particles strikes a thin foil of scattering material, and a detector counts the number of particles scattered at scattering angle θ. The experiment consists in measuring the relative number of scattered particles as a function of θ.

Now consider qualitatively the paths of the α particles traversing the interior of the scattering foil. Any encounter with an electron *is* inconsequential because the α-particle mass greatly exceeds the mass of an electron; the particle is essentially undeflected in such a collision, and a negligible fraction of its energy is transferred to any one electron. An α particle is appreciably deflected or scattered only by a close encounter with a nucleus. The mass of the nucleus of a gold atom is considerably greater than (50 times) the mass of the α particle; it remains essentially at rest. The α particles and nuclei, both positively charged, repel each other. The only force acting between a nucleus and an α particle, both regarded as point charges, is taken to be the Coulomb electrostatic force. This force varies inversely with the square of the distance; therefore, the force on an α particle, although never zero, is strong only when it is close to a nucleus.

Figure 42-2 shows several paths of α particles as they move through the

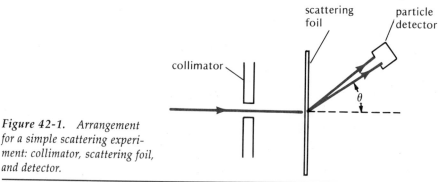

Figure 42-1. Arrangement for a simple scattering experiment: collimator, scattering foil, and detector.

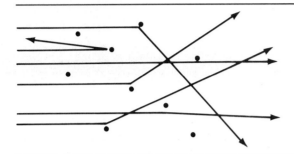

Figure 42-2. Scattering of α particles by nuclei of a material. (The number of α particles scattered through sizable angles is greatly exaggerated.)

interior of a scattering foil. Most pass virtually undeviated; the chance for a close encounter with a scattering center is fairly remote. But those few particles that barely miss a head-on collision can be deflected at sizable angles. Those extremely rare examples of head-on collisions cause the particle's deflection through 180°; that is, the particle is brought to rest momentarily and then returned along its path of incidence. For point charges, most incident particles are scattered only slightly; but a small but significant number are deflected through large angles. (If the positive charge were distributed uniformly throughout the atom rather than being concentrated in nuclei, virtually no particles would be scattered through large angles.) Rutherford's nuclear hypothesis was confirmed in the experiments of Geiger and Marsden. They found that the measured distribution of the scattered α particles agreed with the distribution predicated on the assumption of scattering through a Coulomb force by point charges; the number of scattered particles was found to vary with scattering angle θ according to $1/\sin^4(\theta/2)$.*

Nuclear scattering reveals that when the constituents of nuclei—positively charged protons and electrically neutral neutrons—are separated from one another by less than ~ 1 fm $= 10^{-15}$ m, a strong, attractive *nuclear force* acts between the nuclear constituents. This force, sometimes referred to simply as *the strong interaction*, is substantially stronger than the repulsive Coulomb force between any pair of protons.

Example 42-1. An α particle with an initial kinetic energy of 8.0 MeV is fired head-on at a gold nucleus (79 protons). At what distance from the nucleus is the α particle brought to rest?

The problem is solved most easily by applying energy conservation. The α particle loses kinetic energy K and the system acquires electric potential energy U until the α particle comes to rest momentarily at a distance r from the nucleus, where

$$K = U = \frac{k_e q_1 q_2}{r}$$

We have here used (25-8) for the electric potential energy for point charges $q_1 = 2e$ and $q_2 = 79e$ separated by r.

---

* From the standpoint of quantum theory, a beam of monoenergetic particles is, in effect, a beam of monochromatic waves (Section 41-6). The scattering process consists fundamentally of the diffraction of incident waves by scattering centers. It is remarkable that the wave-mechanical treatment of scattering for an inverse-square force yields precisely the same result as that yielded by a strictly classical analysis. For other types of forces, however, the classical and wave-mechanical results differ.

The relation above can be written as

$$r = \frac{k_e(2e)(79e)}{K}$$

$$= \frac{(9.0 \times 10^9 \text{ N}\cdot\text{m}^2/\text{C}^2)(2)(79)(1.60 \times 10^{-19} \text{ C})^2}{(8.0 \times 10^6 \text{ eV})(1.6 \times 10^{-19} \text{ J/eV})}$$

$$= 2.8 \times 10^{-14} \text{ m} = 28 \text{ fm*}$$

*The distance unit used here is the femtometer = fm = 10^{-15} m, also known as a fermi. The fermi unit, used especially for nuclear distances, honors the Italian-American physicist Enrico Fermi.

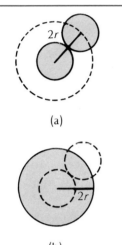

(a)

(b)

Figure 42-3.

Example 42-2. One thousand spheres, each with a radius 1.0 mm, are dispersed at random and held within a Styrofoam cube 1.0 m on an edge. Then another 1000 identical spherical balls are fired at high speed at random over one face of the cube. How many of the incident spheres are expected to make collisions with spheres held within the block and thereby be scattered from their incident paths?

An incident sphere just misses colliding with a sphere at rest when their centers are separated by a distance of just slightly more than $2r$. See Figure 42-3. Each target sphere then presents an effective target area, or *cross section*, of $\sigma = \pi(2r)^2$ to what are effectively point masses in the incident beam. The target spheres are so widely dispersed that no one sphere is likely to "hide" another. The total target area presented to the incident beam is

$$\text{Target area} = N\sigma = N\pi(4r^2) = 10^3(\pi)4(10^{-3} \text{ m})^2 = 1.3 \times 10^{-2} \text{ m}^2$$

The fractional target area exposed for a 1-m^2 surface is then $1.3 \times 10^{-2} = 1.3$ percent. Therefore, about 1.3 percent of the 1000 incident spheres, or about 13, will be scattered from the forward beam.

42-2 The Hydrogen Spectrum

The simplest atomic system is hydrogen; it consists of only two particles—an electron and a much more massive proton as nucleus—interacting by an attractive Coulomb force. Theoretical description of atomic structure has concentrated first on hydrogen and on its observed behavior, primarily the electromagnetic radiation it emits and absorbs.

To observe the spectrum of isolated hydrogen atoms, one must use gaseous *atomic* hydrogen. In a hydrogen gas, the atoms are so far apart that each one behaves as an isolated system (molecular hydrogen H_2 and solid hydrogen radiate different spectra). One can use a prism spectrometer or a diffraction grating. The hydrogen gas may be excited by an electrical discharge or by extreme heating. The dispersed radiation, separated into its various frequency components, falls on a screen or a photographic plate that gives a record of the frequencies and intensities of the emission spectrum.

The spectrum emitted by atomic hydrogen consists of numerous sharp, discrete, bright lines on a black background. See Figure 42-4. In fact, the spectra of all chemical elements in monatomic gaseous form are composed of such bright lines. The spectrum is known as a *line spectrum*. The *emission spectrum* from atomic hydrogen, then, is a bright-line spectrum characteristic of hydrogen. Each chemical element has its own characteristic line spectrum, so that each spectrum is a characteristic "signature" of the particular element and spectroscopy is a particularly sensitive method of identifying the elements.

Atomic hydrogen at room temperature does not, by itself, emit appreciable electromagnetic radiation, but it can selectively absorb electromagnetic radiation, giving an *absorption spectrum.* The absorption spectrum is observed when a beam of white light (all frequencies present) is passed through atomic hydrogen gas and the spectrum of the transmitted light is examined in a spectrometer. The spectrum consists of a series of dark lines superimposed on the spectrum of white light; this is known as a *dark-line spectrum.* The gas is transparent to waves of all frequencies except those corresponding to the dark lines, for which it is opaque; that is, the atoms absorb only waves of certain discrete, sharp frequencies from the continuum of waves passing through the gas. The absorbed energy is very quickly radiated by the excited atoms, but in all directions, not just in the incident direction. The dark lines in the absorption spectrum of hydrogen are at precisely the same frequencies as the bright lines in the emission spectrum are. Hydrogen is a radiator of electromagnetic radiation only at specific frequencies; it is an absorber of radiation only at the same frequencies.

What holds for atomic hydrogen holds also for other elements—a characteristic set of emitted lines when the atoms radiate, and the same set of frequencies for absorption.

It is easy to see that the observed spectrum of hydrogen cannot be accounted for by classical mechanics and classical electromagnetism. Suppose the hydrogen atoms were merely a miniature solar system with the electron orbiting the nucleus like a planet around the sun. Then, with the electrically charged electron accelerated continuously, the atom would radiate continuously. The frequency of the radiation would be the frequency of the orbiting electron. But if the atom were to lose energy continuously by radiation, its energy would decrease continuously, so that the electron would orbit the nucleus in progressively smaller orbits at progressively higher frequencies. In other words, the atom would collapse rapidly, after having radiated a continuous spectrum.

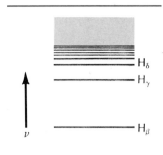

Figure 42-4. Frequency distribution of radiation from atomic hydrogen in the visible region. This particular group of spectral lines is called the Balmer series.

42-3 Bohr Theory of Hydrogen

The first quantum theory of the hydrogen atom was developed in 1913 by Niels Bohr. The photon nature of electromagnetic radiation had been established at that time, but the wave aspects of material particles were not to be recognized until 1924. The Bohr model was only a first step toward a thoroughgoing wave-mechanical treatment of atomic structure. It retains some classical features but introduces some quantum features. Bohr's theory is therefore transitional between classical mechanics and the wave mechanics developed during the 1920s.

In the Bohr theory, we assume the proton to be at rest and the electron to orbit it in a circle. The force between the electron and the proton is the Coulomb force. An electron of mass m and charge e, is in an orbit of radius r about the nucleus, also of electric charge e.

We want the relation for the total energy—kinetic energy of orbiting electron, and electrostatic energy between electron and proton—of the hydrogen atom, considered a sort of junior solar system. We found (Section 15-6) that when a planet of mass m orbits a gravitational force center of mass M in an orbit of radius r, the total energy E is

$$E = -\frac{GmM}{2r} \tag{15-8}$$

To get the corresponding relation for the electric interaction, we merely replace GmM in the equation by $k_e e^2$, where k_e is the Coulomb force constant and e is the magnitude of the electron's and the proton's electric charge.

The total energy of the hydrogen atom then becomes

$$E = -\frac{k_e e^2}{2r} \tag{42-1}$$

The two-particle system has negative total energy with the two particles bound together. Let the electron radius become infinite so that the atom is dissociated into two separate particles; then from (42-1) we have $E = 0$. The ionization energy of hydrogen is 13.6 eV, so that the total energy of the hydrogen in its normal state must be $E = -13.6$ eV. Substituting this value in (42-1) yields $r = 0.53 \times 10^{-10}$ m, roughly the size of an atom.

The quantum condition used by Bohr to select the allowed atomic energies is this: the angular momentum of the electron as it orbits the nucleus is an integral multiple of Planck's constant h divided by 2π. A particle with linear momentum mv in a circular orbit of radius r has angular momentum mvr (Section 14-1). So the quantization rule is

$$mvr = n\frac{h}{2\pi} = n\hbar \tag{42-2}$$

where $n = 1, 2, 3, \ldots$ and $\hbar$ (read as "aitch bar") represents $h/2\pi$. Only those electron orbits are permitted, according to (42-2), for which the angular momentum is an integral multiple of $\hbar$.*

For a particle of mass m orbiting at speed v in a circle of radius r, Newton's second law requires

$$\Sigma F = ma$$

$$\frac{k_e e^2}{r^2} = m\frac{v^2}{r}$$

---

* Equation (42-2) can be cast in a different form, which leads to a simple interpretation of the allowed electron orbits. We have

$$n\frac{h}{mv} = 2\pi r$$

The quantity h/mv is the de Broglie wavelength λ of the electron, so that the equation above may be written as

$$n\lambda = 2\pi r$$

This relation implies that in going the distance $2\pi r$ around the circumference of the circular electron orbit, an integral number n of electron wavelengths may be fitted; that is, an allowed state is one in which an electron, regarded as a wave wrapped around in a self-completing circle, does not cancel itself out by destructive interference. This latter statement cannot be physically correct, however, because an electron does not orbit the nucleus in an atom as a particle; the electron cannot be regarded as existing only around a sharply defined circular orbit. The electron is a wave that extends in all three dimensions. Therefore, we must regard the relation above as a suggestive mnemonic, not as a rigorous application of wave mechanics.

Aha, That Did It!

They had worked long and hard setting up the experiment, and they were finally ready to take data. Then disaster struck. Almost as if by magic, the apparatus went absolutely haywire, the readings were nonsensical, the whole effort seemed a complete wreck. What did it?

A little later the experimenters learned that at just about the time that things had gone so terribly wrong in the laboratory, Wolfgang Pauli happened to be on a train that passed through their city. Here then was still another example of the dreaded "Pauli effect," a kind of malevolent action-at-a-distance.

Just how did Wolfgang Pauli (1900–1958)—noted of course for his formulation of the exclusion principle that, together with quantum theory, explains much of chemistry—get this reputation? He lived at just the time when relativity theory and quantum theory were being developed, and he himself contributed much to that development. He was invited when he was only 19 to write a comprehensive encyclopedia article on relativity theory. One reader said that it was "mature and grandly conceived" and that it had "sureness of mathematical deduction . . . deep physical insight . . . trustworthiness of the critical faculty." That reader was the founder, Albert Einstein.

Note Einstein's remark on "critical faculty." That was what could make mature, highly accomplished physicists tremble: Pauli's ruthless, searing criticism. He hated fuzzy thinking and half truth. One colleague, Paul Ehrenfest, referred to Pauli as "Scourge of God" (*Gottesgeissel*), using the same phrase as was typical for Atilla, the Hun. We must admit that even when it is rightly understood, quantum theory contains uncomfortable paradoxes; to put it more bluntly, quantum physics is a little crazy. At the time it was first emerging, quantum physics included plenty of half-baked ideas. Pauli was determined to rout them out—the ideas that were not merely crazy, but wrong. No wonder theorists stood in terror of Pauli; if there was a flaw in reasoning, Pauli was sure to spot it and say what was wrong with blunt abruptness.

Pauli could be hard on other people, but he was equally critical of his own work. He was prepared, when the occasion required it, to take the bold step. At one point physicists were toying with the idea that momentum, energy, and angular momentum conservation might not apply to radioactive decay. Pauli's proposal: a massless, chargeless, almost undetectable particle that could carry momentum, energy, and angular momentum, and save the conservation laws. That particle—the neutrino (its name, with an Italian diminutive given by Fermi)—plays a central role in elementary-particle physics and is now recognized to come, in fact, in six distinct varieties.

Like almost everyone else who takes physics seriously, Pauli could become frustrated. He wrote at age 25 to a colleague:

> Physics is very muddled again at the moment; it is much too hard for me anyway, and I wish I were a movie comedian or something like that and had never heard of anything about physics!*

Always the critic, always honest.

*See the article on Pauli in *American Journal of Physics* **43**, 205 (1975). It is a translation into English by Ira M. Freeman of the original article in German by Pascual Jordan.

Solving (42-2) for v and substituting in the equation above give

$$\frac{k_e e^2}{r^2} = m \frac{(n\hbar/mr)^2}{r}$$

$$r_n = n^2 \frac{\hbar^2}{k_e m e^2}$$

The smallest allowed radius, the so-called *Bohr radius*, is given by

$$r_1 = \frac{\hbar^2}{k_e m e^2} = 5.291\ 77 \times 10^{-11} \text{ m} \qquad (42\text{-}3)$$

where the values of the known atomic constants have been substituted. The size of the hydrogen atom, ~0.05 nm, is in good agreement with experimental values. The allowed radii given above can be written in simpler form as

$$r_n = n^2 r_1 \tag{42-4}$$

The radii of the stationary orbits are therefore r_1, $4r_1$, $9r_1$, ...

The allowed values of the atom's energy now result from (42-4) substituted in (42-1):

$$E_n = -\frac{k_e e^2}{2r_n} = -\frac{1}{n^2}\frac{k_e e^2}{2r_1} \tag{42-5}$$

We represent the quantity $k_e e^2 / 2r_1$ by E_I, the hydrogen atom's ionization energy, and the equation above becomes

$$E_n = -\frac{E_I}{n^2} \tag{42-6}$$

The only possible energies of the bound electron-proton system that constitutes the hydrogen atom are $-E_I$, $-E_I/4$, $-E_I/9$, ... The permitted energies are discrete; the energy is quantized. The lowest energy (the most negative energy) is that in which the principal quantum number n equals 1; it is called the *ground state*. In the ground state, the energy is $E_1 = -E_I$, and its value from (42-5) and (42-3) is, as computed from fundamental constants,

$$E_1 = -E_I = -\frac{k_e e^2}{2r_1} = -\frac{k_e^2 e^4 m}{2\hbar^2} = -13.605\ 8\text{ eV} \tag{42-7}$$

Figure 42-5 is an *energy-level diagram* for hydrogen. For bound states, E_n is less than zero, and only discrete energies are allowed. As n approaches infinity, the energy difference between adjacent energy levels approaches zero. When n equals infinity, $E_n = 0$, and the hydrogen atom is then dissociated into an electron and a proton; the particles are separated by an infinite distance and both are at rest. In this condition, the atom is ionized, and the energy that must be added to it when it is in its lowest, or ground, state ($n = 1$) to bring its energy

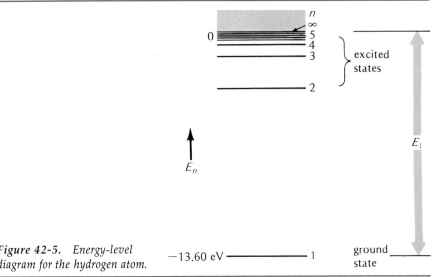

Figure 42-5. Energy-level diagram for the hydrogen atom.

up to $E_n = 0$ is just E_I, the ionization energy. When the system's total energy is positive, the electron and proton are unbound; then all possible energies are allowed, and there is a continuum of energy levels.

Each energy shown in Figure 42-5 corresponds to a stationary state in which the atom can exist without radiating. Stationary states above the ground state ($n = 2, 3, 4, \ldots$) are called *excited states;* an atom in one of them tends to make a transition to some lower stationary state. In a downward transition the electron may, very crudely, be imagined to jump suddenly from one orbit to a smaller orbit. It is better to say that the atom as a whole has made a quantum jump. The amount by which the energy of an atom exceeds the energy of the ground state is called the *excitation energy*. The term *binding energy* denotes the energy that must be added to an atom in an allowed state to free the bound particles and thereby make $E_n = 0$.

Now consider the photons emitted when atoms make downward transitions. An atom is initially in an upper, excited state with energy E_u; it makes a transition to a lower state E_l. In the transition, the atom loses energy $E_u - E_l$. Bohr assumed that in such a transition, a single photon having an energy $h\nu$ is created and emitted by the atom. By energy conservation,

$$h\nu = E_u - E_l \tag{42-8}$$

The Bohr theory, as well as more thoroughgoing wave-mechanical treatments, gives no details of the electron's quantum jump nor of the photon's creation.

Now it is easy to compute the wavelengths of the photons that, according to the Bohr model, are radiated by a hydrogen atom. Using (42-8) and (42-6), we have for the frequency

$$\nu = \frac{E_u - E_l}{h} = \left(-\frac{E_I}{n_u^2 h}\right) - \left(-\frac{E_I}{n_l^2 h}\right) = \frac{E_I}{h}\left(\frac{1}{n_l^2} - \frac{1}{n_u^2}\right) \tag{42-9}$$

where n_u and n_l are the quantum numbers for the upper and lower energy states. The wavelengths $\lambda = c/\nu$ of emitted photons may then be expressed as

$$\frac{1}{\lambda} = \frac{E_I}{hc}\left(\frac{1}{n_l^2} - \frac{1}{n_u^2}\right)$$

This equation can be written more simply as

$$\frac{1}{\lambda} = R\left(\frac{1}{n_l^2} - \frac{1}{n_u^2}\right) \tag{42-10}$$

where R is the *Rydberg constant*, given by

$$R = \frac{E_I}{hc} = \frac{k_e^2 e^4 m}{4\pi \hbar^3 c} = 1.097\,37 \times 10^7 \text{ m}^{-1} \tag{42-11}$$

Actually, (42-10), which gives the wavelengths in the hydrogen spectrum, was known as an empirical relation before Bohr's quantum theory, with the Rydberg constant chosen to fit the observed wavelengths. The fact that Bohr's quantum theory yielded precisely the value of R, and therefore the wavelengths of light emitted and absorbed by hydrogen atoms, is the most emphatic endorsement of Bohr's theory.

The spectral lines of hydrogen can easily be interpreted in terms of the energy-level diagram, Figure 42-6. Vertical lines represent transitions between

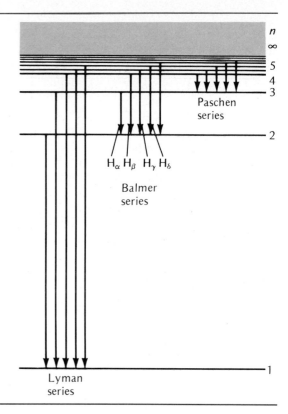

Figure 42-6. Some possible transitions for atomic hydrogen.

stationary states; the length of each line is proportional to the respective photon energy and therefore to the frequency. Spectral lines of the *Lyman series* correspond to those photons created when hydrogen atoms in any of the excited states undergo transitions to the ground state ($n_1 = 1$). Transitions from the unbound states ($E > 0$) to the ground state are responsible for the observed continuous spectrum lying beyond the series limit. In a similar way, the *Balmer series* is produced by downward transitions from excited states to the first excited state ($n_1 = 2$). This series of lines, identified by the labels H_α, H_β, H_γ, H_δ, lie in the visible region of the electromagnetic spectrum. Still further emission series involve downward transitions to $n_1 = 3$, $n_1 = 4$, . . . ; these series fall progressively toward longer wavelengths.

Suppose that the entire emission spectrum from an excited hydrogen gas is observed. We then see the simultaneous emission of many photons produced by downward transitions in many atoms from each of the excited states. To observe the entire emission spectrum, we must have very many hydrogen atoms from each of the excited states making downward transitions to all lower states.

Now consider the absorption spectrum. When white light passes through a gas, those particular photons that have energies equal to the energy difference between stationary states can be removed from the beam. These photons are annihilated, as they give their energy to the internal excitation energy of the atoms. The same set of quantized energy levels participates in both emission and absorption; and the frequencies of their emission and absorption lines are identical. (Because atoms remain in an excited state only very briefly, the Lyman series is the only one observed in absorption.)

We have used the fundamental postulates of the Bohr theory implicitly in developing a model of the hydrogen atom. It is useful, however, to isolate them, since they are retained in their essential forms in more complete wave-mechanical treatments of atomic structure:

- A bound atomic system can exist without radiating, but only in certain discrete stationary states.
- The stationary states are those in which the orbital angular momentum mvr of the atom is an integral multiple of $\hbar$.
- When an atom undergoes a transition from an upper energy state E_u to a lower energy state E_l, a photon of energy $h\nu$ is emitted, where $h\nu = E_u - E_l$. If a photon is absorbed, the atom makes a transition from a low energy state to a higher, according to the same relation.

Example 42-3. Some hydrogen atoms are initially in the second excited state. What are the possible energies of photons emitted by these atoms?

The possible transitions, $3 \rightarrow 2$, $3 \rightarrow 1$, and $2 \rightarrow 1$, are shown in Figure 42-7. The photon energies are, from (42-6):

$3 \rightarrow 2$: $h\nu_{32} = E_3 - E_2 = \left(-\dfrac{13.61 \text{ eV}}{3^2}\right) - \left(-\dfrac{13.61 \text{ eV}}{2^2}\right) = 1.89$ eV

$3 \rightarrow 1$: $h\nu_{31} = (13.61 \text{ eV})\left(1 - \dfrac{1}{3^2}\right) = 12.09$ eV

$2 \rightarrow 1$: $h\nu_{21} = (13.61 \text{ eV})\left(1 - \dfrac{1}{2^2}\right) = 10.21$ eV

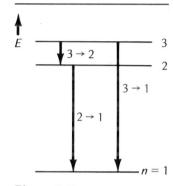

Figure 42-7.

Example 42-4. A hydrogen atom emits a photon corresponding to the Lyman alpha line in the hydrogen spectrum (first excited state to the ground state). (a) With what momentum does the atom recoil on emitting the photon? (b) What is the atom's recoil kinetic energy?

(a) As Example 42-3 shows, the Lyman-alpha transition, $2 \rightarrow 1$, produces a 10.21-eV photon. The photon's momentum is $p = E/c = 10.21$ eV/c. The atom, moving opposite to the photon, has the same momentum magnitude.

(b) The hydrogen atom, with a rest energy of $m_0 c^2 = 0.94$ GeV, recoils at relatively low speed with a kinetic energy of

$$K = \tfrac{1}{2}m_0 v^2 = \dfrac{p^2}{2m_0} = \dfrac{(pc)^2}{2m_0 c^2} = \dfrac{(10.21 \text{ eV})^2}{2(0.94 \times 10^9 \text{ eV})} = 5.5 \times 10^{-8} \text{ eV}$$

The atom's kinetic energy is a very small fraction, only about 5 parts in 10^9, of the photon's energy.

42-4 The Four Quantum Numbers for Atomic Structure

The Bohr quantum theory of hydrogen is only approximately correct. It yields the allowed energies of the hydrogen atom and the hydrogen spectrum fairly satisfactorily. But it does not give the correct quantum relationship for the allowed values of angular momentum. It cannot account for the closely spaced lines (or fine structure) observed in the spectrum. It cannot successfully account for the structure and spectra of atoms with more than one electron. Worst of all, it takes the electron to be a semiclassical particle moving in a

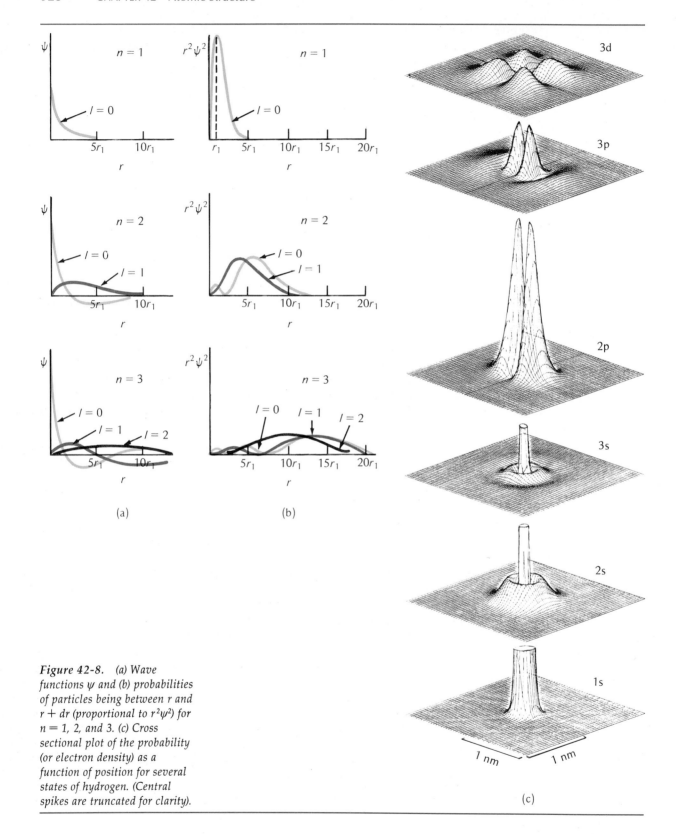

Figure 42-8. (a) Wave functions ψ and (b) probabilities of particles being between r and $r + dr$ (proportional to $r^2\psi^2$) for $n = 1, 2,$ and 3. (c) Cross sectional plot of the probability (or electron density) as a function of position for several states of hydrogen. (Central spikes are truncated for clarity).

well-defined classical orbit, not as a wave extending over three-dimensional space.

A thoroughgoing wave-mechanical analysis of the hydrogen atom involves, as a minimum, finding the allowed wave functions and the corresponding energies by solving for the complete solutions of the Schrödinger wave equation for a particle (the electron) subject to an inverse-square attractive Coulomb force from a fixed force center (the nucleus). The analysis is mathematically sophisticated, and we shall not attempt it here. We shall indicate, in an admittedly approximate and qualitative fashion, the principal features that enter, particularly as they are related to the four quantum numbers characterizing atomic structure.

First recall that for the wave-mechanical problem of a particle confined to a one-dimensional potential well (Section 41-10), a single quantum number emerged. This number was, in effect, the number of half-wave segments that could be fitted between the boundaries of the potential well. In this sense, the problem was altogether analogous to that of a wave on a string attached at both ends. The permitted wave patterns, or allowed standing waves, are those for which an integral number of half-wavelengths can be fitted between the reflecting boundaries (Section 17-6).

For a stationary wave in two dimensions—for example, water waves on the surface of a swimming pool—there are two characteristic integers, or quantum numbers, describing the allowed wave patterns. The allowed wave patterns are, of course, those patterns of standing waves that are consistent with the boundary conditions at the edge of the region over which the waves may extend. For a rectangular swimming pool, two quantum numbers give the number of half-waves that can be fitted along the length and width, respectively, of the pool. For a circular pool, we again have two quantum numbers. One quantum number gives the number of half-waves that can be fitted going radially outward from the center, and the other relates to the number of zeros in the wave function as one goes around a circle.

For a three-dimensional potential well—for example, sound waves trapped inside a rectangular parallelepiped and reflecting from the three sets of parallel side walls—we have three characteristic quantum numbers. Each tells in effect the number of half-waves that can be fitted along each of the three dimensions. In general, the number of characteristic quantum numbers for a wave trapped in a potential well equals the number of dimensions of the well.

And so it is for the electron in a hydrogen atom. Here the particle is in a three-dimensional potential well. The potential energy function, $V = -k_e e^2/r$, depends on the electron's distance r from the nucleus. The electron wave function may extend to infinite distances, but it must be zero there. The potential energy depends on distance from the force center, and so too does the electron wavelength. In fact, as r increases and therefore V increases (becomes less negative), the electron wavelength must also increase. The electron wavelength increases as one goes away from the nucleus.

The simplest class of allowed hydrogen wave functions comprises those that depend only on the radial distance r, and are thereby spherically symmetrical. The first three such s wave functions are illustrated in Figure 42-8. Think of the electron in these states as a diffuse spherical ball centered on the nucleus (for $n = 1$) or as a set of concentric spherical shells (for $n > 1$). In classical terms, the electron wave may be thought of as expanding and contracting radially; the electron wave is reflected at both $r = 0$ and $r = \infty$. In general, however, the three-dimensional electron waves need not be spherically symmetrical; such waves require three quantum numbers to tell how the wave function changes with position.

We give below for each quantum number the symbol, the name, the allowed values, the influence of the atom's energy and angular momentum, and some geometrical characteristics of the associated wave functions for hydrogen. The fourth quantum number that enters naturally in the relativistic wave-mechanical treatment of four-dimensional space-time, the so-called spin quantum number, has no exact classical counterpart. Here results of wave mechanisms are given without proof.

Principal Quantum Number n The principal quantum number n enters in the relation giving the approximate total energy of the atom according to the Bohr formula, $E_n = -E_I/n^2$, (42-6), where E_I is a constant. The allowed values for n are 1, 2, 3, ... As n increases, the energy increases and it approaches zero for infinite n. The wave functions extend toward progressively larger values of r as n increases, in the same way that the size of a classical planetary orbit increases with energy.

Orbital Angular-Momentum Quantum Number l The magnitude of an electron's angular momentum is given by

$$L = \sqrt{l(l+1)}\,\hbar \qquad (42\text{-}12)$$

(Orbital angular momentum classically is that of the orbiting particle around the force center.) Note that the angular-momentum quantization rule is different from $L = n\hbar$, the rule in the Bohr theory.

The orbital angular-momentum quantum number l may take on integral values starting with zero and continuing up to $n - 1$; that is, the l values are restricted to

$$l = 0, 1, 2, 3, 4, \ldots, n-1 \qquad (42\text{-}13)$$
$$\text{s, p, d, f, g, } \ldots$$

(The letter symbols that appear below the numerical values of l are also used to designate the electron state.) For an electron in the state with $n = 1$ (the ground state), the only possible value for l is $l = n - 1 = 1 - 1 = 0$. With $n = 2$, quantum number l may assume the values 0 and 1; the corresponding magnitudes of the orbital angular momentum are, from (42-12), equal to 0 and $\sqrt{2}\hbar$.

Any wave function for which $l = 0$ is spherically symmetrical. The electron has no angular momentum relative to the nucleus (and may be thought of, classically, as passing in an eccentric ellipse through the force center). When $l \neq 0$, the wave functions are not spherically symmetrical but depend on angle. The simplest of these is the p state with $l = 1$. Then the wave function is such that the probability for finding the electron is high at two "knobs" along a line passing through the nucleus, as shown in Figure 42-9. For higher values of l, the wave function shows a more complicated dependence on angle.

When an atom undergoes a transition from one allowed state to another, and a photon is emitted or absorbed, the state and wave functions are changed so that the values of l change by the integer 1, or Δl equals ± 1, for allowed transitions. The permitted energy levels, segregated according to the values of

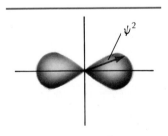

Figure 42-9. Polar plot of the variation of ψ^2 with angle for a p ($l = 1$) state. The pattern is symmetrical with respect to rotation about the symmetry axis.

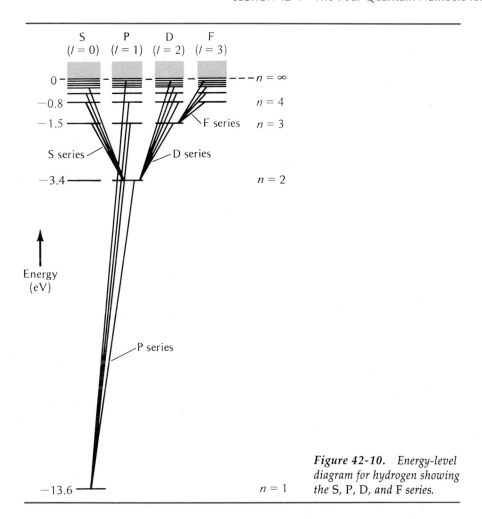

Figure 42-10. Energy-level diagram for hydrogen showing the S, P, D, and F series.

l, together with the allowed transitions, are shown in Figure 42-10. In hydrogen, the several possible levels for a given n have the same total energy; in atoms with more than one electron, however, the S, P, D, F, . . . levels and the corresponding series of emitted lines differ in energy.

Orbital Magnetic Quantum Number m_l The orbital magnetic quantum number may, for a given l, assume positive and negative integral values ranging from $+l$ to $-l$; that is,

$$m_l = l, l-1, l-2, \ldots, 0, \ldots, -l \qquad (42\text{-}14)$$

For example, for a p wave function, with $l = 1$, the allowed values of the orbital magnetic quantum number are 1, 0, and -1. For a d state ($l = 2$), the possibilities are $m_l = 2, 1, 0, -1,$ or -2.

Orbital angular-momentum number l gives, through (42-12), the *magnitude* of the electron's orbital angular momentum. Orbital magnetic quantum number m_l yields the *component* of the electron's orbital angular momentum L_z along some direction z in space. The angular-momentum component L_z is, in fact,

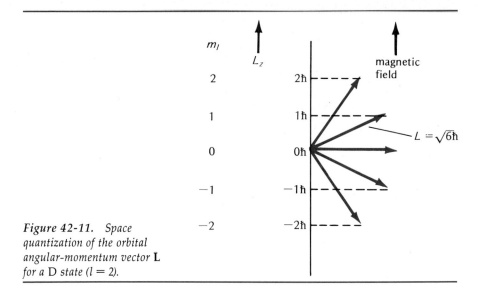

Figure 42-11. Space quantization of the orbital angular-momentum vector **L** for a D state ($l = 2$).

$$L_z = m_l \hbar \tag{42-15}$$

For a d state ($l = 2$), for example, the angular-momentum magnitude is $L = \sqrt{l(l+1)}\hbar = \sqrt{2(3)}\hbar = \sqrt{6}\hbar$, and the allowed projections of this angular-momentum vector along the z axis are $L_z = m_l \hbar = 2\hbar, \hbar, 0, -\hbar,$ and $-2\hbar$. To give a more physical interpretation to m_l and the associated angular-momentum component, imagine the angular-momentum vector **L** to be oriented relative to the z in such directions that its components L_z satisfy (42-15). See Figure 42-11.

What dictates the direction of the z axis? An external magnetic field can do so. The phenomenon whereby the component of the angular momentum is restricted to certain discrete values, and therefore the vector **L** is restricted to certain orientations in space relative to z, is sometimes referred to as *space quantization*. The electron is a negatively charged particle, so that when an electron's wave function indicates nonzero angular momentum, the electron also has an associated magnetic moment. Space quantization implies, then, that the electron magnetic moment is restricted to certain discrete orientations in space. Further, since the orientation of a magnet relative to an external magnetic field controls the energy of the magnet, the energy of an atom (and the photons emitted in transitions between allowed states) can exhibit a multiplicity that is referred to as the *Zeeman effect*, after its discoverer.

Spin Magnetic Quantum Number m_s. The fourth quantum number specifying the state and wave function of an electron is the spin magnetic quantum number. It has just two values:

$$m_s = +\tfrac{1}{2} \text{ and } -\tfrac{1}{2} \tag{42-16}$$

The idea of electron spin is this. Imagine the electron with its charge smeared over a region of space to be spinning perpetually about an internal axis of rotation. The electron has spin angular momentum that is in addition to and independent of the orbital angular momentum associated with quantum number l. Electron spin actually arises as a necessary consequence of a rela-

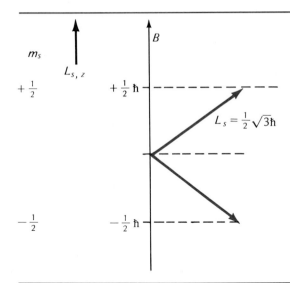

Figure 42-12. *Space quantization of electron-spin angular momentum.*

tivistic treatment of wave mechanics. The magnitude of the spin angular momentum has just one possible value:

$$L_s = \sqrt{s(s+1)}\hbar = \sqrt{(\tfrac{1}{2})(\tfrac{3}{2})}\,\hbar = \tfrac{1}{2}\sqrt{3}\,\hbar \tag{42-17}$$

since the spin quantum number for an electron has the *single* value $s = \tfrac{1}{2}$. Quantum number m_s is related to s as m_l is to l. The component of the spin angular momentum along some direction in space L_{sz} is given by

$$L_{sz} = m_s \hbar = \tfrac{1}{2}\hbar \quad \text{or} \quad -\tfrac{1}{2}\hbar \tag{42-18}$$

The geometrical interpretation of L_{sz} is shown in Figure 42-12.

A perpetually spinning charged particle constitutes in effect an absolutely permanent magnet, so the energy of an atom differs according to the orientation of the spin vector $\mathbf{L}_s$ relative to a magnetic field.

42-5 Pauli Exclusion Principle and the Periodic Table

Specifying the state of an electron in an atom amounts, in the quantum theory, to specifying the values of each of the four quantum numbers n, l, m_l, and m_s. By the procedures of quantum mechanics, it is possible to compute an atom's energy, its angular momentum, and other of its measurable characteristics for each set of quantum numbers. Indeed, it is possible, at least in principle, to predict all properties of the chemical elements from quantum theory. To calculate chemical properties from quantum theory is difficult, however, because of formidable mathematical difficulties that arise with systems having many component particles. Only the problem of the simplest atom, hydrogen, has been solved completely by relativistic quantum theory. Essentially, experiment and theory agree perfectly.

Quantum theory does provide a wealth of information concerning chemical and physical properties. One of its greatest achievements has been to give a fundamental basis for the periodic table of chemical elements. The key is a principle proposed by W. Pauli in 1924, the *Pauli exclusion principle*. This principle together with the quantum theory predicts and accounts for many of the chemical and physical properties of atoms.

Consider again the energy levels available to the single electron in the hydrogen atom. These energy levels are shown schematically (but not to scale) in Figure 42-13. Here each horizontal line corresponds to a particular possible set of values for quantum numbers n, l, and m_l. For each line there are two possible values of the electron-spin quantum number, $m_s = \pm \frac{1}{2}$. The occupancy of an available state by an electron is indicated here by an arrow, whose direction indicates the electron-spin orientation, up for $m_s = +\frac{1}{2}$ and down for $m_s = -\frac{1}{2}$. For brevity only, the energy levels with principal quantum numbers 1, 2, and 3 are shown. For a given value of n, the states with $l = 0$ turn out to be lowest, states with $l = 1$ next, and so on. For a given value of the orbital angular-momentum quantum number l, the possible values of the orbital magnetic quantum number m_l are shown horizontally arranged. Every one of the states (two for each dash) is available to the electron in the hydrogen atom. Some are *degenerate*; they have the *same total energy*, but are nevertheless, distinguishable when a strong magnetic field or other external influence is applied to the atom.

The rules governing the possible values of the quantum numbers and the number of possible values can be summarized as follows:

For a given n: $l = 0, 1, 2, \ldots, n - 1$ (n possibilities)

For a given l: (42-19)
$m_l = l, l - 1, 0, \ldots, -(l - 1), -l$ ($2l + 1$ possibilities)

For a given m_l: $m_s = +\frac{1}{2}, -\frac{1}{2}$ (2 possibilities)

Suppose a hydrogen atom is in its lowest, or ground, state. Then the single electron is in the state in which $n = 1$, $l = 0$, $m_l = 0$, and $m_s = -\frac{1}{2}$. If the

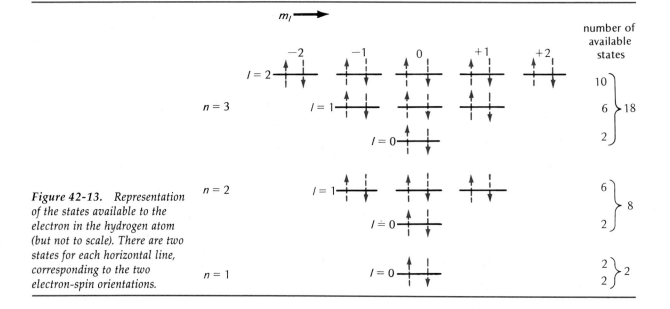

Figure 42-13. Representation of the states available to the electron in the hydrogen atom (but not to scale). There are two states for each horizontal line, corresponding to the two electron-spin orientations.

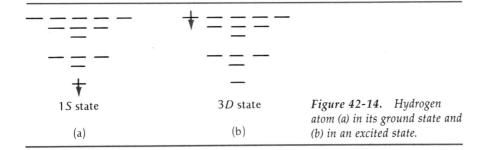

Figure 42-14. Hydrogen atom (a) in its ground state and (b) in an excited state.

hydrogen atom gains energy, this may promote the electron to any one of the higher-lying available states. The atom can then decay to the ground state by downward transitions with the emission of one or more photons. Figure 42-14 depicts a hydrogen atom in its ground state and in an excited 3D state ($l = 2$).

Consider next the element lithium, $_3$Li. This atom has three electrons, to be placed in the levels shown in Figure 42-15. If all three electrons were in the lowest level, that with $n = 0$, $l = 0$, and $m_l = 0$, then two electrons would necessarily have to occupy the state with $m_s = +\frac{1}{2}$ and one would occupy the state with $m_s = -\frac{1}{2}$, or the reverse. This amounts to saying that at least two electrons would have the same set of quantum numbers. But all experimental evidence is that in a lithium atom all three electrons are never simultaneously in the state $n = 1$. The lowest-energy configuration, or ground state, for lithium is this; two electrons, one with $m_s = +\frac{1}{2}$, the other with $m_s = -\frac{1}{2}$, are in the $n = 1$ level, while the third electron occupies a state in the $n = 2$ level. See Figure 42-15. We can interpret this behavior as follows. Two of the three electrons in a lithium atom cannot have the same set of four quantum numbers; that is, two electrons cannot exist in the same state.

The evidence from all elements is the same; atoms simply never occur in nature with two electrons occupying the same state. The Pauli exclusion principle formalizes this experimental fact:

No two electrons in an atom can have the same set of quantum numbers n, l, m_l, and m_s; or no two electrons in an atom can exist in the same state.

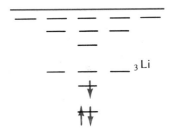

Figure 42-15. Electron configuration of lithium in the ground state.

Exceptions to the exclusion principle, which applies also to systems other than atoms and to particles other than electrons, have never been found. The Pauli principle applies not only to electrons but also to other particles with half-integral spin; the states of particles with integral spin are not limited by it.

Thus, the two electrons in helium in the normal state occupy the two lowest available states indicated in Figure 42-13. No more electrons can be added to the $n = 1$ shell; in helium, $_2$He, the $n = 1$ shell is filled, or closed. The electron spins are then oppositely aligned, and the helium atom has no net angular momentum, either orbital or spin. Furthermore, the two electrons are tightly bound to the nucleus; much energy is required to excite one of them to a higher energy state. That is why helium is chemically inactive.

Suppose that the values of the quantum numbers of each and every electron in an atom are known. Then the electron configuration of the atom is said to be known. A simple procedure is used for specifying an electron configuration. We illustrate it with an example. When a helium atom is in its ground state, each of the two electrons has $n = 1$ and $l = 0$, and their configuration is represented by $1s^2$. The leading number specifies the n value, the lowercase letter s designates the orbital quantum number l of individual electrons, and the postsuperscript gives the number of electrons having the particular values of n and l.

The element lithium, with three electrons, has the electron configuration $1s^2 2s^1$, or two electrons in a completely filled $n = 1$ shell and the third in the $n = 2$ shell. Proceeding in this way—adding one electron as the nuclear charge or atomic number increases by one unit, but always with the restriction that no two electrons within the atom can have the same set of quantum numbers—we can confirm the ground-state configurations of the other atoms at the beginning of the periodic table shown in Table 42-1. We see from Figure 42-13 that two electrons can be accommodated in the s subshell of the $n = 2$ shell and six electrons in the p (or $l = 1$) subshell, after which the $n = 2$ is completely occupied and holds its full quota of eight electrons. With the electron configuration $1s^2 2s^2 2p^6$, corresponding to the rare gas element $_{10}$Ne, the electron wave functions are spherically symmetrical and the atom is chemically inert. In general, a filled subshell for any value of l, with electrons occupying states for all positive, zero, and negative values of m_l and m_s, the atom's net orbital and spin angular momentum is zero and the electron distribution is completely spherical. A closed subshell is effectively a spherical shell of charge.

Chemical properties reflect directly the electron configurations. For example, $_1$H, $_3$Li, $_{11}$Na, and $_{19}$K all have one s electron outside a closed subshell; these elements (the alkali metals) readily relinquish this last s electron to become positive ions, or they may contribute the electron in chemical combinations and thereby exhibit a valence of $+1$. On the other hand, the halogen

Table 42-1

ELEMENT	ELECTRON CONFIGURATION FOR THE GROUND STATE				
$_1$H	$1s^1$				
$_2$He	$1s^2$				
$_3$Li	$1s^2$	$2s^1$			
$_4$Be	$1s^2$	$2s^2$			
$_5$B	$1s^2$	$2s^2$	$2p^1$		
$_6$C	$1s^2$	$2s^2$	$2p^2$		
$_7$N	$1s^2$	$2s^2$	$2p^3$		
$_8$O	$1s^2$	$2s^2$	$2p^4$		
$_9$F	$1s^2$	$2s^2$	$2p^5$		
$_{10}$Ne	$1s^2$	$2s^2$	$2p^6$		
$_{11}$Na	$1s^2$	$2s^2$	$2p^6$	$3s^1$	
$_{12}$Mg	$1s^2$	$2s^2$	$2p^6$	$3s^2$	
$_{13}$Al	$1s^2$	$2s^2$	$2p^6$	$3s^2$	$3p^1$
$_{14}$Si	$1s^2$	$2s^2$	$2p^6$	$3s^2$	$3p^2$
$_{15}$P	$1s^2$	$2s^2$	$2p^6$	$3s^2$	$3p^3$
$_{16}$S	$1s^2$	$2s^2$	$2p^6$	$3s^2$	$3p^4$
$_{17}$Cl	$1s^2$	$2s^2$	$2p^6$	$3s^2$	$3p^5$
$_{18}$Ar	$1s^2$	$2s^2$	$2p^6$	$3s^2$	$3p^6$
$_{19}$K	$1s^2$	$2s^2$	$2p^6$	$3s^2$	$3p^6$ $4s^1$
$_{20}$Ca	$1s^2$	$2s^2$	$2p^6$	$3s^2$	$3p^6$ $4s^2$

elements $_9$F and $_{17}$Cl, both with electron configurations of p^5, lack one electron for completing a p shell; such elements readily acquire an additional electron, to become a negative ion or to form chemical compounds, corresponding to a valence of -1.

42-6 The Laser

The term *laser* is an acronym for Light Amplification by the Stimulated Emission of Radiation. Such a device produces unidirectional, monochromatic, intense, and — most important — coherent visible light.

Consider first the processes by which the energy of an atom can change with the emission or absorption of a photon. (See Figure 42-16.)

- *Spontaneous emission.* An atom is initially in an excited state and decays to a lower state as a photon of energy $h\nu = E_2 - E_1$ is emitted. The decay of unstable atoms is governed by an exponential decay law. Typically, an excited atomic state has a lifetime of the order of 10^{-8} s; on the average, the time for an atom in an excited state to decay spontaneously with the emission of a photon is only 10^{-8} s. Some atomic transitions are much slower, however. For such a metastable state, the atomic lifetime may be as long as 10^{-3} s. (The *spontaneous* transition of an atom from a low energy state to a higher is ruled out by energy conservation.)

- *Stimulated absorption.* An incoming photon stimulates, or induces, an atom to make an upward transition, and the photon is thereby absorbed.

- *Stimulated emission.* An incoming photon stimulates an atom initially in an excited state to make a downward transition. As the atom's energy is lowered, the atom emits a photon, which is *in addition* to the photon inducing the transition. One photon approaches the atom in an excited state, and two photons leave. Afterward, the atom is in the lower energy state. Moreover, the two photons both leave in the same direction as that of the incoming photon,

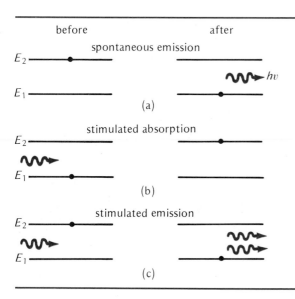

Figure 42-16. Processes in which the quantum state is changed by the absorption or emission of photons: (a) Spontaneous emission. (b) Stimulated absorption. (c) Stimulated emission.

and they are *exactly* in phase relative to one another; that is, they are *coherent* (Section 38-6). We can see that stimulated emission produces coherent radiation as follows. Suppose the two photons were out of phase by some amount; then they would at least partially interfere destructively, violating energy conservation. The stimulated emission produces *light amplification*, or photon multiplication. The trick in constructing a laser is to make the stimulated emission dominate competing processes.

The probability of decay by spontaneous emission can be characterized by the mean life of the excited state. Similarly, one can assign probabilities P_a and P_e to the processes stimulated absorption and stimulated emission. Detailed analysis shows that $P_a = P_e$. That is, for a given photon energy and type of atomic system, stimulated emission is just as probable as stimulated absorption. For example, if a certain number of photons directed at a collection of atoms all initially in a low energy state cause, say, a tenth of the atoms to undergo stimulated absorption, then the same number of photons directed at the same collection of atoms in the upper energy state will cause a tenth of the atoms to undergo stimulated emission.

The three processes—spontaneous emission, stimulated absorption, and stimulated emission—apply to free atoms interacting with photons. If a system consisting of many weakly interacting atoms is in thermal equilibrium, still other so-called relaxation processes may operate to change the quantum state of an atom without, however, emission or absorption of photons. An atom in an excited state may, for example, make a *nonradiative transition* to a lower energy state; the excitation energy goes into the thermal energy of the system rather than into creating a photon. On the contrary, an atom may be raised to a higher energy state as the thermal energy of a system decreases.

Now consider a collection of atoms in thermal equilibrium at some temperature T for which $\epsilon_i > kT$, where ϵ_i is the internal energy. The distribution of the atoms among the available energy states is given to a good approximation by the classical Maxwell-Boltzmann distribution. The number $n(\epsilon_i)$ of atoms with energy ϵ_i is $n(\epsilon_i) \propto e^{-\epsilon_i/kT}$. The relative numbers of atoms in the various possible states are controlled by the system's temperature T according to the Boltzmann factor $e^{-\epsilon/kT}$. The numbers of atoms in progressively higher energy states 1, 2, and 3 are n_1, n_2, and n_3, where $n_1 \propto e^{-E_1/kT}$, $n_2 \propto e^{-E_2/kT}$, and $n_3 \propto e^{-E_3/kT}$. Since $E_1 < E_2 < E_3$, it follows that $n_1 > n_2 > n_3$. The ground state is more heavily populated than the first excited state, and the number of atoms occupying higher states is still lower.

Consider first, for simplicity, a collection of atoms that have only two energy states and are in thermal equilibrium. (Such a collection could be atoms with free or nearly free electrons, whose spin direction is aligned or antialigned with an external magnetic field.) The n_1 atoms in the lower energy state exceed the number n_2 of atoms in the upper energy state; see Figure 42-17(a). Suppose further that a beam of photons with energy $h\nu = E_2 - E_1$ illuminates these atoms. We ignore for the moment spontaneous emission and the relaxation processes within the system, and concentrate on stimulated absorption and stimulated emission only. Stimulated absorption depopulates the lower energy state and reduces the number of photons. Stimulated emission depopulates the upper energy state and increases the number of photons. What is the net effect?

The number of photons disappearing by stimulated absorption is proportional to $P_a n_1$ and the number of additional photons created by virtue of

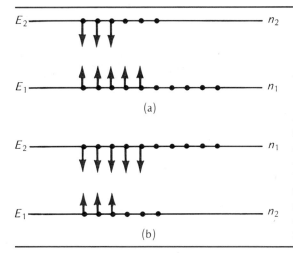

Figure 42-17. Changes in occupancy of quantized states through stimulated absorption and stimulated emission only. (a) In thermal equilibrium, stimulated absorption dominates stimulated emission, and the number of photons is reduced. (b) For a population inversion, stimulated emission dominates stimulated absorption, and the number of photons is enhanced.

stimulated emission is proportional to $P_e n_2 = P_a n_2$. But $n_1 > n_2$, so there is net absorption. Absorption dominates emission simply because more atoms occupy the lower energy state than the upper one. Moreover, the net absorption is accompanied by a tendency toward equalization of the populations of the two states.

Now if we were somehow to produce a *population inversion*, in which the atoms occupying the upper energy state outnumber the atoms in the lower state, then emission would dominate absorption. See Figure 42-17(b). With a population inversion, incoming light would be amplified coherently, since the number of additional photons produced through stimulated emission would more than compensate for the number of photons removed through stimulated absorption. Population inversions have been achieved for lasers in a wide variety of materials by several clever procedures, most of which involve a relatively slow relaxation.

A commonly used, continuously operating laser is the helium-neon laser. The population inversion, the essential condition for photon multiplication, is produced through inelastic collisions between excited (asterisked) helium atoms and neon atoms in the ground state. The process can be written

$$\text{He}^* + \text{Ne} \rightarrow \text{He} + \text{Ne}^*$$

The energy He loses must match the energy Ne gains; that is, the two types of atoms must have excited states with the same or very nearly the same energy. This is indeed the case, as shown in the energy-level diagram of Figure 42-18. The metastable 2s state of helium has an energy of 20.61 eV above the ground state; the 5s state of neon has essentially the same energy, 20.66 eV.* The lifetime of a spontaneous transition of neon from the 5s state to the 3p state is relatively long, but the following transition (3p → 3s) is short. This behavior of the neon atom produces a population inversion between the 5s and 3p states, with the lower state less populated. This transition corresponds to a photon

---

* The states for both helium and neon atoms are indicated by the state of a single excited electron. All other electrons in an atom remain in the ground state. For example, the 5s state of Ne has the electron configuration $1s^2 2s^2 2p^5 5s^1$.

940 CHAPTER 42 Atomic Structure

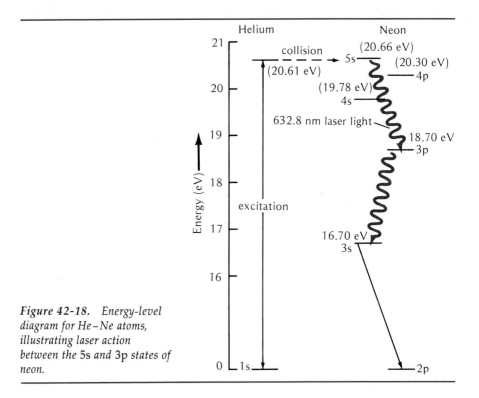

Figure 42-18. Energy-level diagram for He–Ne atoms, illustrating laser action between the 5s and 3p states of neon.

wavelength of 632.8 nm, the orange-red color characteristic of ordinary neon tubes.*

A typical helium-neon laser has a mixture of He–Ne gases enclosed in a sealed tube with parallel silvered mirrors (~99 percent reflecting) at its ends. The gas is excited by a dc voltage source that raises some helium atoms to the 2s metastable state. Inelastic collisions transfer energy to neon atoms and increase the population of the neon atoms in the 5s metastable state. Laser action can then occur, since 632.8-nm light reflecting back and forth between the two ends induces more downward transitions than upward transitions. The coherent, monoenergetic, unidirectional beam is therefore amplified, and laser light emerges from the tube end with the smaller reflectivity.

Energy "lost" through the emission of photons in downward transitions from excited atoms is restored through the continuous excitation of atoms by the dc power supply. Only a small fraction of the energy supplied to the excitation of the atoms is converted to the energy of the output coherent laser beam. A typical working efficiency (output laser power to input excitation power) of a He–Ne gas laser is 1×10^{-3} percent.

Besides the gas laser just described, lasers may involve solids, liquids, and semiconducting materials. Lasers operating from the far infrared to the ultraviolet region of the electromagnetic spectrum have been constructed. In every instance, the condition for laser operation is the existence of a pair of quantized energy levels for which a population inversion has been achieved.

Lasers have many technological applications. They are all possible because, with lasers, one can produce in the visible and nearby regions electromagnetic

---

* Laser action can also take place between other energy states of the neon atom in which population inversion occurs.

radiation that has the coherence properties heretofore available only in radio waves.

Summary

Section 42-4, The Four Quantum Numbers for Atomic Structure, is itself a summary and it is not included here.

Definitions

Alpha particle: helium nucleus.

Rutherford scattering: the scattering of incident charged particles through the Coulomb force by the nucleus as an effective point charge and point mass (in the original experiments suggested by Rutherford, the scattering of alpha particles by gold nuclei).

Cross section: the effective target area presented by a nucleus to a particle.

Types of spectra:

- Bright-line spectrum—collection of sharply defined frequencies in the radiated spectrum, or the emission spectrum.
- Dark-line spectrum—collection of dark lines on the background of the continuous electromagnetic spectrum in the absorption spectrum.

Rydberg constant: the constant R in the empirical relation for the wavelengths λ of the lines in the spectrum of atomic hydrogen

$$\frac{1}{\lambda} = R\left(\frac{1}{n_l^2} - \frac{1}{n_u^2}\right) \quad (42\text{-}10)$$

where n_u and n_l are integral quantum numbers.

Bohr radius: radius of the electron orbit for hydrogen in its ground state in the semiclassical atomic theory of Bohr.

Ground state: lowest-energy quantized state of a system.

Excited state: quantized energy state above the ground state.

Lyman series (for atomic hydrogen): the group of spectral lines originating from or terminating in the ground state.

Balmer series (in the visible region for atomic hydrogen): the group of spectral lines originating from or terminating in the first excited state of hydrogen.

Spontaneous emission: the process in which an atom in an excited state spontaneously decays to a lower energy state with the creation of a single photon.

Stimulated emission: the process in which one photon induces an atom in an excited state to make a downward transition to a lower energy state with the emission of another photon.

Population inversion: the circumstance in which the population of an excited state exceeds the population in a state of lower energy; the essential condition for the operation of a laser.

Fundamental Principles

Basic postulates of the quantum theory of atomic structure:

- Atoms can exist in stationary states without radiating electromagnetic radiation.
- A quantization principle identifies the stationary states. (In the original Bohr theory of hydrogen, the electron's orbital angular momentum mvr was restricted to integral multiples of $\hbar \equiv h/2\pi$, or $mvr = n\hbar$ (42-2); in modern quantum theory the fitting of wave functions to meet boundary conditions produces the allowed states.)
- The absorption or emission of a photon from an atom corresponds to an equal increase or decrease in the energy of the atom.

$$h\nu = E_u - E_l \quad (42\text{-}8)$$

Pauli Exclusion Principle: no two electrons in an atom can have the same set of quantum numbers; equivalently, no two electrons can exist in the same state.

Important Results

Bohr quantum theory of the hydrogen atom:

$$\text{Allowed energies} = E_n = -E_I/n^2 \quad (42\text{-}6)$$

where E_I is the ionization energy and n is the principal quantum number.

$$E_I = k_e^2 e^4 m / 2\hbar^2 = 13.6 \text{ eV} \quad (42\text{-}7)$$

$$\text{Allowed radii} = r_n = n^2 r_1 \quad (42\text{-}4)$$

where Bohr radius $= r_1 = \hbar^2 / k_e m e^2$
$$= 5.29 \times 10^{-11} \text{ m} \quad (42\text{-}7)$$

Problems and Questions

Section 42-1 Nuclear Scattering

· **42-1 Q** The crucial observation in the Rutherford experiment, in which alpha particles were scattered by gold foil, was that

(A) the alpha particles decayed radioactively before reaching the foil.
(B) appreciable numbers of gold nuclei were dislodged from the foil.
(C) no alpha particles were deflected through large angles.
(D) no alpha particles were observed coming from the region of the foil.
(E) some alpha particles were deflected through large angles and some were deflected through small angles.

· **42-2 P** Which particle would get closer to a nucleus as target if the particle were fired directly at the nucleus, a proton accelerated from rest through an electric potential difference V or an alpha particle accelerated from rest through the same potential difference?

: **42-3 Q** Unless it is aimed directly at a nucleus, an alpha particle can never "strike" a nucleus; the positively charged alpha particle is repelled by the positively charged nucleus. But suppose that an energetic negatively charged particle, such as a negative ion, is fired at a nucleus, but not directly head-on. Will it ever "hit" the nucleus?

: **42-4 P** A Styrofoam cube of edge length L contains N_1 steel balls, each of radius r_1, dispersed randomly throughout the interior of the cube. The balls are small enough that no one ball "hides" another. Some other small balls of radius r_2 are fired at random at the cube. What fraction of the incident balls are removed from the forward beam?

: **42-5 P** (a) A 100-eV proton is fired head-on at an electron at rest. What are the kinetic energies of the two particles afterward? (b) A 100-eV electron is fired head-on at a proton at rest. What are the kinetic energies of the two particles afterward?

: **42-6 P** Alpha particles in the Rutherford scattering experiments can be imagined to move effectively in a straight line until they are a distance much less than an atomic radius from the nucleus. As shown in Example 42-1, an 8.0-MeV alpha particle fired head-on at a gold nucleus comes within a distance of 28 fm of the center of the nucleus. (a) What is the acceleration of the alpha particle at this location? (b) Compute the acceleration of the alpha particle resulting from the Coulomb force of the gold nucleus when the alpha particle is an atomic distance away, say, 1.0 nm from the nuclear center.

: **42-7 P** A *neutron star* is an astronomical object that has undergone such severe gravitational collapse that its electrons, protons, and neutrons have been squeezed together to form a highly dense object consisting solely of neutrons in contact; that is, neutrons separated on the average by a distance of the order of 1 fm. In effect, the density of the neutron star is the same as the density of an atomic nucleus. What is the order of magnitude of this density?

: **42-8 P** Since a neutron has no electric charge, it interacts with a nucleus only through the strong, short-range attractive nuclear force. In effect, a neutron feels a force from a nucleus only when it touches it. The radius R of the nucleus is given approximately by $R = r_0 A^{1/3}$ where A is the mass of the nucleus (equal in atomic mass units, u, to the total number of protons and neutrons in the nucleus) and $r_0 = 1.4$ fm. (a) What is the approximate cross section for neutron absorption by lead-206? Cross sections are usually given with the unit *barn*, where 1 barn = 10^{-28} m² = 100 fm². (b) A thin foil of lead has an areal density (an indirect measure of thickness) of 100 mg/cm². This foil is used as a target for a beam of neutrons. What fraction of the incident neutrons is removed from the forward beam by neutron absorption (density of lead, 11.3×10^3 kg/m³)?

: **42-9 P** A beam of 5.0-MeV α particles strikes a target of helium atoms in a gas. (a) What is the minimum distance one helium nucleus comes to another helium nucleus? (b) Show that none of the particles can be scattered by an angle greater than 90°.

: **42-10 P** A particle of charge q_1 with initial kinetic energy K is fired head-on at a massive particle of charge q_2 at rest. (a) What is the minimum distance between the two particles? (b) Now suppose that the particle with charge q_1 and initial kinetic energy K is fired so that, if there were no electric force between the particles, it would miss hitting the massive particle by a distance b. What is now the minimum distance between the two charged particles? (*Big hint:* the Coulomb force is a central force.)

Section 42-3 Bohr Theory of Hydrogen

· **42-11 Q** A key experimental observation led Bohr to postulate what has now become the modern model for the atom. This observation was that
(A) for certain metals, blue light would give rise to a photocurrent but red light would not, independent of the intensity of the light.
(B) isolated atoms emit and absorb discrete light wavelengths, as opposed to a continuum of wavelengths.
(C) 6×10^{23} atoms of any element are always found to have a mass equal in grams to the atomic number of the element.
(D) scattering of alpha particles from gold nuclei indicated that the atom's positive charge is fairly uniformly distributed throughout the atom.
(E) the length of a moving object contracts in its direction of motion.
(F) particles such as electrons have a dual nature, acting in some respects like particles and in other respects like waves.

· **42-12 Q** One of Niels Bohr's brilliant insights, which led to our modern understanding of atomic structure, was that
(A) the electron charge is quantized.
(B) an atom's energy can vary continuously from zero to some maximum cutoff value.

(C) the atom's energy is Planck's constant multiplied by the frequency of the orbiting electron.
(D) the electron's angular momentum is quantized.
(E) the atom's energy is an integral multiple of the electron's rest energy.

· 42-13 P How much energy is required to remove an electron from the first excited state of hydrogen, thereby producing an ion H$^+$?

· 42-14 Q For an atom of hydrogen to emit radiation,
(A) it must be in its ground state.
(B) it must make a transition from the ground state.
(C) it must be in an excited state.
(D) it must simultaneously absorb a photon.
(E) it must be fluorescent.

· 42-15 P The allowed energy levels of a hydrogen atom are characterized by a quantum number n, where $n = 1$ corresponds to the ground state. For which of the transitions listed here would a photon of the shortest wavelength be emitted?
(A) From $n = 100$ to $n = 5$.
(B) From $n = 3$ to $n = 7$.
(C) From $n = 1$ to $n = 5$.
(D) From $n = 4$ to $n = 2$.
(E) From $n = 2$ to $n = 1$.

· 42-16 P What is the wavelength of the photon emitted when a hydrogen atom makes a transition from the $n = 5$ state to the $n = 2$ state?

: 42-17 P Using the Bohr model, determine what energy is required to change a He$^+$ ion into He^{2+}. The experimental value is 54.4 eV.

: 42-18 Q Suppose that an atom had only four distinct energy levels. What is the maximum number of spectral lines of different wavelengths that it could possibly emit?
(A) 1.
(B) 2.
(C) 3.
(D) 4.
(E) 5.
(F) 6.

: 42-19 P A free electron of negligible kinetic energy is captured by a stationary proton to form an excited state of the hydrogen atom. In this process, a photon of energy E is emitted. Shortly thereafter a second photon of energy 10.2 eV is emitted. No further photons are emitted. Deduce the energy E of the first photon emitted under these circumstances.

: 42-20 P Light of wavelength 409 nm is emitted from a hydrogen arc. What transition produced this emission?

: 42-21 P What is the wavelength of a photon that will induce a transition from the ground state to the $n = 4$ state in hydrogen?

: 42-22 P A hydrogen atom can be excited from its ground state to the first excited state when an electron of sufficient kinetic energy hits the atom. What is the minimum electron kinetic energy?

: 42-23 P Two hydrogen atoms, both initially in the ground state, approach one another with the same initial kinetic energy K_i and collide head-on. As a consequence of the collision, one hydrogen atom is excited to the first excited state while the other hydrogen atom remains in the ground state. What is K_i?

: 42-24 P The ion Li^{2+} has a nuclear charge of $+3e$. The ion has a single electron, so that the ion is similar to a hydrogen atom. For this ion, calculate (a) the ground-state energy; (b) the wavelength of the photon emitted when the ion makes a transition from the $n = 2$ state to the $n = 1$ state.

: 42-25 P A free hydrogen atom undergoes a transition from the $n = 3$ state to the $n = 1$ state. A photon is emitted and the atom will necessarily recoil. (a) How much energy, in electron volts, is released in this transition? (b) With what speed does the atom recoil? (c) What is the energy of the photon emitted? (d) What is the kinetic energy of the recoiling atom?

: 42-26 P When a hydrogen atom is excited to the first excited state, it remains in this state for an average time of about 10^{-8} s before making a downward transition to the ground state. (a) Use the uncertainty principle to find the uncertainty in the energy (in eV) of the excited state. (b) What is the fractional uncertainty in the wavelength of the photon emitted in a transition to the ground state? The finite lifetimes of excited atomic states produce a *natural linewidth* in the spectral lines.

: 42-27 P A *muonium* atom consists of a muon, an unstable elementary particle with charge $-e$ and a mass 207 times that of an electron, bound to a proton. What are (a) the ionization energy and (b) Bohr radius for muonium?

: 42-28 P A *positronium* atom consists of an electron and a positron. Each particle may be considered to orbit the atom's center of mass. (a) What is the atom's ionization energy? (b) What is the wavelength of the photon emitted in a transition from the first excited to the ground state?

: 42-29 P Using the fact that the average kinetic energy of a molecule of a gas at temperature T is $3kT/2$, estimate the minimum temperature of a gas of atomic hydrogen that will produce appreciable ionization of the atoms through collisions that will break up the atoms into protons and electrons (a plasma).

: 42-30 P Show that the speed of the electron in the first Bohr orbit is, in units of the speed of light, equal to $k_e e^2/\hbar c$. This combination of fundamental atomic constants, usually abbreviated by the symbol α, is called the *fine-*

structure constant (for reasons now of strictly historical interest). The fine-structure constant plays an important role in quantum theory of electromagnetism because it gives a dimensionless relation among the fundamental constants of electromagnetism (k_e and e), quantum theory (h), and relativity (c).

: **42-31 P** Suppose an "atom" existed with an electron bound to a neutron by the gravitational force. (*a*) From the Bohr model, what would be the energy of the ground state? (*b*) What would be the radius of the first Bohr orbit?

: **42-32 P** In a *positronium* atom, the positively charged proton is replaced by a positively charged positron. For this atom, determine (*a*) the ground-state energy; (*b*) the Bohr radius.

: **42-33 P** Consider a hydrogen atom in a state with very large quantum number n. (*a*) Determine the frequency of a photon emitted in a transition from n to $n - 1$. (*b*) Show that the frequency obtained in (*a*) is approximately equal to the frequency of revolution of the electron in its orbit. This is an illustration of Bohr's correspondence principle, which states that for large n, the results of the quantum theory and classical physics will agree. (*Hint*: $f(n) - f(n-1) \approx df/dn$ for large n.)

: **42-34 Q** Suppose that an electron makes a transition from the $n = 3$ state to the $n = 2$ state in the Bohr model of the hydrogen atom. Which of the following is then correct?
(A) The electron's kinetic energy decreases, and the potential energy of the atom increases, but its total energy remains the same.
(B) Kinetic energy increases, potential energy decreases, but total energy remains the same.
(C) Kinetic and potential energies both increase and so does total energy.
(D) Kinetic and potential energies both decrease, and so does total energy.
(E) Kinetic energy increases by an amount Δ and potential energy decreases by 2Δ, thereby producing a decrease of Δ in total energy.

: **42-35 P** Consider a hypothetical one-electron atom, in which the series of spectral lines corresponding to transitions that end on the $n = 1$ state have wavelengths 130 nm, 110 nm, 95 nm, 86 nm, . . . , 78 nm. The 78-nm wavelength is the shortest that this atom can emit. (*a*) What is the ionization energy, in electron volts, for the atom? (*b*) What are the energies of the first four energy levels, in electron volts? (*c*) What is the wavelength of the photon emitted in a transition from $n = 4$ to $n = 2$? (*d*) What energy must be supplied to induce a transition from $n = 2$ to $n = 3$?

Section 42-4 The Four Quantum Numbers for Atomic Structure

· **42-36 P** Suppose that a relatively heavy atom such as tungsten (atomic number, 74) has all but one of its electrons removed. (*a*) How much energy is required to remove this last electron completely if the ionized atom is initially in its ground state? (*b*) What would be the wavelength of the photon emitted if this single remaining electron were to make a transition from the state with $n = 2$ to the state with $n = 1$? In what part of the electromagnetic spectrum would such a photon be found?

· **42-37 P** When an electron with a kinetic energy of at least 4.88 eV collides with a mercury atom in its ground state, the collision is inelastic and the mercury atom emits radiation. The interpretation of this observation (first done in the *Franck-Hertz experiment* of 1914) is that the kinetic energy lost by the electron brings the mercury atom to its first excited state from which the atom then decays back to the ground state with the emission of a photon. What is the wavelength of this photon?

· **42-38 Q** The chemical behavior of an atom is determined by its
(A) mass number.
(B) binding energy.
(C) atomic weight.
(D) atomic number.
(E) number of isotopes.

· **42-39 Q** A hydrogen atom is in a p state. Therefore
(A) the atom has its lowest possible energy.
(B) the atom is ionized.
(C) the atom is in the ground state.
(D) the atom's orbital angular momentum is not zero.
(E) the electron wave function is spherically symmetrical.

: **42-40 P** The characteristic x-rays from any element result from quantum transitions of the innermost electrons in the atom. The x-ray line of shortest wavelength, called the K_α line, is produced by a transition from the shell for $n = 2$ to a vacancy in the shell for $n = 1$. The remaining electron in the innermost shell shields the electric charge of the nucleus; indeed, the electron making the transition "sees" an effective nuclear charge of $Z - 1$, where Z is the atomic number of the element, or equivalently, the total charge in units of e of the nucleus. What is the wavelength of the K_α x-ray line emitted by iron ($Z = 26$)?

: **42-41 P** Sodium atoms strongly radiate two closely spaced yellow lines (called the sodium D lines) with wavelengths of 589.5944 nm and 588.9977 nm. The two lines originate from transitions from two closely spaced excited energy levels to the single ground state of sodium. What is the energy difference between the two upper energy levels?

: **42-42 P** In one fictitious classical model of the electron, it is assumed to be a sphere of radius 2.8×10^{-15} m with electric charge and mass distributed uniformly throughout its volume. (*a*) At what angular speed would the electron have to spin about a diameter to have spin angular momentum with the magnitude $\sqrt{3}\hbar/2$? (b) What would be the tangential speed of a point on the "equator"?

: 42-43 P The total energy of a hydrogen atom may be written as $E = p^2/2m - ke^2/r$, where p is the orbital linear momentum of the electron and r is its distance from the nucleus. If an electron is confined to a distance r, its momentum is uncertain by an amount governed by the uncertainty principle. (a) Show that the atom's energy is a minimum for some distance r_1, where r_1 is in fact the first Bohr radius. (b) Show that for this distance of electron from nucleus, the atom's energy is that of hydrogen in the ground state.

Section 42-5 Pauli Exclusion Principle and the Periodic Table

· 42-44 Q (a) Write down the quantum numbers for the three outermost electrons in magnesium (atomic number 17). (b) Write down the quantum numbers for the lowest-lying excited state in magnesium.

· 42-45 P For each of the following electron configurations, identify the corresponding element. (a) $1s^22s^22p^63s^1$; (b) $1s^22s^22p^63s^23p^6$; (c) $1s^22s^22p^63s^23p^63d^74s^2$

· 42-46 P (a) What quantum numbers characterize an electron in the $n = 2$ state in the ion He^+? (b) What is the energy of the He^+ ion in this state?

: 42-47 Q Whether or not a particle obeys the Pauli exclusion principle is determined by the particle's
(A) charge.
(B) energy.
(C) spin angular momentum.
(D) orbital angular momentum.
(E) wavelength.

: 42-48 P Suppose three electrons were placed in a one-dimensional box of width L. What minimum energy would be needed to remove one electron (that is, to ionize the system)?

: 42-49 P Seven identical particles (noninteracting) are placed in a cubical box of side 2.5×10^{-10} m. What is the lowest total energy of the system and what are the quantum numbers of the particles if (a) the particles are electrons? (b) the particles have the same mass as the electron but no spin (that is, they are not subject to the Pauli exclusion principle).

Section 42-6 The Laser

· 42-50 Q A process crucial in operating any laser is
(A) spontaneous absorption of radiation.
(B) stimulated emission of radiation.
(C) conversion of photons into electrons.
(D) conversion of electrons into photons.
(E) splitting a single photon into two photons.

· 42-51 P A helium-neon laser operates with a 115-V, 2.0-A power supply. The output is 1.0 mW. What is the efficiency for converting electrical energy to coherent light energy?

: 42-52 P A carbon-dioxide pulsed laser produces pulses of 2.0×10^{11} W with a duration of 1.0 ns. The transverse cross section of the laser beam is 0.50 mm². For each pulse, what is (a) the energy, (b) the energy density, and (c) the linear momentum? (d) Suppose that one pulse were absorbed completely by water at 20° C. How much water would be vaporized?

: 42-53 P A helium-neon laser is most commonly operated to produce light of 633-nm wavelength, but lasing action is also possible at other wavelengths. Use Figure 42-17 to identify the transition for producing coherent light of wavelengths (a) 3.4×10^2 nm and (b) 1.15×10^3 nm.

: 42-54 P A helium-neon laser emits light of wavelength 632.8 nm in a transition between two states of neon. For neon atoms in equilibrium at 300 K, what is the ratio of the population of the upper state to the population of the lower state?

: 42-55 P The particles of a certain system have three possible energies — E_1, E_2, and E_3, where $E_1 < E_2 < E_3$. The corresponding number of particles in the three states are n_1, n_2, and n_3, where for thermal equilibrium $n_1 > n_2 > n_3$. Now suppose that the system is irradiated by strong pumping radiation that equalizes the populations in states 1 and 3. Show that a population inversion must exist between one other pair of states, with either $n_3 > n_2$ or $n_2 > n_1$.

Appendixes

A International System of Units

In the International System of Units (abbreviated SI in all languages) there is one and only one SI unit for each physical quantity, either the appropriate SI base unit itself, defined in the listing below, or the appropriate SI derived unit, formed by multiplication and/or division of two or more SI Base Units, also listed below.

SI Base Units

Meter (m)   The meter is the length equal to the distance traveled in a time interval of 1/299792458 of a second by plane electromagnetic waves in a vacuum.

Kilogram (kg)   The kilogram is the unit of mass; it is equal to the mass of the international prototype of the kilogram.

Second (s)   The second is the duration of 9192631770 periods of the radiation corresponding to the transition between the two hyperfine levels of the ground state of the caesium-133 atom.

Ampere (A)   The ampere is that constant current which, if maintained in two straight parallel conductors of infinite length, of negligible cross section, and placed 1 meter apart in a vacuum, would produce between these conductors a force equal to 2×10^{-7} newtons per meter of length.

Kelvin (K)   The kelvin, unit of thermodynamic temperature, is the fraction 1/273.16 of the thermodynamic temperature of the triple point of water.

Candela (cd)   The candela is the luminous intensity, in the perpendicular direction, of a surface of 1/600000 square meter of a black body at the temperature of freezing platinum under a pressure of 101325 newtons per square meter.

Mole (mol)   The mole is the amount of substance of a system which contains as many elementary entities as there are atoms in 0.012 kilograms of carbon-12. When the mole is used, the elementary entities must be specified and may be atoms, molecules, ions, electrons, other particles, or specified groups of such particles.

SI Derived Units

QUANTITY	UNIT NAME	SYMBOL	BASIC SI UNITS
Frequency	hertz	Hz	s^{-1}
Force	newton	N	$kg \cdot m/s^2$
Pressure	pascal	Pa	N/m^2
Energy	joule	J	$N \cdot m$
Power	watt	W	J/s
Electric charge	coulomb	C	$A \cdot s$
Potential difference	volt	V	W/A
Electric resistance	ohm	Ω	V/A
Conductance	siemens	S	A/V
Capacitance	farad	F	$A \cdot s/V$
Magnetic flux	weber	Wb	$V \cdot s$
Inductance	henry	H	$V \cdot s/A$
Magnetic flux density	tesla	T	Wb/m^2

B SI Prefixes for Factors of Ten

PREFIX	SYMBOL	POWER
exa	E	10^{18}
peta	P	10^{15}
tera	T	10^{12}
giga	G	10^{9}
mega	M	10^{6}
kilo	k	10^{3}
hecto	h	10^{2}
deka	da	10^{1}
deci	d	10^{-1}
centi	c	10^{-2}
milli	m	10^{-3}
micro	μ	10^{-6}
nano	n	10^{-9}
pico	p	10^{-12}
femto	f	10^{-15}

Examples: MW = megawatt = 10^6 watt

nm = nanometer = 10^{-9} meter

C Physical Constants

	SYMBOL	VALUE
Acceleration of gravity	g	9.80665 m/s² (standard value)
Standard atmospheric pressure		1.01325×10^5 Pa
Gravitational constant	G	6.672×10^{-11} N·m²/kg²

	SYMBOL	VALUE
Speed of light	c	2.99792458×10^8 m/s (exact value)
Electron charge	e	1.60219×10^{-19} C
Avogadro's number	N_A	6.0220×10^{23} mol^{-1}
Gas constant	R	8.314 J/mol·K
		8.206 ℓ·atm/mol·K
Boltzmann constant	$k = R/N_A$	1.3807×10^{-23} J/K
		8.617×10^{-5} eV/K
Unified mass unit	u	1.6606×10^{-24} gm (u in gm $= 1/N_A$)
		(1/12)(mass of neutral carbon-12 atom)
Coulomb-law constant	$k_e = 1/4\pi\epsilon_0$	8.98755×10^9 N·m^2/C^2
Permittivity of free space	ϵ_0	8.85419×10^{-12} C^2/N·m^2
Magnetic constant	$k_m = \mu_0/4\pi$	10^{-7} N/A^2 (exact value)
Permeability of free space	μ_0	$4\pi \times 10^{-7}$ N/A^2
Planck's constant	h	6.6262×10^{-34} J·s
		4.1357×10^{-15} eV·s
Mass, electron	m_e	9.1095×10^{-31} kg $= 0.51100$ MeV/c^2
Mass, proton	m_p	1.67265×10^{-27} kg $= 938.26$ MeV/c^2
Mass, neutron	m_n	1.67495×10^{-27} kg $= 939.55$ MeV/c^2

Astronomical Data

Nomenclature: $1.99 \; E \; 30 \equiv 1.99 \times 10^{30}$

OBJECT **Planet** *Satellite*	MASS (kg)	RADIUS (m)	ORBITAL RADIUS (m)	ORBITAL PERIOD
SUN	1.99 E 30	6.95 E 8	—	—
Mercury	3.28 E 23	2.57 E 6	5.8 E 10	88.0 d
Venus	4.82 E 24	6.31 E 6	1.08 E 11	224.7 d
Earth	5.98 E 24	6.38 E 6	1.49 E 11	1.00 y
Synchronous satellite	—	—	4.15 E 7	1.00 d
Moon	7.36 E 22	1.74 E 6	0.38 E 9	27.3 d
Mars	6.34 E 23	3.43 E 6	2.28 E 11	687.0 d
Phobos	2.72 E 16	10.4 E 3	9. E 6	0.318 d
Deimos	1.8 EE 15	5.0 E 3	2.35 E 7	1.26 d
Jupiter	1.90 E 27	7.18 E 7	7.78 E 11	11.86 y
1. Io	7.87 E 22	1.73 E 6	4.22 E 8	1.77 d
2. Europa	4.78 E 22	1.49 E 6	6.71 E 8	3.55 d
3. Ganymede	1.54 E 23	2.53 E 6	1.07 E 9	7.15 d
4. Callisto	7.35 E 22	2.42 E 6	1.88 E 9	16.6 d
5. (Smalthea)	8.3 E 18	8.70 E 4	1.81 E 8	0.498 d
6. (Hestia)	3.8 E 18	6.70 E 4	1.14 E 10	251 d
7. (Hera)	1.0 E 17	2.00 E 4	1.16 E 10	260 d
8. (Poseidon)	7.3 E 16	1.80 E 4	2.35 E 10	2.02 y
9. (Hades)	2.2 E 16	1.20 E 4	2.37 E 10	2.07 y
10. (Demeter)	9.2 E 15	9.0 E 3	1.18 E 10	0.71 y
11. (Pan)	2.2 E 16	1.20 E 4	2.26 E 10	1.91 y
12. (Adrastea)	1.3 E 16	1.00 E 4	2.11 E 10	1.71 y
Saturn	5.68 E 26	6.03 E 7	1.43 E 12	29.46 y
Mimas	4 E 19	2.72 E 5	1.82 E 8	0.94 d
Enceladus	7 E 19	2.99 E 5	2.39 E 8	1.37 d

OBJECT				
Planet			ORBITAL	ORBITAL
Satellite	MASS (kg)	RADIUS (m)	RADIUS (m)	PERIOD
Tethys	4.9 E 20	5.81 E 5	2.95 E 8	1.89 d
Dione	5.4 E 20	5.98 E 5	3.78 E 8	2.74 d
Rhea	1.8 E 21	8.90 E 5	5.28 E 8	4.52 d
Titan	1.2 E 23	2.38 E 6	1.23 E 9	15.95 d
Hyperion	6.8 E 19	2.01 E 5	1.49 E 9	21.28 d
Iapetus	2.3 E 21	6.47 E 5	3.57 E 9	79.33 d
Phoebe	1.9 E 19	1.32 E 5	1.30 E 10	550.4 d
(Janus)	1.2 E 20	2.41 E 5	1.58 E 8	0.749 d
Uranus	8.68 E 25	2.35 E 7	2.87 E 12	84.02 y
Ariel	5.0 E 20	3.11 E 5	1.92 E 8	2.52 d
Umbriel	1.4 E 20	2.01 E 5	2.67 E 8	4.14 d
Titania	2.1 E 21	5.00 E 5	4.39 E 8	8.70 d
Oberon	1.1 E 21	4.01 E 5	5.86 E 8	13.5 d
Miranda	3.0 E 19	1.21 E 5	1.29 E 8	1.41 d
Neptune	1.03 E 26	2.27 E 7	4.49 E 12	164.8 y
Triton	1.46 E 23	2.01 E 6	3.53 E 8	5.88 d
Nereid	5 E 19	1.36 E 5	5.9 E 9	359.4 d
Pluto	1 E 24	5.7 E 6	5.90 E 12	247.7 y

D Conversion Factors

Converting units amounts to *multiplying by the factor 1*, and therefore leaving the quantity (but not its units) unchanged.

Example:

$$60 \text{ mi/h} = ? \text{ m/s}$$

We are given that

$$1 \text{ mi/h} = 0.4470 \text{ m/s}$$

so that

$$1 = \frac{(0.4470 \text{ m/s})}{(1 \text{ mi/h})}$$

Multiplying the quantity above by this conversion factor gives

$$60 \text{ mi/h} \frac{(0.4470 \text{ m/s})}{(1 \text{ mi/h})} = 26.82 \text{ m/s}$$

Note the cancellation of the unwanted units, mi/h.

Example:

$$40 \text{ m/s} = ? \text{ mi/h}$$

This time the units m/s are to cancel, so that the conversion factor must have m/s in its denominator. The required conversion factor in this instance is

$$1 = \frac{(1 \text{ mi/h})}{(0.4470 \text{ m/s})}$$

so that

$$40 \text{ m/s } \frac{(1 \text{ mi/h})}{(0.4470 \text{ m/s})} = 89.49 \text{ mi/h}$$

Length
1 inch = 2.54×10^{-2} m
1 ft = 0.3948 m
1 mi = 1.609344 km
1 Å (Ångstrom unit) = 10^{-10} m = 0.1 nm

Mass
1 u = 1.6606×10^{-27} kg
1 lb (avdp) = 0.4535924 kg

Energy
1 erg = 10^{-7} J
1 ft-lb = 1.355818 J
1 kWh = 3.6000×10^6 J
1 cal = 4.183310 J
1 Btu = 1054.7 J
1 eV = 1.6022×10^{-19} J
1 quad = 10^{15} Btu = 1.0547×10^{18} J
1 ton (nuclear equivalent TNT) = 4.184×10^9 J

Pressure
1 torr = 133.322 Pa
1 mmHg = 133.322 Pa
1 atm = 101325 Pa

Speed
1 mi/h = 0.4470 m/s
1 ft/s = 0.3048 m/s

E References

General Textbooks

These introductory physics textbooks are at a somewhat higher level of sophistication and depth than this one:

- Richard P. Feynman, Robert B. Leighton, and Matthew Sands, *The Feynman Lectures on Physics* (3 vol.). Reading, MA: Addison-Wesley Publishing Company, Inc., 1965. A transcription of lectures given over a two-year period to students at CalTech by Nobel prize winner Richard P. Feynman; the insights are frequently brilliant and the going is sometimes very rough.
- Anthony P. French, *Special Relativity* (1968), *Newtonian Relativity* (1971), *Vibration and Waves* (1971), all published in New York by W. W. Norton Company. These are all parts of the M.I.T. introductory physics series.
- Donald G. Ivey (volume 1) and J. N. Patterson Hume (volume 2) *Physics*. New York, N. Y.: John Wiley and Sons, Inc., 1974. All topics are treated rigorously and in depth; the discussion of classical electromagnetism is particularly insightful.

Biographical References

- Isaac Asimov, *Asimov's Biographical Encyclopedia of Science and Technology*. Garden City, New York: Doubleday & Co., Inc., 1964. A one-volume work with entries for all of the principals and many of the secondary contributors, arranged so as to make it easy to spot contemporaries.

- Charles C. Gillispie, ed., *Dictionary of Scientific Biography* (in 16 volumes). New York: Charles Scribner's Sons, 1970. A thorough, scholarly treatment of the subject's scientific contribution, as well as the contributor's life and times. Information for the biographical sketches appearing in panels in this book comes from this source.

General References

- *McGraw-Hill Encyclopedia of Science and Technology* (15 volumes and annual supplements). New York: McGraw-Hill, Inc., 1960. With a distinguished editorial board and consulting editors, this work not only tells about the basic scientific theory but also describes the applications in technology. Articles differ, of course, but most are direct, clear, easy to follow. The annual supplements keep it up to date.

- Robert C. Weast, ed., *Handbook of Chemistry and Physics.* Cleveland, Ohio: The Chemical Rubber Co., published annually. Usually referred to as the "Chemical Handbook," this book is invaluable to any aspiring scientist or engineer. It has an enormous range of specific data, comprehensive tables on physical and chemical properties, and definitions. As technical books go, it is remarkably cheap.

- *Scientific American,* a monthly with clearly written, beautifully illustrated articles of a relatively nontechnical nature on modern developments in science, and prompt notice on all important advances in physics, published by Scientific American, 415 Madison Avenue, New York, N. Y. 10017. Comprehensive sets of "offprints" of individual articles from past issues are available.

- Jearl Walker, *The Flying Circus of Physics.* New York, N. Y.; John Wiley & Sons, 1975. A fun book, with a variety of questions based on more-or-less ordinary observation and whose explication involves more-or-less ordinary basic physics. Get the version that includes answers and detailed references. Some of the questions are easy; some are very tough.

- Weber, Robert L. *A Random Walk in Science* (1973) and *More Random Walks in Science* (1982). London, Institute of Physics. Stories, anecdotes, satire, fun items; this is where I learned about the *N. Y. Times* comment in 1920 on Goddard's work.

Answers to Selected Problems

Chapter 23   Point Electric Charges
23-1 A
23-3 Speeds up
23-5 E
23-7 E
23-9 A
23-11 8.43 N directed at 25.5° above the axis drawn from q_2 to q_1
23-13 14.5 cm left of the origin
23-15 $6.36 \times 10^9 \left(\dfrac{Q}{a}\right)^2$ N, directed downward
23-17 Give each sphere the charge $\frac{1}{2}Q$
23-19 $0.100 \dfrac{kQ^2}{a^2} (\mathbf{i} + \mathbf{j} + \mathbf{k})$ with the particular charge ($+$ or $-$) at the origin (0, 0, 0). The nearest neighbor charges are at $(a, 0, 0)$, $(0, a, 0)$, and $(0, 0, a)$; the next nearest charges are at $(a, a, 0)$, $(a, 0, a)$ and $(0, a, a)$; the most remote charge is at (a, a, a).
23-21 At $z = \pm 0.707a$
23-23 6.32×10^5 m/s²
23-25 11.2×10^{-3} g
23-27 C
23-29 1.81×10^3 N
23-31 $\Delta q/q = 9.00 \times 10^{-19}$
23-33 $2.0 \times 10^{-19} \equiv$ fraction of electrons
23-35 6.54 N/C
23-37 C
23-39 9.00 mm from the 4 μC charge and 11.0 mm from the 6 μC charge
23-41 F
23-43 A
23-45 E
23-47 (a) **E** is normal to the side with the two negative charges and directed toward it; $|\mathbf{E}| = 4\sqrt{2}\,\dfrac{kQ}{a^2}$ where $4\sqrt{2}\,k = 5.09 \times 10^{10}$ N·m²/C²

Chapter 24   Continuous Distributions of Electric Charge
24-1 9.0×10^5 N/C
24-3 (a) $\sigma = 3.98 \times 10^{-4}$ C/m² (b) $\sigma = \rho \Delta r = 2.98 \times 10^{-2} \Delta r$ in C/m³
24-5 3.60×10^{10} N/C normal to the sheet and directed toward it
24-7 $\theta = 13.0°$
24-9 $E(z) = \dfrac{kQ}{z^2 - a^2}$ for $z < -a$ and for $z > a$
24-11 $\dfrac{2kQ}{R^2}\left(1 - \dfrac{1}{\sqrt{1 + (R/z)^2}}\right)$ directed away from the disk and along the z axis (the axis of symmetry)
24-13 $E_x(x) = 4k\sigma \tan^{-1}(a/x)$
24-15 8.22×10^{-9} C
24-17 (a) 9.60×10^{-17} N (b) $F_e = 2.30 \times 10^{-18}$ N (c) $mg = 8.94 \times 10^{-3}$ N
24-19 (a) 56.9×10^{-9} s (b) 5.69 cm
24-21 C
24-23 5.20×10^2 N·m²/C
24-25 $(\phi_E)_{max} = 1.01$ N·m²/C
24-27 $(1.88 \times 10^{10}\,Q)$ N·m²/C
24-29 D
24-31 B

A-7

Answers to Selected Problems

24-33 D

24-35 $E = \left(\dfrac{\rho}{3\epsilon_0}\right) r = \left(\dfrac{kQ}{R^3}\right) r$ for $r \leq R$;

$E = \left(\dfrac{\rho R^3}{3\epsilon_0}\right) \dfrac{1}{r^2} = \dfrac{kQ}{r^2}$ for $r \geq R$

Note: $E = \left(\dfrac{\rho}{3\epsilon_0}\right) r = \dfrac{kQ}{R^2}$ for $r = R$; $Q \equiv \rho(\tfrac{4}{3}\pi R^3)$

24-37 $g(r) = \left(\dfrac{GM}{R^2}\right) \dfrac{r}{R}$ for $r < R$; $g(r) = \dfrac{GM}{r^2}$ for $r < R$;

$g(r = R) = \dfrac{GM}{R^2}$

24-39 $E = \left(\dfrac{\rho}{2\epsilon_0}\right) r = 2\pi k\rho r$ for $r \leq R$; $E =$

$\left(\dfrac{\rho R^2}{2\epsilon_0}\right) \dfrac{1}{r} = (2\pi k\rho R^2) \dfrac{1}{r}$ for $r \geq R$. Note: $E(r=R) = 2\pi k\rho R$

24-41 Inside: $E = 4k\rho t \left[1 - \cos\pi\dfrac{x}{t}\right]$; outside: $E = 8k\rho t$

where $4k = 3.60 \times 10^{10}$ N·m²/C²

24-45 $\sigma = 26.6\ \mu$C/m²

24-47 C

24-49 B

24-51 $Q = 4.53 \times 10^5$ C

24-53 D

Chapter 25 Electric Potential

25-1 E

25-3 86.9 kg of ice

25-5 (a) -18.0 kV (b) -12.7 kV (c) -3.00 kV (d) -3.37 kV

25-7 (a) For $r < R$: $V = \pi k\rho r^2$ (b) For $r > R$: $V = 2\pi k\rho R^2$ ln r/R where ρ is the volume charge density. Note that $\pi k = 2.83 \times 10^{10}$ N·m²/C² and $2\pi k = 5.65 \times 10^{10}$ N·m²/C²

25-9 A

25-11 $U = -5.40$ V

25-13 D

25-15 A

25-17 (a) $r_a = \dfrac{4}{5} R = \dfrac{32}{15} \dfrac{kZe^2}{mv_0^2}$ (b) $r_b = \dfrac{3}{4} R = 2 \dfrac{kZe^2}{mv_0^2}$

25-19 $U = \dfrac{3}{5} \dfrac{GM_E^2}{R_E} = 2.24 \times 10^{32}$ J where M_E and R_E are earth's mass and radius, respectively

25-21 E

25-23 $\Delta K = 40.0$ eV $= 6.40 \times 10^{-18}$ J

25-25 zero volts

25-27 0.885 mm

25-29 (a) C (b) G (c) B and H

25-31 No.

25-33 E

25-37 A

25-39 B

25-41 B

25-45 (a) $V_L = 2.52 V_0$ (b) $\Delta V = V_L - 4V_0 = -1.48 V_0$. In the large droplet, the ratio of charge to radius is reduced.

25-47 (a) 2.22×10^{-4} C (b) 2.22×10^{-5} C (c) $E_1 = 2.00 \times 10^6$ N/C for 1-m radius; $E_{0.1} = 20.0 \times 10^6$ N/C for 10-cm radius. For a given charge, a larger radius of curvature means a smaller value of E.

25-49 $Q_1 = 64.8\ \mu$C, $Q_2 = 132.4\ \mu$C and $Q_3 = 202.8\ \mu$C, where the subscripts refer to the sphere's radius in meters.

25-51 (a) $E = 2.09 \times 10^6$ N/C (b) $E = 1.74 \times 10^4$ N/C (c) $\lambda = 1.16 \times 10^{-8}$ C/m

Chapter 26 Capacitance and Dielectrics

26-1 B

26-5 B

26-7 (a) 4.43×10^{-7} m (b) 16.6 V

26-9 F

26-11 (a) $q_1 = q_2 = q_4 = 68.6 \times 10^{-12}$ C; $V_1 = 68.6$ V, $V_2 = 34.3$ V, $V_4 = 17.1$ V (b) $q_1 = 120 \times 10^{-12}$ C, $q_2 = 240 \times 10^{-12}$ C, $q_4 = 480 \times 10^{-12}$ C; $V_1 = V_2 = V_4 = 120$ V where the subscripts refer to the capacitance in pF.

26-17 B

26-19 (a) $q_3 = 120\ \mu$C, $V_3 = 40$ V; $q_2 = 40\ \mu$C, $V_2 = 20$ V; $q_4 = 80\ \mu$C, $V_4 = 20$ V (b) $q_3 = 180\ \mu$C, $V_3 = 60$ V; $q_2 = 0$, $V_2 = 0$. Note: the subscripts refer to the capacitance in μF.

26-21 $q_1 = q_2 = 6.55\ \mu$C; $q_3 = 29.45\ \mu$C where the subscripts refer to the capacitance in μF.

26-23 3 in series: $C = C_0/3$; 3 in parallel: $C = 3C_0$; 1 in series with 2 in parallel: $C = 2C_0/3$; 1 in parallel with 2 in series: $C = 3C_0/2$

26-25 Still Q_1 on C_1, Q_2 on C_2 and Q_3 on C_3

26-27 $\left(\dfrac{\kappa + 3}{4}\right) C_0$

26-29 B

26-31 (a) remain unchanged (b) decrease (c) increase (d) increase (e) remain unchanged (f) increase

26-35 E

26-37 5.74 cm³ per cm length. Put a dielectric sheath around the inner conductor.

26-39 1.96×10^{-9} J

26-41 Subscripts refer to capacitance in μF. Note $C_{12} = \kappa(4\ \mu$F$) = 12\ \mu$F
(a) Before: $V_2 = 10.3$ V, $V_4 = 1.71$ V $= V_8$; after: $V_2 = 10.91$, $V_8 = 1.091$ V $= V_{12}$
(b) Before: $q_2 = 20.6\ \mu$C, $q_4 = 6.86\ \mu$C, $q_8 = 13.7\ \mu$C
After: $q_2 = 21.8\ \mu$C, $q_8 = 8.73\ \mu$C, $q_{12} = 13.1\ \mu$C
(c) Before: $U_2 = 105.8\ \mu$J, $U_4 = 5.88\ \mu$J, $U_8 = 11.76\ \mu$J
After: $U_2 = 119.0\ \mu$J, $U_8 = 4.76\ \mu$J, $U_{12} = 7.14\ \mu$J

26-43 75%

26-47 (a) $\left[1 + (\kappa - 1)\dfrac{x}{L}\right]^{-1} V_0$ (b) $\left[1 + (\kappa - 1)\dfrac{x}{L}\right]^{-1}$ $(\tfrac{1}{2}C_0 V_0^2)$ where L is the dimension of the plate parallel to x.

26-49 $r = \dfrac{ke^2}{2mc^2} = 1.41 \times 10^{-15}$ m

26-51 (a) $\dfrac{3}{5}\dfrac{kQ^2}{R}$ (b) $\dfrac{1}{2}\dfrac{kQ^2}{R}$ (c) $U_a/U_b = 6/5 = 1.20$

26-53 (a) $U_{\text{Total}} = \left[\dfrac{\kappa}{\kappa - (\kappa - 1)\alpha}\right] U_0$; $U_A/U_{\text{Total}} = \dfrac{\kappa(1 - \alpha)}{\kappa - (\kappa - 1)\alpha}$; $U_D/U_{\text{Total}} = \dfrac{\alpha}{\kappa - (\kappa - 1)\alpha}$ where $U_0 \equiv (\tfrac{1}{2}C_0 V_0^2)$ with $C_0 \equiv \epsilon_0(A/d)$; note that C_0 is the capacitance when $\alpha = 0$ (i.e., when the dielectric is absent).
(b) $V_A = \left[\dfrac{\kappa(1 - \alpha)}{\kappa - (\kappa - 1)\alpha}\right] V_0$, $V_D = \left[\dfrac{\alpha}{\kappa - (\kappa - 1)\alpha}\right] V_0$
(c) $U_{\text{Total}} = 13.3 \times 10^{-3}$ J; $U_A/U_{\text{Total}} = 2/3$; $U_D/U_{\text{Total}} = 1/3$; $V_A = 66.7$ V; $V_D = 33.3$ V. Note that subscript A refers to the air gap and D refers to the dielectric.

Chapter 27 Electric Current and Resistance
27-1 600 C
27-3 12.4 A/(mm)^2 = 12.4 × 10^6 A/m^2
27-5 36 mm/s
27-7 (a) 3.93 × 10^{-6} A (b) 1.96 A/m^2
27-9 30.2 mA
27-11 37.5 s
27-13 C
27-15 C
27-17 (a) 2.22 × 10^6 A (b) 8.88 W (c) 1.37 × 10^3 years
27-19 (a) 14.1 V (b) 141 mA
27-21 D
27-23 D
27-25 (a) 223 W (b) 1.86 A
27-27 (a) 0.326 Ω (b) 4.75 × 10^6 A/m^2 (c) 8.15 V (d) 81.5 × 10^{-3} V/m (e) 204 W
27-29 2.50 W
27-31 (a) 2.88 Ω (b) 41.7 A (c) 4.30 × 10^3 kcal (d) 4050 W. The actual power will be greater because the actual resistance will be less than 2.88 Ω.
27-33 C
27-35 25.9 × 10^{-3} Ω
27-37 11.0 Ω
27-39 (a) ~14.5 s (b) 2.42 A. It was 2.40 A at 20 °C
27-41 (a) 26.7 × 10^6 Ω·m (b) 11.9 × 10^9 pores/m^2 (c) ~9.15 × 10^{-6} m
27-43 (a) 4 s (b) 0.347 RC
27-45 A
27-47 (a) 5.31 × 10^{-9} F (b) 1.7 × 10^{10} Ω = 17 kMΩ (c) 62.6 s

Chapter 28 DC Circuits
28-1 0.150 Ω
28-3 11.5 kJ
28-5 C
28-7 7.50 V
28-9 32.0 W
28-11 (a) 0.50 A counterclockwise (b) 1.5 W
28-13 7.20 V
28-17 C
28-19 8 Ω
28-21 A
28-23 4R in series, 37.5 W; 2R in series with 2R in parallel, 60 W; R in series with the combination of R in parallel with 2R in series, 90 W; R in series with 3R in parallel, 112.5 W; 2R in series in parallel with 2R in series, 150 W; R in parallel with 3R in series, 200 W; R in parallel with R in series with 2R in parallel, 250 W; 2R in series in parallel with 2R in parallel, 375 W; 4R in parallel, 600 W; where $R = V^2/P = 96$ Ω = resistance of a single coil
28-25 B
28-27 (a) 10 Ω (b) 30 V
28-29 $\tfrac{5}{6}R$
28-31 D
28-35 C
28-37 E
28-39 (a) Use 2.50 × 10^3 ohm series resistor (b) Use 4.80 × 10^{-3} ohm shunt
28-41 (a) 3.53 V (b) 17.14 V (c) 29.56 V (d) 29.98 V
28-43 (a) $R_a = 50$ Ω (b) $R_b = 11\,250$ Ω (c) $R_c = 112\,500$ Ω
28-45 E
28-47 5 V
28-49 (a) 1.5 W with three in series; (b) 2.9 W with two in parallel and in series with third; (c) 1.8 W with three in parallel.

Chapter 29 The Magnetic Force
29-1 C
29-3 1.89 × 10^{-4} Wb
29-5 (a) 7.52 × 10^{-8} Wb (b) 2.74 × 10^{-8} Wb (c) 0
29-9 (a) Normal to page, directed inward (b) Normal to page, directed outward (c) normal to $\mathbf{v}$, parallel to page, directed downward (d) normal to $\mathbf{v}$, parallel to page, directed to the right (e) force is zero (f) normal to $\mathbf{v}$, parallel to page, directed downward
29-11 C
29-13 C
29-15 (a) 14.4 m (b) 1.52 kHz
29-17 (a) 2.80 × 10^6 Hz (b) 0.337 m
29-19 45°
29-21 5.08 mm
29-25 F
29-27 A

29-29 33.5 kV with upper plate positive
29-31 0.312 N into the paper
29-33 6.13×10^{-3} N/m
29-35 (a) 2.65×10^8 A/m² (b) Orient along east-west line with conventional current from west to east
(c) 199 kW would be the power needed to suspend a 1-m length of wire with a 1-cm² cross section. Clearly the wire would melt in a very short time.
29-37 IBR
29-39 (a) zero (b) 2.60×10^{-3} N·m
29-41 2.00×10^{-3} N·m
29-43 1.98 T
29-45 $\tfrac{1}{12}\omega L^2 QB \sin\theta$

Chapter 30 Sources of the Magnetic Field
30-1 D
30-3 C
30-5 B
30-7 B
30-9 B
30-11 $\lambda = 3.34 \times 10^{-9}$ C/m
30-17 6.28×10^{-6} T
30-19 $B = \dfrac{\mu_0 I}{4R} = 3.14 \times 10^{-7}\,\dfrac{I}{R}$
30-21 $B_z = \dfrac{\mu_0}{2}\dfrac{R^2 I}{z^3} = 6.28 \times 10^{-7}\,\dfrac{R^2 I}{z^3}$ where $z \gg R$
30-23 (a) $B_a = 5.66 \times 10^{-7}\, I/a$ (b) $B_b \simeq 5.36 \times 10^{-7}\, I/a$ so $B_a/B_b = 1.055$ (c) No. The approximation is independent of a.
30-25 C
30-27 A
30-29 (a) 0 for $r < R_1$ (b) $\dfrac{\mu_0 I(r^2 - R_1^2)}{2\pi r(R_2^2 - R_1^2)}$ for $R_1 < r < R_2$ where $\mu_0/2\pi = 5.00 \times 10^{-8}$ Wb/A·m (c) $\left(\dfrac{\mu_0}{4\pi}\right)\dfrac{2I}{r}$ for $r > R_2$ where $\left(\dfrac{\mu_0}{4\pi}\right)(2) = 2 \times 10^{-7}$ Wb/A·m
30-31 $\tfrac{1}{2}\mu_0 j$
30-33 C
30-35 15.7×10^{-3} T
30-37 F
30-39 3.20×10^{-3} T

Chapter 31 Electromagnetic Induction
31-1 A, B, C, D
31-3 1.42 Hg
31-5 0.533 V
31-7 B
31-9 1.88×10^3 V
31-11 $|\mathcal{E}| = 126$ V
31-13 0.136 V
31-15 0.711 A
31-17 $|\mathcal{E}| = \mu_0 x\,\dfrac{di}{dt}\,(1 - \sqrt{1 - (R/x)^2})$
31-19 0.815 G
31-21 $\left|\dfrac{d\phi_B}{dt}\right| = \tfrac{1}{3}\mathcal{E}_0$. B either up from page and increasing or down and decreasing.
31-23 $\Delta\phi = 4.80 \times 10^{-2}$ Wb
31-25 G
31-29 $\dfrac{mgR}{(aB)^2}\,(= \text{constant})$
31-31 (a) 24.0 m/s² (b) 150 m/s $\simeq$ 336 mph
31-33 R_1/R_2
31-37 B
31-39 B
31-41 D
31-45 (a) zero current (b) $\tfrac{1}{2}\omega a^2 B$

Chapter 32 Inductance and Electric Oscillations
32-1 1 mH
32-3 Change current at rate of 40.0 A/s
32-5 D
32-7 $\phi = 1.0$ Wb/N where N is the number of turns
32-9 $|\mathcal{E}| = 1.97 \times 10^{-6}\,(R_2^2/R_1)\,di/dt$
32-13 4.61 µs
32-19 (a) 6.91 (L/R) (b) 7.60 (L/R)
32-21 (b) $\left(\dfrac{V_0}{6.00 \times 10^5}\right)\tau = (1.67 \times 10^{-6}\,V_0)\,\tau$
32-23 D
32-25 (a) $i_1 = i_2 = 3.53$ A (b) $i_1 = 3.14$ A, $i_2 = 0.923$ A (c) $i_1 = 0$, $i_2 = -2.22$ A (d) $i_1 = i_2 = 0$
32-27 (a) 43.3 W (b) 45.6 W (c) 2.27 W
32-29 (a) 197 mH (b) 98.7 mJ
32-31 (a) 4.00 mJ (b) 0.750 mJ
32-33 D
32-35 $1.59 \times 10^{-8}(i/r)^2$ in J/m³
32-37 $1.59 \times 10^{-8}(Ni/R)^2$ in J/m³
32-39 0.504 T
32-41 (a) 0.989 pF (b) 9.02 pF
32-43 From $\tfrac{1}{2}f_0$ to $\sqrt{2}f_0$ where $f_0 = \dfrac{1}{2\pi}\dfrac{1}{\sqrt{LC}}$. Note: C is the *minimum* capacitance of a single capacitor and L is the fixed inductance
32-45 (a) Resonant frequency doubled. (b) Total resistance doubled.
32-47 An object that is subject to both a spring force ($F = -kx$) and a damping force ($F_D = -kv$).

Chapter 33 AC Circuits
33-1 377 rad/s
33-3 A, B, C

33-5 (a) 2.53 A (c) 6.40×10^{-3} J
33-7 A
33-9 (a) 0.159 pF (b) 1.59×10^{-9} H
33-11 31.7 pF
33-13 (a) $\dfrac{1}{\sqrt{1 + (1/\omega CR)^2}}$
33-15 B
33-17 4 μs
33-19 B
33-21 (a) 20 Ω (b) 286 kΩ (c) 106 kHz
33-25 (a) 16.9 mH (b) 28.1 Ω (c) At 100 Hz, $\phi = 20.7°$; at 500 Hz, $\phi = 62.1°$
33-27 (a) 497 Hz
33-29 G
33-31 (a) 2 A (b) 1.41 A (c) 144 Hz (d) $\theta = 288.375\,\pi$
33-33 D
33-35 (a) 5.00 V (b) 1.77 V (c) 2.89 V
33-37 84.9 W
33-39 (a) $I = 5.00$ A, $\phi = 0°$ (b) $I = 2.73$ A, $\phi = -56.9°$ (c) $I = 3.81$ A, $\phi = +40.4°$
33-41 (a) I_{max} may either increase or decrease; both ω_0 and $\Delta\omega$ decrease. (b) I_{max} may either increase or decrease; ω_0 decreases; $\Delta\omega$ does not change. (c) I_{max} decreases; ω_0 does not change; $\Delta\omega$ increases.
33-43 (a) $R = 33.9$ Ω, $L = 60.0$ mH, $C = 39.1$ μF (b) $P_R = 150$ W, $P_L = 0 = P_C$
33-45 B

Chapter 34 Maxwell's Equations
34-1 D
34-3 C
34-7 (a) $I_R = 75.2$ mA $= I_d$ (b) $I_R = 1.00$ A $= I_d$
34-9 C
34-11 D
34-13 D
34-15 (a) A·m

Chapter 35 Electromagnetic Waves
35-1 C
35-3 D
35-5 C
35-7 A
35-9 (a) 470 nm (b) Blue
35-11 5.26×10^{14} Hz
35-13 E
35-15 3.33×10^{-8} J/m³
35-17 (a) 6.43×10^7 W/m² (b) 2.20×10^5 N/C (c) 7.34×10^{-4} T
35-19 (a) 1.0×10^5 J (b) 3.3×10^{-4} kg·m/s (c) 3.3×10^{-5} N (d) 3.0×10^9 m
35-23 (a) 24 nm (b) 0.20 μm (c) 6.5×10^{-7} J
35-25 $0.875\, I/c$
35-27 (a) 2.41×10^{-23} kg·m/s (b) 7.23×10^{-15} J, 0.072 μW

Chapter 36 Ray Optics
36-1 610 cm
36-3 90.0 cm
36-7 (b) Doubles the range
36-15 A
36-17 1.63
36-19 2.26×10^8 m/s
36-21 C
36-25 4.26 cm
36-27 40.6°
36-29 $\theta = \cos^{-1}\left(\dfrac{n^2 - 2}{2}\right)$
36-31 A
36-38 1.21 m
36-43 78.5°
36-45 If index $n \geq \sqrt{2}$, then $\theta \leq 45°$ so that $\left(90° - \sin^{-1}\dfrac{1}{n}\right) \geq \theta \geq \sin^{-1}\dfrac{1}{n}$.
36-47 C (If A were true, n could be less than 1.414.)

Chapter 37 Thin Lenses
37-1 E
37-3 5.01 mm
37-7 C
37-11 (a) 1.33 m (b) 3.00 m (c) 0.200 m
37-15 (a) Light intensity cut in half (b) Successive steps of 2
37-17 D
37-21 E
37-23 At $s = 35.7$ cm from her eye
37-25 B
37-27 Far-sighted
37-29 E
37-31 +10.0 cm
37-33 (a) 2.40 m (b) 2.42 m
37-35 1.11 cm
37-37 S (Recall Problem 37-31.)
37-39 (a) 52 cm (b) 25 (c) 51.85 cm
37-41 B
37-43 A
37-45 (a) Converging lens (b) 150 cm using $n_{21} = 1.20$
37-47 $f = \left(\dfrac{1}{n-1}\right)\left(\dfrac{d^2 + t^2}{8t}\right)$
37-49 E

Chapter 38 Interference
38-3 (a) R_2 (b) Signal strength the same at R_1 and R_2, (c) R_2, (d) R_2

A-12   Answers to Selected Problems

38-5 17.3 m
38-9 A
38-11 0.37 mm
38-13 (a) 1.49 mm (b) 0.435 mm
38-17 A
38-19 (a) 18.7° (b) 39.8° (c) 73.7°
38-21 (a) 17.9° (b) 38.0° (c) 67.4°
38-23 3rd order (corresponds to $\theta = 49.9°$)
38-25 C
38-27 $\lambda = 648$ mm
38-29 B
38-33 If the two beams depart significantly from absolutely parallel, the phase relationship between them would vary with distance from the source.
38-35 0.443 mm
38-37 40.0 μm
38-39 8.21×10^{-9} m
38-45 0.181 mm
38-47 $n = 1.000\ 288$
38-49 $n = 1.000\ 300$

Chapter 39 Diffraction
39-1 E
39-3 (a) 51.5° (b) 25.1°
39-7 D
39-9 (a) 90.0° (b) 30.0° (c) 11.5° (d) 5.74°
39-11 (a) no (b) no (c) yes (d) no
39-13 (a) 3.15° $\simeq 5.50 \times 10^{-2}$ rad (b) 6.60 cm
39-15 (a) $\phi = 6.12$ rad $= 351°$ (b) 2.66×10^{-2} (c) 7.09×10^{-4}
39-17 A
39-19 The maximum for which $\dfrac{d \sin \theta}{\lambda} = m = 8$ at angle $\theta_8 = 5.00 \times 10^{-3}$ rad $= 0.287°$
39-21 B
39-23 Yes. The reflecting surface acts as an aperture in that it reflects (and thereby isolates) only a portion of the incident light.
39-25 1.21 km
39-27 3.95 mm
39-31 60° and 180°

Chapter 40 Special Relativity
40-1 A
40-3 0.96 c
40-5 (a) 0.198c, (b) 0.200c
40-7 D
40-9 E
40-11 0.6c
40-13 0.87c
40-15 A
40-17 900
40-19 600 km

40-21 (a) $L_0 [\sin^2 \theta_0 + \cos^2 \theta_0 (1 - v^2/c^2)]^{1/2}$
(b) $\tan^{-1} [\tan \theta_0 / \sqrt{1 - v^2/c^2}]$
40-23 (a) 9.39 MeV/c (b) 94.5 MeV/c (c) 542 MeV/c (d) 1.93×10^3 MeV/c
40-25 0.563
40-27 (a) 51.1 keV (b) 0.417c
40-29 4.53×10^{-14} J
40-31 (a) 100 MeV/c (b) 10 MeV/c
40-33 (a) 0.659 MeV (b) 2.45 MeV
40-35 (a) 0.302c (b) 2.7 percent
40-37 (a) 100 000 y (b) 297 s
40-39 (a) No (b) Yes (c) No (d) Yes (e) No
40-41 (a) $0.405\ m_0 c^2$ (b) $1.29\ m_0 c^2$
40-45 4.7×10^{-12} kg
40-47 0.25 kg
40-49 42 μg
40-51 3.1×10^{13}
40-53 9×10^6

Chapter 41 Quantum Theory
41-1 D
41-3 7.5×10^{31}
41-5 544 nm
41-7 (a) 7.7×10^{31} (b) 9.2×10^{21}
41-9 (a) 685 nm (b) 0.61 eV (c) 3.47×10^5 m/s
41-11 (a) 5 min (b) 4.88 eV (c) 3.44×10^{14}
41-15 (a) 17.6 MeV/c (b) 21 keV
41-17 A
41-19 1 μm
41-21 4.86 pm
41-23 0.058 fm for Na, 0.048 fm for Cl
41-27 B
41-29 0.82 pm
41-31 0.85 MeV
41-35 71°
41-37 C
41-39 E
41-41 (a) same (b) particle greater (c) photon greater
41-43 7.8 pm
41-47 C
41-49 0.21 MeV
41-51 (a) 6.65×10^{-27} J (b) 7.9×10^{-9} s (c) 2.38 m
41-57 E
41-59 D
41-61 0.33 pJ, 1.32 pJ, 2.96 pJ
41-63 (a) 0 (b) 1/100 (c) 1/50 (d) 1/100 (e) 0

Chapter 42 Atomic Structure
42-1 E
42-3 No, because of angular momentum conservation
42-5 (a) 0.22 eV (b) 0.22 eV
42-7 10^{17} kg/m^3
42-9 (a) 2.3 fm
42-11 B

42-13 10.2 eV
42-15 E
42-17 54.4 eV
42-19 3.4 eV
42-21 12.8 eV
42-23 6.8 eV
42-25 (a) 12.1 eV (b) 4.8×10^4 m/s (c) 12.1 eV
(d) 7.8×10^{-8} eV
42-27 (a) 2.82 keV, (b) 256 fm
42-29 5×10^4 K
42-31 (a) -2.6×10^{-78} eV (b) 1.2×10^{29} m
42-35 (a) 15.9 eV (b) -6.4 eV, -4.6 eV, -2.9 eV, -1.5 eV (c) 3.5 eV (d) 1.8 eV
42-37 254 nm
42-39 D
42-41 2.1×10^{-3} eV
42-45 (a) $_{11}$Na (b) $_{18}$Ar (c) $_{27}$Co
42-47 C
42-49 (a) 106 eV (b) 16.9 eV
42-51 4.3×10^{-6}
42-53 (a) $4p \rightarrow 3s$ (b) $3p \rightarrow 4s$

Index

Aberration:
  astigmatism, 807
  coma, 807
  chromatic, 807
  distortion, 807–808
  in lenses, 807–808
  spherical, 807
Absorption, electromagnetic waves, 763–764
Absorption spectrum, 921, 926
Acceleration of particle, and electric field, 506, 517
AC circuit, 710–730
  basic characteristics, 710–713
  phasors, 719–724
  power factor, 726
  power, 725–728
  rms values, 724–725
  series RLC arrangement, 713–719, 721–724
  transformers, 729–730
Accommodation in focusing, 798
AC generator, emf for, 711
Allowed energy, 924–927
Allowed radius, 923–924
Alpha particles, nuclear scattering, 917–920
Alternating current, 583, 710 (*see also* AC circuit)
Ammeter, 583
  circuit symbol, 606
  in dc circuits, 606–608
Ampère (unit), 496, 583
Ampère, André M., 583
Ampèrian current, 659
Ampère's law:
  general form, 737–741

  magnetic fields, 652–656, 737–741, 745
Analog computer and Kirchhoff's rules, 611
Anderson, C. D., 895
Angle of deviation, ray optics, 783
Angle of incidence:
  lens maker's formula, 805–806
  ray optics, 775–778, 782–784
Angle of reflection, ray optics, 775–778, 784–785
Angle of refraction:
  critical, 784–785
  ray optics, 775–776, 782–783
Angular frequency:
  electrical free oscillation, 700
  series RLC circuit, 728
Angular magnification:
  astronomical telescope, 802–803
  in microscope, 804
Angular momentum, particle, 922
Angular speed, charged particle, 626
Antineutron, 500
Antiparticle, 499, 895
Antiproton, 499
Antiquark, 500
Apex angle, prism, 783
Apparent depth, immersed object, 783
Astigmatism, 807
Astronomical telescope:
  diffraction effects, 848–849
  lens arrangement, 801–803
Atom (*see also* Atomic structure):
  electric charge, 492
  size, 917
  structure, 917–941
Atomic model and diamagnetism, 682

Atomic nuclear forces, 491
Atomic number and charge quantization, 501
Atomic structure, 917–941
  allowed radius, 923–924
  Bohr radius, 923
  Bohr theory, 921–927
  energy, 921–927
  hydrogen spectrum, 920–921
  nuclear scattering, 917–920
  Pauli exclusion principle, 933–937
  quantum numbers, 927–933
  quantum theory, 917–941
Attractive force, point charges, 497

Back emf, 689, 691
Balmer series, 921, 926
Baseline geodesy, 779
Baseline interferometry, 849
Battery:
  dc circuit, 598–611
  internal resistance, 600
Bell, Alexander Graham, 667
Binding energy, 879
  of atoms, 925
  of deuteron, 890–891
  of electron, 891
Binocular lens arrangement, 802
Biot, J. B., 644
Biot-Savart law, 647–650, 652
Blackbody radiation, 887
Bohr, Niels, 903, 921
Bohr formula, 924
Bohr radius, 923
Bohr theory:
  of hydrogen, 921–927

A-15

Bohr theory (continued)
  postulates, 927
Born, Max, 901
Bound systems, special relativity, 879
Bradley, James, 760
Bragg, W. L., 852
Bragg plane, 851–852
Bragg relation, 852–853
Bremsstrahlung, 891–892, 900
Bremsstrahlung collision, 891
Bright-line spectrum, 920–921
Bubble chamber, 622, 895–896

Camera lens, 798
Capacitance, 560–574
  capacitors in parallel, 566–567
  capacitors in series, 565–566
  coaxial capacitor, 563
  definition, 560–561
  electric field and, 571–572
  equation, 561
  LC circuit, 697–702
  parallel-plate capacitor, 561–562, 568, 569
  RC circuit, 589–591
  series RLC circuit, 713–719
  of Van de Graaff generator, 570
Capacitive reactance:
  audio amplifier, 718
  definition, 715
Capacitor, 560–563
  in ac circuit, 712
  characteristics, 690
  charging, 561, 569–571
  circuits, 564–567
  circuit symbol, 561
  coaxial, 563
  dielectric, 568–569
  electric potential energy, 569–571
  energy, 569–571
  free charge on, 573
  paper, 568
  in parallel, 564, 566–567
  parallel-plate, 561–562, 568, 569, 571, 572–574, 761
  plates for, 560
  polarization, 572–574
  potential, 560–561
  in series, 564–566
  work done in charging, 569–570
Cathode ray tube, 518
Cavendish, Henry, 526
Center-of-mass frame, 873–875
Central force, electric force as, 495, 496, 539
Charge (see Electric charge)
Charge carriers, 583
Charge conservation, 499, 500, 564, 609–610
Charged conductors (see Conductor; Electric conductor)
Charge distributions, 495, 512–533
Charged objects, 491–494
Charged particle:
  angular speed, 626
  energy, 875–878

frequency, 626
and generation of electronmagnetic waves, 757–759
interactions between, 666
kinetic energy, 629–630
momentum, 626, 629–630
mass, 629–630
path, in magnetic field, 625–626
period, 626
in quantum theory, 886–911
in special relativity, 875
speed, 628, 629, 875–878
in uniform magnetic field, 625–630
Charging an object, 493
Charging by induction, 493–494
Chemical force in electrochemical cell, 599
Chemical properties and quantum theory, 933–937
Chromatic aberration, 807
Circularly polarized waves, 766
Classical conservation laws, 499
Classical mechanics, and wave mechanics, 906
Coaxial cable (see Coaxial conductor)
Coaxial capacitor, 563
Coaxial conductor, 693–694
Coherent radiation, 938
Coherent sources of waves, 829–830
Collisions between nuclei, forces in, 491
Coma, 807
Complementarity principle, 902–903
Compton, Arthur H., 894, 900
Compton collision, 893
Compton effect, 892–894, 900–901
Compton wavelength, 893–894
Concave-convex lens, 794
Conductance, 586
Conducting rod, induced emf, 675
Conducting surface:
  arbitrary shape, 527–530
  electric field, 525–530
  spherical, 524–525, 527
Conduction electrons:
  forces on, 671, 672–674
  in photoelectric effect, 887
Conductivity, 492–493, 588
Conductivity ratio, 492
Conductor:
  cylindrical coaxial, 655–656
  electric field, 503, 525–530
  electric potential, 552–554
  electrons in, 492
  electrostatic effects, 493
  examples of, 492
  grounding, 494
  induced emf, 669–675
  loops, 667–671, 672–674, 675–680
  metallic, and resistivity, 588
  resistance, 591–592
  resistivity, 587–588
  straight, and magnetic field, 644–647, 650, 653
Confined particle, 908–911, 929
Conjugate points in lenses, 798
Conservation law:
  charge, 499, 500, 564, 609–610

classical, 499
energy, 541–542, 545, 564, 585, 600, 601–602, 609, 610, 896, 901
mass-energy, 876–880
momentum, 874–875, 901
Conservative force:
  electric force as, 496
  requirement, 540
Constructive interference:
  definition, 817
  diffraction grating, 828
  two point sources, 818–824, 829
Continuous charge, 495, 512–533
  electric field, 512–519, 522–530
  electric flux, 519–525
  Gauss's law, 521–525, 533
Contracted length, special relativity, 871
Converging lens (see Lenses, converging)
Convex lenses, 794, 805
Corner reflector, 778
Cosmic radiation, 624
Coulomb (unit), 496
Coulomb, C. A. de, 495
Coulomb constant, 495–496, 521
Coulomb force:
  atomic, 920, 921
  between charged particles, 660
  as conservative force, 539–541
  definition, 495
  electric flux and, 521–522
  superposition principle and, 497
Coulomb's law, 495–498, 652
  definition, 495
  electric field, 501–503
  electromagnetic induction, 678
Critical angle of refraction, 784–785
Crystalline solid and x-ray diffraction, 850–853
Current (see Electric current)
Current-carrying conductors:
  magnetic field from, 651, 653–656
  magnetic force between, 646–647, 648
  magnetic force on, 630–633
Current-carrying inductor, 694–696, 697
Current-carrying loop, 676–677
Current-carrying solenoid, induced emf, 670
Current-density vector, 584–585, 592
Current loop:
  magnetic field at center, 649–650
  magnetic field from, 653–654
  magnetic torque on, 633–637
Current rule for dc circuits, 610
Current-voltage-impedance relation, 715–716, 722, 726
Cyclotron, 627–628
Cyclotron frequency, 626

Dark-line spectrum, 921
Davisson, C., 898
DC circuits, 598–611
  emf, 598–601
  instruments for measurements, 606–609
  Kirchhoff's rules, 609–611

measurements, 606–609
multiloop, 609–611
resistors in parallel, 604–606
resistors in series, 603–604
single-loop, 601–603
de Broglie, Louis, 897
de Broglie relation for wavelength, 897
de Broglie wavelength, 922
Decay:
atoms, 937–941
probability of, 938
unstable particles, 491
Degenerate state of electron, 934
Destructive interference:
definition, 817
two point sources, 819, 822
Detector loop, 667–668
Deuteron, 890–891
Diamagnetism, 681–682
Dielectric constant, 567–569
Dielectrics, 560–574
behavior, in electric field, 572–574
electric charge and, 492
electrostatic effects, 493
microscopic properties, 572–574
nonpolar, 573–574
polarized, 573
Dielectric strength:
air, 554
insulating material, 568–569
Diffraction:
strong, 852
of x-rays, 897–898
Diffraction effects for lenses, 807
Diffraction grating, 855–856
interference of waves, 826–829
wavelength measurements, 783
Diffraction of waves, 773, 841–856
definition, 841
double-slit, 846–847
Fraunhofer, 845, 853–855
Fresnel, 845
multiple point sources, 841–844, 853–855
and intensity zero, 842–847, 853–857
resolution, 847–850
single slit, 844–846, 853–855
by X-rays, 850–853
Diffraction patterns, 842–847, 856
Diffuse reflection, 776
Diopter (unit), 798
Dipole (see Electric dipole)
Dirac, P. A. M., 895
Direct current, 583 (also see DC circuits)
Dispersion of waves, 781
Displacement:
electric potential and, 540
measuring, by using interference of waves, 834–835
Displacement current, 739
Dissipative circuit elements, 585–586
Distortion aberration, 807–808
Diverging Lens (see lenses, diverging)
Doppler shift, 761–762
Double-slit diffraction, 846–847
Double-slit interference, 824–825, 846–847

Drift velocity:
charged particle, 630–631
electrons, 583–584, 592

East-west effect of earth's magnetic field, 624
Eddy currents, 680–681
Einstein, Albert, 673, 860, 923
Einstein relation, 876
Electric charges:
on atom, 492
capacitance and, 560–568
characteristics, 498–501
conservation, 499, 500, 564, 609–610
continuous, 495, 512–533
Electric charge:
Coulomb's law for, 495–498, 501–503, 652, 678
definition, 491–492, 500–501
detection of, 493
discreteness, 501
electric potential and, 539–554
on electron, 492, 498, 500
electrostatic effects, 493
field for, 501–507
field lines for, 505–507
fundamental particles, 492
like, 492
magnetic force and, 492
magnitude, 498
in motion, 492, 582
on neutron, 492
on nonmetallic objects, 492
point, 491–508 (see also Point charge)
on proton, 492
quantization of, 500–501, 887
RC circuit, 590
at rest, 491–508
separation, 493
unlike, 492
Electric-charge conservation, 499, 500, 564, 609–610
Electric circuits:
ac (see also AC circuits)
capacitors, 564–567
capacitors in parallel, 566–567
capacitors in series, 565–566
dc (see also DC circuits)
energy of inductor and, 696–697
LC arrangement, 697–702
LR arrangement, 694–696
Ohm's law and, 585–586
open, 600
oscillations, 697–702
RC arrangement, 589–591
resistance, 586
self-inductance, 689–694
short, 600
Electric conductance, 586
Electric conductivity (see Conductivity)
Electric conductor (see Conductor)
Electric current, 582–592
alternating, 583
definition, 582
density vector, 583–584, 592
direct, 583

direction (conventional), 582–583
energy conservation and, 585
equation, 582
induced (see Induced current)
instantaneous, in series RLC circuit, 714–716, 721–722
magnetic field and, 630–636, 647–650, 653–655
magnetic force and, 646–647
measurement, 583, 606–608, 634–635
Oersted effect, 643–646
peak value, in series RLC circuit, 716, 718
RC circuit, 589–591
rms value, 724
sinusoidally varying, 710–711
Electric dipole:
compared with magnetic dipole, 635–637
definition, 530
in electric field, 530–531
equipotential lines for, 548
field lines, 506–507
oscillating, 757–759
potential energy, 531
torque on, 530
Electric-dipole moment:
dielectrics and, 573–574
equation, 530
permanent, 531
Electric-dipole oscillator:
electric and magnetic fields, 757–759
as generator of electromagnetic waves, 757–759
radiation pattern, 759
Electric energy density, 571–572, 697
Electric field, 501–505
between two charges, 504–505
capacitance and, 571–572
from changing magnetic flux, 677–680
charged conductors and, 525–530, 591–592
compared with magnetic field, 619
conservative, 678
from continuous line of charge, 512–515
critical value, 554
definition, 503
in diffraction, 842, 853–855
direction, 503–504
disturbances, 504
effect on dielectrics, 572–574
effect on electron, 517–519
electric dipole and, 530–531
for electric-dipole oscillator, 757–759
electromagnetic waves and, 751, 752, 754–755, 756, 816–824
energy density, 571–572, 754–755
equation, 503
equipotential surface and, 548–549
from flat sheet of continuous charge, 512, 515–516
fringing, 517
induced (see Induced electric field)
for infinite uniform line of charge, 514
for infinite uniformly charged sheet of charge, 516

Electric field (continued)
  for interior of conducting surface, 526
  loops, 522
  magnitude, 503, 505–506
  Maxwell's equations and, 737–749
  nonconservative, 678
  normal component, 516
  for photon, 899–902
  for point charge, 503–505, 513, 540, 545
  rectangular components, 551
  representation, 505–507
  for short wire, 589
  from single point charge, 512–513
  for spherical shell of charge, 524–526
  from static charges, 678
  strength, 503
  surface-charge density and, 527–529
  transverse component, 514
  uniform, 516–521, 549–552, 628–630
  from uniformly charged infinite sheet, 523–524
  of uniformly charged infinite wire, 522–523
  for uniformly charged ring, 451–452
Electric-field lines, 505–507, 622, 651
  continuity, 506
  density, 506
  electric flux and, 520, 522
  equipotential lines and, 548
  properties, 506
Electric-field vector, 550
Electric flux, 519–525, 531–533, 651
  in creating magnetic field, 739–740
  definition, 520
  dot product, 519
  equation, 520
  electromagnetic waves and, 747
  for point charge, 521
  signs for, 520, 522
  zero, 522
Electric force, 619, 621
  between nonmetallic objects, 492
  compared with gravitational force, 496–497, 539, 546
  electric field and, 501–505
  electric-field lines and, 506
  equation, 495
  like charges, 497
  properties, 495
  qualitative features, 491–494
  ratio of, to gravitational force, 501
  per unit positive charge, 502
  work done by, 539–551
  zero, 492
Electric-force constant, 495–496
Electric generator, 601
  definition, 671
  induced emf, 671
Electricity and magnetism (see Electromagnetism)
Electric load, 585
Electric oscillation(s), 689–705
  angular frequency, 700
  frequency, 700
  LC circuit, 697–702
  radio receiver, 700–701
  series RLC circuit, 713–719, 721–724

Electric oscillator:
  analogy with mechanical oscillator, 702–704
  for generating electromagnetic waves, 777
Electric permittivity:
  of dielectric, 568
  of vacuum, 496, 521
Electric polarization, 572–574
Electric potential, 539–554
  conductors and, 552–554
  definition, 541–543
  equation, 542
  for point charges, 543–545
  for sheet of charge, 451
  signs for, 544
Electric potential difference, 541–543
  capacitor, 560
  capacitors in parallel, 567
  capacitors in series, 565
  circuit, 585
  circuits, 598–611
  emf and, 601–603
  Hall effect and, 632–633
  instantaneous, in series RLC circuit, 721–722
  measurement, 606–608
  moving conductor, 674–675
  short wire, 589
  signs for, 602
Electric potential drop, 549
  electric circuit, 564, 711
  instantaneous, in series RLC circuit, 722
  peak, in series RLC circuit, 724
Electric potential energy, 541–543, 545–547
  charged capacitors, 569–571
  electric dipole, 531, 636
  more than two charges, 546–547
  signs for, 546
  two charges, 545–546
  Van de Graaff generator, 570
Electric potential increase, 549
  circuit, 586
Electric power:
  definition, 585
  single-loop dc circuit, 601–609
  time-average, 725, 726–727
Electric shielding and conducting shell, 527–528, 530
Electrochemical cell, 599
Electrolyte as conductor, 492
Electromagnetic force, 491, 901
Electromagnetic induction, 666–682
  definition, 667
  diamagnetism and, 681–682
  eddy currents and, 680–681
  effects, 601
  emf and, 667–675
  Faraday's law, 669, 671–672, 677–680, 737, 740, 744–745
  induced current and, 667–671
  induced electric field and, 677–680
  Lenz's law, 675–677
Electromagnetic spectrum, 752–754
Electromagnetic waves, 751–768
  absorption, 763–764
  basic properties, 751

diffraction effects, 841–856
electric field and, 751, 752, 754–755, 756
electromagnetic spectrum and, 752–754
energy density, 754–755, 756
frequency, 752, 759
generation, 757–759, 841–844
intensity, 755–756
  and diffraction effects, 842–847, 853–856
  in quantum theory, 899–900
interference and, 816–835
linear momentum, 762, 765
magnetic field and, 751, 752, 754–755, 756
from Maxwell's equations, 745–749
path difference, 818–824
phase differences in, 818, 820–824, 830–834
polarization, 752, 759, 765–768
Poynting vectors and, 756–757
quantization, 887
in quantum theory, 886–911
radiation force, 762–764
radiation pressure, 762, 764–765
ray optics and, 772–785
reflection of, 764
representation, 752
sinusoidal, 752–754
in special relativity, 860–861
speed, 749, 751
speed of light and, 759–762
time-average intensity for, 817–820, 822–832
transverse, 751
wavelength, 752
wave number, 752
Electromagnetism, 491, 643–646, 666
  definition, 501
  electromagnetic waves and, 751–768
  Maxwell's equations, 737–749
  speed of light and, 759–762
Electromotance, 600
Electromotive force, 600
Electron:
  atomic structure and, 921–937
  charge on, 492, 498, 500
  complementarity principle and, 903
  in conductors, 492
  in dielectric materials, 492
  drift speed, 583–584
  effects on, in electric field, 517–519
  energy levels, 934
  ground-state energy, 910
  kinetic energy, 519, 547
  magnetic force on, 645–646
  orbits, 682, 922–937
  positron and, 499, 895–897
  in semiconductors, 493
  thermal speed, 583–584
  uncertainty principle and, 903–908
  velocity in electric field, 518
  as wave, 904
  zero-point energy, 911
Electron beam, 518
Electron configuration, 934–937
Electron-positron pair, 499, 895–897

Electron spin, 659–660, 932–933
Electron-spin magnetic moment, 659
Electron-spin orientation, 934
Electron state, 930, 931
Electron volt (unit), 546–547
Electrostatic effects, 493
Electroscope, 493, 527
Electrostatics, 492
Elementary particles, 871
Elliptically polarized wave, 766
EMF, 598–601
  ac circuit, 711
  back, 689, 691
  comparison of, 606, 608–609
  dc circuit, 601–603
  induced (see Induced emf)
  instantaneous, in series RLC circuit, 714–715, 721
  as phasor, 719
  signs for, 602
Emission spectrum, 920, 926
Energy:
  binding, 879, 925
  charged capacitor, 569–571
  electric (see Electric potential energy)
  excitation, 925
  ground state, 910
  hydrogen atom, 921–927
  inductor, 696–697
  kinetic (see Kinetic energy)
  in LC circuit, 700
  photon, 887–888, 890, 891–893
  quantization of, 910
  in quantum theory, 887–897, 909–911
  relativistic, 875–878
  rest, 876–878
  zero-point, 911
Energy conservation:
  electric circuits, 564, 600, 601–602, 609–610
  electric current, 585
  electric potential, 541–542, 545
  pair annihilation, 896
  in quantum electrodynamics, 901
Energy density:
  electric field, 571–572, 754–755
  electromagnetic waves, 754–755
  magnetic field, 697, 754–755, 756
Energy-level diagram, 910, 924, 925–926
Energy levels for electron, 934
Equipotential surfaces, 547–549
Ether as medium for propagating light, 861
Ether-drift experiment, 861
Exchange force, 660
Excitation energy of atoms, 925
Excited state of atoms, 925
Eye, as lens, 798, 801
Eyepiece:
  astronomical telescope, 801
  microscope, 804

Faller, J. E., 526
Farad (unit), 561
Faraday, Michael, 526, 561, 667, 743
Faraday's law of electromagnetic induction, 669, 671–672, 677–680, 737, 740, 744–745
Far-sightedness and lenses, 798, 801
Fermi (unit), 871, 920
Fermi, Enrico, 870–871, 920
Fermi-Dirac statistics, 871
Fermi energy, 871
Fermilab, 871
Fermions, 871
Fermi surface, 871
Ferromagnetic materials, 60
Field concept, 501–505
Field theory, 900
Fizeau, A. H. L., 760
Flux linkage, 669, 691
Focal length:
  in basic lens equation, 796–798
  converging thin lens, 792–795, 798
  definition, 794
  diverging lens, 800, 801
Focal plane, converging lens, 795
Focal point, 794
Focal point, principal, 794
Focus, principal, 794
Foucault, J. B. L., 782
Franklin, Benjamin, 500, 526
Fraunhofer, J., 826
Fraunhofer diffraction, 845, 853–855
Free charge on capacitor, 573
Free electron in conductors, 492, 591–592
Frequency:
  angular, 700, 728
  charged particle, 626
  electric oscillation, 700
  electromagnetic waves, 752, 759
  photoemission, 888
  radiation, in photoelectric effect, 887–891
  resonance, 716–717, 718, 727–728
  wave, 779–781, 887
Fresnel diffraction, 845
Fringing of electric field, 517
Fundamental interaction forces, 491
Fundamental magnetic interaction, 646
Fundamental particles, 492

Galilean telescope, 804
Galvani, Luigi, 599
Galvanometer:
  circuit symbol, 607
  in dc circuits, 607–609
  magnetic torque and, 634–635
Gamma rays, 754
Gamma-ray microscope, 903
Gauss (unit), 621
Gaussian surface:
  closed, 519, 521
  cylindrical, 523
  electric flux and, 519–533
  magnetism and, 651
  shape, in uniform electric field, 520–521
  spherical, 521, 524
Gauss's law, 519, 521–525, 526
  electric fields, 737, 744
  electricity, 651, 652
  magnetic fields, 737, 744
  magnetism, 651
  proof, 531–533
Germer, L. H., 898
Gradient of electric potential, 550
Gravitational-field effect, on particle motion, 517
Gravitational force, 491
  compared with electric force, 496–497, 539, 546
  ratio of, to electric force, 501
Gravitational mass, 496
Grounding, 494
Ground state of atoms, 924, 934, 935
Ground-state energy, 910

Hall effect, 632–633
Hall potential difference, 632
Heisenberg, Werner, 901, 903
Heisenberg uncertainty principle (see Uncertainty principle)
Helium electron configuration, 935
Helium-neon laser, 939–940
Henry (unit), 691
Henry, Joseph, 667, 691
Hertz, Heinrich, 701, 887
High-frequency oscillator, 701–702
High-pass filter, 713
Hill, H. A., 526
Holes in semiconductors, 633
Hooke, Robert, 834
Huygen's principle and reflected plane wave fronts, 777
Hydrogen:
  absorption spectrum, 921, 926
  electron configuration, 934
  emission spectrum, 920, 926
  ground state, 924
  ionization energy, 924–925
  spectrum, 920–921, 926
  wave functions, 928–930
  wavelengths in spectrum of, 925
  wave-mechanical analysis, 929–930

Ice-pail experiment, 526–527
Ideal particle, 889
Ideal wave, 889
Image:
  real, 795–798, 804, 809
  retinal, 802–803, 804, 849–850
  virtual, 798–799, 800, 801, 804, 809
Image distance, 796–798
Impedance, 715–716, 718, 722–723
  vector diagram, 723
Incoherent sources of waves, 829–830
Index of refraction, 775, 779–781
  relative, 780–781, 804–807
  vacuum, 780
Induced charge on capacitor, 573
Induced current, 667–671, 675–677
Induced electric field, 677–680
Induced emf, 667–671
  electric field and, 677–680
  electric generator, 671

Induced emf *(continued)*
  inductor, 691
  moving conductor, 671–675
  multiturn coil, 689–690
  mutual inductor, 705
  solenoid, 670
Inductance, 689–705
  coaxial conductor, 693–694
  coil, 692, 695–696
  definition, 691
  electric oscillations and, 697–704
  energy and, 696–697
  LC circuit, 697–702
  LR circuit, 694–696
  mutual, 704–705
  self-, 689–694
  series RLC circuit, 713–719
  transmission line, 718
Inductive reactance, 715
Inductor:
  ac circuit, 712, 726–727
  characteristics, 690–691
  circuit symbol, 690
  definition, 690
  energy, 696–697
  flux linkage, 691
  induced emf for, 691
  magnetic-energy density, 697
  mutual, 704–705
Inertial frame of reference, 873–875
Infrared radiation, 752
Insulating materials:
  dielectric strength of, 568–569
  electric charge and, 492
Insulators, 492
Intensity:
  electromagnetic waves, 755–756, 842–847, 853–856, 899–900
  light, 767, 768
  time-average, 817–820, 822–832
  wave, 755
Intensity zero in diffraction, 842–847, 853–856
Interference fringes, 824–825
  peaks, 827–828, 854, 856
  of waves, 816–835
    from coherent sources, 829–830
    constructive, 817–835
    destructive, 817, 819, 822
    diffraction effects, 841–856
    in diffraction grating, 826–829
    from incoherent sources, 829–830
    Michelson interferometer and, 834–835
    phase change and, 830–832
    in quantum theory, 902
    reflection and, 830–832
    superposition and, 816–818
    thin films, 832–834
    from two point sources, 818–824
    Young's experiment, 824–825
Interferometer, 834
Internal reflection (total), 784–785
Inverse-square force, electric force as, 496, 539
Ionization energy of hydrogen atom, 924–925

Joule heating, 586
Junction in dc circuits, 609
Junction rule in dc circuits, 609–610

Kinetic energy:
  for charges, 546–547
  charged particle, 629–630
  confined particle, 909–911
  electron, 519, 547
  free electrons, 591
  orbiting electron, 921–922
  photon, 888
  recoil, 927
  relativistic, 875–878
Kinetic-energy selector, 629
Kirchhoff's rules:
  first, 610
  second, 610

Laboratory frame in special relativity, 873–875
Lasers, 829–830, 937–941
Lateral magnification in lenses, 797
Latitude effect of earth's magnetic field, 624
Lattice spacing in crystalline solid, 852
Law:
  Ampère's, 652–656, 737–741, 745
  Biot-Savart, 647–650, 652
  charge conservation, 499, 500, 564, 609–610
  classical conservation, 499
  Coulomb's, 495–498, 501–503, 652, 678
  electromagnetic induction, 669, 671–672, 677–680, 737, 740, 744–745
  energy conservation, 541–542, 545, 564, 585, 600, 601–602, 609–610, 896, 901
  Faraday's, 669, 671–672, 677–680, 737, 740, 744–745
  Gauss's 519, 521–525, 526, 531–533, 651, 652, 737, 744
  Lenz's, 675–677, 681–682
  of Malus, 767
  mass conservation, 874–875
  mass-energy conservation, 876–880
  momentum conservation, 874–875, 901
  Ohm's, 585–586, 592
  Snell's, 775, 777, 781–783, 805–806
Lawrence, E. O., 627
LC circuit, 697–702
Lebedev, P. N., 762
Length:
  contracted, 871
  proper, 871
Lenses:
  aberrations in, 807–808
  astronomical telescope, 801–803
  camera, 798
  coated, 834
  combinations of, 801–804
  concave-convex, converging, 794
  conjugate points for, 798

  converging, 792–799, 801–807, 809
  diffraction effects, 807
  diverging, 779–801, 810
  double convex, converging, 794, 805
  equation, 797, 800, 806
  focal length, 792–795, 800
  focal point, 794
  Galilean telescope, 804
  lens maker's formula, 804–807
  microscope, 803–804
  objective, 801, 804
  plano-convex, converging, 794, 805
  radius of curvature, 804–807
  real image and, 795–798
  spherical mirrors and, 808–810
  with spherical surfaces, 793–807
  thin, 792–810
  virtual image and, 798–799, 800
Lens maker's formula, 804–807
Lens power, 798
Lenz, H. F. E., 676
Lenz's law:
  diamagnetism, 681–682
  electromagnetic induction, 675–677
Light (*see also* Visible light):
  as electric field, 504
  intensity, 767, 768
  monochromatic, in quantum theory, 888–889
  propagation of, 861
  ray optics for, 772–785
  speed of, 759–762
  unpolarized, 766–767, 768
  as wave phenomenon, 772–774
Light amplification, 938
Light pipe, 785
Like charges:
  Columb's law and, 495
  electric field lines for, 506–507
  electric force between, 497
  interactions between, 492
Linear charge density, 513–515
Linearly polarized wave, 766
Linear momentum of electromagnetic waves, 762, 765
Line of charge, 512–515
Line spectrum, 920–921
Lithium electron configuration, 935, 936
Livingston, M. S., 627
Lloyd, H., 831
Load (*see* Electric load)
Loop of conductor, 667–671, 672–674, 675–680, 689–690, 737–740
Loop equation for dc circuits, 610
Loop rule for dc circuits, 610
Lorentz, H. E., 628, 880
Lorentz force relation, 628
Lorentz transformations, 880–881
Low-pass filter, 713
Low-speed physics versus relativistic physics, 862, 864–865, 880
LR circuit, 694–696
Lyman series, 926

Magnet(s):
  behavior of, 636

magnetic field for, 651, 658
permanent, 660
Magnetic dipoles, 651
Magnetic dipole moment, 635–637
Magnetic domains, 660
Magnetic effect, 619
Magnetic-energy density, 697, 754–755, 756
Magnetic field, 619
  Ampère's law, 652–656
  Biot-Savart law, 647–650
  from capacitor plates, 740–741
  center of circular current loop, 649–650
  from changing electric flux, 739–740
  compared with electric field, 619
  from current element, 647–650
  cylindrical coaxial conductor, 655–656
  definition, 620–621
  direction, 621
  earth, 624
  effect on moving charge, 619–637
  electric-dipole oscillator, 757–759
  electric field and, 501
  electromagnetic waves and, 751, 752, 754–755, 756
  energy density, 754–755
  energy in inductor, 697
  from long, straight conductor, 644–647, 650
  magnitude, 621, 624
  Maxwell's equations and, 737–749
  Oersted effect and, 643–646
  from one moving charge, 643–660
  solenoid, 656–658
  sources, 643–660
  in special relativity, 875
  uniform, 623, 625–630
  velocity dependence, 625
Magnetic-field intensity, 622
Magnetic-field lines, 622–625
  induction and, 668–671
  symbols, 623
Magnetic flux, 622–625
  definition, 623
  in detector loop, 668–669
  electric field and, 677–680
  electromagnetic waves and, 746
  Gauss's law and, 651
  from moving conductor, 671
Magnetic-flux density, 622, 623
Magnetic force, 619–637
  between charged particles, 666
  between current-carrying conductors, 646–647, 648
  characteristics, 620–621
  on current-carrying conductor, 630–633
  on current loop, 634
  definition, 620–621
  direction, 620
  electric charges in motion and, 492
  on electron, 645–646
  Hall effect and, 632–633
  magnitude, 620
Magnetic induction, 622
Magnetic-induction field, 622
Magnetic-interaction constant, 644

Magnetic materials, 658–660
Magnetic monopoles, 651
Magnetic potential energy, 636
Magnetic torque, 633–637
Magnetism and electron spin, 659–660
Magnifying glass, 799
Magnifying power of astronomical telescope, 802–803
Malus, E. L., 767
Maser, 779
Mass:
  charged particle, 629–630
  conservation of, 874–875
  relativistic, 874–875, 877
  rest, 875, 876–877, 878–879, 894
Mass conservation and special relativity, 874–875
Mass-energy, conservation of in special relativity, 876–880
Mass spectrometer, 630
Matter waves in quantum theory, 897–899
Maxwell, James Clerk, 737, 742–743, 745
Maxwell demon, 742–743
Maxwell's equations:
  electric and magnetic fields, 737, 741–745
  electromagnetic waves, 745–749
  in quantum theory, 901–902
  ray optics, 776
Mechanical oscillator, analogy with electric oscillator, 702–704
Metastable state of atoms, 937
Mho (unit), 586
Michelson, A. A., 760, 835, 861
Michelson interferometer, 834–835
Michelson-Morley experiment, 861
Microscope lens arrangement, 803–804
Microwaves, 752, 767
Microwave oscillator, 702
Momentum:
  angular, 922
  charged particle, 626, 629–630
  confined particle, 909
  conservation of, 874–875, 901
  of electron-positron pair, 895, 896
  measuring, 875
  photon, 892–894, 927
  proton, 626–627, 877
  relativistic, 873–875, 877
Momentum conservation:
  quantum electrodynamics, 901
  special relativity, 874–875
Momentum selector, 629
Monochromatic radiation, 761, 937–941
Morley, E. W., 861
Motion of particle in electric field, 517
Multiloop circuits, 609–611
Muon, in relativity experiments, 868–869
Mutual inductance, 704–705

Near-sightedness and lenses, 798
Neutral Intermediate Vector Boson, (neutrino), 923
Neutron:
  and antineutron, 500

charge on, 492
thermal, 898
wavelength, 898–899
Newton, Isaac, 772, 834
Newton's rings, 834
Node, in interference of waves, 817, 822
Nonmagnetic materials, 659
Nonpolar dielectrics, 573–574
Nonpolar molecule, 531, 573–574
Nonradiative transitions, 938
Normal components of electric field, 516
n-type semiconductor, 633
Nuclear explosion, first, 870
Nuclear force, 492, 919
Nuclear scattering, 917–920
Nucleus, 492, 918
Null instrument, 608

Object distance, in lens equation, 796–798, 800
Objective lens, 801, 804
Observation angle:
  diffraction, 842–846, 847–856
  interference of waves, 818–820, 822–824, 827–829
Ocular, in astronomical telescope, 801
Oersted, H. C., 643
Oersted effect, 643–646
Ohm (unit), 586
Ohm, Georg S., 586
Ohm's law, 585–586, 592
Open circuit, 600
Opera glass, 804
Optical fiber, 785
Optical path length, 833–834
Optical pumping, 830
Optical reversibility principle, 774
Optical system design, 776
Optics (see Ray optics)
Orbital angular-momentum component of electron, 931–932
Orbital angular-momentum quantum number, 930, 931
Orbital magnetic quantum number, 931–932
Orbiting satellite, 779
Oscillations, electric (see Electric oscillations)

Pair annihilation, 896–897, 901
Pair production, 895–896, 901
Paper capacitor, 568
Parabolic reflector as transmitter and receiver, 774
Parallel capacitors, 54, 566–567
Parallel-plate capacitor, 561–562, 568, 569, 571, 572–574, 761
Parallel resistors, 604–606
Paramagnetic materials, 659–660
Paraxial rays, 795, 796, 800, 805
Particle(s):
  acceleration, in electric field, 517
  confined, 908–911, 929
  motion, in electric field, 517
  motion, in gravitational field, 517

Particle(s) (continued)
  wave character, 897–902
Particle-antiparticle pairs, 499
Path difference:
  diffraction, 842–844, 852, 856
  electromagnetic waves, 818–824
Pauli, Wolfgang, 923, 934
Pauli exclusion principle, 933–937
Perfect insulator (vacuum), 492
Period:
  charged particle, 626
  electric oscillation, 700
Periodic table, and Pauli exclusion principle, 933–937
Permanent electric-dipole moment, 531
Permanent magnet, 660, 680–681
Permeability of free space, 644
Permittivity of free space, 521
Phase angle in series RLC circuit, 715, 717, 718, 722–723
Phase constant in ac circuit, 711
Phase difference:
  electromagnetic waves, in diffraction, 842–844, 848–850, 853–855
  electromagnetic waves, in interference, 818, 820–824, 830–834
  LC circuit, 700
  series RLC circuit, 717
Phasors, 719–724
Photodisintegration, 890
Photoelectric effect, 887–891, 900
Photoelectrons, 887–891
Photoemission, 888–889
Photoemitter, 888
Photon(s):
  absorption, 937–941
  electric charge and, 499
  electric field, 899–902
  emission, 887–890, 925–927, 937–941
  energy, 887–888, 890, 891–893, 897, 927
  momentum, 892–894, 927
  in photoelectric effect, 887–891
  speed, 888
  wavelength, 890–891, 893–894, 925
Photon multiplication, 938, 939
Photovoltaic cell, 601
π meson, 499, 500
Pion, 500
Planck, Max, 673, 887
Planck's constant, 887–888
Plane of charge, 516
Plane wave fronts:
  lenses and, 793, 801
  reflection and, 777
Plano-convex lens, 794, 805, 834
Plate tectonics, 779
Point charge(s), electric, 491–508 (see also Electric charge)
  Coulomb's law and, 495–498
  as electric dipole, 530–531
  electric field between, 504–505
  electric field for, 503–507, 512–513, 540, 545
  electric-field lines for, 505–507
  electric flux for, 521

  electric force between, 495–498, 539–540
  electric force for, 495
  electric potential for, 543–545
  electric potential energy for, 545–547
  equipotential surfaces for, 547–549
  Gauss's law and, 521, 531–532
  magnitude, 498–501
  total energy, 547
  work done in moving, 51, 542–544
Point image in ray optics, 778
Point source(s):
  diffraction effects and, 841–844, 853–855
  interference in waves from, 818–824
  for measuring small shifts in position, 779
Polar dielectrics, 574
Polarization:
  of electromagnetic waves, 765–768
  law of Malus and, 767
  by scattering, 767–768
  types, 765–766
Polarization charge on capacitor, 573
Polar molecule, 531, 573–574
Polaroid material, 767, 768
Population inversion in laser, 939–940
Positron, 499, 895–897
Positronium, 896
Potential (see Electric potential)
Potential difference (see Electric potential difference)
Potential drop (see Electric potential drop)
Potential energy, electric (see Electric potential energy)
Potentiometer, 606, 608–609
Power:
  average value, in ac circuit, 726
  in electric circuit, 585
  inductor, 696
  instantaneous, in ac circuit, 725–728
  in loop of conductor, 673–674
  radio antenna, 821
  time-average in circuit, 725, 726–727
Power factor, 726
Poynting, J. H., 756
Poynting vector for electromagnetic wave, 756
Primary of transformer, 729
Principal focus for lenses, 794, 800
Principal quantum number, 930
Principle of indeterminacy, 903
Prism:
  dispersion of visible light and, 783
  in refraction, 782–783, 792–795
Prism spectrometer, 783
Probability:
  decay of atoms and, 938
  wave function and, 899–902, 906, 909
Propagation speed of wave, 755
Proper length in special relativity, 871
Proper time in special relativity, 868
Proton:
  antiproton and, 499
  atomic structure and, 920–921
  charge, 492
  cyclotron frequency, 626–628

  momentum, 626–627, 877
  path in magnetic field, 877–878
p-type semiconductor, 633

Quality, in series RLC circuit, 728
Quanta, 887
Quantization, 886–887
  charge, 500–501, 887
  energy, 924
Quantization rule, 922
Quantum electrodynamics, 900–901
Quantum jump, 925
Quantum mechanics, 901
Quantum number, 909
  for atomic structure, 927–937
  orbital angular-momentum, 930–931
  orbital magnetic, 931–932
  principal, 930
  spin magnetic, 932–933, 934
Quantum-state changes, 937–941
Quantum theory, 673, 886–911
  of atomic structure, 917–941
  Bohr's model, 921–927
  Bremsstrahlung effect, 891–892
  chemical properties and, 933–937
  complementarity principle, 902–903
  Compton effect, 892–894
  for confined particle, 908–911
  electromagnetism and, 649
  energy, 887–897, 909–911
  of hydrogen atom, 921–927
  matter waves, 897–899
  pair annihilation, 896–897
  pair production, 895–896
  Pauli exclusion principle, 933–937
  photoelectric effect, 887–891
  uncertainty principle, 903–908
  wave function, 899–902, 904–905, 908–909
  work function, 888
  x-ray production and, 891–892
Quark, 500
Quasar, as point source, 779

Radiation force of electromagnetic waves, 762–764
Radiation pressure of electromagnetic waves, 762, 764–765
Radio antenna:
  as electromagnetic wave generator, 756–757
  interference of waves and, 821
Radio interferometry, 779
Radio receiver, electric oscillator in, 700–701
Radio telescope and diffraction effects, 849
Radio transmitter, photon emission, 889
Radio waves, 752, 816–835
Radius of curvature of lenses, 804–807
Ray:
  deviation, through lenses, 792–810
  paraxial, 795, 796, 800, 805
Rayleigh, Lord, 849

Rayleigh's criterion for resolution, 849–850
Ray optics:
  diffraction effects, 841–856
  electromagnetic waves of small wavelength, 772–785
  interference, 816–835
  reciprocity principle, 774–775
  reflection, 775–779, 784–785
  refraction, 775–777, 779–783
  Snell's law, 775, 777, 781–783
  thin lenses and, 792–810
  wave optics and, 772–774, 906
Ray tracing:
  astronomical telescope, 801–803
  converging lenses, 795–799
  diverging lenses, 800–801
  real image, 795–798
  spherical mirror, 809–810
  virtual image, 798–799, 800
RC circuit, 589–591
Reactance, 715
  capacitive, 715
  inductive, 715
  in series RLC circuit, 714–715
Real image, 795–798, 804, 809
Receiver in ray optics, 774
Reciprocity principle for ray optics, 774, 775, 794
Recoil kinetic energy of atom, 927
Reductio ad absurdum proof, 513
Reference frames in special relativity, 862–865, 866–875, 878–879, 880–881
Reflected ray, 775
Reflection:
  definition, 775
  diffuse, 776
  electromagnetic waves, 764
  interference of waves and, 830–832
  internal, 784–785
  plane wave fronts, 777
  rules of, 775–777
  Snell's law, 775, 777
  specular, 776
  spherical wave fronts, 777–778
Reflection-transmission grating, 826
Reflection-transmission ration, 776
Refracted ray, 775
Refraction:
  definition, 775
  index of, 775, 779–781
  lenses, 792–795, 804–807
  rules of, 775–777, 781–783
  Snell's law, 775, 777, 781–783
  through nonparallel-face plate, 782–783, 792–795
  through parallel-face plate, 782, 792
  through prism, 782–783, 792–795
  through window glass, 782
Refractive index:
  lenses, 804–807
  thin-film interference, 833–834
Relative index of refraction, 780–781, 804–807
Relativistic dynamics, 877
Relativistic energy, 875–878, 893

Relativistic kinetic energy, 875–878
Relativistic mass, 874–875, 877, 897
Relativistic momentum, 873–875, 877, 897
Relativistic space, 869–873
Relativistic speed, 877
Relativistic time, 865–871
Relativistic transformations:
  space and time, 880
  velocity, 861–865, 874, 881
Relativity, 860–861
  electromagnetism and, 649
  quantum theory and, 886–911
  of simultaneity, 880–881
  special (*see* Special relativity)
Relaxation process, 938
Repulsion of like charges, 492
Repulsive force for point charges, 497
Resistance, 582–592
  comparison, with Wheatstone bridge, 606, 608
  equation, 586
  inductor, 695–696
  infinite, in dc circuit, 600
  LR circuit, 694–696
  measurement, 608
  microscopic view, 591–592
  Ohm's law, 585–586
  open circuit, 600
  parallel resistors in dc circuits, 604–606
  RC circuit, 589–591
  series resistors, 603–606
  series RLC circuit, 713–719
  short circuit, 600
  shunt, 607
  temperature coefficient of, 588
Resistance thermometer, 588
Resistivity, 587–589
Resistor(s):
  ac circuits, 711, 727
  circuit symbol, 586
  dc circuits, 598–611
  in parallel in dc circuits, 604–606
  in series in dc circuits, 603–604
Resolution:
  diffraction and, 847–850
  in photography, 849–850
  Rayleigh's criterion for, 849–850
Resonance, in series RLC circuit, 717, 727–728
Resonance frequency, in series RLC circuits, 716–717, 718, 727–728
Rest energy, 876–878
Rest mass, 875, 876–877, 878–879, 894
Rest time, 868
Resultant electric force, 501–505
Retinal image:
  astronomical telescope, 802–803
  microscope, 804
  resolution and, 849–850
Rheostat circuit, 586
Right-hand rule:
  current, 653
  magnetic-dipole moment, 636
  magnetic field, 621, 644
RLC circuit, series (*see* Series *RLC* circuit)
RMS value:

  current, 724
  voltage, 724
Roemer, Ole, 760
Röntgen, Wilhelm C., 673, 891
Root-mean-square value (*see* RMS value)
Rotating vector (*see* Phasor)
Rubbia, Carlo, 758
Rutherford, Ernest, 918
Rutherford scattering, 917–920
Rydberg constant, 925

Savart, F., 644
Sawtooth wave from RC circuit, 591
Scalar equation, 495
Scattering:
  in Compton effect, 892–894
  nuclear, 917–920
  polarization and, 767–768
  Rutherford's experiments, 917–918
  of x-rays, 894
Scattering angle, 894, 918–919
Schrödinger, Erwin, 901
Schrödinger equation:
  for particle, 929
  in quantum theory, 901–902
Secondary of transformer, 729
Self-inductance, 689–694
Semiconductor(s), 493
  electric charge and, 493
  Hall effect and, 633
  resistivity and, 588
Series capacitors, 564–566
Series resistors, 603–604
Series RLC circuit, 713–719, 721–724, 725–728
Short circuit, 600
Shunt, 607
Shunt resistance, 607
Sieman (unit), 586
Simple harmonic oscillator in LC circuit, 699–700
Single-loop dc circuits, 601–603
Single-slit diffraction, 844–846, 853–855
Sinusoidal variation in ac circuit, 710–722, 725–728
Sinusoidal waves, electromagnetic, 752–754
Snell, W., 775
Snell's law:
  in lens maker's formula, 805–806
  reflection and refraction of rays, 775, 777, 781–783
Sodium-lamp photon emission, 890
Solenoid:
  induced electric field, 678–680
  induced emf, 670
  magnetic field, 656–658
Solids, arrangement of atoms in, 492
Source charges, 501–503
Source phase difference for electromagnetic waves, 818, 820
Sound waves, speed of, 861
Space contraction in special relativity, 869–873, 881
Space quantization, 932

Special relativity, 673, 860–881
  bound systems, 879
  constancy of speed of light and, 860–861
  dynamics, 877
  energy, 875–878
  first postulate, 860
  Lorenz transformations, 880–881
  mass-energy conservation, 876–880
  momentum, 873–875
  muon experiments, 868–869
  reference frames, 862–865, 866–875, 878–879, 880–881
  rest energy, 876–878
  rest mass, 875, 876–877
  rest time, 868
  second postulate, 873
  space, 869–873, 881
  time, 865–869, 881
  twin paradox, 869
  velocity transformations, 861–865, 874, 881
  unbound systems, 878–879
Spectroscopy, 920
Spectrum:
  absorption, 921, 926
  Balmer series, 921, 926
  bright-line, 920–921
  dark-line, 921
  electromagnetic, 752–754
  emission, 920–926
  hydrogen, 920–921, 925–926
  line, 920–921
  Lyman series, 926
Specular reflection, 776
Speed of light:
  constancy of, 761, 860–861
  Doppler effect 761–762
  electromagnetic waves and, 759–762
  measurement, 760–761, 861
  relative to observer in motion, 861–865
  special relativity theory and, 860–881
  standard meter and, 760
Speed of waves, 779–780
Spherical aberration, 807
Spherical mirrors, 808–810
  concave, 808–810
  convex, 810
Spherical shell of charge:
  electric field, 524–525
  electric potential, 553–554
Spherical wave front:
  lenses and, 794, 800
  reflection of, 777–778
Spin magnetic quantum number, 932–933, 934
Spontaneous emission of photon, 937
Spring, motion of, 702–703
Static charges and Coulomb's law, 496
Stellar aberration, 760
Step-down transformer, 730
Step-up transformer, 730
Stimulated absorption of photon, 937
Stimulated emission of photon, 937–938
Strength of electric field, 503
Strong diffraction, 852
Strong interaction, 919

Strong nuclear force, 491
Superconductors:
  magnetic field and, 657–658
  resistivity and, 588–589
Superposition principle:
  for coulomb force, 497
  for electric field, 515, 522
  interference of waves and, 816–818
  for magnetic field, 621
Surface charge, 526
Surface-charge density, 515–516, 523, 527–529, 541
Surface integral for electric flux, 520

Telescope:
  astronomical, 801–803
  Galilean, 804
Temperature coefficient of resistance, 588
Tesla (unit), 621
Tesla, N., 621
Test charge, 501–503, 541–542
Thermal coefficient of resistivity, 588
Thermal energy dissipated in moving conductor, 675
Thermal neutron, 898
Thermal speed of electrons, 583–584, 592
Thermocouple, 600
Thin films, interference for, 832–834
Thomson, J. J., 630
Threshold frequency for photoemission, 888
Time:
  proper, 868
  rest, 868
  in special relativity, 865–872
Time average of energy density, electromagnetic field, 755
Time-average intensity for electromagnetic waves, 817–820, 822–832
Time-average power for electric circuit, 725, 726–727
Time constant:
  LR circuit, 695
  RC circuit, 590
Time dilation, 865–869, 881
Toroid inductance, 692–693
Torque:
  electric dipole, 530
  magnetic, 633–635
Transformer:
  ac circuits, 729–730
  circuit symbol, 729
  definition, 729
  as mutual inductor, 704
  step-down, 730
  step-up, 730
Transistor radio:
  current and resistance, 586
  energy and energy costs, 601
Transitions in electron energy levels, 924–927, 930–932, 934–936, 938–940
Transmission diffraction grating, 826
Transmitter in ray optics, 774

Transverse component of electric field, 514
Transverse waves, electromagnetic, 751
Twin paradox, 869

Ultraviolet radiation, 752
Unbound system, in special relativity, 878–879
Uncertainty:
  particle energy, 904
  particle location, 903–904, 906
  particle momentum, 903–904, 906–907
  time interval, 904
Uncertainty principle, 903–908
Unified atomic mass unit, 877
Uniform charge, 512–533
Uniform circular motion in magnetic field, 625
Uniform electric field, 516–519, 628–630
  electric flux, 520–521
  field lines and equipotential, 549–552
Uniformly charged sheets, electric field from, 516–517, 523–524
Uniformly charged wire, electric field for, 522–523
Uniform magnetic field, 623, 625–630
Unlike charges:
  electric-field lines, 506–507
  interactions between, 492
Unpolarized light, 766–767, 768

Vacuum, as perfect insulator, 492
Valence, and charge quantization, 501
Van de Graaff generator, 493, 570
Van der Meer, Simon, 758
Variable resistor circuit symbol, 586
Vector equation for electric force, 495
Velocity:
  electron, in electric field, 518
  falling object, 703–704
Velocity selector, 629
Velocity transformations in special relativity, 861–865, 874, 881
Virtual image:
  converging lenses, 798–799, 801, 804, 809
  definition, 799
  diverging lens, 800
  ray optics, 778
Visible light, 752, 766–767 (see also Light)
  diffraction, 844–850
  dispersion in prism, 783
  generated by laser, 937–941
  interference analysis, 816–835
  ray optics, 772–785
  sources, 829–830
  wave length, 781, 825
Volt (unit), 542
Volta, Alessandro, 542, 599
Voltage rms value in ac circuit, 724
Voltage-current plots, 585–586
Voltage divider in dc circuits, 609
Voltmeter in dc circuits, 606–608

Wave(s):
  behavior, 772–774
  diffraction, 841–856
    effects, 773
  dispersion, 781
  electromagnetic (see Electromagnetic waves)
  frequency, 779–781
  intensity, 755
  interference, 816–835, 902
  propagation speed, 755
  ray optics, 772–785
  reciprocity principle, 774–775
  reflection, 775–779, 784–785
  refraction, 775–777, 779–783
  sinusoidal (see Sinusoidal waves)
  sound, 861
  speed, 779–780
  wavelengths, 772–774, 781
Wave front(s):
  plane, 777, 793, 801
  in ray optics, 772–785
  spherical, 777–778, 794, 800
Wave function:
  electron, 929, 930
  hydrogen, 929–930
  quantum theory, 899–902, 904–905, 908–909
Wavelength:
  de Broglie relation, 897
  electromagnetic waves, 752
  in hydrogen spectrum, 925
  neutron, 898–899
  photon, 890–891, 893–894, 925
  point sources, 818–824
  visible light, 781, 825
Wave mechanics, 901, 906
Wave number, for electromagnetic waves, 752
Wave optics and ray optics, 906
Wave packet, 904
Weak interaction force, 491
Weber (unit), 624
Weber, W., 624
Wheatstone, C., 608
Wheatstone bridge, 606, 608
Williams, E. R., 526
Work:
  charging capacitors, 569–570
  in coil by magnetic field, 636
  electric current, 585
  electric force between two point charges, 539–540
  loop of conductor, 673
  moving a charge, 541, 542–544
Work function, 888

X-rays, 752–753
  diffraction, 850–853
    effects, 897–898
  discovery, 673, 891
  production, 891–892

Young, Thomas, 824
  interference experiment, 824–825

Zeeman effect, 932
Zero-point energy, 911

Credits *(Continued from copyright page)*
ator Laboratory. **Figure 30-22:** After photo by R. W. de Blois in D. Halliday and R. Resnick, *Fundamentals of Physics.* New York: John Wiley, 1981. **Pages 742, 743:** The Institution of Electrical Engineers, London. **Figure 36-10:** Educational Development Center. **Figure 36-19:** Bell Laboratories. **Figure 38-6:** Educational Development Center. **Figure 38-10:** Addison-Wesley Publishing Company. **Page 826:** Ewing Galloway, New York. **Page 827:** Young, Thomas. *A Course of Lectures on Natural Philosophy and the Mechanical Arts.* London: Taylor and Walton, 1845. Courtesy of AIP Niels Bohr Library. **Figure 38-18b:** Bausch and Lomb. **Figures 39-5, 39-7:** Addison-Wesley Publishing Company. **Figure 39-8:** Allyn and Bacon file photo. **Figures 39-9b, 39-10c:** M. Cagnet, M. Francon, J. C. Thrier, *Atlas of Optical Phenomena.* Berlin: Springer, 1962. Courtesy of Springer-Verlag, Heidelberg. **Figure 39-11:** Dan McCoy/Rainbow. **Figure 39-15b:** Eastman Kodak Company. **Figure 41-8:** G. D. Rochester and J. G. Wilson, *Cloud Chamber Photographs of the Cosmic Radiation.* Pergamon Press, 1952. **Figure 41-10:** Courtesy RCA Laboratories. **Figure 42-8c:** John R. Van Wazer and Ilyas Absar, *Electron Densities in Molecules and Molecular Orbitals.* New York: Academic Press, 1975.

Constants

Acceleration due to gravity (g)	9.81 m/s²
Gravitational constant (G)	6.67×10^{-11} N·m²/kg²
Gas constant (R)	8.31 J/mol·K
Avogadro's number (N_A)	6.02×10^{23} mol⁻¹
Boltzmann constant (k)	1.38×10^{-23} J/K
Earth: Mass	5.98×10^{24} kg
Radius (mean)	6.38×10^{6} m
Distance (mean) to moon	3.84×10^{8} m
Distance (mean) to sun	1.49×10^{11} m
Density of water (20 °C)	1.00×10^{3} kg/m³
Standard atmospheric pressure	1.01×10^{5} Pa
Volume of 1 mole ideal gas at STP	22.4 ℓ
Absolute zero temperature	-273 °C
Coulomb-law constant ($k_e = 1/4\pi\epsilon_0$)	9.0×10^{9} N·m²/C²
Permittivity of free space (ϵ_0)	8.85×10^{-12} C²/N·m²
Magnetic interaction constant ($k_m = \mu_0/4\pi$)	10^{-7} T·m/A
Permeability of free space (μ_0)	$4\pi \times 10^{-7}$ N/A²
Electron charge (e)	1.60×10^{-19} C
Speed of light (vacuum) (c)	3.00×10^{8} m/s
Planck's constant (h)	6.63×10^{-34} J·s
Electron mass	9.11×10^{-31} kg = 0.511 MeV/c²
Proton mass	1.67×10^{-27} kg = 938 MeV/c²

Values are to three significant figures. See Appendix C for further information.